Selected Works

V.A. Fock

Quantum Mechanics and Quantum Field Theory

Selected Works

V.A. Fock

Quantum Mechanics and Quantum Field Theory

Edited by

L.D. Faddeev
Steklov Mathematical Institute
St. Petersburg, Russia

L.A. Khalfin
Steklov Mathematical Institute
St. Petersburg, Russia

I.V. Komarov
St. Petersburg State University
St. Petersburg, Russia

CRC Press is an imprint of the
Taylor & Francis Group, an informa business
A CHAPMAN & HALL BOOK

CRC Press
Taylor & Francis Group
6000 Broken Sound Parkway NW, Suite 300
Boca Raton, FL 33487-2742

First issued in paperback 2019

ISBN-13: 978-0-415-30002-5 (hbk)
ISBN-13: 978-0-367-39430-1 (pbk)

Library of Congress Cataloging-in-Publication Data

Fock, V. A. (Vladimir Aleksandrovich), 1898-1974
[Selections. English. 2004]
V.A. Fock--selected works : quantum mechanics and quantum field theory / by L.D. Faddeev, L.A. Khalfin, I.V. Komarov.
p. cm.
Includes bibliographical references and index.
ISBN 0-415-30002-9 (alk. paper)
1. Quantum theory. 2. Quantum field theory. I. Title: Quantum mechanics and quantum field theory. II. Faddeev, L. D. III. Khalfin, L. A. IV. Komarov, I. V. V. Title.

QC173.97.F65 2004
530.12--dc22 2004042806

Library of Congress Card Number 2004042806

Visit the Taylor & Francis Web site at
http://www.taylorandfrancis.com

and the CRC Press Web site at
http://www.crcpress.com

Contents

Preface

On December 22, 1998 we celebrated the centenary of Vladimir Aleksandrovich Fock, one of the greatest theoretical physicists of the XX-th century. V.A. Fock (22.12.1898–27.12.1974) was born in St. Petersburg. His father A.A. Fock was a silviculture researcher and later became an inspector of forests of the South of Russia. During all his life V.A. Fock was strongly connected with St. Petersburg. This was a dramatic period of Russian history — World War I, revolution, civil war, totalitarian regime, World War II. He suffered many calamities shared with the nation. He served as an artillery officer on the fronts of World War I, passed through the extreme difficulties of devastation after the war and revolution and did not escape (fortunately, short) arrests during the 1930s. V.A. Fock was not afraid to advocate for his illegally arrested colleagues and actively confronted the ideological attacks on physics at the Soviet time.

In 1916 V.A. Fock finished the real school and entered the department of physics and mathematics of the Petrograd University, but soon joined the army as a volunteer and after a snap artillery course was sent to the front. In 1918 after demobilization he resumed his studies at the University.

In 1919 a new State Optical Institute was organized in Petrograd, and its founder Professor D.S. Rozhdestvensky formed a group of talented students. A special support was awarded to help them overcome the difficulties caused by the revolution and civil war. V.A. Fock belonged to this famous student group.

Upon graduation from the University V.A. Fock was already the author of two scientific publications — one on old quantum mechanics and the other on mathematical physics. Fock's talent was noticed by the teachers and he was kept at the University to prepare for professorship. From now on his scientific and teaching activity was mostly connected with the University. He also collaborated with State Optical Institute, Physico-Mathematical Institute of the Academy of Sciences (later split into the Lebedev Physical Institute and the Steklov Mathematical Institute), Physico-Technical Institute of the Academy of Sciences (later the Ioffe Institute), Institute of Physical Problems of the Academy of Sciences and some other scientific institutes.

Fock started to work on quantum theory in the spring of 1926 just after the appearance of the first two Schrödinger's papers and in that same year he published his own two papers on this subject (see [26-1, 2]). They attracted attention and in 1927 he received the Rockefeller grant for one year's work in Göttingen and Paris. His scientific results of this period (see [28-1, 2, 3, 4]) placed him at once in the rank of the most active theorists of the world. The outstanding scientific achievements of V.A. Fock led to his election to the USSR Academy of Sciences as a corresponding member in 1932 and as an academician in 1939. He was awarded the highest scientific domestic prizes. The works by V.A. Fock on a wide range of problems in theoretical physics — quantum mechanics, quantum field theory, general relativity and mathematical physics (especially the diffraction theory), etc. — deeply influenced the modern development of theoretical and mathematical physics. They received worldwide recognition. Sometimes his views differed from the conventional ones. Thus, he argued with deep physical reasons for the term "theory of gravitation" instead of "general relativity." Many results and methods developed by him now carry his name, among them such fundamental ones as the Fock space, the Fock method in the second quantization theories, the Fock proper time method, the Hartree–Fock method, the Fock symmetry of the hydrogen atom, etc. In his works on theoretical physics not only had he skillfully applied the advanced analytical and algebraic methods but systematically created new mathematical tools when the existing approaches were not sufficient. His studies emphasized the fundamental significance of modern mathematical methods for theoretical physics, a fact that became especially important in our time.

In this volume the basic works by Fock on quantum mechanics and quantum field theory are published in English for the first time. A considerable part of them (including those written in co-authorship with M. Born, P.A.M. Dirac, P. Jordan, G. Krutkov, N. Krylov, M. Petrashen, B. Podolsky, M. Veselov) appeared originally in Russian, German or French. A wide range of problems and a variety of profound results obtained by V.A. Fock and published in this volume can hardly be listed in these introductory notes. A special study would be needed for the full description of his work and a short preface cannot substitute for it.

Thus without going into the detailed characteristics we shall specify only some cycles of his investigations and some separate papers. We believe that the reader will be delighted with the logic and clarity of the original works by Fock, just as the editors were while preparing this edition.

In his first papers on quantum mechanics [26-1, 2] Fock introduces the concept of gauge invariance for the electromagnetic field, which he called "gradient invariance," and, he also presents the relativistic generalization of the Shrödinger equation (the Klein–Fock–Gordon equation) that he obtained independently and simultaneously with O. Klein and earlier than W. Gordon. In a series of works [29-3, 4] on the geometrization of the Dirac equation Fock gives the uniform geometrical formulation of gravitational and electromagnetic fields in terms of the general connection defined not only on the space–time, but also on the internal space (in modern terms). In the most direct way these results are connected with modern investigations on Yang–Mills fields and unification of interactions.

Many of Fock's works [30-2, 3; 33-2; 34-1, 2; 40-1, 2] are devoted to approximation methods for many-body systems based on the coherent treatment of the permutational symmetry, i.e., the Pauli principle. Let us specifically mention the pioneer publication [35-1], where Fock was the first to explain the accidental degeneracy in the hydrogen atom by the symmetry group of rotations in 4-space. Since then the dynamical symmetry approach was extensively developed. In the work [47-1] an important statement of the quantum theory of decay (the Fock–Krylov theorem) was formulated and proved, which has become a cornerstone for all the later studies on quantum theory of unstable elementary (fundamental) particles.

A large series of his works is devoted to quantum field theory [32-1, 2, 3, 4, 5; 34-3; 37-1]. In those works, Fock establishes the coherent theory of second quantization introducing the Fock space, puts forward the Fock method of functionals, introduces the multi-time formalism of Dirac–Fock–Podolsky etc. The results of these fundamental works not only allowed one to solve a number of important problems in quantum electrodynamics and anticipated the approximation methods like the Tamm–Dankov method, but also formed the basis for subsequent works on quantum field theory including the super multi-time approach of Tomonaga–Schwinger related to ideas of renormalizations. It is particularly necessary to emphasize the fundamental work [37-2] in which Fock introduced an original method of proper time leading to a new approach to the Dirac equation for the electron in the external electromagnetic field. This method played an essential role in J. Schwinger's study of Green's functions in modern quantum electrodynamics.

The new space of states, now called the Fock space, had an extraordinary fate. Being originally introduced for the sake of consistent analysis of the second quantization method, it started a new independent life in modern mathematics. The Fock space became a basic tool for studying stochastic processes, various problems of functional analysis, as well as in the representation theory of infinite dimensional algebras and groups.

Besides theoretical physics, Fock also worked in pure mathematics. For this edition we have chosen two such works most closely related to quantum physics. In [29-1], published only one year after delta-functions were introduced by Dirac, Fock obtained the rigorous mathematical background for these objects unusual for classical analysis, and thus preceded the further development of the theory of generalized functions. In [43-1] Fock gave an original representation of an arbitrary function by an integral involving Legendre's functions with a complex index. Later, this work entered the mathematical background of the well-known Regge method.

As time goes on, the significance of works of classics of science — and Fock is undoubtedly such a classic — becomes more and more obvious. The fame of brilliant researchers of new particular effects, sometimes recognized by contemporaries higher than that of classics, is perhaps less lasting. This is not surprising, for in lapse of time more simple and more general methods appear to deal with particular effects, while the classical works lay in the very basis of the existing paradigms. Certainly, when a paradigm changes (which happens not so often), new classics appear. However, it does not belittle the greatness of the classics as to who founded the previous paradigm. So the discovery of quantum theory by no means diminished the greatness of the founders of classical physics. If in the future the quantum theory is substituted for a new one, it by no means will diminish the greatness of its founders, and in particular that of Fock. His name will stay forever in the history of science.

As a real classic of science Fock was also interested in the philosophical concepts of new physics. In this volume we restricted ourselves to only two of his papers on the subject [47-1, 57-1]. Fock fought against the illiterate attacks of marxist ideologists on quantum mechanics and relativity. His philosophical activity helped to avoid in physics a pogrom of the kind suffered by Soviet biology.

The works by Fock were translated into English and prepared for this edition by A.K. Belyaev, A.A. Bolokhov, Yu.N. Demkov, Yu.Yu. Dmitriev, V.V. Fock, A.G. Izergin, V.D. Lyakhovsky, Yu.V. Novozhilov, Yu.M. Pis'mak, A.G. Pronko, E.D. Trifonov, A.V. Tulub, and V.V. Vechernin. Most of them knew V.A. Fock, worked with him and were affected by his outstanding personality. They render homage to the memory of their great teacher.

Often together with the Russion version Fock published its variant in one of the European languages, mainly in German. We give references to all variants. Papers are in chronological order and are enumerated by double numbers. The first number indicates the year of first publication. To distinguish papers published the same year we enumerated them by the second number. Hence the reference [34-2] means the second paper in this issue originally published in 1934. In 1957 the collected papers by Fock on quantum field theory were published by Leningrad University Press in *V.A. Fock, Raboty po Kvantovoi Teorii Polya, Izdatel'stvo Leningradskogo Universiteta, 1957.* The articles for the book were revised by the author. In the present edition papers taken from that collection are shown with an asterisk.

The editors are grateful to A.G. Pronko who has taken on the burden of the LaTeX processing of the volume.

We believe that the publication of classical works by V.A. Fock will be of interest for those who study theoretical physics and its history.

L.D. Faddeev, L.A. Khalfin, and I.V. Komarov
St. Petersburg

The following abbreviations are used for the titles of Russian editions:

TOI — Trudy Gosudarstvennogo Opticheskogo Instituta (Petrograd-Leningrad)

JRPKhO — Journal Russkogo Fiziko-Khemicheskogo Obshchestva, chast' fizicheskaja

JETP — Journal Eksperimentalnoi i Teoreticheskoi Fiziki

DAN — Doklady Akademii Nauk SSSR

Izv. AN — Izvestija Akademii Nauk SSSR, serija fizicheskaja

UFN — Uspekhi Fizicheskikh Nauk

Vestnik LGU — Vestnik Leningradskogo Gosudarstvennogo Universiteta, serija fizicheskaja

UZ LGU — Ucheniye Zapiski Leningradskogo Gosudarstvennogo Universiteta, serija fizicheskikh nauk

OS — Optika i Spektroskopija

Fock57 — V.A. Fock, Raboty po Kvantovoi Teorii Polya. Izdatel'stvo Leningradskogo Universiteta, Leningrad, 1957

23-1
On Rayleigh's Pendulum

G. KRUTKOV AND V. FOCK

Petrograd

Received 12 December 1922

Zs. Phys. **13**, 195, 1923

The importance of Ehrenfest's "Adiabatic Hypothesis" for the present and future of the quantum theory makes very desirable an exact examination of its purely mechanical meaning. Some years ago one of the authors[1] found a general method to look for the adiabatic invariants, whereas the other author[2] investigated the case of a degenerated, conditionally periodic system which had not been considered in the first paper (see also the paper by Burgers[3]). An objection which can be attributed to this theory is that in the course of calculations at some point a simplifying assumption was made in the integrated differential equations, namely that its right-hand side is subject to an averaging process; to explain this approximation, arguments connected with the slowness of changes of the system parameters were used. This shortcoming makes it difficult to use the ordinary methods and to check the adiabatic invariance of several quantities. Therefore it seems reasonable to consider a very simple example which we can integrate without any additional assumptions and only then use the slowness of parameter changes.

As such an example we chose the Rayleigh pendulum, which is "classic" for the "adiabatic hypothesis," i.e., a pendulum the length of which is changing continuously but the equilibrium point remains fixed. As is well known the adiabatic invariant here is the quantity

$$v = \frac{E}{\nu},$$

the relation of the energy of the pendulum to the frequency. This

[1] G. Krutkow, Verslag Akad. Amsterdam **27**, 908, 1918 = Proc. Amsterdam **21**, 1112; Verslag **29**, 693, 1920 = Proc. **23**, 826; TOI **2**, N12, 1–89, 1921.

[2] V. Fock, TOI **3**, N16, 1–20, 1923.

[3] J.M. Burgers, Ann. Phys. **52**, 195, 1917.

quantity can be also written in the form of the action integral

$$\int_{t_0}^{t_0+\tau} 2T\, dt \qquad \text{or} \qquad \oint p\, dq\,.$$

One can propose at least four different methods to prove the adiabatic invariance of v:

1. The calculations of Lord Rayleigh. The already-mentioned averaging is performed here, also.[4]

2. and 3. The general proofs of the adiabatic invariance of the "phase integral" and the above mentioned general theory prove the v-invariance as a special case.[5,6] Here in the course of calculations we neglected some terms, too; and finally

4. The variational principle

$$\delta \int_{t_0}^{t_0+\tau} 2\,T\, dt = 0\,,$$

which is already less sensitive to our objection.[7] The following considerations give the fifth proof. It is more complicated than all the previous ones but does not contain their defects. Moreover, we hope that in the course of the proof we shall be able to find how the adiabatic invariants behave during the transitions through the instants of the degeneration of states.

---

[4]Lord Rayleigh, Papers **8**, 41. See also H. Poincaré, Cosmogonic Hypothesis, p. 87 (in French) and A. Sommerfeld, Atomic Structure and Spectral Lines, 3d edition, p. 376 (in German). (*Authors*)

[5]J.M. Burgers, l.c., A. Sommerfeld, l.c. 718.

[6]G. Krutkow, l.c. p. 913 (corr. p. 1117).

[7]P. Ehrenfest, Ann. Phys. **51**, 346, 1916. One of us learnt from a private conversation with Professor Ehrenfest that the proof needed an improvement: the formula (i) on p. 347 should be replaced by a more general one $A = \left[\frac{\partial L}{\partial a} - \frac{d}{dt}\frac{\partial L}{\partial \dot{a}}\right]_{\dot{a}=0}$ in which the second term generally does not vanish. However the second term is a complete derivative and therefore vanishes after integration over the period so the former result is still valid. (*Authors*)

1 The General Method. Establishing the Differential Equations of the Problem

A system has f degrees of freedom. Its Hamilton function[8] depends on the (generalized) coordinates q_r, momenta p_r, and on parameter a

$$H = H(q_1, \ldots, q_f, p_1, \ldots, p_f; a) . \qquad (a)$$

In the Hamilton differential equations

$$\dot{p}_r = -\frac{\partial H}{\partial q_r}, \quad \dot{q}_r = \frac{\partial H}{\partial p_r} \quad (r = 1, 2, \ldots, f) \qquad (b)$$

we put $a =$ const and integrate the resulting *isoparametric problem*, then we obtain f integrals of motion

$$H_1 = c_1 \ , \ H_2 = c_2 \ , \ldots, H_f = c_f \qquad (c)$$

which are in involution. Then we solve them relative to p_r

$$p_r = K_r(q_1, \ldots, q_f, c_1, \ldots, c_f; a) , \qquad (c')$$

and form the Jacobi characteristic function

$$V = \int \sum_r K_r \, dq_r \ , \qquad (d)$$

which gives us another set of f integrals

$$\frac{\partial V}{\partial c_1} = \vartheta_1 \ , \ \frac{\partial V}{\partial c_2} = \vartheta_2 \ , \ \ldots, \ \frac{\partial V}{\partial c_f} = \vartheta_f \ , \qquad (e)$$

needed for complete solution. Here $\vartheta_1 = t + \tau$ and $\tau, \vartheta_2, \ldots, \vartheta_f$ are considered as constants.

Now we turn from the variables p_r, q_r to the "elements" c_r, ϑ_r. This is the "contact (canonical) transformation" with the transformation function $V(q_1, \ldots, q_f, c_1, \ldots, c_f; a)$. Now we remove the condition $a =$ const; a can be an arbitrary function of time t. We come then to the *rheoparametric problem*. According to a known theorem the differential equations for the "elements" remain canonical with the new Hamilton function

$$\mathcal{H} = c_1 + \left(\frac{\partial V}{\partial a} \, \dot{a} \right) , \qquad (f)$$

[8] V.A. Fock avoided the use of the currently common word "Hamiltonian" saying that it sounded to him like an Armenian name. (*Editors*)

where the brackets mean that the derivative of V should be expressed through c_r, ϑ_r. Thus the "*rheoparametric equations*" are

$$\left.\begin{aligned} \dot{c}_r &= -\frac{\partial \mathcal{H}}{\partial \vartheta_r} = -\frac{\partial}{\partial \vartheta_r}\left(\frac{\partial V}{\partial a}\,\dot{a}\right) \\ \dot{\vartheta}_1 &= \frac{\partial \mathcal{H}}{\partial c_1} = 1 + \frac{\partial}{\partial c_1}\left(\frac{\partial V}{\partial a}\,\dot{a}\right) \\ \dot{\vartheta}_s &= \frac{\partial \mathcal{H}}{\partial c_s} = \frac{\partial}{\partial c_s}\left(\frac{\partial V}{\partial a}\,\dot{a}\right) \end{aligned}\right\} \quad \begin{pmatrix} r = 1, 2, \ldots, f \\ s = 2, 3, \ldots, f \end{pmatrix}. \tag{g}$$

If $f = 1$ and one puts $\dot{a} = \text{const}$, one obtains

$$\dot{c} = -\frac{\partial}{\partial \vartheta}\left(\frac{\partial V}{\partial a}\right)\cdot \dot{a}\,, \quad \dot{\vartheta} = 1 + \frac{\partial}{\partial c}\left(\frac{\partial V}{\partial a}\right)\cdot \dot{a}\,. \tag{g'}$$

The next step, the averaging process, should not be performed.

Now we turn to the Rayleigh pendulum. We make a preliminary condition that we stay in the region of small oscillations which is a restriction for the change of the pendulum length; actually by a large enough shortening of the length we shall come to the non-small elongation angles. However this restriction is not essential because we shall further assume that the velocity of the length decrease is small. For the pendulum length λ we put

$$\lambda = l - \alpha t\,, \tag{1}$$

with $\alpha = \text{const}$, i.e., *we consider the case of a constant velocity of the parameter change.*

If the mass of a heavy point is equal to 1, φ is the angle of elongation, $p = \lambda^2\dot{\varphi}$ is the angular momentum and g is the gravity acceleration, then we have the Hamilton function

$$H = \frac{1}{2\lambda^2}\,p^2 + \frac{1}{2}\,g\,\lambda\,\varphi^2\,. \tag{2}$$

If we put $H = c$, we have

$$p = \sqrt{2\lambda^2 c - g\lambda^3\varphi^2} \tag{$2'$}$$

and

$$V = \int p\;d\varphi = \frac{\lambda\varphi}{2}\sqrt{2c - g\lambda\varphi^2} + c\sqrt{\frac{\lambda}{g}}\,\sin^{-1}\varphi\,\sqrt{\frac{g\lambda}{2c}}\,; \tag{3}$$

and further

$$\left.\begin{aligned} \vartheta &= \frac{\partial V}{\partial c} = \sqrt{\frac{\lambda}{g}}\ \sin^{-1}\varphi\,\sqrt{\frac{g\lambda}{2c}}\,, \\ \varphi &= \sqrt{\frac{2c}{g\lambda}}\ \sin\vartheta\sqrt{\frac{g}{\lambda}}\,. \end{aligned}\right\} \tag{4}$$

Now we find $\left(\frac{\partial V}{\partial \lambda}\right)\dot{\lambda}$:

$$\left(\frac{\partial V}{\partial \lambda}\right)\dot{\lambda} = -\left(\frac{\partial V}{\partial \lambda}\right)\alpha = -\left[\frac{c\ \vartheta}{2\ \lambda} + \frac{3\ c}{4\sqrt{g\lambda}}\ \sin\vartheta\sqrt{\frac{g}{\lambda}}\right]\alpha\,. \tag{5}$$

The rheoparametric equations for our problem are

$$\left.\begin{aligned} \frac{dc}{dt} &= \alpha\left(\frac{c}{2\lambda} + \frac{3\ c}{2\ \lambda}\cos\vartheta\ 2\sqrt{\frac{g}{\lambda}}\right), && (A) \\ \frac{d\ \vartheta}{dt} &= -\alpha\left(\frac{\vartheta}{2\ \lambda} + \frac{3}{4\sqrt{g\lambda}}\ \sin\vartheta\ 2\sqrt{\frac{g}{\lambda}}\right) + 1\,, && (B) \end{aligned}\right\} \tag{$*$}$$

where for λ one must substitute $\lambda = l - \alpha\ t$.

2 Integration of the Differential Equations (∗)

Because equation (B) does not contain the variable c it can be considered separately. We put

$$\left.\begin{aligned} x &= \frac{2\sqrt{g}}{\alpha}\ \sqrt{\lambda}\,, \\ \vartheta &= \frac{\alpha\ x}{2\ g}\ \tan^{-1} y \end{aligned}\right\} \tag{6}$$

and then get a simple equation

$$\frac{dy}{dx} - \frac{3}{x}y + 1 + y^2 = 0\,,$$

i.e., the Ricatti differential equation. Now we put

$$y = \frac{1}{u}\,\frac{du}{dx}\,,$$

after that the differential equation will have the form:

$$\frac{d^2u}{dx^2} - \frac{3}{x}\frac{du}{dx} + u = 0\,. \tag{**}$$

Now we introduce the general Bessel functions with the k-index:

$$Z_k = A\ J_k(x) + B\ Y_k(x)\,, \tag{7}$$

where A, B are constants. Then the general solution of (**) is

$$u = x^2\ Z_2(x)\,,$$

and using the known formulas for Bessel functions,[9]

$$y = \frac{Z_1(x)}{Z_2(x)}\,,$$

we have

$$\vartheta = \frac{\alpha x}{2g}\tan^{-1}\frac{Z_1(x)}{Z_2(x)}\,.$$

As can be easily seen this expression for ϑ does not contain both constants A, B but only one, namely, their ratio.

If we put the value of ϑ into equation (A), we easily get:

$$\frac{\log xc}{dx} + \frac{3}{x}\frac{Z_2^2(x) - Z_1^2(x)}{Z_2^2(x) + Z_1^2(x)} = 0 \tag{8}$$

or, using again the well-known formulas,

$$\frac{d\ \log xc}{dx} = \frac{d\ log x[Z_1^2(x) + Z_2^2(x)]}{dx}\,,$$

and therefore

$$c = \text{const}\ [Z_1^2(x) + Z_2^2(x)]\,. \tag{9}$$

3 Equation of Motion for the Angle φ. Its Integration

Before we go further we shall check the results obtained by establishing the equation of motion for the deviation angle φ of the pendulum and by integration of this equation.

[9]See, e.g., P. Schafheitlin, Die Theorie der Besselschen Funktionen, p. 123, formulas **4(1)** and **4(2)**. (*Authors*)

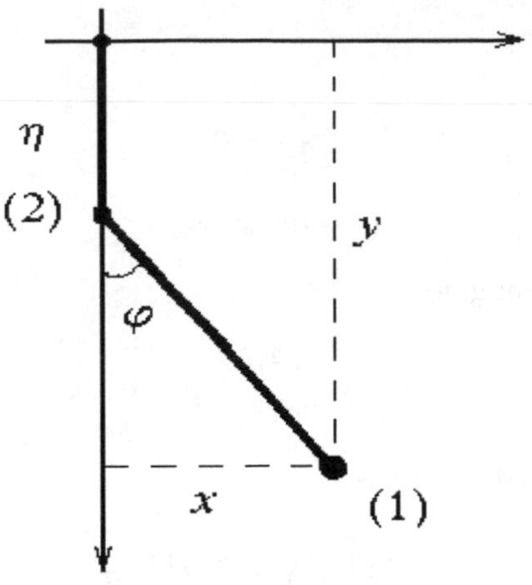

Fig. 1

For the coordinates of the point 1 which remains in the (x, y)-plane and keeps the distance $l - \eta$ from point 2 (see Fig. 1) which in turn lies on the y axis at a distance η from the origin, we have

$$x = (l - \eta)\ \sin\varphi\ , \qquad y = (l - \eta)\ \cos\varphi + \eta$$

and consequently for the kinetic energy T and potential energy Π

$$\left.\begin{aligned} T &= \frac{1}{2}[(l-\eta)^2\dot{\varphi}^2 + 2(1-\cos\varphi)\dot{\eta}^2 - 2(l-\eta)\sin\varphi\cdot\dot{\varphi}\dot{\eta}]\,, \\ \Pi &= -g(l-\eta)\cos\varphi - g\eta\,. \end{aligned}\right\} \tag{10}$$

The equation of motion for φ is

$$(l-\eta)\,\ddot{\varphi} - 2\dot{\varphi}\dot{\eta} - \sin\varphi\cdot\ddot{\eta} + g\sin\varphi = 0\,. \tag{11}$$

As before we restrict ourselves to small oscillations. Thus we have

$$\ddot{\eta} = 0\ , \quad \dot{\eta} = \alpha\ , \quad \eta = \alpha\, t \tag{12}$$

and $\eta = 0$ at $t = 0$. Then the differential equation has the form

$$(l - \alpha t)\,\ddot{\varphi} - 2\alpha\dot{\varphi} + g\varphi = 0\,. \tag{11$'$}$$

We introduce now a new independent variable

$$\tau = \sqrt{\frac{g}{l}}\ t \tag{13}$$

and put

$$\frac{\alpha}{\sqrt{lg}} = \sigma \,, \tag{14}$$

$$(1 - \sigma\tau)\varphi = y \,. \tag{15}$$

A simple calculation gives

$$(1 - \sigma\tau)\frac{d^2y}{d\tau^2} + y = 0 \,.$$

Returning to the old variables

$$x = \frac{2\sqrt{g}}{\alpha}\sqrt{\lambda} = \frac{2}{\sigma}\sqrt{1 - \sigma\tau} \,,$$

we have

$$\frac{d^2y}{dx^2} - \frac{1}{x}\,\frac{dy}{dx} + y = 0 \,.$$

The solution, as one easily finds, is:

$$y = x[A\ J_1(x) + B\ Y_1(x)] = x\ Z_1(x) \,.$$

For the angle φ it follows:

$$\varphi = \frac{4}{\sigma^2 x}\ Z_1(x) \,, \tag{16}$$

and for the angular velocity $\dot{\varphi}$, using the known formulas for Bessel functions, we have

$$\dot{\varphi} = \sqrt{\frac{g}{l}}\ \frac{8}{\sigma^3\ x^2}\ Z_2(x) \,. \tag{17}$$

Now we form the expressions

$$\frac{\lambda^2}{2}\ \dot{\varphi}^2 = \frac{(l - \alpha t)^2}{2}\ \varphi^2 \quad \text{and} \quad \frac{\lambda g}{2}\ \varphi^2 = \frac{(l - \alpha t)g}{2}\ \varphi^2 \,,$$

the sum of which by definition is the quantity c (see (2)) which we can consider as the energy of the pendulum:

$$c = \frac{2gl}{\sigma^2}\ [Z_1^2(x) + Z_2^2(x)] \,, \tag{18}$$

in complete accordance with formula (9).

A simple calculation allows us to express the constants A and B through the initial values of φ_0 and $\dot{\varphi}_0$:

$$\left.\begin{aligned} A &= \frac{\pi}{2}\left[\varphi_0\, Y_2\left(\frac{2}{\sigma}\right) - \dot{\varphi}_0\sqrt{\frac{l}{g}}\, Y_1\left(\frac{2}{\sigma}\right)\right], \\ B &= \frac{\pi}{2}\left[\dot{\varphi}_0\,\sqrt{\frac{l}{g}}\, J_1\left(\frac{2}{\sigma}\right) - \varphi_0\, J_2\left(\frac{2}{\sigma}\right)\right], \end{aligned}\right\} \tag{19}$$

where one has to use the known formula[10]

$$Y_k J_{k-1} - J_k Y_{k-1} = \frac{2}{\pi x}\,.$$

4 The Adiabatic Invariance of $\mathbf{v} = \dfrac{\mathbf{c}}{\boldsymbol{\nu}}$

To prove the adiabatic invariance relative to v, we assume that the length of the pendulum decreases *slowly*, i.e., α is *small*, whereas σ and x are *large*; the latter assumption demands $\sigma\tau \ll 1$. We can now replace in $Z_k(x)$ $(k = 1, 2)$ the $J_k(x)$ and $Y_k(x)$ and $J_k\left(\frac{2}{\sigma}\right)$, $Y_k\left(\frac{2}{\sigma}\right)$, which enter formulas (19) for A and B, by their asymptotic expressions:

$$\left.\begin{aligned} J_k(x) &= \sqrt{\frac{2}{\pi x}}\,\sin\left(x - \frac{2k-1}{4}\right) \\ Y_k(x) &= \sqrt{\frac{2}{\pi x}}\,\cos\left(x - \frac{2k-1}{4}\right) \end{aligned}\right\} \quad x \to \infty\,.$$

After some simple calculations we have:

$$\left.\begin{aligned} Z_2(x) &= \sqrt{\frac{\sigma}{2x}}\left[\dot{\varphi}_0\sqrt{\frac{l}{g}}\cos\left(\frac{2}{\sigma} - x\right) - \varphi_0\sin\left(\frac{2}{\sigma} - x\right)\right] \\ Z_1(x) &= \sqrt{\frac{\sigma}{2x}}\left[\dot{\varphi}_0\sqrt{\frac{l}{g}}\sin\left(\frac{2}{\sigma} - x\right) + \varphi_0\cos\left(\frac{2}{\sigma} - x\right)\right] \end{aligned}\right\} \tag{20}$$

and for c:

$$c = \frac{2\,g\,l}{\sigma^2}\cdot\frac{\sigma}{2x}\left(\frac{l}{g}\dot{\varphi}_0^2 + \varphi_0^2\right), \tag{21}$$

[10] Schafheitlin, l.c., p. 124, formula 13(4). (*Authors*)

or, if we denote $\frac{1}{2}(l^2\dot{\varphi}_0^2 + lg\varphi_0^2)$ by c_0:

$$c = \frac{2\, c_0}{\sigma x}. \tag{21$'$}$$

According to the definition the vibrational number ν is equal to

$$\nu = \frac{1}{2\pi}\sqrt{\frac{g}{l-\alpha t}} = \frac{1}{2\pi}\sqrt{\frac{g}{l}\,\frac{2}{\sigma x}}. \tag{22}$$

The relation v is then equal to

$$v = \frac{c}{\nu} = 2\pi c\,\sqrt{\frac{\lambda}{g}} = 2\pi c_0\,\sqrt{\frac{l}{g}} = \frac{c_0}{\nu_0}; \tag{23}$$

by that *the adiabatic invariance relative to* v is proven.

Petrograd,
Physical Institute of the University,
Summer 1922

Translated by Yu.N. Demkov

26-1
On Schrödinger's Wave Mechanics

V. Fock

Leningrad

5 June 1926

Zs. Phys. **38**, 242, 1928

In his extremely important paper [1] E. Schrödinger proposed a wave equation which is the basic equation of the "undulatorical" mechanics and can be considered as a substitution to the Hamilton–Jacobi partial differential equations (HJ) of the ordinary (classical) mechanics. The wave equation was established on condition that the Lagrange function contains no terms linear in velocities. Schrödinger writes (footnote on p. 514 l.c.):

> "In relativistic mechanics and in the case of a magnetic field the expression of the (HJ) is more complicated. In the case of a single electron this equation means the constancy of the four-dimensional gradient diminished by a given vector (the four-dimensional potential). The wave mechanical translation of this Ansatz meets some difficulties."

In this paper we will try to remove some of these difficulties and to find the corresponding wave equation for the more general case when the Lagrange function contains the linear (in velocities) terms.

Our paper consists of two parts. In part I the wave equation is formulated; part II contains examples of the Schrödinger quantization method. Schrödinger has already obtained some of these results, but only the results themselves; the calculations were not submitted.

Part I

The Hamilton–Jacobi differential equation for a system with f degrees of freedom is

$$H\left(q_i, \frac{\partial W}{\partial q_i}\right) + \frac{\partial W}{\partial t} = 0\,. \tag{1}$$

The left-hand side of this equation is a quadratic function of the derivatives of the function of action W with respect to coordinates.[1]

We replace here

$$\left.\begin{aligned} \frac{\partial W}{\partial t} \quad \text{by} \quad -E &= -E\frac{\frac{\partial\psi}{\partial t}}{\frac{\partial\psi}{\partial t}}\,, \\ \frac{\partial W}{\partial q_i} \quad \text{by} \quad -E &= \frac{\frac{\partial\psi}{\partial q_i}}{\frac{\partial\psi}{\partial t}} \quad (i = 1, 2, \ldots, f)\,, \end{aligned}\right\} \tag{2}$$

where E is the energy constant of the system. After multiplication by $\left(\frac{\partial\psi}{\partial t}\right)^2$ we have the uniform quadratic function of the first derivatives of ψ relative to the coordinates and time:

$$Q = \frac{1}{2}\sum_{k=1}^{f}\sum_{i=1}^{f} Q^{ik}\,\frac{\partial\psi}{\partial q_i}\,\frac{\partial\psi}{\partial q_k} + \frac{\partial\psi}{\partial t}\sum_{i=1}^{f} P^i\,\frac{\partial\psi}{\partial q_i} + R\left(\frac{\partial\psi}{\partial t}\right)^2 . \tag{3}$$

To find the wave equation we consider the integral

$$J = \int Q\,d\Omega\,dt\,. \tag{4}$$

Here by $d\Omega$ we denote the volume element of the multidimensional coordinate space; in the case of a system with n mass-points with coordinates x_i, y_i, z_i, one can understand under $d\Omega$ the product of the proper volume elements

$$d\tau_i = dx_i\;dy_i\;dz_i$$

and then

$$d\Omega = d\tau_1\;d\tau_2\ldots d\tau_n\,.$$

The product $d\Omega\,dt$ is then *not* the volume element of the space–time domain, where $2Q$ is the square of the gradient of the function ψ.

The integration over the coordinates should be extended over the whole coordinate spaces and over an arbitrary time interval $t_2 > t > t_1$.

[1]This is in classical mechanics. The relativistic mechanics of *a single* point-like mass (at least without a magnetic field) allows the equation to be written in this form; however it looks like the operations connected with the transformed equation are not completely inarguable. (*V. Fock*)

One can obtain the required wave equation by putting the first variation of the integral (functional) J equal to zero:

$$\delta \int Q \, d\Omega \, dt = 0 \,. \tag{5}$$

Here we can postulate the vanishing of the variation $\delta\psi$ either at the borders of the whole integration region or only at instants t_1 and t_2.

The explicit presentation of the wave equation is rather bulky; we will better consider other examples.

If one looks for periodic solutions and puts

$$\psi = e^{2\pi i\nu t}\psi_1 = e^{i\frac{E}{\hbar}t}\psi_1 \,, \tag{6}$$

one obtains for ψ_1 an equation which does not contain time. The energy E enters here as a parameter and at least in the nonrelativistic mechanics E enters linearly. In the special case of vanishing P^i equation (3) coincides with that established by Schrödinger. If P^i does not vanish, the coefficients of the (time-independent) wave equation are complex.

The special energy values can be found using the uniqueness, finiteness and continuity conditions of the solutions.

For pure periodic solutions (6) one can put directly the expressions in (2) equal to $\dfrac{\partial W}{\partial q_i}$. Then one has

$$\psi = \text{const}\; e^{-\frac{i}{\hbar}W} \,, \tag{7}$$

which clarifies the meaning of the action function W as a phase of the wave process.

Part II

1 The Kepler Motion in Magnetic Field

Let us consider a uniform magnetic field of strength $\mathcal{H}$ oriented along the z-axis. The Lagrange function

$$L = \frac{1}{2}\, m\, (\dot{x}^2 + \dot{y}^2 + \dot{z}^2) + \frac{e\mathcal{H}}{2c}\, (x\dot{y} - y\dot{x}) + \frac{e^2}{r} \tag{8}$$

and the (HJ) equation

$$\frac{1}{2m}\Big\{(\text{grad } W)^2 + \frac{e\mathcal{H}}{c}\Big(y\,\frac{\partial W}{\partial x} - x\,\frac{\partial W}{\partial y}\Big) +$$

$$+\frac{e^2\mathcal{H}^2}{4c^2}(x^2+y^2)\Big\} - \frac{e^2}{r} + \frac{\partial W}{\partial t} = 0 \tag{9}$$

are well known.

The quadratic form Q is

$$Q = \frac{E^2}{2m}(\text{grad }\psi)^2 - \frac{e\mathcal{H}E}{2mc}\left(y\frac{\partial\psi}{\partial x} - x\frac{\partial\psi}{\partial y}\right)\frac{\partial\psi}{\partial t} - \\ -\left[E + \frac{e^2}{r} - \frac{e^2\mathcal{H}^2}{8mc^2}(x^2+y^2)\right]\left(\frac{\partial\psi}{\partial t}\right)^2, \tag{10}$$

and the wave equation is

$$\Delta\psi - \frac{e\mathcal{H}}{Ec}\left(y\frac{\partial^2\psi}{\partial x\,\partial t} - x\frac{\partial^2\psi}{\partial y\,\partial t}\right) - \\ -\frac{2m}{E^2}\left[E + \frac{e^2}{r} - \frac{e^2\mathcal{H}^2}{8mc^2}(x^2+y^2)\right]\frac{\partial^2\psi}{\partial t^2} = 0 \tag{11}$$

(Δ is the Laplace operator). Introducing the function ψ_1

$$\psi = \psi_1 e^{i\frac{E}{\hbar}t},$$

and denoting for brevity

$$\frac{e\mathcal{H}}{2mc} = \omega, \quad \frac{\hbar^2}{e^2 m} = a, \tag{12}$$

we obtain for ψ_1 the equation

$$\Delta\psi_1 - \frac{2im}{\hbar}\omega\left(y\frac{\partial\psi_1}{\partial x} - x\frac{\partial\psi_1}{\partial y}\right) + \\ +\left[\frac{2E}{ae^2} + \frac{2}{ar} - \frac{m^2\omega^2}{\hbar^2}(x^2+y^2)\right]\psi_1 = 0. \tag{13}$$

Introducing the spherical coordinates and choosing a as a unit of length, we have

$$\Delta\psi_1 + 2\,i\,\omega_1\frac{\partial\psi_1}{\partial\varphi} + \left[\alpha + \frac{2}{r} - \omega_1^2 r^2 \sin\vartheta^2\right]\psi_1 = 0, \tag{14}$$

with the abbreviations

$$\frac{m}{\hbar}\omega a^2 = \omega_1, \quad \frac{2Ea}{e^2} = \alpha. \tag{15}$$

If we neglect ω_1^2 (weak magnetic field), the equation can be solved if we look for ψ_1 in the form (separation in spherical coordinates)

$$\psi_1 = e^{in_1\varphi}\, P_n^{n_1}(\cos\vartheta)\, r^n\, \psi_2(r)\,, \tag{16}$$

where $P_n^{n_1}(\cos\vartheta)$ are the "adjoint spherical functions." Then for $\psi_2(r)$ we have the equation

$$r\,\frac{d^2\psi_2}{dr^2} + 2(n+1)\frac{d\psi_2}{dr} + [\,2 + (\alpha - 2n_1\omega_1)r\,]\,\psi_2 = 0\,, \tag{17}$$

the eigenvalues of which were already considered by Schrödinger. We have then

$$\alpha = 2\,n_1\omega_1 - \frac{1}{(n+p)^2}\,. \tag{18}$$

For the shift of spectral lines, we get the value

$$\Delta\nu = n_1 \cdot \frac{\omega}{2\pi} = n_1 \cdot \frac{e\mathcal{H}}{4\pi mc} \tag{19}$$

which agrees with the old quantum theory.

2 The Motion of an Electron in the Electrostatic Field of the Nucleus and in the Magnetic Field of the Dipole which Coincides with the Nucleus[2]

Solving this problem we meet the difficulty of a general nature, which we cannot overcome here. The example is chosen just to draw attention to the possibility of these difficulties.

Let the z-axis coincide with the direction of the dipole momentum of the value M. The Lagrange function is

$$L = \frac{1}{2}\,m\,(\dot{x}^2 + \dot{y}^2 + \dot{z}^2) - \frac{eM}{cr^3}\,(x\dot{y} - y\dot{x}) + \frac{e^2}{r}\,, \tag{20}$$

and the (HJ) equation in spherical coordinates after neglecting M^2 is

$$\frac{1}{2m}\,(\text{grad}\,W)^2 + \frac{eM}{m\,c\,r^3}\,\frac{\partial W}{\partial\varphi} - \frac{e^2}{r} + \frac{\partial W}{\partial t} = 0\,. \tag{21}$$

If we form the wave equation

$$\Delta\psi + \frac{2eM}{Ecr^3}\,\frac{\partial^2\psi}{\partial t\,\partial\varphi} - \frac{2\,m}{E^2}\left(E + \frac{e^2}{r}\right)\frac{\partial^2\psi}{\partial t^2} = 0\,, \tag{22}$$

[2] See [2] on the treatment of this problem according to the old quantization recipes. (*V. Fock*)

we see that the point $r = 0$ is an essentially singular point (the uncertainty position) for all integrals. To understand this more clearly, let us choose quantity a (12) as a unit of length

$$\beta = \frac{e^2}{\hbar^3} \cdot \frac{M\,m}{c} \tag{23}$$

and, making the assumption about the form of ψ

$$\psi = r^n\ P_n^{n_1}(\cos\vartheta)\ e^{i\frac{E}{\hbar}t + in_1\varphi}\ F\ (r)\,, \tag{24}$$

we reduce equation (22) to the ordinary differential equation

$$\frac{d^2F}{dr^2} + \frac{2n+2}{r}\ \frac{dF}{dr} + \left(\alpha + \frac{2}{r} - \frac{2n_1\beta}{r^3}\right) F = 0\,. \tag{25}$$

The essentially singular character of point $r = 0$ comes from the term $\frac{2n_1\beta}{r^3}$; from the physical point of view this term must play the role of a small correction[3] and by no means could be essentially important for the solution. This difficulty is characteristic not only of the chosen example, but occurs in all cases where we consider the approximate presentation of forces; in the theory of the Schrödinger wave equation we have to consider these forces not only in the region of electronic orbits, but in the whole space. In a "natural" mechanical system (electrons and nuclei) this difficulty would possibly arise. It is now not clear how to overcome it. Probably we have to use different approximations of forces in different regions of space and put some continuity conditions for the wave function ψ at the borders of these regions. Whether any ambiguity in the evaluation of the eigenenergies could then be excluded remains unclear. Anyway, the question touched upon here needs a deeper investigation.

3 The Relativistic Kepler Motion[4]

The (HJ) equation (cleared from the square roots) has the form

$$(\text{grad}\ W)^2 = \frac{1}{c^2}\left(\frac{\partial W}{\partial t}\right)^2 - 2\left(m + \frac{e}{c^2 r}\right)\frac{\partial W}{\partial t} + 2m\frac{e^2}{r} + \frac{e^4}{c^2 r^2}\,, \tag{26}$$

[3]We have already neglected the squares of β. (*V. Fock*)

[4]See footnote [2]. (*V. Fock*)

and the corresponding wave equation is

$$\Delta\psi = \frac{1}{E^2}\left[2mE + \frac{E^2}{c^2} + 2\left(m + \frac{E}{c^2}\right)\frac{e^2}{r} + \frac{e^4}{c^2r^2}\right]\frac{\partial^2\psi}{\partial t^2}\,. \tag{27}$$

We denote

$$\left.\begin{aligned} \frac{1}{a_1} = \frac{e^2}{\hbar^2}\left(m + \frac{E}{c^2}\right)\,, \quad \alpha_1 = \frac{a_1^2}{\hbar^2}\left(2mE + \frac{E^2}{c^2}\right)\,, \\ \gamma = \frac{e^2}{\hbar c} \quad \text{(the fine-structure constant)}\,. \end{aligned}\right\} \tag{28}$$

We introduce the quantity a_1 as a unit of length and make the periodicity assumption (6). Then the equation for ψ will be

$$\Delta\psi_1 + \left(\alpha_1 + \frac{2}{r} + \frac{\gamma^2}{r^2}\right)\psi_1 = 0\,. \tag{29}$$

Separating the variables

$$\psi_1 = r^{n'} \cdot Y_n(\vartheta, \varphi) \cdot F(r) \tag{30}$$

with

$$n' = -\frac{1}{2} + \sqrt{\left(n + \frac{1}{2} + \gamma\right)\left(n + \frac{1}{2} - \gamma\right)} \tag{31}$$

(so that n' is *not* an integer) we obtain the differential equation for F

$$r\,\frac{d^2F}{dr^2} + 2\,(n' + 1)\,\frac{dF}{dr} + (\,2 + \alpha_1 r\,)\,F = 0\,, \tag{32}$$

i.e., again equation (17). Thus, we have

$$\alpha_1 = -\frac{1}{(n' + p)} \quad (p = 1, 2, \ldots)\,. \tag{33}$$

Calculating the energy from (33) we obtain

$$E = mc^2\left(\frac{n' + p}{\sqrt{(n' + p)^2 + \gamma^2}} - 1\right)\,, \tag{34}$$

which is the Sommerfeld formula, with the only difference being that the quantum numbers are half-integers, as already mentioned by Schrödinger on p. 372 l.c.

4 The Stark Effect

Again we orient the direction of the electric uniform field with the strength D along the z-axis. We introduce the parabolic coordinates

$$z + i\rho = \frac{a}{2}(\xi + i\eta)^2, \tag{35}$$

use the notations

$$a = \frac{\hbar^2}{me^2}, \quad \alpha = \frac{2Ea}{e^2}, \quad \varepsilon = D \cdot \frac{a^2}{e} \tag{36}$$

and obtain a time-independent wave equation for ψ_1 :

$$\begin{aligned} &\frac{\partial^2 \psi_1}{\partial \xi^2} + \frac{\partial^2 \psi_1}{\partial \eta^2} + \frac{1}{\xi}\frac{\partial \psi_1}{\partial \xi} + \frac{1}{\eta}\frac{\partial \psi_1}{\partial \eta} + \\ &\quad + \left(\frac{1}{\xi^2} + \frac{1}{\eta^2}\right)\frac{\partial^2 \psi_1}{\partial \varphi^2} + [4 + \alpha(\xi^2 + \eta^2) - \varepsilon(\xi^4 - \eta^4)]\psi_1 = 0\,. \end{aligned} \tag{37}$$

Separating the variables

$$\psi_1 = X(\xi)\; Y(\eta)\; (\xi\eta)^n\; e^{in\varphi} \tag{38}$$

we obtain the equations for X and Y

$$\left.\begin{aligned} &\frac{d^2X}{d\xi^2} + \frac{2n+1}{\xi}\frac{dX}{d\xi} + (2 + A + \alpha\xi^2 - \varepsilon\xi^4)\; X = 0\,, \\ &\frac{d^2Y}{d\eta^2} + \frac{2n+1}{\eta}\frac{dY}{d\eta} + (2 - A + \alpha\eta^2 + \varepsilon\eta^4)\; Y = 0\,. \end{aligned}\right\} \tag{39}$$

We introduce new variables x, y and new parameters $\lambda^{(1)}$, $\lambda^{(2)}$, μ

$$\left.\begin{aligned} &\xi^2 = \frac{2}{\sqrt{-\alpha}}x\,, \qquad \eta^2 = \frac{2}{\sqrt{-\alpha}}y\,, \\ &2\epsilon = \mu(\sqrt{-\alpha})^3\,, \qquad \frac{1}{\sqrt{-\alpha}} = \frac{\lambda^{(1)} - \lambda^{(2)}}{2} \end{aligned}\right\} \tag{40}$$

and obtain instead of (39):

$$\left.\begin{aligned} &x^2\frac{d^2X}{dx^2} + (n+1)x\frac{dX}{dx} + (\lambda^{(1)}x - x^2 - \mu x^3)\; X = 0\,, \\ &y^2\frac{d^2Y}{dy^2} + (n+1)y\frac{dY}{dy} + (-\lambda^{(2)}y - y^2 + \mu y^3)\; Y = 0\,. \end{aligned}\right\} \tag{41}$$

Both equations have the form

$$t^2 \frac{d^2F}{dt^2} + (n+1)t\frac{dF}{dt} + (\lambda t - t^2 - \mu t^3)\, F = 0\,, \tag{42}$$

and in the first equation of (41) we have to choose the parameter λ in the first equation of (41) so that $F(t)$ is finite and continuous for $t \geq 0$ and for the second equation of (41) the same must be valid for $t \leq 0$.

Now we use the Laplace transformation

$$F(t) = \int e^{tz}\; f(z)\; dz\,, \tag{43}$$

and have the differential equation for $f(z)$

$$\mu f^{''}(z) + (z^2-1)f^{'}(z) - [(n-1)z + \lambda]f(z) = 0\,. \tag{44}$$

Here μ is a small parameter of the order of the electric field strength. We look now for the expansions of $\lambda, F(t), f(z)$ in powers of μ:

$$\left.\begin{aligned} \lambda &= \lambda_0 + \mu\lambda_1 + \mu^2\lambda_2 + \dots\,,\\ F(t) &= F_0 + \mu F_1 + \mu^2 F_2 + \dots\,,\\ f(z) &= f_0 + \mu f(z) + \mu^2 f_2 + \dots\,. \end{aligned}\right\} \tag{45}$$

Anyway, the series for $f(z)$ is divergent but is useful as an asymptotic expansion. For f_0 the expression

$$f_0(z) = (z-1)^{\overline{\frac{n-1+\lambda_0}{2}}} \cdot (z+1)^{\overline{\frac{n-1+\lambda_0}{2}}} \tag{46}$$

is obtained and for f_1 we get the differential equation

$$\frac{d}{dz}\frac{f_1(z)}{f_0(z)} = \frac{\lambda_1}{z^2-1} - \frac{f_0^{''}(z)}{(z^2-1)f_0(z)}\,. \tag{47}$$

Now from Schödinger's analysis of equation (17) it follows that f_0 must be a rational function, therefore the function F_0 must be an integer-transcendent one, so that in the first equation (41) λ_0 must be equal to

$$\lambda_0^{(1)} = n - 1 + 2p_1 \qquad (p_1 = 1, 2, \dots)\,, \tag{48a}$$

and in the second equation (41)

$$\lambda_0^{(2)} = -n + 1 - 2p_2 \qquad (p_2 = 1, 2, \dots)\,. \tag{48b}$$

An analogous consideration shows that f_1 must be also rational. This can be possible only if the residue of the right side of (47) vanishes at $z = \pm 1$. Simple calculation shows that this condition is valid if

$$\lambda_1 = \frac{1}{8}\,(3\lambda_0^2 - n^2 + 1). \qquad (49)$$

If we restrict ourselves to the first approximation, we can write:

$$\left.\begin{aligned} \lambda^{(1)} &= n - 1 + 2p_1 + \frac{\mu}{8}\,[3(n-1+2p_1)^2 - n^2 + 1]\,, \\ \lambda^{(2)} &= -n + 1 + 2p_2 + \frac{\mu}{8}\,[3(-n+1-2p_2)^2 - n^2 + 1]\,. \end{aligned}\right\} \qquad (50)$$

Calculating the value α from (50) and (40), we have

$$-\alpha = \frac{1}{(n-1+p_1+p_2)^2} - 3\epsilon(p_1 - p_2)(n - 1 + p_1 + p_2), \qquad (51)$$

which agrees with the Epstein formula.

References

1. E. Schrödinger, *Quantisiering als Eigenwertproblem*, Ann. Phys. **79**, 361 *(I. Mitteilung)* and **79**, 489 (*II. Mitteilung*), 1926.
2. G. Krutkow, *Adiabatic Invariants and their Application in Theoretical Physics*, TOI **2**, N12, 38, 1922 (in Russian).

Leningrad,
Physical Institute of the University

Translated by Yu.N. Demkov

26-2
On the Invariant Form of the Wave Equation and of the Equations of Motion for a Charged Massive Point[1]

V. Fock

Leningrad

Received 30 July 1926

Zs. Phys. **39**, N2–3, 226–232, 1926

In his not yet published paper H. Mandel[2] uses the notion of the five-dimensional space for considering the gravity and the electromagnetic field from a single point of view. The introduction of the fifth coordinate parameter seems to us very suitable for representing both the Schrödinger wave equation and the mechanical equations in the invariant form.

1 The special relativity

The Lagrange function for the motion of a charged massive point is, in easily understandable notations,

$$L = -mc^2\sqrt{1-\frac{\mathfrak{v}^2}{c^2}} + \frac{e}{c}\mathfrak{A}\mathfrak{v} - e\varphi, \tag{1}$$

[1]The idea of this work appeared during a discussion with Prof. V. Fréederickcz, to whom I am also obliged for some valuable pieces of advice. (*V. Fock*)

Remark at proof. When this note was in print, the excellent paper by Oskar Klein (Zs. Phys. **37**, 895, 1926) was received in Leningrad where the author obtained the results which are identical in principle with the results of this note. Due to the importance of the results, however, their derivation in another way (a generalization of the Ansatz used in my earlier paper) may be of interest. (*V. Fock*)

For history of the subject see Helge Kragh *Equation with many fathers. Klein-Gordon equation in 1926.* Am. J. Phys. **52**, N 11, 1024–1033, 1984. (*Editors*)

[2]The author kindly gave me a possibility to read his paper in manuscript. (*V. Fock*)

and the corresponding Hamilton–Jacobi equation (HJ) reads

$$(\operatorname{grad} W)^2 - \frac{1}{c^2}\left(\frac{\partial W}{\partial t}\right)^2 - \frac{2e}{c}\left(\mathfrak{A}.\operatorname{grad} W + \frac{\varphi}{c}\frac{\partial W}{\partial t}\right) + \\ + m^2c^2 + \frac{e^2}{c^2}(\mathfrak{A}^2 - \varphi^2) = 0. \tag{2}$$

Analogously to the Ansatz used in our earlier paper[3] we put here

$$\operatorname{grad} W = \frac{\operatorname{grad}\psi}{\frac{\partial\psi}{\partial p}}; \quad \frac{\partial W}{\partial t} = \frac{\frac{\partial\psi}{\partial t}}{\frac{\partial\psi}{\partial p}}, \tag{3}$$

where p denotes a new parameter with the dimension of the quantum of action. Multiplying by $\left(\frac{\partial\psi}{\partial p}\right)^2$ we get a quadratic form

$$Q = (\operatorname{grad}\psi)^2 - \frac{1}{c^2}\left(\frac{\partial\psi}{\partial t}\right)^2 - \frac{2e}{c}\frac{\partial\psi}{\partial p}\left(\mathfrak{A}\operatorname{grad}\psi + \frac{\varphi}{c}\frac{\partial\psi}{\partial t}\right) + \\ + \left[m^2c^2 + \frac{e^2}{c^2}(\mathfrak{A}^2 - \varphi^2)\right]\left(\frac{\partial\psi}{\partial p}\right)^2. \tag{4}$$

We note that the coefficients at the zeroth, first, and second powers of $\frac{\partial\psi}{\partial p}$ are four-dimensional invariants. Further, the form Q remains invariant if one puts

$$\left.\begin{aligned} \mathfrak{A} &= \mathfrak{A}_1 + \operatorname{grad} f, \\ \varphi &= \varphi_1 - \frac{1}{c}\frac{\partial f}{\partial t}, \\ p &= p_1 - \frac{e}{c}f, \end{aligned}\right\} \tag{5}$$

where f denotes an arbitrary function of the coordinates and of time. The latter transformation also leaves invariant the linear differential form[4]

$$d'\Omega = \frac{e}{mc^2}(\mathfrak{A}_x dx + \mathfrak{A}_y dy + \mathfrak{A}_z dz) - \frac{e}{mc}\varphi dt + \frac{1}{mc}dp. \tag{6}$$

[3]V. Fock, *Zur Schrödingerschen Wellenmechanik*, Zs. Phys. **38**, 242, 1926 (see [26-1] in this book).

[4]The symbol d' means that $d'\Omega$ is not a complete differential. (*V. Fock*)

We will now represent the form Q as the squared gradient of the function ψ in the five-dimensional space (R_5) and look for the corresponding space interval. One can easily find

$$ds^2 = dx^2 + dy^2 + dz^2 - c^2 dt^2 + (d'\Omega)^2. \tag{7}$$

The Laplace equation in (R_5) says

$$\nabla\psi - \frac{1}{c^2}\frac{\partial^2\psi}{\partial t^2} - \frac{2e}{c}\left(\mathfrak{U}.\operatorname{grad}\frac{\partial\psi}{\partial p} + \frac{\varphi}{c}\frac{\partial^2\varphi}{\partial t\partial p}\right) - \\ -\frac{e}{c}\frac{\partial\psi}{\partial p}\left(\operatorname{div}\mathfrak{U} + \frac{1}{c}\frac{\partial\varphi}{\partial t}\right) + \left[m^2c^2 + \frac{e^2}{c^2}(\mathfrak{U}^2 - \varphi^2)\right]\frac{\partial^2\psi}{\partial p^2} = 0. \tag{8}$$

It remains, as well as (7) and (4), invariant under the Lorentz transformations and under the transformations (5).

Since the coefficients of equation(8) do not contain parameter p, the dependence of function ψ on p can be assumed in the form of the exponential factor. Namely, to ensure correspondence with the experiment, we have to put[5]

$$\psi = \psi_0\, e^{2\pi i \frac{p}{h}}. \tag{9}$$

Equation for ψ_0 is invariant under the Lorentz transformations but not under the transformations (5). The meaning of the additional coordinate parameter p is exactly that it ensures the invariance of the equations with respect to the addition of an arbitrary gradient to the four-potential.

It is to be noted here that the coefficients of the equation for ψ_0 are complex-valued in general.

Assuming further that these coefficients do not depend on t and putting

$$\psi_0 = e^{-\frac{2\pi i}{h}(E+mc^2)t}\,\psi_1, \tag{10}$$

one gets for ψ_1 a time-independent equation that is identical to the generalization of the Schrödinger wave equation given in our earlier paper. Those values of E for which function ψ_1 exists satisfying some requirements of finiteness and continuity are then the Bohr energy levels. It follows from the above-mentioned considerations that the addition of a gradient to the four-potential cannot affect the energy levels. Both functions ψ_1 and $\overline{\psi}_1$ corresponding to the vector potentials $\mathfrak{U}$ and

[5]The appearance of the parameter p, connected with the linear form, in the exponent can be possibly associated with some relations observed by E. Schrödinger (Zs. Phys. **12**, 13, 1923). (*V. Fock*)

$\overline{\mathfrak{U}} = \mathfrak{U} - \operatorname{grad} f$ differ, namely, by a factor $e^{\frac{2\pi ie}{ch}f}$ of modulus 1, thus only possessing (at very general assumptions for the function f) the same continuity properties.

2 The General Relativity

A. The wave equation. For the space interval in the five-dimensional space we put

$$\left.\begin{aligned} ds^2 &= \sum_{i,k=1}^{5} \gamma_{ik}\, dx_i\, dx_k = \\ &= \sum_{i,k=1}^{4} g_{ik}\, dx_i\, dx_k + \frac{e^2}{m^2}\left(\sum_{i=1}^{5} q_i dx_i\right)^2 . \end{aligned}\right\} \tag{11}$$

The quantities g_{ik} here are the components of the Einstein fundamental tensor, the quantities q_i ($i = 1, 2, 3, 4$) being the components of the four-potential divided by c^2, so that

$$\sum_{i=1}^{5} q_i dx_i = \frac{1}{c^2}(\mathfrak{U}_x dx + \mathfrak{U}_y dy + \mathfrak{U}_z dz - \varphi c dt). \tag{12}$$

The quantity q_5 is a constant, and x_5 is the additional coordinate parameter. All the coefficients are real-valued being independent of x_5.

The quantities g_{ik} and q_i depend on the fields only but not on the properties of the massive point, the latter being represented by the factor $\frac{e^2}{m^2}$. For brevity, we will, however, introduce the quantities depending on $\frac{e}{m}$:

$$\frac{e}{m} q_i = a_i \quad (i = 1, 2, 3, 4, 5) \tag{13}$$

introducing also the following agreement: at the summation from 1 to 5, the sign of the sum is written explicitly, and at the summation from 1 to 4 it is suppressed.

Using these notations we find

$$\gamma_{ik} = g_{ik} + a_i a_k; \quad g_{i5} = 0, \quad (i, k = 1, 2, 3, 4, 5), \tag{14}$$

$$\gamma = \| \gamma_{ik} \| = a_5^2\, g, \qquad (i, k = 1, 2, 3, 4, 5), \tag{15}$$

$$\left.\begin{aligned}\gamma^{lk} &= g^{lk},\\ \gamma^{5k} &= -\frac{1}{a_5}g^{ik}a_i = -\frac{a^i}{a_5},\\ \gamma^{55} &= \frac{1}{(a_5)^2}\cdot(1+a_i a^i),\end{aligned}\right\} \quad (i,k,l=1,2,3,4). \tag{16}$$

The wave equation corresponding to equation (8) reads

$$\sum_{i,k=1}^{5}\frac{\partial}{\partial x_i}\left(\sqrt{-\gamma}\,\gamma^{ik}\frac{\partial\psi}{\partial x_k}\right)=0 \tag{17}$$

or, written in more detail,

$$\frac{1}{\sqrt{-g}}\frac{\partial}{\partial x_i}\left(\sqrt{-g}\,g^{ik}\frac{\partial\psi}{\partial x_k}\right)-\frac{2}{a_5}a^i\frac{\partial^2\psi}{\partial x_i\partial x_5}+ \\ +\frac{1}{(a_5)^2}(1+a_i a^i)\frac{\partial^2\psi}{\partial x_5^2}=0. \tag{18}$$

Finally, introducing the function ψ_0 and the potentials q_i enables one to rewrite these equations as

$$\frac{1}{\sqrt{-g}}\frac{\partial}{\partial x_i}\left(\sqrt{-g}\,g^{ik}\frac{\partial\psi_0}{\partial x_k}\right)-\frac{4\pi}{h}\sqrt{-1}\,c\,e\,q^i\frac{\partial\psi_0}{\partial x_i}- \\ -\frac{4\pi^2c^2}{h^2}(m^2+e^2q_iq^i)\,\psi_0=0. \tag{19}$$

B. The equations of motion. We will now represent the equations of motion of a charged massive point as the equations of the geodesic line in R_5.

To this end, we must now calculate the Christoffel symbols. We denote the five-dimensional symbols $\left\{{kl \atop r}\right\}_5$, the four-dimensional symbols being denoted as $\left\{{kl \atop r}\right\}_4$. Further, we introduce the covariant derivative of the four-potentials:

$$A_{lk}=\frac{\partial a_l}{\partial x_k}-\left\{{kl \atop r}\right\}_4 a_r, \tag{20}$$

and split the tensor $2A_{ik}$ to its symmetric and antisymmetric parts:

$$\left.\begin{aligned}B_{lk} &= A_{lk}+A_{kl},\\ M_{lk} &= A_{lk}-A_{kl}=\frac{\partial a_l}{\partial x_k}-\frac{\partial a_k}{\partial x_l}.\end{aligned}\right\} \tag{21}$$

Then we have

$$\left.\begin{aligned}
\left\{ {kl \atop r} \right\}_5 &= \left\{ {kl \atop r} \right\}_4 + \frac{1}{2}(a_k g^{ir} M_{il} + a_l g^{ir} M_{ik}), \\
\left\{ {kl \atop 5} \right\}_5 &= \frac{1}{2a_5} B_{lk} - \frac{1}{2a_5}(a_k a^i M_{il} + a_l a^i M_{ik}), \\
\left\{ {k5 \atop 5} \right\}_5 &= -\frac{1}{2} a^i M_{ik}, \\
\left\{ {55 \atop k} \right\}_5 &= 0, \\
\left\{ {55 \atop 5} \right\}_5 &= 0.
\end{aligned}\right\} \tag{22}$$

The equations of the geodesic line in R_5 are then

$$\frac{d^2 x_r}{ds^2} + \left\{ {kl \atop r} \right\}_4 \frac{dx_k}{ds}\frac{dx_l}{ds} + \frac{d'\Omega}{ds} \cdot g^{ir} M_{il} \frac{dx_l}{ds} = 0, \tag{23}$$

$$\frac{d^2 x_5}{ds^2} + \frac{1}{2a_5} B_{lk} \frac{dx_k}{ds}\frac{dx_l}{ds} - \frac{1}{a_5}\frac{d'\Omega}{ds} a^i M_{il} \frac{dx_l}{ds} = 0. \tag{24}$$

Here $d'\Omega$ denotes, as earlier, the linear form

$$d'\Omega = a_i dx_i + a_5 dx_5. \tag{25}$$

Multiplying the four equations (23) by a_r, the fifth equation (24) by a_5 and summing up, one obtains an equation which can be written in the form

$$\frac{d}{ds}\frac{d'\Omega}{ds} = 0. \tag{26}$$

Hence, it is true that

$$\frac{d'\Omega}{ds} = \text{const.} \tag{27}$$

Multiplying equation (23) by $g_{r\alpha}\dfrac{dx_\alpha}{ds}$ and summing up over r and α, one gets, due to the antisymmetry of M_{ik},

$$\frac{d}{ds}\left(g_{r\alpha} \frac{dx_r}{ds}\frac{dx_\alpha}{ds} \right) = 0 \tag{28}$$

which, after introducing the proper time τ by the formula

$$g_{ik} dx_i dx_k = -c^2 d\tau^2, \tag{29}$$

gives

$$\frac{d}{ds}\left(\frac{d\tau}{ds}\right)^2 = 0. \tag{30}$$

Equation (28) (or (30)) is, however, a corollary of (26) due to the relation

$$ds^2 = -c^2 d\tau^2 + (d'\Omega)^2. \tag{31}$$

It follows from what is said that equation (24) is a sequence of (23) and can be omitted. After introducing the proper time as an independent variable in (23), the fifth parameter drops out completely; we also suppress the subscript 4 at the bracket symbol:

$$\frac{d^2 x_r}{d\tau^2} + \left\{ {kl \atop r} \right\} \frac{dx_k}{d\tau}\frac{dx_l}{d\tau} + \frac{d'\Omega}{d\tau} g^{ir} M_{il} \frac{dx_l}{d\tau} = 0. \tag{32}$$

The last term in the left-hand side represents the Lorentz force. In the special relativity, the first of these equations can be written as

$$m\frac{d}{dt}\frac{dx}{d\tau} + \frac{1}{c}\frac{d'\Omega}{d\tau}\left[\frac{e}{c}\left(\dot{z}H_y - \dot{y}H_z + \frac{\partial \mathfrak{A}_x}{\partial t}\right) + e\frac{\partial \varphi}{\partial x}\right] = 0. \tag{33}$$

For the agreement with the experiment, the factor in the square brackets should be equal to unity. Thus

$$\frac{d'\Omega}{d\tau} = c \tag{34}$$

and

$$ds^2 = 0. \tag{35}$$

The paths of the massive point are hence geodesic lines of zero length in the five-dimensional space.

To get the Hamilton–Jacobi equation, we put the square of the five-dimensional gradient of function ψ equal to zero,

$$g^{ik}\frac{\partial \psi}{\partial x_i}\frac{\partial \psi}{\partial x_k} - \frac{2}{a_5}\frac{\partial \psi}{\partial x_5} a^i \frac{\partial \psi}{\partial x_i} + (1 + a_i a^i)\left(\frac{1}{a_5}\frac{\partial \psi}{\partial x_5}\right)^2 = 0. \tag{36}$$

Putting here

$$m c a_5 \frac{\dfrac{\partial \psi}{\partial x_i}}{\dfrac{\partial \psi}{\partial x_5}} = \frac{\partial W}{\partial x_i}, \tag{37}$$

and introducing the potentials q_i instead of a_i, we obtain an equation

$$g^{ik}\frac{\partial W}{\partial x_i}\frac{\partial W}{\partial x_k} - 2\,e\,c\,q^i\,\frac{\partial W}{\partial x_i} + c^2(m^2 + e^2\,q_i q^i) = 0. \qquad (38)$$

It can be considered as a generalization of our equation (2) which was our starting point.

Leningrad,
Physical Institute of the University

Translated by A.G. Izergin and A.G. Pronko

28-1
A Comment on Quantization of the Harmonic Oscillator in a Magnetic Field

V. Fock[1]

temporarily in Göttingen

Received 12 January 1928

Zs. Phys. **47**, 446, 1928

The isotropic planar harmonic oscillator in a uniform magnetic field perpendicular to the plane of the oscillator presents a simplest example of the application of the perturbation theory to the degenerated unperturbed system. Therefore the exact solution of this simple problem would be of some interest. The purpose of this remark is to present this solution.

The Hamilton function for this problem is

$$H = \frac{1}{2m}(p_x^2 + p_y^2) + \frac{1}{2}\, m\, \omega_0^2(x^2 + y^2) + \frac{e\mathcal{H}}{2mc}\,(yp_x - xp_y) + \frac{e^2\mathcal{H}^2}{8mc^2}\,(x^2 + y^2)\,.$$

Here m is the mass, e the charge and ω_0 the frequency of the oscillator, $\mathcal{H}$ the strength of the magnetic field. Introducing the Larmor frequency

$$\omega_1 = \frac{e\mathcal{H}}{2mc}\,,$$

one can rewrite the Hamilton function

$$H = \frac{1}{2m}(p_x^2 + p_y^2) + \omega_1(yp_x - xp_y) + \frac{1}{2}\, m\, (\omega_0^2 + \omega_1^2)\, (x^2 + y^2)\,.$$

The corresponding Schrödinger amplitude equation is

$$\Delta\psi \mp \frac{2im}{\hbar}\omega_1\left(y\frac{\partial\psi}{\partial x} - x\frac{\partial\psi}{\partial y}\right) + \frac{2m}{\hbar^2}\left[E - \frac{1}{2}m(\omega_0^2 + \omega_1^2)(x^2 + y^2)\right]\psi = 0.$$

We introduce the new unit of length

$$b = \left(\frac{\hbar}{m}\right)^{1/2}(\omega_0^2 + \omega_1^2)^{-1/4}\,,$$

[1] International Education Board Fellow.

and denote

$$\frac{\omega_1}{\sqrt{\omega_0^2+\omega_1^2}}=\omega, \quad \frac{1}{\hbar}\,\frac{E}{\sqrt{\omega_0^2+\omega_1^2}}=W\,.$$

Using the polar coordinates

$$x=b\,r\cos\varphi\,,\quad y=b\,r\sin\varphi\,,$$

we will have the amplitude equation

$$\frac{\partial^2\psi}{\partial r^2}+\frac{1}{r}\frac{\partial\psi}{\partial r}+\frac{1}{r^2}\frac{\partial^2\psi}{\partial\varphi^2}\pm 2i\omega\frac{\partial\psi}{\partial\varphi}+(2W-r^2)\psi=0\,.$$

This equation easily admits the separation of variables. If one puts

$$\psi=e^{in\varphi}R(r)\,,$$

$$W\mp n\omega=W_1\,,$$

we have an equation for $R(r)$,

$$\frac{d^2R}{dr^2}+\frac{1}{r}\frac{dR}{dr}+\left(2W_1-\frac{n^2}{r^2}-r^2\right)R=0\,,$$

i.e., exactly the equation that one gets for a flat isotropic oscillator *without a magnetic field* in polar coordinates. By substitution

$$r^2=\rho\,,$$

one finds

$$\frac{d^2R}{d^2\rho}+\frac{1}{\rho}\,\frac{dR}{d\rho}+\left(-\frac{1}{4}+\frac{W_1}{2\rho}-\frac{n^2}{4\rho^2}\right)R=0\,.$$

This equation was already investigated by Schrödinger [1]. The energy levels here are

$$W_1=2k+n+1\quad(k=0,1,\ldots)\,,$$

and the eigenfunctions

$$R_{n,k}=\rho^{n/2}e^{-\rho/2}L_{n+k}^{n}(\rho)\,,$$

where

$$L_{n+k}^{n}(\rho)=\frac{d^n}{d\rho^n}L_{n+k}(\rho)$$

$$[L_{n+k}(\rho)\quad \text{the Laguerre polynomial}]\,.$$

Now calculating the energy E and returning to frequencies ω_0, ω_1 one gets

$$E = \pm n\hbar\omega_1 + (2k + n + 1)\hbar\sqrt{\omega_0^2 + \omega_1^2}\,,$$

or

$$E = (n_1 - n_2)\hbar\omega_1 + (n_1 + n_2 + 1)\hbar\sqrt{\omega_0^2 + \omega_1^2}\,,$$

where n_1 and n_2 are two nonnegative integers.

I am cordially thankful to the International Educational Board for enabling my stay in Göttingen.

References

1. E. Schrödinger, *Quantisiering als Eigenwertproblem, III. Mitteilung,* Ann. Phys. **80**, 484, 1926.

Translated by Yu.N. Demkov

28-2

On the Relation between the Integrals of the Quantum Mechanical Equations of Motion and the Schrödinger Wave Equation

V. FOCK[1]
temporarily in Göttingen

Received 1 May 1928

Zs. Phys. **49**, N5-6, 323-338, 1928

1. The mathematical part of the solution of a quantum mechanical problem, also in the case when the Hamilton function contains the time explicitly, is essentially reduced to finding a complete system of normalized orthogonal functions satisfying the Schrödinger equation

$$H\psi + \frac{h}{2\pi i}\frac{\partial\psi}{\partial t} = 0. \tag{1}$$

In what follows such a system will be called a "basic system." A basic system having been found, the most difficult part of the problem from the mathematical point of view is overcome. First, by means of the basic system one can calculate the matrices satisfying the equations of motion; second, the transition probabilities; and finally, as will be shown in the subsequently published paper of the author, it is possible to find the general solution of the Dirac statistical equation.

For the energy operator H not containing the time explicitly, one can write a basic system at once, if only the eigenfunctions of H are known. Denote these functions as $\varphi_1(q), \varphi_2(q), \ldots,$[2] then $\varphi_n(q)$ satisfies the equation

$$H\varphi_n = W_n\varphi_n, \tag{2}$$

[1]International Education Board Fellow.

[2]All formulae also remain valid in the case of several variables; writing $\varphi_n(q)$ etc. chosen here is to be regarded as an abbreviation only. (*V. Fock*)

and the functions

$$\psi_n(q,t) = e^{-\frac{2\pi i}{h}W_n t}\varphi_n(q) \tag{3}$$

evidently form a basic system. Now one can also construct a basic system if the energy operator H contains the time explicitly. Namely, it is possible to prove the following proposition:

Proposition 1. Let a complete normalized orthogonal system

$$\varphi_1(q), \varphi_2(q), \ldots$$

be given. Denote as $\psi_n(q,t)$ the solution of the Schrödinger equation (1), which reduces itself to $\varphi_n(q)$ at $t = 0$. Then for all t the system of functions $\psi_n(q,t)$ is complete, normalized and orthogonal, i.e., the functions $\psi_n(q,t)$ form a basic system.

Let us prove first that the functions $\psi_n(q,t)$ remain normalized and orthogonal for all t. To this end, consider the integral

$$a_{mn} = \int \overline{\psi}_m(q,t)\psi_n(q,t)\varrho(q)\,dq \tag{4}$$

[$\varrho(q)$ is the density function in the q-space]

and calculate its derivative with respect to time t. We have

$$\begin{aligned}\frac{da_{mn}}{dt} &= \int \overline{\psi}_m \frac{\partial\psi_n}{\partial t}\varrho\,dq + \int \frac{\partial\overline{\psi}_m}{\partial t}\psi_n\varrho\,dq = \\ &= \frac{2\pi i}{h}\Big\{-\int \overline{\psi}_m H\psi_n\varrho\,dq + \int \psi_n\overline{H\psi_m}\varrho\,dq\Big\} = 0,\end{aligned} \tag{5}$$

since the functions ψ_n satisfy equation (1), and the operator H is self-adjoint. But one has for $t = 0$

$$a_{mn} = \delta_{mn}, \tag{6}$$

hence equation (6) holds true for all t.

The proof of the completeness of the system of functions ψ_n is somewhat more difficult.

We put

$$S_t(f,g) = \sum_{n=1}^{\infty} \int \overline{\psi}_n f\varrho\,dq \int \psi_n\overline{g}\varrho\,dq, \tag{7}$$

where f and g are two quadratically integrable functions. For $t = 0$, the system of functions ψ is complete, and expression (7) is equal to

$$Q(f,g) = \int f\overline{g}\varrho\,dq. \tag{8}$$

We denote the complete derivative of (7) as $\dfrac{dS_t(f,g)}{dt}$, while $\dfrac{\partial S_t}{\partial t}(f,g)$ denotes the "partial" derivative, i.e., the derivative taken at the assumption that f and g do not depend on time. Then we have

$$\frac{dS_t(f,g)}{dt} = \frac{\partial S_t}{\partial t}(f,g) + S_t\left(\frac{\partial f}{\partial t}, g\right) + S_t\left(f, \frac{\partial g}{\partial t}\right). \tag{9}$$

We will calculate the expression $\dfrac{\partial S_t}{\partial t}(f,g)$. We have

$$\begin{aligned}\frac{\partial S_t}{\partial t}(f,g) &= \sum_{n=1}^{\infty}\Big\{\int \frac{\partial \overline{\psi}_n}{\partial t} f\varrho dq \int \psi_n \overline{g}\varrho dq + \int \overline{\psi}_n f \varrho dq \int \frac{\partial \psi_n}{\partial t}\overline{g}\varrho dq\Big\}\\ &= \frac{2\pi i}{h}\sum_{n=1}^{\infty}\Big\{\int f\overline{H\psi}_n \varrho dq \int \psi_n \overline{g}\varrho dq - \int \psi_n f \varrho dq \int H\psi_n \overline{g}\varrho dq\Big\}\\ &= \frac{2\pi i}{h}\sum_{n=1}^{\infty}\Big\{\int \overline{\psi}_n H f \varrho dq \int \psi_n \overline{g}\varrho dq - \int \overline{\psi}_n f \varrho dq \int \psi_n \overline{Hg}\varrho dq\Big\},\end{aligned}$$

i.e.,

$$\frac{\partial S_t}{\partial t}(f,g) = \frac{2\pi i}{h}\Big\{S_t(Hf,g) - S_t(f,Hg)\Big\}. \tag{10}$$

We must prove that

$$S_t(f,g) = Q(f,g) \tag{11}$$

holds for all t.

Before passing to the proof, let us make the following remark. If equation (11) holds true for arbitrary time-independent functions f, g, then it holds also in the case where f and g are time-dependent, since time enters here as a parameter only. The same remark applies to all other equations which do not contain the derivatives of functions f, g with respect to time.

To prove this, we will show that all "partial" (in the sense explained above) derivatives of $S_t(f,g)$, which we denote now as $S_t^{(k)}(f,g)$ instead of $\dfrac{\partial^k S_t}{\partial t^k}(f,g)$, vanish for $t=0$. If $S_t(f,g)$ is an analytic function of t,[3] it follows that equation (11) is fulfilled identically in t.

---

[3]This assumption is quite necessary for the proof. That it always appears is not, however, evident since the functions $\psi_n(q,t)$ are not, in general, analytic functions of t. They are such only if the given functions $\varphi_n(q) = \psi_n(q,0)$ satisfy certain conditions, e.g., if they are entire transcendental functions, and H is a second-order differential operator depending on time analytically. We assume here that the necessary conditions are fulfilled. (*V. Fock*)

It follows from equation (10) that the first derivative vanishes, since $S_t = Q$ for $t = 0$, and the right-hand side of (10) vanishes due to the self-adjointness of H. Denote as $H^{(l)}$ the operator that is obtained by differentiating H with respect to time l times. Then the operator $H^{(l)}$ is also self-adjoint, due to the validity of the following relation for any time-independent functions u and v:

$$\int (H^{(n)}u)\overline{v}\varrho\, dq - \int n\overline{H^{(n)}v}\varrho\, dq = \frac{d^n}{dt^n}\{Q(Hu,v) - Q(u,Hv)\} = 0. \quad (12)$$

Let us calculate the $(n+1)$-th derivative of S_t,

$$S_t^{(n+1)}(f,g) = \frac{2\pi i}{h}\sum_{k=0}^{n}\frac{n!}{k!(n-k)!}\{S_t^{(k)}(H^{(n-k)}f,g) - S_t^{(k)}(f,H^{(n-k)}g)\}. \quad (13)$$

Assume that vanishing of all derivatives $S_t^{(k)}$ up to the n-th one is proved, and show that then the $(n+1)$-th derivative also should vanish. Indeed, we get from (13) for $t = 0$

$$[S_t^{(n+1)}(f,g)]_{t=0} = \frac{2\pi i}{h}\{Q(H^{(n)}f,g) - Q(f,H^{(n)}g)\} = 0, \quad (14)$$

due to the self-adjointness of $H^{(n)}$.

Vanishing of the first derivative being established already, it follows that all derivatives at $t = 0$ equal zero.

If now, as assumed, $S_t(f,g)$ is analytic in t, then $S_t(f,g)$ is constant being equal to $Q(f,g)$. Thus, the completeness for all t of the system of functions $\psi_n(q,t)$ is proved.

2. Any complete normalized orthogonal system of functions $\varphi_s(q,t)$ is known to enable one representing a given operator F as a matrix with the elements

$$F_{mn} = \int \overline{\varphi}_m(q,t)F\varphi_n(q,t)\varrho\, dq.$$

If the matrix elements are calculated by means of functions $\varphi_s(q,t)$, we say that the operator is represented "in the scheme of $\varphi_s(q,t)$." If the corresponding system of functions is a basic system, we say that the operator is represented in its basic scheme.

In the basic scheme of $\varphi_s(q,t)$, a matrix with the elements

$$F_{mn} = \int \overline{\psi}_m F\psi_n\varrho\, dq \quad (15)$$

corresponds to the operator F.

We calculate the time derivative of the matrix elements (15). Making use of the differential equation (1) and of the self-adjointness of the operators F and H, we get the equations

$$\frac{d}{dt}\int \overline{\psi}_m F \psi_n \varrho \, dq = \int \overline{\psi}_m \left[\frac{\partial F}{\partial t} + \frac{2\pi i}{h}(HF - FH)\right] \psi_n \varrho \, dq \tag{16}$$

or

$$\frac{dF_{mn}}{dt} = \left[\frac{\partial F}{\partial t} + \frac{2\pi i}{h}(HF - FH)\right]_{mn}. \tag{17}$$

Denoting as **F** and **H** the matrices corresponding in the basic scheme to the operators F and H, we can write equation (17) as a matrix equation

$$\frac{d\mathbf{F}}{dt} = \frac{\partial \mathbf{F}}{\partial t} + \frac{2\pi i}{h}(\mathbf{HF} - \mathbf{FH}). \tag{18}$$

Substituting for **F** in (18) the matrices **q** and **p** corresponding to the coordinates and to the momenta, one gains the quantum mechanical equations of motion

$$\left.\begin{aligned} \frac{d\mathbf{p}}{dt} &= \frac{2\pi i}{h}(\mathbf{Hp} - \mathbf{pH}), \\ \frac{d\mathbf{q}}{dt} &= \frac{2\pi i}{h}(\mathbf{Hq} - \mathbf{qH}). \end{aligned}\right\} \tag{19}$$

Thus, the equations of motion are valid for the matrices constructed by means of our basic systems (the matrices in the basic scheme), and, as noticed already by Dirac,[4] also if the Hamilton function H depends on time explicitly.

The integral of the equation of motion is understood as an operator whose matrix in the basic scheme is constant. For such an operator F, the right-hand side of (16) vanishes for all m and n. The expression

$$\left[\frac{\partial F}{\partial t} + \frac{2\pi i}{h}(HF - FH)\right] \psi_n(q,t), \tag{20}$$

being orthogonal to all $\psi_m(q,t)$, must vanish due to the completeness of the system of functions ψ_n for all values of n.

Expanding any function $f(q)$ in the functions $\psi_n(q,t)$, we can come from (20) to the conclusion that the operator $\dfrac{\partial F}{\partial t} + \dfrac{2\pi i}{h}(HF - FH)$

[4] P.A.M. Dirac, *Physical Interpretation in Quantum Dynamics*, Proc. Roy. Soc. **A 113**, 621, 1926.

being applied to any functions gives zero. Hence, we can also say that the operator itself equals zero:

$$\frac{\partial F}{\partial t} + \frac{2\pi i}{h}(HF - FH) = 0. \tag{21}$$

This equation is thus fulfilled if the operator F is the integral of the quantum mechanical equations of motion, and equation (21) can be regarded as the definition of the notion "integral of the equations of motion."[5]

3. We consider an operator F with the following properties. Its eigenfunctions are assumed to be defined up to factors that can depend only on time, and not on the coordinates being chosen so that any quadratically integrable function in the q-space can be expanded in these functions. Such an operator P is said to be "complete." The energy operator H can serve as an example of a complete operator. On the contrary, the operator

$$yp_x - xp_y$$

corresponding to the surface integral of the equations of motion of a free particle in the axially symmetric field is not complete since its eigenfunctions are of the form

$$\varphi_n = f(z, x^2 + y^2, t)e^{in \arctan \frac{y}{x}} \qquad (n \text{ is integer})$$

containing an arbitrary factor depending on the coordinates.

Now we will state the following proposition:

Proposition 2. If an integral F of the equations of motion is a complete operator, then its eigenfunctions can be normalized so that they satisfy the Schrödinger equation also forming a basic system.

We will prove this proposition for the eigenfunctions of the operator F belonging to the point spectrum. Evidently, the proposition also holds true for the continuous spectrum.

Consider the integral

$$\int \overline{\varphi}_m \left[\frac{\partial F}{\partial t} + \frac{2\pi i}{h}(HF - FH)\right] \varphi_n \varrho \, dq = 0, \tag{22}$$

[5] The equivalence of both definitions was already pointed out by Dirac [Proc. Roy. Soc. **A 112**, 661, 1926]. (*V. Fock*)

where φ_n is an eigenfunction[6] of the operator F belonging to the point spectrum

$$F\varphi_n = \lambda_n \varphi_n, \tag{23}$$

and φ_m can belong both to the point and continuous spectrum; in the latter case it should be understood as being replaced by the corresponding differential.

Differentiating equation (23) with respect to time and assuming for a while that λ_n can also be time-dependent, one gets

$$\frac{\partial F}{\partial t}\varphi_n = \lambda_n \frac{\partial \varphi_n}{\partial t} - F\frac{\partial \varphi_n}{\partial t} + \frac{d\lambda_n}{dt}\varphi_n$$

and hence

$$\int \overline{\varphi}_m \frac{\partial F}{\partial t}\varphi_n \varrho \, dq = (\lambda_n - \lambda_m)\int \overline{\psi}_m \frac{\partial \varphi_n}{\partial t}\varrho \, dq + \frac{d\lambda_n}{dt}\delta_{nm},$$

where the self-adjointness of F was also used.

Further,

$$\int \overline{\varphi}_m HF\varphi_n \varrho \, dq = \lambda_n \int \overline{\varphi}_m H\varphi_n \varrho \, dq$$

and

$$\int \overline{\varphi}_m FH\varphi_n \varrho \, dq = \lambda_m \int \overline{\varphi}_m H\varphi_n \varrho \, dq.$$

Substituting these expressions in (22) one obtains

$$(\lambda_n - \lambda_m)\int \overline{\varphi}_m \Big[\frac{\partial \varphi_n}{\partial t} + \frac{2\pi i}{h}H\varphi_n\Big]\varrho \, dq + \frac{d\lambda_n}{dt}\delta_{nm} = 0. \tag{24}$$

From this one sees that, first,

$$\frac{d\lambda_n}{dt} = 0 \tag{25}$$

(which was to be expected), and, second, that the expression

$$\frac{h}{2\pi i}\frac{\partial \varphi_n}{\partial t} + H\varphi_n$$

[6]The function φ_n contains an arbitrary phase factor of modulus 1 which can also depend on time. In some cases, this factor should be defined so that φ_n considered as a function of time would oscillate possibly less. The exact formulation of this requirement and a method of defining the corresponding phase factors are given in the Appendix to this paper. (*V. Fock*)

is orthogonal to all the eigenfunctions whose eigenvalue differs from λ_n. If this eigenvalue is simple, it follows that

$$\frac{h}{2\pi i}\frac{\partial \varphi_n}{\partial t} + H\varphi_n = l_n \varphi_n, \tag{26}$$

where

$$l_n = \frac{h}{2\pi i}\int \overline{\varphi}_n \frac{\partial \varphi_n}{\partial t}\varrho\, dq + \int \overline{\varphi}_n H \varphi_n \varrho\, dq \tag{26*}$$

is a real constant, and the function

$$\psi_n(q,t) = e^{-\frac{2\pi i}{h}\int l_n dt}\varphi_n(q,t) \tag{27}$$

satisfies the Schrödinger equation (1). If, on the contrary, λ_n is a multiple eigenvalue with the multiplicity m, we denote the corresponding eigenfunctions (which can be regarded as normalized and orthogonal to each other) as

$$\varphi_n^{(1)}, \varphi_n^{(2)}, \cdots, \varphi_n^{(m)}.$$

Instead of (27), one obtains the system of equations

$$\left.\begin{aligned} H\varphi_n^{(1)} + \frac{h}{2\pi i}\frac{\partial \varphi_n^{(1)}}{\partial t} &= c_{11}^{(n)}\varphi_n^{(1)} + \cdots + c_{m1}^{(n)}\varphi_n^{(m)} \\ &\cdots\cdots\cdots\cdots \\ H\varphi_n^{(m)} + \frac{h}{2\pi i}\frac{\partial \varphi_n^{(m)}}{\partial t} &= c_{1m}^{(n)}\varphi_n^{(1)} + \cdots + c_{mm}^{(n)}\varphi_n^{(m)} \end{aligned}\right\} \tag{28}$$

with the Hermitian matrix $c_{kl}^{(n)}$ of the coefficients

$$c_{kl}^{(n)} = \int \overline{\varphi}_n^{(k)} H \varphi_n^{(l)} \varrho\, dq + \frac{h}{2\pi i}\int \overline{\varphi}_n^{(k)} \frac{\partial \varphi_n^{(l)}}{\partial t}\varrho\, dq. \tag{28*}$$

The eigenfunctions belonging to the multiple eigenvalue λ_n are defined up to an orthogonal transformation whose coefficients may depend on time. One can introduce new functions $\psi_n^{(l)}$ instead of $\varphi_n^{(k)}$ by means of the formulae

$$\left.\begin{aligned} \varphi_n^{(k)} &= \sum_{l=1}^{m} y_{lk}\psi_n^{(l)}, \\ \psi_n^{(l)} &= \sum_{k=1}^{m} \overline{y}_{lk}\varphi_n^{(k)}. \end{aligned}\right\} \tag{29}$$

Due to this, it is possible to ensure that the new functions $\psi_n^{(l)}$ satisfy the Schrödinger equation (1). The coefficients y_{lk} must then satisfy the system of equations

$$\frac{h}{2\pi i}\frac{dy_{lk}}{dt} = \sum_{s=1}^{m} y_{ls}c_{sk}^{(n)}. \tag{30}$$

We can choose conditions

$$y_{lk}(0) = \delta_{lk} \tag{31}$$

as the initial conditions. The functions $\psi_n^{(l)}$, $l = 1, 2 \ldots m$ are now defined up to an orthogonal transformation with constant coefficients. They are, like functions φ_n, eigenfunctions of the operator F, but also satisfy the Schrödinger equation (1), q.e.d.

We would like to make here a remark concerning integrals of motion that are not complete operators.

If G is such an integral, then the equality

$$\frac{h}{2\pi i}\frac{\partial G}{\partial t}\varphi + (HG - GH)\varphi = 0 \tag{32}$$

is valid for any function φ. If φ is an eigenfunction of the operator G

$$G\varphi = \lambda\varphi \quad (\lambda = \text{const}), \tag{33}$$

one can write equation (32) in the form

$$G\left(\frac{h}{2\pi i}\frac{\partial\varphi}{\partial t} + H\varphi\right) = \lambda\left(\frac{h}{2\pi i}\frac{\partial\varphi}{\partial t} + H\varphi\right). \tag{34}$$

The function

$$\varphi' = \frac{h}{2\pi i}\frac{\partial\varphi}{\partial t} + H\varphi, \tag{35}$$

thus, satisfies the same equation (33) as the function φ, being an eigenfunction belonging to the same eigenvalue of the operator G. In some cases that are not considered here, it is possible to conclude that φ' is proportional to φ. Further, in some cases it is possible to normalize the eigenfunction φ so that the normalized function satisfies the Schrödinger equation (1). Equation (34) is also obviously valid for the continuous spectrum; we do not want, however, to discuss the questions arising here in more detail.

4. As an example of application of Proposition 2, we consider a harmonic oscillator with its instantaneous period being a quadratic function of time.[7]

The Hamilton function $\mathbf{H}$ as a matrix function of the canonical variables $\mathbf{p}$ and $\mathbf{q}$ is

$$\mathbf{H} = \frac{1}{2m}\mathbf{p}^2 + \frac{2\pi^2 m}{T^2}\mathbf{q}^2, \tag{36}$$

the quantum mechanical equations of motion being

$$\left.\begin{aligned} \dot{\mathbf{p}} &= -\frac{4\pi^2 m}{T^2}\mathbf{q}, \\ \dot{\mathbf{q}} &= \frac{\mathbf{p}}{m}. \end{aligned}\right\} \tag{37}$$

It is easy to see that the expression

$$\frac{T}{m}\mathbf{p}^2 + \frac{4\pi^2 m}{T^2}\mathbf{q}^2 - \frac{1}{2}\dot{T}(\mathbf{qp} + \mathbf{pq}) + \frac{1}{2}\ddot{T}m\mathbf{q}^2 \tag{38}$$

is an integral of the equations of motion (37) if, as assumed, the period T is a quadratic function of time so that its third derivative $\dddot{T}$ vanishes identically. Now we pass from the matrices to the operators.

Let us denote time as t', then the Schrödinger equation is, in our case,

$$-\frac{h^2}{8\pi^2 m}\frac{\partial^2\psi}{\partial q^2} + \frac{2\pi^2 m}{T^2}q^2\psi + \frac{h}{2\pi i}\frac{\partial\psi}{\partial t'} = 0, \tag{39}$$

and the operator F corresponding to expression (38) is

$$F\psi = -\frac{h^2 T}{4\pi^2 m}\cdot\frac{\partial^2\psi}{\partial q^2} + \frac{4\pi^2 m}{T}q^2\psi - \frac{h}{4\pi i}\dot{T}\left(q\frac{\partial\psi}{\partial q} + \frac{1}{2}\psi\right) + \frac{1}{2}\ddot{T}mq^2\psi. \tag{40}$$

The period T as a function of time t' is given as

$$T = T_0 + 4\pi\beta t' + \frac{4\pi^2\gamma}{T_0}t'^2, \tag{41}$$

[7]This example is also suitable for explaining the adiabatic rule carried over by M. Born to the new quantum mechanics. [A new proof of this rule (by M. Born and the author) will appear in this journal (see [28-4], this book).] Compare also with the paper by G. Krutkow and the author: *Über das Rayleighsche Pendel*, Zs. Phys. **13**, 195, 1923 (see [23-1], this book), where a similar problem is considered within the framework of the old quantum mechanics. (*V. Fock*)

where T_0 is a constant with the dimension of time, β and γ being dimensionless constants. We introduce the dimensionless quantities x and t' instead of q and t', writing

$$\left.\begin{aligned} 2\pi\sqrt{\frac{m}{hT_0}}\,q &= x, \\ 2\pi\frac{t'}{T_0} &= t. \end{aligned}\right\} \tag{42}$$

The expression for the period T is now

$$T = T_0(1 + 2\beta t + \gamma t^2). \tag{43}$$

In terms of the new variables, the Schrödinger equation becomes

$$\frac{\partial^2\psi}{\partial x^2} - \frac{x^2}{(1 + 2\beta t + \gamma t^2)^2}\psi + 2i\frac{\partial\psi}{\partial t} = 0 \tag{44}$$

and the equation for the eigenfunctions φ of the operator F is

$$\begin{aligned} F\varphi = &-(1 + 2\beta t + \gamma t^2)\frac{\partial^2\varphi}{\partial x^2} + 2i(\beta + \gamma t)\left(x\frac{\partial\varphi}{\partial x} + \frac{1}{2}\varphi\right) + \\ &+ \left(\frac{1}{1 + 2\beta t + \gamma t^2} + \gamma\right)x^2\varphi = \lambda\varphi. \end{aligned} \tag{45}$$

To remove the term with the first derivative we put

$$\varphi = \exp i\left(\frac{T'}{T}\cdot\frac{x^2}{4}\right)\cdot\varphi_1 = \exp i\left(\frac{(\beta + \gamma t)x^2}{2(1 + 2\beta t + \gamma t^2)}\right)\cdot\varphi_1$$

obtaining for φ_1 the equation

$$(1 + 2\beta t + \gamma t^2)\frac{\partial^2\varphi_1}{\partial x^2} - \frac{1 + \gamma - \beta^2}{1 + 2\beta t + \gamma t^2}x^2\varphi_1 + \lambda\varphi_1 = 0. \tag{46}$$

Finally, we introduce, instead of x and λ, the quantities ξ and λ_1

$$\xi = \frac{\sqrt[4]{1 + \gamma - \beta^2}}{\sqrt{1 + 2\beta t + \gamma t^2}}\cdot x; \quad \lambda_1 = \frac{\lambda}{\sqrt{1 + \gamma - \beta^2}}, \tag{47}$$

so that the equation for φ_1 is written as

$$\frac{\partial^2\varphi}{\partial\xi^2} + (\lambda_1 - \xi^2)\varphi_1 = 0. \tag{48}$$

This is a well-known equation for the Hermit orthogonal functions. The eigenvalues here are $\lambda_1 = 2n+1$ $(n = 0, 1 \ldots)$. The eigenfunctions must be normalized so that

$$\int_{-\infty}^{+\infty} |\varphi_1|^2 dx = \frac{\sqrt{1+2\beta t+\gamma t^2}}{\sqrt[4]{1+\gamma-\beta^2}} \int_{-\infty}^{+\infty} |\varphi_1|^2 d\xi = 1. \tag{49}$$

The normalized eigenfunctions of the operator F (45) are, hence,

$$\varphi_n(x,t) = \frac{(1+\gamma-\beta^2)^{1/8}}{\sqrt{\sqrt{\pi}\cdot 2^n \cdot n!}}(1+2\beta t+\gamma t^2)^{-1/4} e^{i\frac{(\beta+\gamma t)\xi^2}{2\sqrt{1+\gamma-\beta^2}}} e^{-\frac{1}{2}\xi^2} H_n(\xi), \tag{50}$$

where $H_n(\xi)$ denotes a Hermit polynomial and ξ is the brief notation for expression (47). Following Proposition 2, these functions can differ from the solutions of the Schrödinger equation (basic system) only by a factor modulus 1 depending on time.

The calculation following the method used while proving Proposition 2 gives for this factor the value

$$e^{-i(n+1/2)\tau},$$

where, for brevity, we put

$$\tau = \int_0^t \frac{\sqrt{1+\gamma-\beta^2}}{1+2\beta t+\gamma t^2} dt. \tag{51}$$

The functions

$$\psi_n(x,t) = e^{-i(n+1/2)\tau} \varphi_n(x,t) \tag{52}$$

satisfy the Schrödinger equations also forming a basic system.

This result can be verified by substituting the expression

$$\psi(x,t) = e^{i\frac{(\beta+\gamma t)x^2}{2\sqrt{1+2\beta t+\gamma t^2}}} \cdot (1+2\beta t+\gamma t^2)^{-1/4} f(\xi,\tau) \tag{53}$$

for ψ in equation (44) and going to the variables ξ and τ. In terms of these variables, one gets the following equation for the function $f(\xi,\tau)$ in (53):

$$\frac{\partial^2 f}{\partial \xi^2} - \xi^2 f + 2i\frac{\partial f}{\partial \tau} = 0, \tag{54}$$

i.e., the Schrödinger equation for an oscillator with a constant period. The solution of this equation is obvious:

$$f_n(\xi,\tau) = C_n e^{-i(n+1/2)\tau} \cdot e^{\frac{1}{2}\xi^2} \cdot H_n(\xi). \tag{55}$$

The latter expression, being put into (53), results in equation (52).

5. As a simple example of the applicability of Proposition 2 in the region of the continuum spectrum (in which case it was not proved here) let us consider the free motion of a particle. The equations are written with respect to the reduced dimensionless quantities. The Hamilton function is

$$\mathbf{H} = \tfrac{1}{2}(\mathbf{p}_x^2 + \mathbf{p}_y^2 + \mathbf{p}_z^2), \tag{56}$$

the Schrödinger equation being

$$\Delta\psi + 2i\frac{\partial\psi}{\partial t} = 0, \tag{57}$$

where Δ denotes the ordinary Laplace operator. The equations of motion

$$\left.\begin{aligned} \dot{\mathbf{p}}_x &= 0, \quad \dot{\mathbf{p}}_y = 0, \quad \dot{\mathbf{p}}_z = 0, \\ \dot{\mathbf{q}}_x &= \mathbf{p}_x, \; \dot{\mathbf{q}}_y = \mathbf{p}_y, \; \dot{\mathbf{q}}_z = \mathbf{p}_z \end{aligned}\right\} \tag{58}$$

possess six integrals, three of the kind

$$p_x = \lambda = \text{const} \tag{59}$$

and three of the kind

$$x - p_x t = \alpha = \text{const}. \tag{60}$$

The operators

$$\frac{1}{i}\frac{\partial}{\partial x} \quad \text{or} \quad x - \frac{t}{i}\frac{\partial}{\partial x}$$

correspond to integrals (59) or (60). Their eigenfunctions satisfy the equations

$$\frac{\partial\psi(x,\lambda)}{\partial x} = i\lambda\,\psi(x,\lambda) \tag{61}$$

and

$$\frac{\partial\psi(x,\alpha)}{\partial x} - i\frac{x}{t}\varphi(x,\alpha) + \frac{i\alpha}{t}\varphi(x,\alpha) = 0. \tag{62}$$

We will introduce these eigenfunctions by the requirement

$$\frac{1}{\Delta\lambda}\int\limits_{-\infty}^{+\infty} dx \left| \int\limits_{\lambda}^{\lambda+\Delta\lambda} \psi(x,\lambda)\, d\lambda \right|^2 = 1 \tag{63}$$

and

$$\frac{1}{\Delta\alpha}\int\limits_{-\infty}^{+\infty} dx \left| \int\limits_{\alpha}^{\alpha+\Delta\alpha} \varphi(x,\alpha)\, d\alpha \right|^2 = 1. \tag{64}$$

The eigenfunctions normalized in this way are

$$\left.\begin{aligned} \psi(x,\lambda) &= c_1 \frac{1}{\sqrt{2\pi}} e^{i\lambda x}, \\ \varphi(x,\alpha) &= c_2 \frac{1}{\sqrt{2\pi t}} e^{\frac{i(x-\alpha)^2}{2t}}, \end{aligned}\right\} \tag{65}$$

where c_1 and c_2 are phase factors of modulus 1 that can depend only on time. If one puts

$$c_1 = e^{-\frac{i\lambda^2 t}{2}}, \quad c_2 = 1,$$

then the functions

$$\left.\begin{aligned} \psi(x,\lambda) &= \frac{1}{\sqrt{2\pi}} e^{i\lambda x - i\lambda^2 \frac{t}{2}}, \\ \varphi(x,\alpha) &= \frac{1}{\sqrt{2\pi t}} e^{\frac{i(x-\alpha)^2}{2t}} \end{aligned}\right\} \tag{66}$$

satisfy the Schrödinger equation (57), which can be verified in a direct way.

Appendix

The normalization of the eigenfunctions of the operators depending on time explicitly

As is known, the eigenfunctions of an operator are defined only up to a factor of modulus 1 (phase factor), or — in the case of a multiple eigenvalue — up to an orthogonal substitution. If the operator is time-dependent, then the normalization elements (e.g., the phase factor, or the coefficients of the substitution) also can be time-dependent. Determining the normalization elements properly, one can ensure that the time derivatives of the eigenfunctions satisfy certain requirements. It is natural to require that the first derivative becomes as small as possible. This normalization proves to be particularly advisable in the case of the eigenfunctions of the time-dependent energy operator for constructing the approximate solutions of the Schrödinger equation.

We wish to demonstrate here how one can normalize the eigenfunctions of a time-dependent operator so that they oscillate as weakly as possible.

In the case of a simple eigenvalue we will fix the time-dependent phase factor by imposing the condition that the integral

$$\int \left| \frac{\partial \varphi_n}{\partial t} \right|^2 \varrho \, dq = \min \tag{A1}$$

is as small as possible. Then the function φ_n is defined up to a phase factor that does not depend on time any more. In the case of a multiple eigenvalue, let the eigenfunctions belonging to it be

$$\varphi_n^{(1)}, \varphi_n^{(2)}, \cdots, \varphi_n^{(m)}.$$

These are defined up to an orthogonal transformation, whose coefficients are functions of time. We choose this transformation so that the sum of the integrals

$$\sum_{r=1}^{m} \int \left| \frac{\partial \varphi_n^{(r)}}{\partial t} \right|^2 \varrho \, dq = \min \tag{A2}$$

is as small as possible. Then the functions $\varphi_n^{(r)}$ are defined up to an orthogonal transformation with constant coefficients. The requirement (1a) is fulfilled if φ_n satisfies the equation

$$\int \overline{\varphi}_n \frac{\partial \varphi_n}{\partial t} \varrho \, dq = 0. \tag{A3}$$

Indeed, for some other permissible function

$$\psi_n = e^{i\omega_n} \varphi_n \tag{A4}$$

the corresponding integral is equal to

$$\int \left| \frac{\partial \psi_n}{\partial t} \right|^2 \varrho \, dq = \dot{\omega}_n^2 + \int \left| \frac{\partial \varphi_n}{\partial t} \right|^2 \varrho \, dq, \tag{A5}$$

so that it is always greater than (1a) if only ω_n is not constant.

If the function φ_n does not satisfy equation (3a), one can choose the quantity ω_n in (4a) so that (3a) is satisfied by ψ_n. To this end, one has to solve the equation

$$\dot{\omega}_n = i \int \overline{\varphi}_n \frac{\partial \varphi_n}{\partial t} \varrho \, dq, \tag{A6}$$

ω_n thus obtained being obviously real.

In the case of a multiple eigenvalue, condition (2a) is fulfilled if the functions $\varphi_n^{(r)}$ satisfy the equations

$$\int \frac{\partial \varphi_n^{(k)}}{\partial t} \overline{\varphi}_n^{(l)} \varrho \, dq = 0 \qquad (k, l = 1, 2, \ldots, m). \tag{A7}$$

Then one introduces, via the orthogonal substitution

$$\left.\begin{aligned} \psi_n^{(l)} &= \sum_{k=1}^{m} \overline{y}_{lk}(t) \varphi_n^{(k)}, \\ \varphi_n^{(k)} &= \sum_{l=1}^{m} y_{lk}(t) \psi_n^{(l)}, \end{aligned}\right\} \tag{A8}$$

new functions $\psi_n^{(l)}$ and calculates for these functions the expression analogous to (2a). Taking into account equation (7a) and the orthogonality of the substitution, one gets the expression

$$\sum_{r=1}^{m} \int \left| \frac{\partial \psi_n^{(r)}}{\partial t} \right|^2 \varrho \, dq = \sum_{r=1}^{m} \int \left| \frac{\partial \varphi_n^{(r)}}{\partial t} \right|^2 \varrho \, dq + \sum_{s,r=1}^{m} \left| \frac{dy_{rs}}{dt} \right|^2 . \tag{A9}$$

Since the second term is necessarily positive vanishing only for constant y_{rs}, the functions $\varphi_n^{(k)}$ possess the minimal value searched for if they satisfy equation(4a).

If equation (7a) is not satisfied, however, by the functions $\varphi_n^{(k)}$, one can choose the coefficients of the substitution in (8a) so that an analogous equation will be satisfied by the functions $\psi_n^{(l)}$. These coefficients must be solutions of the system of differential equations

$$\frac{dy_{sr}}{dt} i \sum_{k=1}^{m} y_{sk} b_{kr}, \tag{A10}$$

where the matrix with the elements

$$b_{kr} = i \int \frac{\partial \overline{\varphi}_n^{(k)}}{\partial t} \varphi_n^{(r)} \varrho \, dq \tag{A11}$$

is obviously Hermitian. To fulfill the orthogonality relation of y_{rs}, it is sufficient to choose the values

$$y_{sr}(0) = \delta_{sr} \tag{A12}$$

as an initial condition.

If the functions $\varphi_n^{(r)}$ are normalized in the way described here, the coefficients $c_{kl}^{(n)}$ in equation (28) have a simple meaning of the matrix elements of the energy matrix.

I would like to express my heartfelt gratitude to the International Education Board for the possibility of working in Göttingen.

Translated by A.G. Izergin and A.G. Pronko

28-3*
Generalization and Solution of the Dirac Statistical Equation

V. Fock[1]
temporarily in Göttingen

Received 1 May 1928

Zs. Phys. **49**, 339, 1928
Fock57, pp. 9–24

In his paper "Emission and Absorption of Radiation" [1] Dirac considers the Bose–Einstein statistics of an ensemble of mechanical systems by a completely new method. He admits that the perturbation of the system by an external force takes place and considers the changes caused by this perturbation in the given probability distribution of the energy levels in the ensemble.

What is essentially new in the Dirac method is that he considers the number N_s of systems on the s-th energy level as a canonical variable. In the space of such variables (which we'll call the Dirac space) Dirac establishes the wave equation; its solutions are the functions of N_s and time. The square modulus of the solution defines the probability of the corresponding distribution of the energy levels in the ensemble.

However an algorithm on how to solve this equation is not given in the Dirac paper.

In the present paper[2] the problem is generalized and we look for the probability distribution of any arbitrary mechanical quantity (not only of energy), whereas the probability amplitudes for the distribution of another (or the same) arbitrary mechanical quantity at time $t = 0$ are given as initial conditions.

---

[1] International Education Board Fellow.

[2] In the original text an attempt has been made to apply the method developed in this paper to the Fermi statistics. This attempt was unsuccessful, and therefore the parts related to the Fermi statistics (including §8) are omitted. The erratum was published in the paper "*On Quantum Electrodynamics*" Sow. Phys. **6**, 5, 428 (1934). The difficulties, connected with the application of this method to the Fermi statistics, were mentioned already in the original paper (*V. Fock, 1957*). The present translation uses the revised version in Fock57. (*Translator*)

The generalized Dirac equation is solved generally by the use of the generating function for which an explicit expression can be given.

§1. We consider an ensemble of identical mechanical systems. The energy operator H of each system can contain time. Together with energy we also consider two other mechanical quantities a and b with operators A and B.

Further we need operator A only for a fixed time $t_0 = 0$, so we can assume that A does not depend on time explicitly. On the contrary, we consider operator B at a variable time and correspondingly we assume that it can explicitly contain time.[3]

The Schrödinger equation for a single system is

$$H\psi + \frac{\hbar}{i}\frac{\partial\psi}{\partial t} = 0\,. \tag{1}$$

The eigenfunctions of operators A and B satisfy the equations

$$A\ \psi_s(q) = a_s\ \psi_s(q)\,, \tag{2}$$

$$B\ \varphi_s(q,t) = \beta_s\ \varphi_s(q,t)\,. \tag{3}$$

The arbitrary time-dependent phase factors of functions $\varphi_s(q,t)$ can be chosen by the prescription[4] proposed by the author in [2].

We will also consider a system of solutions $\psi_s(q,t)$ of the Schrödinger equation

$$H\psi_s(q,t) + \frac{\hbar}{i}\frac{\partial\psi_s}{\partial t} = 0 \tag{4}$$

that satisfy the initial conditions

$$\psi_s(q,0) = \psi_s(q)\,. \tag{5}$$

According to the theorem proved in the quoted paper by the author the system of functions $\psi_s(q,t)$ will be complete, normalized and orthogonal for any t. Further we will call it the basic system.

§2. Each solution $f(q,t)$ of the Schrödinger equation (1) is defined uniquely by its initial value $f(q,0)$. If we expand the initial value into the set of functions $\psi_s(q) = \psi_s(q,0)$:

$$f(q,0) = \sum_s g_s\ \psi_s(q,0)\,, \tag{6}$$

[3]In Dirac's paper both operators A and B coincide and are equal to the energy operator of the unperturbed system. (*V. Fock*)

[4]Each eigenfunction normalized along this prescription should be orthogonal to its time derivative. (*V. Fock*)

then at time t the solution will be

$$f(q,t) = \sum_s g_s\ \psi_s(q,t)\,. \tag{7}$$

The same solution can be expanded into eigenfunctions of operators A and B:

$$f(q,t) = \sum_s x_s\ \psi_s(q,0)\,, \tag{8}$$

$$f(q,t) = \sum_s y_s\ \varphi_s(q,t)\,, \tag{9}$$

where coefficients x_s and y_s are functions of time, whereas g_s in (7) were constants.[5] The infinite-dimensional space of all sets of expansions' coefficients

$$\left.\begin{array}{l} g_1,\ g_2, \dots\ g_s, \dots \\ x_1,\ x_2, \dots\ x_s, \dots \\ y_1,\ y_2, \dots y_s, \dots \end{array}\right\} \tag{10}$$

is called the complex Hilbert space. The elements of each line in (10) will be the "coordinates" in this space. Transition from one set (line) to another corresponds to a linear orthogonal (unitary) transformation of coordinates, which can be considered as a rotation of a coordinate system in the Hilbert space.

The physical meaning of y_s is the following. Let the operator b with eigenvalues β_s and eigenfunctions $\varphi_s(q,t)$ correspond to the physical quantity b. If we expand the solution of the Schödinger equation into functions $\varphi_s(q,t)$, the square modulus $|y_s|^2$ of the expansion coefficient y_s gives a relative probability for the quantity b to have the value β_s at time t.

Instead of saying "the quantity b is equal to the eigenvalue β_s of operator B" we can say shorter "the system is in state s" each time when it is clear by which operator the states we are talking about are defined.

Because the sum of the squared moduli $|y_s|^2$ is equal to the corresponding sum of g_s and therefore is a constant, we can put it equal to the number of systems N in the whole ensemble

$$\sum_s |y_s|^2 = \sum_s |g_s|^2 = N. \tag{11}$$

[5] If we consider $\psi_s(q,0)$ as eigenfunctions of an unperturbed system, then x_s are the same values that Dirac calls b_s. (*V. Fock*)

The square modulus

$$|y_s|^2 = N_s^\beta \tag{12}$$

is then the probable number N_s^β of systems in state s (i.e., with the eigenvalue β_s). The same is naturally true also for operator A.

§3. Let us establish the differential equations which are satisfied by the expansion coefficients y_s.

If we multiply each of expansions (7) and (9) by

$$\overline{\varphi}_r(q,t)\ \varrho(q)\,dq$$

[$\varrho(q)$ is the density function in q-space], integrate over q-space and put both results equal, then we obtain the expression for y_r through constants g_s:

$$y_r = \sum_s Y_{rs}\ g_s\,, \tag{13}$$

where for brevity we denoted

$$Y_{rs} = \int \overline{\varphi}_r(q,t)\ \psi_s(q,t)\ \varrho dq\,. \tag{14}$$

The values Y_{rs} satisfy the equations

$$\left.\begin{aligned} \sum_l Y_{rl}\overline{Y}_{sl} = \delta_{rs} \\ \sum_l Y_{lr}\overline{Y}_{ls} = \delta_{rs} \end{aligned}\right\} \tag{15}$$

and therefore are the entries of the unitary matrix.

Differentiating (14) by time and taking into account that $\psi_s(q,t)$ satisfies the Schrödinger equation (4), we obtain

$$\frac{dY_{rs}}{dt} = \int \frac{\partial\overline{\varphi}_r}{\partial t}\psi_s\ \varrho dq - \frac{i}{\hbar}\int \overline{\varphi}_r H\psi_s\ \varrho dq\,. \tag{16}$$

Expanding here $\psi_s(q,t)$ into $\varphi_l(q,t)$:

$$\psi_s(q,t) = \sum_l Y_{ls}\ \varphi_l(q,t)\,, \tag{17}$$

and denoting for brevity

$$K_{rl} = \int \overline{\varphi}_r(q,t)\ H\varphi_l(q,t)\ \varrho dq + \frac{\hbar}{i}\int \overline{\varphi}_r(q,t)\frac{\partial\varphi_l}{\partial t}\varrho dq\,, \tag{18}$$

we get the differential equations

$$\frac{\hbar}{i}\frac{dY_{rs}}{dt} = -\sum_l K_{rl} Y_{ls} \,. \tag{19}$$

The coefficients matrix K_{rl} is evidently Hermitian. Because y_s are the linear functions of Y_{rs} with constant coefficients g_s, the y_s satisfy the same differential equations (19). This statement can be presented as a theorem:

Theorem 1. The expansion coefficients y_s of the solution of the Schrödinger equation into an orthogonal system of $\varphi_s(q,t)$ are solutions of the differential equations

$$\frac{\hbar}{i}\frac{dy_r}{dt} = -\sum_l K_{rl}\; y_l \,, \tag{20}$$

with the definition (18) of K_{rl}.

This theorem is evidently valid for any complete orthogonal system. Particularly for the basic system (e.g., for $\psi_r(q,t)$) expressions (18) are equal to zero and the expansion coefficients are constants.

§4. Together with the system of differential equations (20) we consider its complex conjugate one

$$\frac{\hbar}{i}\frac{d\overline{y}_r}{dt} = \sum_l K_{lr}\; \overline{y}_l \,. \tag{20*}$$

Denoting the bilinear form by F,

$$F = \sum_{sr} K_{sr}\; \overline{y}_s y_r \,, \tag{21}$$

we can write equations (20) and (20*) as

$$\frac{\hbar}{i}\frac{dy_r}{dt} = -\frac{\partial F}{\partial \overline{y}_r} \tag{22}$$

or

$$\frac{\hbar}{i}\frac{d\overline{y}_r}{dt} = \frac{\partial F}{\partial y_r} \,. \tag{22*}$$

If we consider $\overline{y}_r$ (or y_r) as a canonical coordinate and $\frac{\hbar}{i} y_r$ (or $-\frac{\hbar}{i}\overline{y}_r$) as a canonical momentum, i.e., if we put $\overline{y}$

$$Q_r = \overline{y}_r \,, \quad P_r = \frac{\hbar}{i}\; y_r \tag{23}$$

or

$$Q_r = y_r \,, \qquad P_r = -\frac{\hbar}{i}\,\overline{y}_r \,, \tag{23*}$$

equations (22) and (22*) can be considered as canonical equations of motion in the Hilbert space with the Hamilton function F.

Now following Dirac we will consider that canonical variables are operators (matrices, q-numbers) and establish the Schrödinger equation corresponding to the Hamilton operator F. We can use here either the space of y_r or the space of $\overline{y}_r$.

In the space of y_r the operator y_r means "multiplication by y_r" and the operator $\overline{y}_r$ means "changing the sign and taking the derivative by y_r":

$$y_r \to y_r \,; \qquad \overline{y}_r \to -\frac{\partial}{\partial y_r} \,. \tag{24}$$

In the space of $\overline{y}_r$ the operator $\overline{y}_r$ means "multiplication by $\overline{y}_r$" and operator y_s means "taking derivative by $\overline{y}_r$":

$$\overline{y}_r \to \overline{y}_r \,; \qquad y_r \to \frac{\partial}{\partial \overline{y}_r} \,. \tag{25}$$

This follows from the well-known general formulae

$$P_r = \frac{\hbar}{i}\,\frac{\partial}{\partial Q_r} \,; \qquad Q_r = -\frac{\hbar}{i}\,\frac{\partial}{\partial P_r} \,, \tag{26}$$

and is equally valid whether we define the canonical variables explicitly according to (23) or (23*).

Denoting the wave function in space of y_r by Ω we get

$$\frac{\hbar}{i}\,\frac{\partial \Omega}{\partial t} - \sum_{sr} K_{sr}\, y_r \frac{\partial \Omega}{\partial y_s} = 0 \,. \tag{27}$$

We have to remark here that the sequence of operators in the diagonal terms of (27) is not essential. Indeed, if we apply the operators y_r and $\frac{\partial}{\partial y_r}$ in a different sequence, and, e.g., write

$$\frac{\partial}{\partial y_r}\,(y_r \Omega) \qquad \text{or} \qquad \frac{1}{2}\left[\frac{\partial}{\partial y_r}(y_r \Omega) + y_r \frac{\partial \Omega}{\partial y_r}\right] ,$$

then instead of zero in the right-hand side of (27) we would have a term of the form $c(t)\Omega$, where $c(t)$ is a real function of time. However it is easily seen that the solution of the new equation differs from that of

(27) only by a factor with the absolute value equal to unity, i.e., by an inessential phase factor.

Equation (27) represents a wave equation in space of y_r. To obtain the corresponding equation in the Hilbert space with $\overline{y}_r$ as coordinates we have to substitute expressions (24) for operators $y_r, \overline{y}_r$ in the Hamilton function (21) by the expressions (25). If we denote the wave function in the space $\overline{y}_r$ as Ω, we find

$$\frac{\hbar}{i}\,\frac{\partial\overline{\Omega}}{\partial t} + \sum_{sr} K_{sr}\,\overline{y}_s\,\frac{\partial\overline{\Omega}}{\partial\overline{y}_r} = 0\,. \tag{28}$$

Due to $K_{sr} = \overline{K}_{rs}$ this equation is exactly complex conjugate to (27).

Thus, to get the wave equation in the Hilbert space it is unimportant whether we consider the expressions (23) or (23*) as "coordinates" and "momenta." As the sequence of operators in the Hamilton function is also unimportant, we can state that the wave equation in Hilbert space is established uniquely.

Now we return to equation (27). This is a linear partial differential equation of the first order. If we form the system of ordinary differential equations corresponding to (27) we will exactly get the system (20). So this system and equation (27) are adjoint in the sense of the theory of partial differential equations. From this it follows that to make the transformation of wave equation (25) to a different coordinate system in the Hilbert space we need only to change the independent variables according to the ordinary rules of differential calculus. The wave function Ω is covariant relative to such transformation.

This remark allows one to find the general solution of wave equation (27) without any calculations. Indeed, if we choose the expansion coefficients of g_s into the basic system of functions as the coordinates in the Hilbert space, then in these coordinates the wave equation has a simple form

$$\left(\frac{\partial\Omega}{\partial t}\right)_{g_s} = 0\,, \tag{29}$$

where the subscript g_s means that the time derivative is taken at constant g_s. The general solution of this equation is an arbitrary function of variables g_s:

$$\Omega = \Omega(g_1,\ g_2,\ \ldots\ g_s\ \ldots)\,, \tag{30}$$

where in the case of equation (25) we have to express g_s through y_r:

$$g_s = \sum_r \overline{Y}_{rs}\, y_r\,. \tag{31}$$

The general solution of equation (28) is complex conjugate to (30). We can summarize the results of this section as a theorem.

Theorem 2. The Dirac statistical equation in the Hilbert space with the coordinates y_r is a first-order partial differential equation adjoint to the system of ordinary differential equations (20) of Theorem 1. Its general solution will be an arbitrary function of the expansion coefficients into the basic system of functions.

§5. As we mentioned in §2, not y_s themselves but their square moduli $|y_s|^2 = \overline{y}_s y_s$ (namely, the probable number of systems in state s) have the physical meaning. Therefore using a canonical transformation we introduce new variables n_s and θ_s in a way that the operator

$$\overline{y}_s y_s \to \overline{y}_s \frac{\partial}{\partial \overline{y}_s} \to n_s \tag{32}$$

means simply the multiplication by a nonnegative integer n_s. Such a canonical transformation will be

$$\left.\begin{aligned} \overline{y}_s &= \frac{\Phi(n_s)}{\Phi(n_s-1)} e^{\frac{i}{\hbar}\theta_s} = e^{\frac{i}{\hbar}\theta_s} \frac{\Phi(n_s+1)}{\Phi(n_s)}, \\ y_s = \frac{\partial}{\partial \overline{y}_s} &= \frac{(n_s+1)\Phi(n_s)}{\Phi(n_s+1)} e^{-\frac{i}{\hbar}\theta_s} = e^{-\frac{i}{\hbar}\theta_s} \frac{n_s \Phi(n_s-1)}{\Phi(n_s)}, \end{aligned}\right\} \tag{33}$$

where $\Phi(n)$ is an arbitrary function of the integer n for which we demand that at $n = 0$ it will be equal to unity and for negative n is infinite:[6]

$$\Phi(0) = 1; \quad \frac{1}{\Phi(-k)} = 0 \quad (k = 1, 2, \ldots). \tag{34}$$

The values θ_s should be considered as canonical variables and the values n_s as corresponding momenta. Then the meaning of θ_s will be:

$$\theta_s \to -\frac{\hbar}{i} \frac{\partial}{\partial n_s},$$

and the operator

$$e^{\frac{i}{\hbar}\theta_s} = e^{-\frac{\partial}{\partial n_s}}$$

will mean "the decrease of the number n_s by unity," whereas the operator

$$e^{-\frac{i}{\hbar}\theta_s} = e^{\frac{\partial}{\partial n_s}}$$

[6] In the original text the form of the function $\Phi(n)$ was related to the type of statistics, whereas in fact it is connected with the normalizing condition; see further the formula (*). In the present edition this mistake is corrected. (*V. Fock, 1957*)

means "the increase of the number n_s by unity." The transformation formulae can also be written in the form

$$\left.\begin{aligned}\overline{y}_s &= \Phi(n_s)e^{-\frac{\partial}{\partial n_s}}\frac{1}{\Phi(n_s)},\\ y_s &= \frac{\partial}{\partial \overline{y}_s} = \frac{\Phi(n_s)}{n_s!}e^{\frac{\partial}{\partial n_s}}\frac{n_s!}{\Phi(n_s)}.\end{aligned}\right\} \tag{33*}$$

We must require that the operator y_s is conjugate to $\overline{y}_s$. This gives

$$|\Phi(n)|^2 = n!\,.$$

Considering $\Phi(n)$ to be real, we can put

$$\Phi = \sqrt{\Gamma(n+1)} = \sqrt{n!}\ . \tag{35}$$

This function evidently satisfies conditions (34). Substituting (35) into (33) we get the transformation used by Dirac, namely,

$$\left.\begin{aligned}\overline{y}_s &= \sqrt{n_s}e^{\frac{i}{\hbar}\theta_s} = e^{\frac{i}{\hbar}\theta_s}\sqrt{n_s+1},\\ y_s &= \sqrt{n_s+1}e^{-\frac{i}{\hbar}\theta_s} = e^{-\frac{i}{\hbar}\theta_s}\sqrt{n_s},\end{aligned}\right\} \tag{36}$$

or

$$\left.\begin{aligned}\overline{y}_s &= \sqrt{n_s}\ e^{-\frac{\partial}{\partial n_s}} = e^{-\frac{\partial}{\partial n_s}}\sqrt{n_s+1},\\ y_s &= \sqrt{n_s+1}\ e^{\frac{\partial}{\partial n_s}} = e^{\frac{\partial}{\partial n_s}}\ \sqrt{n_s}.\end{aligned}\right\} \tag{36*}$$

Besides, we have to show that transformations (33) or (36) are really canonical, i.e., that the next commutation relations between operators $\overline{y}_s$ and $\dfrac{\partial}{\partial \overline{y}_s}$ hold

$$\left.\begin{aligned}\frac{\partial}{\partial \overline{y}_s}\overline{y}_r - \overline{y}_r\frac{\partial}{\partial \overline{y}_s} &= \delta_{rs},\\ \frac{\partial}{\partial \overline{y}_s}\frac{\partial}{\partial \overline{y}_r} - \frac{\partial}{\partial \overline{y}_r}\frac{\partial}{\partial \overline{y}_s} &= 0,\\ \overline{y}_s\overline{y}_r - \overline{y}_r\overline{y}_s &= 0.\end{aligned}\right\} \tag{37}$$

At $r = s$ the first of these relations follows from equations

$$\frac{\partial}{\partial \overline{y}_s}\ \overline{y}_s = n_s + 1; \qquad \overline{y}_s\ \frac{\partial}{\partial \overline{y}_s} = n_s\,, \tag{37*}$$

which can be obtained from (33). The other relations are also evidently valid. So transformation (33) is canonical.

§6. Now we have to investigate the transition given by formulae (33) and (36) from the Hilbert space of variables $\overline{y}_r$ to the Dirac space of variables n_r. Let us consider first the case of one variable $\overline{y}_r$, which we will denote as z.

The function $c(n)$ of an integer number n in the Dirac space corresponds to the function $F(z)$ of variable z in the Hilbert space. According to the Dirac general theory of representations, the transition from $F(z)$ to $c(n)$ is realized by the complete system of functions $f(n, z)$:

$$F(z) = \sum_n c\,(n)\; f(n, z)\,. \tag{38}$$

Before we go further, it should be recalled that expression (35) for $\Phi(n)$ was defined from the requirement that both operators (33) are mutually conjugated. But the form of the conjugation condition depends on the form of the weight function in the normalizing condition. Therefore the function $\Phi(n)$ is connected with the weight function. Namely, the normalizing condition

$$\sum_n \frac{n!}{|\Phi(n)|^2}\,|c(n)|^2 = 1 \tag{*}$$

corresponds to the arbitrary $\Phi(n)$ and so condition (35) means that the normalizing condition has an ordinary form

$$\sum_n |c(n)|^2 = 1\,. \tag{**}$$

Further in this section we keep $\Phi(n)$ arbitrary and put $\Phi(n) = \sqrt{n!}$ only in final formulae.[7]

We will write equations (33) which define the considered transformation in the form

$$\begin{aligned} z \;\rightarrow\; S &= \frac{\Phi(n)}{\Phi(n-1)}\,e^{-\frac{\partial}{\partial n}}\,, \\ \frac{\partial}{\partial z} \rightarrow T &= \frac{(n+1)\Phi(n)}{\Phi(n+1)}\,e^{\frac{\partial}{\partial n}}\,. \end{aligned} \tag{39}$$

[7]The text between the formulae (38) and (39) is added in this edition. (*V. Fock, 1957*)

They mean that the result of the action of the operator z (or $\frac{\partial}{\partial z}$) on the left-hand side $F(z)$ in formula (38) can be expanded in functions $f(n, z)$, and the expansion coefficients $Sc\ (n)$ (or $Tc\ (n)$) will be results of action of operator S (or T) on the function $c\ (n)$. Thus,

$$zF(z) = \sum_n [Sc\ (n)]\ f(n, z)\,, \tag{40}$$

$$\frac{\partial F(z)}{\partial z} = \sum_n [Tc\ (n)]\ f(n, z)\,, \tag{41}$$

where, according to (39), the coefficients $Sc\ (n)$ (or $Tc\ (n)$) have the following values:

$$Sc\ (n) = \frac{\Phi(n)}{\Phi(n-1)}\ c\ (n-1)\,, \tag{42}$$

$$Tc\ (n) = \frac{(n+1)\Phi(n)}{\Phi(n+1)}\ c\ (n+1)\,. \tag{43}$$

Replacing n in (40) by $n+1$ and in (41) by $n-1$, we can rewrite these formulae in the form:

$$zF(z) = \sum_n \frac{\Phi(n+1)}{\Phi(n)}\ c\ (n)\ f(n+1, z)\,, \tag{44}$$

$$\frac{\partial F(z)}{\partial z} = \sum_n \frac{n\Phi(n-1)}{\Phi(n)} c\ (n)\ f(n-1, z)\,. \tag{45}$$

On the other hand we can take directly expansion (38) and multiply it by z or take the z-derivative. Then we have

$$zF(z) = \sum_n c\ (n)\ zf(n, z)\,, \tag{46}$$

$$\frac{\partial F(z)}{\partial z} = \sum_n c\ (n)\ \frac{f(n, z)}{z}\,. \tag{47}$$

Expressions (44) and (46) and also (45) and (47) should be equal to each other and identically equal relative to the function $F(z)$ and therefore also relative to $c\ (n)$, i.e., term by term. For that the function $f(n, z)$ must satisfy the following functional equations:

$$zf(n, z) = \frac{\Phi(n+1)}{\Phi(n)}\ f(n+1, z)\,, \tag{48}$$

$$\frac{\partial f(n,z)}{\partial z} = \frac{n\Phi(n-1)}{\Phi(n)} f(n-1,z). \tag{49}$$

Multiplying both parts of (49) by z and expressing the product $zf(n-1,z)$ according to (48) through $f(n,z)$, we get a differential equation

$$z\frac{\partial f(n,z)}{\partial z} = nf(n,z), \tag{50}$$

the validity of which could also be seen directly from (37*). Its solution is

$$f(n,z) = f(n)z^n. \tag{51}$$

Putting (51) in (48) we get

$$f(n)\Phi(n) = f(n+1)\Phi(n+1). \tag{52}$$

Because the quantity (52) does not depend on n, we can simply put it equal to unity.

Thus, we defined the function $f(n,z)$ up to a factor independent of n and z:

$$f(n,z) = \frac{z^n}{\Phi(n)}. \tag{53}$$

For the ordinary normalizing condition (**) we have $\Phi(n) = \sqrt{n!}$ and consequently

$$f(n,z) = \frac{z^n}{\sqrt{n!}}. \tag{53*}$$

Transition to many variables proceeds without any difficulties; the eigenfunctions are the products of eigenfunctions of a single variable. We write down only the final formula for the canonical transformation of the function $\overline{\Omega}\,(\overline{y}_1, \overline{y}_2, \ldots)$ in the Hilbert space into the function $\psi(n_1, n_2, \ldots)$ in the Dirac space

$$\overline{\Omega}\,(\overline{y}_1, \overline{y}_2, \ldots) = \sum_{n_1, n_2, \ldots} \psi(n_1, n_2, \ldots)\frac{\overline{y}_1^{n_1}\overline{y}_2^{n_2}\ldots}{\Phi(n_1)\Phi(n_2)\ldots} \tag{54}$$

or for the usual normalization of the function ψ

$$\overline{\Omega}\,(\overline{y}_1, \overline{y}_2, \ldots) = \sum_{n_1, n_2, \ldots} \psi(n_1, n_2, \ldots)\frac{\overline{y}_1^{n_1}\overline{y}_2^{n_2}\ldots}{\sqrt{n_1!}\sqrt{n_2!}\ldots}. \tag{54*}$$

§7. The theory of canonical transformation (33) or (36) considered in the previous two sections allows us not only to establish the wave equation in the Dirac space, but also to get at once its solution if the solution in the Hilbert space is known. The transformation of the equation itself is actually not needed; nevertheless, we shall perform it to make easier the comparison with the Dirac formulae.

In formula (21) for the Hamilton function we must replace the operators $\overline{y}_s$ and $y_r = \dfrac{\partial}{\partial \overline{y}_r}$ by their expressions (33) and (36). Then we get

$$F = \sum_s K_{ss} n_s + \sum_{r \neq s} K_{rs} \frac{\Phi(n_r)\Phi(n_s)(n_s+1)}{\Phi(n_r-1)\Phi(n_s+1)}\, e^{\frac{i}{\hbar}(\theta_r - \theta_s)} \tag{55}$$

and in the case of usual normalization when $\Phi(n) = \sqrt{n!}$

$$F = \sum_s K_{ss} n_s + \sum_{r \neq s} K_{rs} \sqrt{n_r}\sqrt{n_s+1}\, e^{\frac{i}{\hbar}(\theta_r - \theta_s)} . \tag{55*}$$

Let us denote the wave function in the Dirac space as $\psi(n_1, n_2, \dots)$. This function satisfies the wave equation

$$\frac{\hbar}{i}\frac{\partial}{\partial t}\psi(n_1, n_2, \dots) + F\psi(n_1, n_2, \dots) = 0 . \tag{56}$$

In an explicit form this equation can be written as

$$\frac{\hbar}{i}\frac{\partial}{\partial t}\psi(n_1, n_2, \dots) + \sum_s K_{ss} n_s \psi(n_1, n_2, \dots) +$$
$$+ \sum_{r \neq s} K_{rs} \frac{\Phi(n_r)\Phi(n_s)(n_s+1)}{\Phi(n_r-1)\Phi(n_s+1)} \psi(n_1, \dots n_r - 1, \dots n_s + 1, \dots) = 0 . \tag{57}$$

If the function $\psi(n_1, n_2, \dots)$ is normalized by the formula

$$\sum_{n_1, n_2, \dots} |\psi(n_1, n_2, \dots)|^2 = \text{const} , \tag{58}$$

then $\Phi(n) = \sqrt{n!}$ and equation (57) gets the form

$$\frac{\hbar}{i}\frac{\partial}{\partial t}\psi(n_1, n_2, \dots) + \sum_s K_{ss} n_s \psi(n_1, n_2, \dots) +$$
$$+ \sum_{r \neq s} K_{rs} \sqrt{n_r}\sqrt{n_s+1}\,\psi(n_1, \dots n_r - 1, \dots n_s + 1, \dots) = 0 . \tag{57*}$$

The solution of this equation is already known: $\psi(n_1, n_2, \dots)$ is the expansion coefficient in (54) or (54*) if $\overline{\Omega}(\overline{y}_1, \overline{y}_2, \dots)$ satisfies the differential equation (28). This can be checked directly if one inserts (54) or

(54*) in the differential equation and makes all coefficients zero at all products of any powers of $\overline{y}_s$; then one gets exactly equation (57) and correspondingly (57*).

§9.[8] Now we are in a position to formulate the statistical problem in a general form and get its solution.

Let us consider two mechanical quantities a and b with operators A and B having discrete eigenvalues α_s and β_s. The probability amplitudes are given at the initial moment $t = 0$ for the distribution of the systems over the eigenstates of operator A. They are

$$\psi_0^{(\alpha)}(n_1, n_2, \ldots) . \tag{59}$$

We have to calculate the probability amplitudes

$$\psi_t^{(\beta)}(n_1, n_2, \ldots) \tag{60}$$

for the distribution of systems over the eigenstates of operator B at time t.

We will get the problem considered by Dirac if operators A and B coincide and if they are equal to the energy operator of the unperturbed system. The solution of the given problem can be found in the following way.

Let us build up, as it was shown in §1, the basic system of functions which are equal to the eigenfunctions of operator A at $t = 0$. Consider also the eigenfunctions of operator B. Using these systems of functions we form the matrix Y_{rs} according to (14).

Using the quantities conjugate to the given probability amplitudes (59) and introducing arbitrary parameters $g_1, g_2, \ldots$, we form the generating function

$$\Omega^0(g_1, g_2, \ldots) = \sum_{n_1, n_2, \ldots} \overline{\psi}_0^{(\alpha)}(n_1, n_2, \ldots) \frac{g_1^{n_1} g_2^{n_2} \ldots}{\sqrt{n_1!}\sqrt{n_2!}\ldots} , \tag{61}$$

where the summation extends over all nonnegative values of $n_1, n_2, \ldots$, satisfying the condition

$$n_1 + n_2 + \ldots = N \qquad \text{(the number of systems)}. \tag{62}$$

So the function Ω^0 is a uniform function of an N-th power relative to parameters g_s. Now let us substitute the entering Ω^0 parameters g_s

[8] §8 containing an attempt to apply theory to the Fermi statistics is omitted here and §§ 9, 10 are slightly shortened. (*V. Fock, 1957*)

with linear forms

$$g_s = \sum_r \overline{Y}_{rs} y_r \,, \tag{63}$$

where y_r are new parameters. We denote the function of y_r, obtained in this way, by Ω so that

$$\Omega(y_1, y_2, \ldots) = \Omega^0(g_1, g_2, \ldots) \,. \tag{64}$$

The function Ω satisfies wave equation (27) in the Hilbert space. Let us expand Ω in powers of y_s and write down the expansion as

$$\Omega(y_1, y_2, \ldots) = \sum_{n_1, n_2, \ldots} \overline{\psi}_t^{\beta}\,(n_1, n_2, \ldots) \frac{y_1^{n_1} y_2^{n_2} \ldots}{\sqrt{n_1!}\sqrt{n_2!}\ldots} \,. \tag{65}$$

Then the quantities $\psi_t^{(\beta)}$, conjugate to the expansion coefficients, satisfy the wave equation in the Dirac space and are the wanted probability amplitudes.

§10. Now we consider special cases of the general problem. Let A be the energy operator at time $t = 0$ and B be the energy operator at time t:

$$A = H(0), \qquad B = H(t). \tag{66}$$

In this case the functions $\psi_s(q, t)$ and $\varphi_s(q, t)$ coincide at time $t = 0$,

$$\psi_s(q, 0) = \varphi_s(q, t), \tag{67}$$

and the matrix element Y_{rs} turns into δ_{rs} at $t = 0$, so that at $t = 0$ the quantities y_s coincide with g_s:

$$y_s(0) = g_s, \qquad Y_{rs}(0) = \delta_{rs}. \tag{68}$$

The initial values of the expansion coefficients $\overline{\psi}_t^{(\beta)}(n_1, n_2, \ldots)$ (65) will be the values $\psi_0^{(\alpha)}(n_1, n_2, \ldots)$ from formula (61). Hence the method described in §9 allows one to find the solution of the Dirac equation when the initial conditions are given.

Now let us consider the case when the ensemble consists of a single system which is in state s at time $t = 0$. Then the generating function Ω^0 is equal to

$$\Omega^0(g_1, g_2, \ldots) = g_s \tag{69}$$

and the function $\Omega(y_1, y_2, \ldots)$ takes the form

$$\Omega^0(y_1, y_2, \ldots) = \sum_r \overline{Y}_{rs}\; y_r \,. \tag{70}$$

Comparing this formula with (65), we get

$$\overline{\psi}_t\left(0,0,\ldots 1^{(r)},0,\ldots\right)=\overline{Y}_{rs} \tag{71}$$

(the unity in the left-hand side stands on the r-th place).

The square modulus of this value

$$|\psi_t|^2=|Y_{rs}|^2 \tag{72}$$

is the probability that in state r at time t there is one system (in this case the only one). In other words $|\psi_t|^2$ is the transition probability from state s to state r, which coincides with the definition given by Born [3].

As another example, let us consider an ensemble of N systems; all of them are in the same state at $t=0$. Then functions Ω^0 and Ω will be

$$\left.\begin{aligned}\Omega^0&=\frac{(g_s)^N}{\sqrt{N!}},\\ \Omega&=\frac{\left(\sum_r\overline{Y}_{rs}y_r\right)^N}{\sqrt{N!}}.\end{aligned}\right\} \tag{73}$$

The probability amplitude $\psi_t(n_1,n_2,\ldots)$ for the distribution $(n_1,n_2,\ldots)$ at time t will be

$$\overline{\psi}_t(n_1,n_2,\ldots)=\frac{\sqrt{N!}}{\sqrt{n_1!}\sqrt{n_2!}\ldots}Y_{1s}^{n_1}Y_{2s}^{n_2}\ldots\,. \tag{74}$$

The square modulus of this value, which represents the probability of the given distribution, is equal to

$$|\psi_t(n_1,n_2,\ldots)|^2=\frac{N!}{n_1!n_2!\ldots}|Y_{1s}|^{2n_1}|Y_{2s}|^{2n_2}\ldots\,. \tag{75}$$

Because the values $|Y_{rs}|^2$ give us the probabilities for a single system (see formula (72)), this expression corresponding to the Bose–Einstein statistics coincides with that calculated by the ordinary probability theory.

In conclusion I would like to thank cordially Professor M. Born for his interest in my work and the International Education Board, which made this work possible.

References

1. P.A.M. Dirac, *The Quantum Theory of Emission and Absorption of Radiation*, Proc. Roy. Soc. **A 144**, 243, 1926.
2. V. Fock, *Über die Beziehung zwischen den Integralen der quantenmechanischen Bewegungsgleichungen und der Schrödingerschen Wellengleichung*, Zs. Phys. **49**, 323, 1928 (and [28-2] in this book (*Editors*)).
3. M. Born, *Das Adiabatenprinzip in der Quantenmechanik*, Zs. Phys. **40**, 107, 1926.

Translated by Yu.N. Demkov

References

28-4
Proof of the Adiabatic Theorem

M. Born and V. Fock
Göttingen

Received 1 August 1928

Zs. Phys. **51**, 165, 1928

In old quantum mechanics the adiabatic theorem established by Ehrenfest had the meaning that the quantized action variables $J = n\hbar$ are invariant relative to an infinitely slow (adiabatic) change of a mechanical system.[1] From this one can guess that if the system is in a state with definite quantum numbers before the adiabatic change started, then after the changes it will be characterized by the same quantum numbers.

The adiabatic theorem in new quantum mechanics has analogous meaning. If we enumerate the states of a system with numbers corresponding to the energy levels, the adiabatic theorem holds claiming that if the system was initially in a state with a definite number, then after the adiabatic change the transition probability of the system to be in the state with a different number would be infinitely small, although the energy levels after the adiabatic interaction could differ from the initial levels by finite values.

The adiabatic theorem was transferred into new quantum mechanics by one of the authors[2] already in 1926. Nevertheless, the proof given there and also the proof by Fermi and Persico[3] are mathematically not indisputable. Both proofs consider only the case when at an adiabatic change none of the frequencies vanish, i.e., no degeneration takes place. So there was no generalization, which was already performed for old quantum mechanics by M. von Laue.[4]

In this paper we try to give a proof which is more general and more satisfactory from the mathematical standpoint.

---

[1]The literature about the adiabatic theorem in old quantum mechanics is given at the end of this paper. (*M. Born and V. Fock*)

[2]M. Born, *Das Adiabatensatz in der Quantenmechanik*, Zs. Phys. **40**, 167, 1926.

[3]E. Fermi and F. Persico, *Il prinzipio delle adiabatiche e la nozione de forza vivo nella nuova meccanica ondulatoria*. Lincei Rend. (6) **4**, 452–457, 1926.

[4]M. v. Laue, Ann. Phys. **76**, 619, 1925.

§1. We consider a mechanical system with the energy operator containing time explicitly and changing with time slowly. We express this slowness in a way that the time dependence will be

$$H = H(s); \qquad s = \frac{t}{T}, \tag{1}$$

where T is a large parameter with the dimension of time, and the derivatives by s of the coefficients in the operator H and in its eigenfunctions must be finite.

The eigenfunctions of the operator $H(s)$ are

$$\varphi_1(q,s), \quad \varphi_2(q,s), \dots .$$

They satisfy the equations

$$H(s)\ \varphi_n(q,s) = W_n(s)\ \varphi_n(q,s). \tag{2}$$

Because this equation contains time (as a parameter), we can multiply the eigenfunctions by arbitrary phase factors containing time. We can fix these phases by the condition that each eigenfunction is orthogonal to its time derivative

$$\int \varphi_n^*(q,s)\ \frac{\partial \varphi_n}{\partial s}\ \varrho(q)\ dq = 0, \tag{3}$$

($\varrho(q)$ is the density function in q-space). Then the eigenfunctions are defined up to a constant phase factor.

Together with the eigenfunctions of the energy operator let us consider a system of functions

$$\psi_1(q,t), \quad \psi_2(q,t), \ \dots,$$

which satisfy the Schrödinger equation

$$H\ \psi_n + \frac{\hbar}{i}\ \frac{\partial \psi_n}{\partial t} = 0 \tag{4}$$

and coincide with φ_n at $t = 0$:

$$\psi_n(q,0) = \varphi_n(q,0). \tag{5}$$

The systems of functions $\psi_n(q,t)$ are normalized and orthogonal for each t.[5]

[5]V. Fock, *Über die Beziehung zwischen den Integralen der quantenmechanischen Bewegungsgleichungen und der Schrödingerschen Wellengleichung*, Addenda, Zs. Phys. **49**, 323, 1928. (See also [28-2] in this book. (*Editors*))

§2. Now we go over from eigenfunctions to matrices. We can construct the matrices using three complete sets of eigenfunctions:

$$\psi_n(q,0) = \varphi_n(q,0)\,, \tag{a}$$

or

$$\varphi_n(q,s)\,, \tag{b}$$

or

$$\psi_n(q,s)\,. \tag{c}$$

We indicate the matrices formed by the set (a) by the superscript ^{0}, those formed by the set (b) by the superscript ^{s} and those formed by the set (c) by no superscript. For instance, we write

$$H^0_{mn} = \int \varphi^*_m(q,0)\; H(0)\; \varphi_n(q,0)\; \varrho\; dq\,, \tag{6a}$$

$$W_n\;\delta_{nm} = H^s_{mn} = \int \varphi^*_m(q,s)\; H(s)\; \varphi_n(q,s)\; \varrho\; dq\,, \tag{6b}$$

$$H_{mn} = \int \psi^*_m(q,t)\; H(s)\; \psi_n(q,t)\; \varrho\; dq\,. \tag{6c}$$

The representation (a) has the property that time-independent matrices will correspond to constant (explicitly independent on time) operators; in the representation (b) the energy matrix has the diagonal form and in the representation (c) the equations of motion have the form:

$$\left.\begin{aligned} \dot q &= \frac{i}{\hbar}\,(H\,q - q\,H)\,,\\ \dot p &= \frac{i}{\hbar}\,(H\,p - p\,H)\,. \end{aligned}\right\} \tag{7}$$

The transition from one representation to another can be done by a unitary matrix that we shall call U, V and Y. (As usual we will call $U^\dagger$ the "transposed-conjugate" or "adjoint" matrix

$$U^\dagger_{mn} = U^*_{nm}\,,$$

so the matrix U is unitary if it satisfies the conditions

$$U\,U^\dagger = 1; \qquad U^\dagger\,U = 1.$$

The matrix representation of H^0, W and H of the operator H is connected by the equations

$$H = U^\dagger \; H^0 \; U\,, \tag{8ac}$$

$$W = V^\dagger \; H^0 \; V\,, \tag{8ab}$$

$$H = Y^\dagger \; W \; Y\,. \tag{8bc}$$

The matrix Y can be expressed through U and V as follows:

$$Y = V^\dagger \; U\,. \tag{9}$$

All these relations can be easily verified using explicit expressions for the matrix elements

$$\left.\begin{aligned} U_{mn} &= \int \varphi_m^*(q,0)\; \psi_n(q,t)\; \varrho\; dq\,,\\ V_{mn} &= \int \varphi_m^*(q,0)\; \varphi_n(q,s)\; \varrho\; dq\,,\\ Y_{mn} &= \int \varphi_m^*(q,s)\; \psi_n(q,t)\; \varrho\; dq\,. \end{aligned}\right\} \tag{10}$$

§3. As it was indicated by one of the authors,[6] the entries Y_{mn} have the following physical meaning. The square modulus $|Y_{mn}|^2$ is the probability that the mechanical system, being at $t = 0$ in state n (having the energy level $W_n(0)$), will be in state m at time t (having the energy level $W_n(t/T)$). This meaning of $|Y_{mn}|^2$ follows also from the Dirac statistical equation, which was recently shown by the second author.[7] The adiabatic theorem states now that at an infinitely slow change of the system, i.e., at an infinitely large value of the parameter T in (1), the transition probability $|Y_{mn}|^2$ $(m \neq n)$ (which is the function of time) remains infinitely small even for finite values of $s = \dfrac{t}{T}$.

We will prove this statement under some restrictions investigating the differential equation to which the values Y_{lm} satisfy.

§4. First let us establish the differential equation for the transformation matrix U.

The matrices q and p satisfy the equations of motion

$$\left.\begin{aligned} \dot q &= \frac{i}{\hbar}(H\; q - q\; H)\,,\\ \dot p &= \frac{i}{\hbar}(H\; p - p\; H) \end{aligned}\right\} \tag{7}$$

[6] M. Born, *The Adiabatic Principle in Quantum Mechanics*, Zs. Phys. **40**, 167, 1926.

[7] V. Fock, *Verallgemeinerung und Lösung der Diracshen statistischen Gleichung*, Zs. Phys. **49**, 339, 1928. (See [28-3] of this book. (*Editors*))

and the constant matrices q^0 and p^0 are connected with p and q by the relations

$$\left.\begin{aligned} q &= U^\dagger\, q^0\, U\,, \\ p &= U^\dagger\, p^0\, U\,. \end{aligned}\right\} \tag{11}$$

The connection between the representations of the energy operators H and H^0 is

$$H = U^\dagger\, H^0\, U\,. \tag{12}$$

Now we use the expressions for q, p and H through q^0, p^0 and H^0 and calculate the derivatives $\dot{q}$ and $\dot{p}$:

$$\begin{aligned} \dot{q} &= \dot{U}^\dagger\, q^0\, U + U^\dagger\, q^0\, \dot{U}\,, \\ \dot{p} &= \dot{U}^\dagger\, p^0\, U + U^\dagger\, p^0\, \dot{U}\,. \end{aligned}$$

Because $U^\dagger \dot{U} = 1$, we have

$$\dot{U}^\dagger = -U^\dagger\, \dot{U}\, U^\dagger$$

and, therefore,

$$\left.\begin{aligned} \dot{q} &= U^\dagger\ -\dot{U}\, U^\dagger\, q^0 + q^0\, \dot{U}\, U^\dagger\, U\,, \\ \dot{p} &= U^\dagger\ -\dot{U}\, U^\dagger\, p^0 + p^0\, \dot{U}\, U^\dagger\, U\,. \end{aligned}\right\} \tag{13}$$

Otherwise, due to the equations of motion,

$$\left.\begin{aligned} \dot{q} &= \frac{i}{\hbar}\ U^\dagger\, H^0\, q^0 - q^0\, H^0\, U\,, \\ \dot{p} &= \frac{i}{\hbar}\ U^\dagger\, H^0\, p^0 - P^0\, H^0\, U\,. \end{aligned}\right\} \tag{14}$$

Let us denote by K^0 the Hermitian matrix

$$K^0 = H^0 + \frac{\hbar}{i}\, \dot{U}\, U^\dagger\,, \tag{15}$$

and compare expressions (13) and (14); then we come to the equations

$$\left.\begin{aligned} K^0\, q^0 - q^0\, K^0 &= 0\,, \\ K^0\, p^0 - p^0\, K^0 &= 0\,. \end{aligned}\right\} \tag{16}$$

The matrix K^0 commutes with q^0 and p^0. Then, if q^0 and p^0 form an irreducible system of matrices, the matrix K^0 must be a multiple of the unity matrix, and we can put it simply equal to zero. Then for the transformation matrix U, we get the equation

$$\frac{\hbar}{i}\,\dot{U} + H^0\,U = 0\,, \tag{17}$$

which is nothing else but the Schödinger equation in a matrix representation.

If the matrices q^0 and p^0 are not constants as stated before, they would satisfy the equations of motion

$$\left.\begin{aligned} \dot{q}^0 &= \frac{i}{\hbar}\,(K^0\,q^0 - q^0\,K^0)\,, \\ \dot{p}^0 &= \frac{i}{\hbar}\,(K^0\,p^0 - p^0\,K^0) \end{aligned}\right\} \tag{18}$$

with the "Hamilton matrix" K^0. The considerations of this section also contain the theory of general (time-independent) canonical transformations of the quantum mechanical equations of motion.

§5. From the differential equation (17) for the matrix U it is easy to get a corresponding equation for the matrix Y. We have

$$Y = V^\dagger\,U$$

and

$$\dot{Y} = V^\dagger\,\dot{U} + \dot{V}^\dagger\,U\,. \tag{19}$$

If we now express H^0 in (17) using ($8ab$) through the diagonal matrix W, then we have

$$\frac{\hbar}{i}\,\dot{U} + V\,W\,V^\dagger\,U = 0$$

or

$$\dot{U} = -\frac{i}{\hbar}\,V\,W\,Y\,. \tag{20}$$

On the other hand,

$$U = V\,Y\,. \tag{21}$$

Putting expressions (20) and (21) for $\dot{U}$ and U into (19), we obtain

$$\dot{Y} = -\frac{i}{\hbar}\,W\,Y + \dot{V}^\dagger\,V\,Y\,, \tag{22}$$

i.e., the needed differential equation.

Here the superscript points mean (as usual) the time derivative. Now instead of time we want to introduce the quantity $s = \frac{t}{T}$ from formula (1):

$$\frac{dY}{ds} = -\frac{iT}{\hbar} W\, Y + \frac{dV^\dagger}{ds} V\, Y\,. \tag{22'}$$

Further we introduce:

$$Q = -i\frac{dY^\dagger}{ds} V = i\, V^\dagger \frac{dY}{ds}\,. \tag{23}$$

Because the transformation matrix V is unitary, the matrix Q just introduced is Hermitian. Its entries can be expressed through the eigenfunctions of the energy operator:

$$Q_{mn} = i \int \varphi^*_m(q,s)\, \frac{\partial \varphi_n(q,s)}{\partial s}\, \varrho\, dq\,. \tag{24}$$

We note that due to normalization (3) of the eigenfunctions $\varphi_n(q,t)$ all diagonal elements of the matrix Q vanish.

Now we want to find another expression for the elements of Q. Let us differentiate equation (8ab) by parameter s. Then we get

$$\frac{dW}{ds} = \frac{dV^\dagger}{ds} H^0\, V + V^\dagger\, H^0\, \frac{dV}{ds} + V^\dagger\, \frac{dH^0}{ds}\, V\,. \tag{25}$$

In the representation where the energy matrix is diagonal, the expression $V^\dagger\, \frac{dH^0}{ds}\, V$ is the matrix for s-derivative of the energy operator; for brevity we denote it H':

$$H' = V^\dagger\, \frac{dH^0}{ds}\, V\,. \tag{26}$$

Now from (8ab) it follows

$$H^0\, V = V\, W\,; \qquad V^\dagger\, H^0 = W\, V^\dagger\,. \tag{27}$$

If one puts (27) into (25) and takes into account the differential equations (23) and (26) for Q and H', then

$$\frac{dW}{ds} = i(Q\, W - W\, Q) + H'. \tag{28}$$

We consider now the nondiagonal element of matrix (28). Because W is diagonal, from (28) for $m \neq n$ it follows

$$iQ_{mn}(W_n - W_m) + H'_{mn} = 0$$

or

$$Q_{mn} = -\frac{i\,H'_{mn}}{W_m - W_n}\,, \tag{29}$$

where H'_{mn} has the meaning

$$H'_{mn} = \int \varphi^*_m \,\frac{\partial H}{\partial s}\, \varphi_n \,\varrho\, dq\ . \tag{30}$$

For $m = n$ as we already established, $Q_{nn} = 0$.

It should be noticed here that Q_{mn} remains finite also in the case when for some special value of s the difference $W_m(s) - W_n(s)$ vanishes; this follows from expression (24) for Q_{mn}.

Let us return to the differential equation (22$'$) for the matrix Y, which we will write now in the form

$$\frac{dY}{ds} = -\frac{iT}{\hbar}\, W\, Y + i\, Q\, Y\ . \tag{31}$$

If we rewrite (31) for matrix elements, then we get a system of equations

$$\frac{dY_{mn}}{ds} = -\frac{i\,T}{\hbar}\, W_m\, Y_{mn} + i \sum_k Q_{mk}\, Y_{kn}\ . \tag{32}$$

We consider also the system of equations[8]

$$\frac{dy_m}{ds} = -\frac{iT}{\hbar} W_m\, y_m + i \sum_k Q_{mk}\, y_k\ . \tag{33}$$

Let us take into account that due to (5) and (10) the matrix Y at $t = 0, s = 0$ is a unity matrix; then we can consider the matrix elements of a column

$$Y_{1n}, Y_{2n}, \ldots Y_{mn} \ldots$$

as such a solution

$$y_1, y_2, \ldots y_m \ldots$$

of the system of equations (33), which satisfies the initial conditions

$$y_m = Y_{mn} = \delta_{mn} \qquad \text{for} \qquad s = 0\ . \tag{34}$$

The quantities y_m are uniquely defined by the differential equations (33) and the initial conditions (34).

[8]See footnote[7]; compare formulas (18) and (19) and theorem 1. (*M. Born and V. Fock*)

§6. Now we want to indicate the method for solution of equations (33). We put for brevity

$$\omega_k(s) = \frac{1}{\hbar}\int_0^s W_k(s)\, ds\,, \tag{35}$$

and introduce in (33), instead of y_k, new variables

$$c_k = y_k\ e^{iT\omega_k}\ . \tag{36}$$

The quantities c_k satisfy the differential equations

$$\frac{dc_m}{ds} = i\sum_k P_{mk}\ c_k\ , \tag{37}$$

where for brevity we denote

$$P_{mk} = Q_{mk}\ e^{iT(\omega_m - \omega_n)}\,. \tag{38}$$

The difference of new equations from the original ones is first that now the coefficient at c_m is equal to zero, whereas the coefficient at y_m is proportional to a large parameter T; second, that P_{mk} loaded with the large T in the exponent oscillate rapidly, whereas Q_{mk} are slowly varying quantities.

Now we denote by $c_{mn}(s)$ those solutions of (37) that satisfy the initial conditions

$$c_m(0) = \delta_{mn}\ , \tag{39}$$

i.e., the quantities

$$c_{mn}(s) = Y_{mn}\ e^{iT\omega_m}\ . \tag{40}$$

Their square moduli are equal to those of Y_{mn}, so they are transition probabilities.

As it can be easily checked, the differential equations (37) with the initial conditions (39) are equivalent to the system of integral equations

$$c_{mn}(s) = \delta_{mn} + i\sum_k \int_0^s P_{mk}(\sigma)\ c_{kn}(\sigma)\ d\sigma\ . \tag{41}$$

One can solve these integral equations by iterations. As the zero approximation, we can put

$$c^{(0)}_{mn} = \delta_{mn}\ ,$$

and in the first approximation as the result of inserting the zero approximation into the right-hand side of (41), one gets

$$c_{mn}^{(1)} = \delta_{mn} + i \int_0^s P_{mn}(\sigma)\, d\sigma \ ,$$

and generally

$$c_{mn}^{(l)} = \delta_{mn} + i \sum_k \int_0^s P_{mn}(\sigma)\, c_{kn}^{(l-1)}(\sigma)\, d\sigma \ . \tag{42}$$

As the final result, we obtain an infinite series

$$c_{mn}(s) = \delta_{mn} +$$

$$+ \sum_{k=1}^{\infty} i^k \int_0^s ds_k \int_0^{s_k} ds_{k-1} \ldots \int_0^{s_2} ds_1 \left[P(s_k)\, P(s_{k-1}) \ldots P(s_1)\right]_{mn} . \tag{43}$$

Up to now we did not take into account the convergency considerations. To ensure the convergency of the method, we have to introduce preliminary requirements on the matrix $P(s)$ to be absolutely restricted[9] for all s and a constant-restricted matrix M can be found that is a majorant:

$$|P_{mn}(s)| = |Q_{mn}(s)| \leq M_{mn}; \quad (M_{mn}) \ \text{ restricted.} \tag{44}$$

Then[10] the majorant system of equations

$$\frac{db_{mn}}{ds} = \sum_k M_{mk}\, b_{kn} \ , \tag{45}$$

[9]A matrix (P_{mn}) is called restricted if for each system of numbers x_n, y_n that satisfies the normalization conditions

$$\sum_n |x_n|^2 = 1; \quad \sum_n |y_n|^2 = 1 \ ,$$

the double sum

$$\sum_{mn} P_{mn}\, x_m\, y_n$$

converges and its absolute value stays within some limit independent of the choice of x_n, y_n. The matrix is called absolutely restricted if the matrix consisting of absolute values $|P_{mn}|$ is also restricted. (*M. Born and V. Fock*)

[10]W.L. Hart, Amer. Journ. **39**, 407–424, 1917.

with the initial conditions

$$b_{mn}(0) = \delta_{mn} \, ,$$

possesses the solution

$$b_{mn}(s) = (e^{sM})_{mn} = \delta_{mn} + \sum_{k=1}^{\infty} \frac{s^k}{k!} \, (M^k)_{mn} \, , \tag{46}$$

which presents the uniformly convergent series in powers of s. It is easy to see, replacing in (43) P_{mn} by M_{mn}, that the absolute value of each term of the series (43) is no larger than the corresponding term in the series (46). From this it is immediately clear that conditions (44) are sufficient for the convergence of series (43).

Whether the matrix Q_{mn} in some problem is really restricted, one can decide using the following (sufficient) criteria: according to the theorem by Schur[11] it is really the case if the series

$$z_m = \sum_k |Q_{mk}| \tag{47}$$

converges and remains independent of m within a limit. According to (29) for Q_{mn}, this series is equal to

$$z_m = {\sum_k}' \frac{|H'_{mk}|}{|W_m - W_n|} \, , \tag{48}$$

where the prime means that the term with $k = m$ is omitted.

Now if we accept that the set

$$\alpha_m = {\sum_k}' \frac{1}{(W_m - W_k)^2} \tag{49}$$

converges, and denote by β_m the expression

$$\beta_m = \sum_k \left| H'_{mk} \right|^2 = \int \left| \frac{\partial H}{\partial s} \, \varphi_m \right|^2 \, \varrho \, dq \, , \tag{50}$$

then we can estimate the sum z_m using the Schwarz inequality

$$z_m = \sqrt{\alpha_m \, \beta_m} \, . \tag{51}$$

[11] J. Schur, *Restricted Bilinear Forms*, Crelles Journ. **140**, 1, 1911 (Theorem I).

Thus, we obtain the following sufficient condition for the matrix Q to be absolutely bounded: the product $\alpha_m \, \beta_m$ should lie within a limit A independent of m:

$$\alpha_m \, \beta_m \leq A \, . \tag{52}$$

When the eigenvalues W_n grow proportionally to n, which is the case for a harmonic oscillator, the set (49) converges and its sum remains smaller than a number independent of m. Then for Q to be absolutely bounded it is sufficient for β_m (50) to be finite, which always will be the case if the time derivative of the perturbation energy is a limited function.

If a mechanical system is restricted by a volume so that the q-space is finite, then for a single degree of freedom the eigenvalues W_n grow proportionally to n^2. Then α_m decrease as $1/m^2$, and for Q to be absolutely bounded it is sufficient to admit that β_m increase not faster than proportionally to m^2, which is the case for a very general assumption about the perturbation energy.

§7. Now we come to the initial problem: the proof of the adiabatic theorem. We must prove the following: if the parameter T [formulas (1), (27), (32) and (34)] is sufficiently large, then the square moduli $|Y_{mn}|^2 = |c_{mn}|^2$ for finite s differ arbitrarily little from their initial values δ_{mn}. The exact conditions needed for the theorem to be valid will be formulated later.

Next we formulate the Lemma:

Lemma. If in the interval $0 \leq s \leq s'$ the following assumptions are fulfilled:

1. The following inequality is valid:

$$|Q_{mn}(s)| = |P_{mn}(s)| \leq M_{mn} \, .$$

2. Within the interval each function (frequency)

$$\frac{d\omega_m}{ds} - \frac{d\omega_n}{ds} = 2\pi\nu_{mn}(s)$$

has maximally N_1 zeroes of maximally the r-th order (i.e., for degenerated states of a mechanical system) and in the vicinity of the zero point s_0 the estimation is valid

$$\frac{1}{|2\,\pi\,\nu_{mn}(s)|} < \frac{A}{|s - s_0|^r} \, .$$

3. The real and imaginary parts of the function

$$\frac{Q_{mn}(s)}{\nu_{mn}(s)}$$

are piecewise monotonous; the largest number of segments where they are monotonous is N_2.

Then the estimation is valid:

$$\left| \int_0^{s'} P_{mn}(s)\, ds \right| =$$

$$= \left| \int_0^{s'} Q_{mn}(s)\, e^{iT(\omega_m - \omega_n)}\, ds \right| < 4M_{mn}(N_1 + N_2) \sqrt[r+1]{\frac{4A}{T}}. \tag{53}$$

The proof of this statement will be given in the Appendix.

Using this Lemma it is easy to prove the adiabatic theorem.

We perform the first integration (over s_1) in the k-th term of the set (43) and estimate the result using formula (53).[12]

We do the remaining integrations replacing P_{mn} by M_{mn}. Then we get

$$|c_{mn} - \delta_{mn}| < 4\,(N_1 + N_2) \sqrt[r+1]{\frac{4A}{T}} \sum_{k=1}^{\infty} \frac{s^{k-1}}{(k-1)!} (M^k)_{mn} =$$

$$= 4(N_1 + N_2) \sqrt[r+1]{\frac{4A}{T}} \frac{db_{mn}}{ds}\,, \tag{54}$$

where $\dfrac{db_{mn}}{ds}$ is the derivative of the solution (46) of the auxiliary equations (45). This quantity is finite for finite s just as the factor of the radical in (54); however the radical tends to zero for infinite T.

Therefore, we have proven a mathematical theorem:

Theorem. *If the matrix Q is absolutely restricted and all the conditions for finite s and infinitely large T are valid, the difference $c_{mn} - \delta_{mn}$ has the order $T^{-\frac{1}{r+1}}$:*

$$c_{mn} = \delta_{mn} + O\left(T^{-\frac{1}{r+1}}\right). \tag{55}$$

Thus, this difference tends to zero when T grows infinitely.[13]

It follows directly from this theorem that the probability of the transition $n \to m$ to another energy level is of the order of magnitude $T^{-\frac{2}{r+1}}$:

$$|Y_{mn}|^2 = |c_{mn}|^2 = O(T^{-\frac{2}{r+1}})\,, \tag{56}$$

[12] One should keep in mind that Q_{nn} is zero. (*M. Born and V. Fock*)

[13] The notation $x = O(\alpha)$ means that x is of the order of α. (*M. Born and V. Fock*)

or, e.g., of the order $\frac{1}{T^2}$ if none of frequencies ν_{mn} vanish in the course of adiabatic evolution.

Using the normalization relation

$$\sum_n |Y_{mn}|^2 = 1$$

for the probability $|Y_{mn}|^2$ for the system to stay in the same state m, we obtain the expression

$$|Y_{mm}|^2 = 1 - \sum_n{}' |Y_{mn}|^2 = 1 - O(T^{-\frac{2}{r+1}}) . \tag{57}$$

This probability differs from unity by the quantity of the same order $T^{-\frac{2}{r+1}}$.

Up to now we considered as the initial state the "sharp" (pure) one, i.e., that at time $t = 0$ the system is in the state W_n with the probability 1, whereas all the other states have the zero probability. If, on the contrary, at time $t = 0$ all the energy levels W_n are populated with the probabilities $|b_n|^2$, then we calculate the probabilities $|b'_m|^2$ of different levels at time t using the formula

$$b'_m = \sum_n c_{mn}\, b_n . \tag{58}$$

From (55), we have

$$b'_m = b_m + O\left(T^{-\frac{1}{r+1}}\right) \tag{59}$$

and, therefore,

$$\left.\begin{aligned} |b'_m|^2 = |b_m|^2 + O\left(T^{-\frac{1}{r+1}}\right) \qquad & \text{if} \quad b_m \neq 0 , \\ |b'_m|^2 = O\left(T^{-\frac{2}{r+1}}\right) \qquad & \text{if} \quad b_m = 0 . \end{aligned}\right\} \tag{60}$$

Thus, the deviation of the probability $|b'_m|^2$ of state m from its initial value $|b_m|^2$ is of different order whether its initial value is zero or nonzero, and actually in the first case it is generally[14] smaller, i.e., of a higher order of $1/T$.

[14]Compare on the contrary formula (57). (*M. Born and V. Fock*)

Finally, we would mention that the adiabatic theorem can be valid in cases for which it was not proven here. As an example we can present a perturbed harmonic oscillator considered by one of the authors,[15] where matrix Q is not restricted and the method considered in §**6** is not applicable.

Appendix

Proof of the Lemma in §7

To estimate the integral

$$\int_0^{s'} Q_{mn}(s)\, e^{iT(\omega_m-\omega_n)}\, ds$$

we denote briefly the real or imaginary part of the function $Q_{mn}(s)$ by $f(s)$ and the difference $\omega_m(s)-\omega_n(s)$ as $g(s)$ and consider the integral

$$J=\int_0^{s'} f(s)\, e^{iTg(s)}\, ds \; .$$

We split the integration interval in two groups E_1 and E_2 of segments, namely, the first group is the vicinities

$$\alpha_k-\varepsilon < s < \alpha_k+\varepsilon$$

of the zeroes α_k of the derivative $g'(s)$ and the second group is the other parts of the segment $(0, s')$.

Evidently, the integral over E_1

$$J_1=\int_{E_1} f(s)\, e^{iTg(s)}\, ds$$

satisfies the inequality

$$|J_1| < M\int_{E_1} ds = 2MN_1\varepsilon \; ,$$

[15] V. Fock, *Über die Beziehung zwischen den Integralen der quantenmechanischen Bewegungsgleichungen und der Schrödingerschen Wellengleichung*, Zs. Phys. **49**, N5–6, 323–338, 1928 (this book [28-2]). Here the perturbation energy for $x \to \infty$ diverges as x^2. (*M.B. and V.F.*)

where N_1 is the number of zeroes α_k of the $g'(s)$ and M is the maximum of the absolute value of $f(s)$.

We write the integral over E_2 in the form

$$J = \int\limits_{E_2} \frac{f(s)}{g'(s)}\, e^{iTg(s)}\, g'(s)\, ds \ .$$

In E_2 the $\dfrac{1}{g'(s)}$ is finite, whereas in the vicinity of the zeroes α_k we can estimate

$$\frac{1}{|g'(s)|} < \frac{A}{\varepsilon^r} \, .$$

Now we apply the second averaging theorem of the integral calculus

$$\int\limits_{\alpha}^{\beta} \varphi(s)\psi(s)\, ds = \varphi(\alpha) \int\limits_{\alpha}^{\sigma} \psi(s)\, ds + \varphi(\beta) \int\limits_{\sigma}^{\beta} \psi(s)\, ds \ ,$$

$$\alpha \leq \sigma \leq \beta$$

with

$$\varphi(s) = \frac{f(s)}{g'(s)}$$

and

$$\psi(s) = g'(s) \cos\left[T\ g(s)\right]$$

or

$$\psi(s) = g'(s) \sin\left[T\ g(s)\right]$$

to each of the N_2 intervals where $\dfrac{f(s)}{g'(s)}$ are monotonous, then due to

$$\frac{f(s)}{g'(s)} < \frac{M\ A}{\varepsilon^r}$$

and

$$\left| \int\limits_{\sigma_1}^{\sigma_2} g'(s) {\sin \atop \cos}\ [T\ g(s)]\ ds \right| = \left| \int\limits_{g_1}^{g_2} {\sin \atop \cos}\ [T\ g]\ dg \right| < \frac{2}{T}$$

we come to the inequality

$$|J_2| < \frac{8\ M\ A}{\varepsilon^r\ T}\ N_2 \ .$$

Together with the inequality for the first integral we get

$$|J| < 2\ M\ N_1\varepsilon + \frac{8\ M\ A\ N_2}{\varepsilon^r\ T}\ .$$

Up to now the choice of ε remained arbitrary (it should only be small).

Now we choose

$$\varepsilon = \left(\frac{4\ A}{T}\right)^{\frac{1}{r+1}}\ ,$$

and then we obtain the estimate

$$|J| < 2M(N_1 + N_2)^{r+1}\sqrt{\frac{4\ A}{T}}\ .$$

The imaginary part of $Q_{mn}(s)$ can be taken into account simply by multiplication of this formula by a factor 2. So formula (53) is proven.

List of Papers on the Adiabatic Theorem in Quantum Mechanics

1. P. Ehrenfest, *Adiabatische Invarianten und Quantentheorie*, Ann. Phys. **51**, 327, 1916.
2. J.M. Burgers, *Die adiabatischen Invarianten bedingt periodischer Systeme*, l.c. **52**, 195, 1917; further: Verslagen Amsterdam **25**, 25, 918 and 1055, 1917.
3. G. Krutkow, *Contribution to the Theory of Adiabatic Invariants*, Proceedings Akad. Amsterdam **21**, 1112, 1919; further: *On the Determination of Quantum Conditions by Means of Adiabatic Invariants*, l.c. **23**, 826, 1920.
4. P. Ehrenfest, *Adiabatische Transformationen in der Quantentheorie und ihre Behandlung dureh Niels Bohr*, Naturwissensch. **11**, 543, 1923.
5. V. Fock, *Conditionally Periodic Systems with Commensurabilities and Their Adiabatic Invariants*, Transactions of the Optical Institute in Petrograd **3**, N16, 1–20, 1923.
6. G. Krutkow and V. Fock, *Über das Rayleighsche Pendel*, Zs. Phys. **13**, 195, 1923. (See [23-1] in this book. (*Editors*))
7. H. Kneser, *Die adiabatische Invarianz der Phasenintegrals bei einem Freiheitsgrad*, Math. Ann. **91**, 155, 1924.
8. M. v. Laue, *Zum Prinzip der mechanischen Transformierbarkeit (Adiabatenhypothese)*, Ann. Phys. **76**, 619, 1925.

9. P.A.M. Dirac, *The Adiabatic Invariance of the Quantum Integrals*, Proc. Roy. Soc. **A 107**, 725, 1925; further: *The Adiabatic Hypothesis for Magnetic Fields*, Proc. Cambridge Phil. Soc. **23**, 69, 1925.
10. A.M. Mosharrafa, *On the Quantum Dynamics of Degenerate Systems*, Proc. Roy. Soc. **A 107**, 237, 1925.

Translated by Yu.N. Demkov

29-1
On "Improper" Functions in Quantum Mechanics

V. Fock

Göttingen

Received 22 November 1928

JRPKhO **61**, 1, 1929

1. The application of the so-called improper functions in quantum mechanics, e.g., the function $\delta(x-y)$, originally introduced by Dirac [1], does not meet minimal requirements of mathematical rigor. Sometimes it is not even clear what is meant by a particular mathematical relation expressed in terms of these functions. Certainly, the simplest way to gain some level of rigor is to avoid the use of improper functions, which is quite possible due to the notion of the Stieltjes integral. It was the way that J. von Neumann [3] followed. On the other hand, it is sometimes very convenient to apply improper functions, and the authors of many valuable works make wide use of them. The correct use of the improper functions, despite the lack of rigor, leads to true results. This can be explained only by the fact that manipulations with these functions are essentially legitimate and contain nothing paradoxical, and, in an appropriate formulation, they can be made rigorous in the mathematical sense.

2. Let K be a linear operator. If the action of K on a function f can be represented in the form

$$Kf(x) = \int K(x,y)f(y)dy,$$

we call K an integral operator and $K(x,y)$ the kernel of K.

Any "improper" function in quantum mechanics appears as the kernel $T(x,y)$ of an integral operator T which has no true kernel. The statement that an improper function $T(x,y)$ is the limit of a sequence $T_n(x,y)$ of usual functions, strictly speaking, makes no sense. The absence of the limit is just the reason why the function $T(x, y)$ is "improper."

If we are given a sequence $T_n(x, y)$ of usual functions such that the limit

$$\lim_{n\to\infty} T_n(x, y)$$

does not exist, but the expression

$$\lim_{n\to\infty} \int T_n(x, y) f(y) dy \tag{1}$$

makes sense for a certain class of functions (e.g., for the square integrable functions, continuous functions or functions with a finite number of discontinuities, or the functions with continuous derivatives up to a certain order, etc.), we denote the latter expression by

$$\int T(x, y) f(y) dy,$$

where $T(x, y)$ will be an "improper" kernel or "improper" function.

Thus, the equation

$$\lim_{n\to\infty} \int T_n(x, y) f(y) dy = \int T(x, y) f(y) dy \tag{2}$$

is *the definition* of the improper kernel $T(x, y)$. Consequently, the expression $T(x, y)$ is not a function but only a concise notation of a certain passage to the limit.

If the result of the action of an operator T on a function $f(x)$ can be represented in the form (1),

$$\lim_{n\to\infty} \int T_n(x, y) f(y) dy = T f(x), \tag{3}$$

then we call the symbol $T(x, y)$ in (2) the "improper kernel of T" and the functions $T_n(x, y)$ the "approximate functions" for this kernel. From the above discussion it is clear that the latter notation should not be taken literally.

Obviously, the sequences $T_n(x, y)$ can be chosen in different ways.

Suppose we are given two sequences $T_n(x, y)$ and $T'_n(x, y)$. Let these sequences define the same improper function. This means that, for any admissible function $f(y)$, the following relation holds:

$$\lim_{n\to\infty} \int T_n(x, y) f(y) dy = \lim_{n\to\infty} \int T'_n(x, y) f(y) dy. \tag{4}$$

The improper functions $T(x, y)$ and $T'(x, y)$ should be regarded as equal only if the expressions of the form (3) (or the corresponding operators) are equal but not if the limit

$$\lim_{n\to\infty} [T_n(x, y) - T'_n(x, y)]$$

equals zero. The latter limit may even make no sense. For example, the symbolic equation

$$(x - y)\delta(x - y) = 0 \tag{5}$$

does not imply that $\delta(x - y) = 0$ for $x \neq y$ (since δ is not equal to any quantity at all, δ is not a function) or that

$$\lim_{n\to\infty} \delta_n(x - y) = 0,$$

but follows only that, for any admissible $f(y)$,

$$\int (x - y)\delta(x - y)f(y)dy = 0, \tag{6}$$

or, in more detail,

$$\lim_{n\to\infty} \int (x - y)\delta_n(x - y)f(y)dy = 0. \tag{7}$$

Similarly,

$$\left(\frac{\partial}{\partial x} + \frac{\partial}{\partial y}\right)\delta(x - y) = 0 \tag{8}$$

does not mean that the approximate functions $\delta_n(x, y)$ actually depend only on the difference $(x - y)$ but means only that

$$\lim_{n\to\infty} \int \left[\frac{\partial\delta_n(x, y)}{\partial x} + \frac{\partial\delta_n(x, y)}{\partial y}\right] f(y)dy = 0. \tag{9}$$

If expressions (5) and (8) are regarded simply as a short notation for quite cumbersome formulae (7) and (9), then it is possible to use them safely, without any risk of going into contradiction. However, if we want to make these arguments rigorous, every time we must justify the passage to the limit $n \to \infty$. For this, it is necessary to study the specific features of the sequence that defines the improper function in question and to determine the functions for which the expressions of type (9) and (2) make sense.

3. The kernel of an operator can be obtained by applying this operator to the kernel of the identity operator, i.e., to the Dirac function $\delta(x-y)$ [4]. A sequence of approximate functions $T_n(x,y)$ for the improper kernel $T(x,y)$ can be constructed by applying the operator T to approximate functions of the identity kernel $\delta(x,y)$. Therefore, different types of approximate functions for the Dirac function δ are of special interest.

The function δ can be defined by the following equations:

$$\lim_{n\to\infty}\int_a^b \delta_n(x,y)f(y)dy = 0\,, \tag{10}$$

if x does not belong to the interval (a,b), and

$$\lim_{n\to\infty}\int_a^b \delta_n(x,y)f(y)dy = f(x), \qquad a < x < b \tag{11}$$

for any function $f(x)$ satisfying certain general conditions.[1]

Consider the sequence of functions

$$\delta_n(x,y) = \sum_{k=0}^{n} \varphi_k(x)\overline{\varphi}_k(y), \tag{12}$$

where $\varphi_k(y)$ is a closed orthonormal system of functions and $\overline{\varphi}_k(y)$ is a complex quantity conjugate to $\varphi_k(y)$. This sequence satisfies conditions (10) and (11). Therefore, expression (12) can be regarded as an n-th approximate function for the Dirac function $\delta(x,y)$.

If φ_k are the eigenfunctions of the operator T with eigenvalues λ_k,

$$T\varphi_k = \lambda_k\varphi_k, \tag{13}$$

then expression (12) is the kernel of the identity operator corresponding to the Neumann's "partition of unity" (Zerlegung der Einheit) for the operator T.

The expression

$$T_n(x,y) = \sum_{k=0}^{n} \lambda_k\varphi_k(x)\overline{\varphi}_k(y) \tag{14}$$

[1]These conditions can precisely be stated only if a specific sequence $\delta_n(x,y)$ is known. For the sequences of type (12), these are the conditions under which the function $f(x)$ can be expanded into the corresponding series. (*V. Fock*)

is then the n-th approximate function for the improper kernel $T(x, y)$ of T. Due to this property, sequence (12) is especially convenient since this sequence can be adapted to any specific problem. Note that, in some instances, the limit of expression (14) can exist as $n \to \infty$, and then it makes sense to speak of the kernel of T in the true meaning. For example, if T is the inverse to a differential operator L, so that

$$F(x) = Tf(x)$$

implies that

$$f(x) = LF(x),$$

then, in general, the kernel of T exists and is equal to the Green's function for the differential operator L. In this case, expression (14) coincides with the first terms of the expansion of the Green's function in the eigenfunctions.

If the operator has a continuous spectrum in addition to a discrete one, then expression (12) contains an integral along with the sum. For example, if a continuous spectrum is $0 \leq \lambda < \infty$, then we can take

$$\delta(x, y; n, \lambda) = \sum_{k=0}^{n} \varphi_k(x)\overline{\varphi_k}(y) + \int_0^\lambda \varphi(x, \lambda)\overline{\varphi}(y, \lambda)d\lambda. \tag{12*}$$

After integration in (12) and (3), not only n but also λ must be increased to infinity. In this case, the expression

$$T(x, y; n, \lambda) = \sum_{k=0}^{n} \lambda_k \varphi_k(x)\overline{\varphi_k}(y) + \int_0^\lambda \lambda\varphi(x, \lambda)\overline{\varphi}(y, \lambda)d\lambda \tag{14*}$$

plays the role of an approximate kernel of the operator T.

4. Now we consider several examples of approximate functions $\delta_n(x, y)$. For the operator

$$T = \frac{h}{2\pi i}\frac{d}{dx} \tag{15}$$

with a continuous spectrum, we obtain

$$\varphi(x, \lambda) = \frac{1}{\sqrt{h}} e^{\frac{2\pi i\lambda}{h}x}. \tag{16}$$

The expression

$$\delta(x, y; \lambda) = \int_{-\lambda}^{+\lambda} \varphi(x, \lambda)\overline{\varphi}(y, \lambda)d\lambda = \frac{\sin\frac{2\pi\lambda}{h}(x-y)}{\pi(x-y)} \tag{17}$$

has all properties of an approximate function for the Dirac δ since it is known from the theory of the Fourier integral that

$$\lim_{\lambda\to\infty}\frac{1}{\pi}\int_a^b \frac{\sin\frac{2\pi\lambda}{h}(x-y)x-y}{f}(y)dy = \begin{cases} 0 & \text{if } x \text{ is outside of } (a,b). \\ f(x) & \text{if } \quad a < x < b. \end{cases}$$

For the Fourier series in the interval $(0, 1)$,

$$\varphi_n(x) = e^{2\pi i n x}, \tag{18}$$

we have

$$\delta(x, y) = \sum_{-n}^{+n} \varphi_k(x)\overline{\varphi}_k(y) = \sum_{-n}^{+n} e^{2\pi i k(x-y)} = \frac{\sin{(2n+1)\pi(x-y)}}{\sin(x-y)}. \tag{19}$$

The next example is given by the following orthonormal system of functions on $(-\infty, +\infty)$:

$$\varphi_n(x) = \frac{1}{\sqrt[4]{\pi}}\frac{e^{-\frac{1}{2}x^2}H_n(x)}{\sqrt{2^n n!}}, \tag{20}$$

where $H_n(x)$ is the Chebyshev–Hermite polynomial.

The finite sum

$$\delta_n(x, y) = \sum_{n=0}^{n-1} \varphi_n(x)\overline{\varphi}_n(y) \tag{21}$$

can be represented in the following integral form:

$$\delta_n(x, y) = \frac{1}{\sqrt{\pi}}\frac{1}{2\pi i}\int e^{-\frac{(x-y)^2}{4}v - \frac{(x+y)^2}{4v}} \left(\frac{v+1}{v-1}\right)^n \frac{dv}{2\sqrt{v}} \tag{22}$$

taken along the contour going around the point $v = 1$ in the positive direction. For infinitely increasing n and finite x and y, or, more precisely, for

$$x^2 \ll 2n, \qquad y^2 \ll 2n \qquad \text{as} \quad n \to \infty,$$

this implies the asymptotic expression

$$\delta_n(x,y) = \frac{\sin\sqrt{2n}(x-y)}{\pi(x-y)} + O\left(\frac{1}{\sqrt{n}}\right), \tag{23}$$

which obviously has the required properties (10) and (11) [5].

In all examples in question, approximate functions for $x \neq y$ and $n \to \infty$ do not tend to zero but only oscillate increasingly stronger, which confirms all that has been said about the sense of formula (5).

References

1. P.A.M. Dirac, *Physical Interpretation of Quantum Dynamics*, Proc. Roy. Soc. London **113 A**, 621, 1926.
2. V. Fock, Izv. AN **18**, 161, 1935. (See [35-1] in this book. (*Editors*))
3. J. von Neumann, *Mathematishe Begründung der Quantenmechanik*, Göttingen Nachr. Math. Phys. Klasse 1927, S. 1.
4. D. Hilbert, J. von Neumann and L. Nordheim, *Über die Grundlagen der Quantenmechanik*, Math. Ann. **98**, 1, 1927.
5. Further examples of asymptotic expressions for sums of type (12) can be found in the dissertation of Prof. A.A. Adamov "On Expansions of Arbitrary Function of One Real Variable into Series over the Functions of a Certain Type," St. Petersburg, 1907. See also his papers in Izv. SPb Polytechn. Inst., 1908.

Translated by L.A. Khalfin

29-2
On the Notion of Velocity in the Dirac Theory of the Electron

V. Fock

Leningrad

Zs. Phys. **55**, N2, 127, 1929

1. The wave equation for an electron according to Dirac reads as[1]

$$(P_0 + \alpha_1 P_1 + \alpha_2 P_2 + \alpha_3 P_3 + \alpha_4 mc)\psi = 0\,, \tag{1}$$

where

$$\left.\begin{aligned} &P_k = p_k + \tfrac{e}{c}A_k = \frac{h}{2\pi i}\frac{\partial}{\partial x_k} + \frac{e}{c}A_k \qquad (k = 0, 1, 2, 3),\\ &x_0 = ct, \quad x_1 = x, \quad x_2 = y, \quad x_3 = z. \end{aligned}\right\} \tag{1*}$$

The electron charge is denoted here by $-e$; A_1, A_2, A_3 are the components of the vector potential, A_0 is minus the scalar potential.

Equation (1) can also be rewritten in the form

$$H\psi + \frac{h}{2\pi i}\frac{\partial\psi}{\partial t} = 0, \tag{2}$$

where H is the Hamilton operator

$$H = eA_0 + c(P_0 + \alpha_1 P_1 + \alpha_2 P_2 + \alpha_3 P_3 + \alpha_4 mc). \tag{3}$$

We assume that the four matrices $\alpha_1, \alpha_2, \alpha_3, \alpha_4$ are Hermitian and therefore the operator H is self-adjoint.

The so-called equation of motion for an operator F is

$$\frac{dF}{dt} = \frac{\partial F}{\partial t} + \frac{2\pi i}{h}(HF - FH). \tag{4}$$

[1]P.A.M. Dirac, *The Quantum Theory of the Electron*, Proc. Roy. Soc. London **A 177**, 610, 1928; **118**, 351, 1928.

This equation expresses the following mathematical fact.[2] If one constructs a matrix $||F_{mn}||$ using a complete system of solutions of equation(2), the time derivatives of the matrix elements

$$F_{mn} = \int \overline{\psi}_m F \psi_n dxdydz \tag{5}$$

are equal to the corresponding matrix elements of the right-hand side of (4), i.e.,

$$\frac{dF_{mn}}{dt} = \left\{ \frac{\partial F}{\partial t} + \frac{2\pi i}{h}(HF - FH) \right\}_{mn} . \tag{6}$$

Equation (4) also has the following physical meaning. If F is an operator corresponding to a classical quantity, equation (4) yields the operator corresponding to its time derivative.

2. One may ask why time plays a preselected role in our formulae, while in the relativity theory coordinates and time enter formally symmetrically. This speculation can be easily clarified.

The relativistic symmetry between coordinates and time is valid as far as one does not distinguish between real and imaginary quantities. In fact, time (and all time-like variables) differs from coordinates by the sign in the fundamental metric form and plays a rather preselected role.

We would like to discuss this problem in more detail by considering the eigenfunctions problem for the operator F.

In formulae (2–6) we have chosen time as an independent variable and all operators and their eigenfunctions depend on it only as on a parameter. The domain of action for the operators is therefore an infinite space, i.e., a certain domain defined by the space variables; in other words, it is the integration domain in formula (5). On the boundary of the domain the eigenfunctions should satisfy certain boundary conditions. If we choose the coordinate x as an independent variable, all the operators should be expressed via y, z and t, while x should be considered as a parameter. The action domain of the operators turns into a certain space-time domain. But now the boundary conditions for such a domain are of completely different character than the ones for a purely spatial domain. For example, they can be given only on part of the boundary and the eigenfunction problem cannot be formulated in this

[2]See *Uber die Beziehung zwischen den Integralen der quantenmechanichen Bewegungsgleichungen und der Schrödingerischen Wellengleichung*, Zs. Phys. **49**, 323, 1928 (and [28-2] in this book (*Editors*)). The results of this paper are directly applicable to the Dirac wave equation as far as the operator H is self-adjoint. (*V. Fock*)

situation. In other words, this difference is the difference between the elliptic and hyperbolic differential equations.

We see therefore that time plays a role quite different from the role of coordinates both in the relativity theory and in quantum relativity theory.

Here we are going to make a short remark about the notion of proper time. Classically the proper time is defined as the length of the world line of a particle. In the quantum theory the notion of the path of a particle makes no sense and the definition of the proper time is therefore impossible.[3]

3. Consider now the equations of motion. According to formula (4) the operator corresponding to the derivative of a coordinate x_k by time, i.e., the velocity $\dfrac{dx_k}{dt}$, is

$$\frac{dx_k}{dt} = c\alpha_k \qquad (k = 1, 2, 3)\,. \tag{7}$$

As is generally known from the Dirac theory, the components of the current density vector are

$$j_k = ec\overline{\psi}\alpha_k\psi. \tag{8}$$

This formula is in complete agreement with the interpretation of the operators $c\alpha_k$ as representing the three-dimensional (but not four-dimensional) velocity. On the other hand, this interpretation is related with an essential difficulty. The eigenvalues of this operator are $\pm c$. One comes therefore to a paradox that the measurement of each velocity component can give only the values $\pm c$. Breit[4] tried to give a physical meaning to this result, which he obtained in another way;[5] however, his speculations do not seem to be very convincing to the author.

---

[3]The statements by Eddington (*The Charge of an Electron*, Proc. Roy. Soc. London **A 122**, 358, 1929), taken in his eqn. (11), are obviously based on a mistake. (*V. Fock*)

[4]G. Breit, *An Interpretation of Dirac's Theory of the Electron*, Proc. Nat. Acad. Amer. **14**, 553, 1928.

[5]Breit interpreted α_4 as $\sqrt{1-\frac{v^2}{c^2}}$, which surely cannot be justified; cf. formula (16). (*V. Fock*)

4. Consider now the four-dimensional velocity, which has classical components

$$\left.\begin{aligned} v_1 &= \frac{\dot{x}}{\sqrt{1-\frac{v^2}{c^2}}}, & v_2 &= \frac{\dot{y}}{\sqrt{1-\frac{v^2}{c^2}}}, \\ v_3 &= \frac{\dot{z}}{\sqrt{1-\frac{v^2}{c^2}}}, & v_4 &= \frac{1}{\sqrt{1-\frac{v^2}{c^2}}}. \end{aligned}\right\} \tag{9}$$

According to classical relativistic mechanics these components can be expressed in terms of the components of the momentum p_k and of the four-potential A_k:

$$v_k = \frac{1}{m}\left(p_k + \frac{e}{c}A_k\right) \quad (k = 1, 2, 3), \tag{10}$$

$$v_0 = -\frac{1}{m}\left(p_k + \frac{e}{c}A_0\right). \tag{11}$$

They satisfy the equations of motion (the charge is $-e$):

$$\left.\begin{aligned} m\tfrac{dv_1}{dt} &= -\tfrac{e}{c}(\dot{y}\mathfrak{H}_z - \dot{z}\mathfrak{H}_y) - e\mathfrak{E}_x, \\ m\tfrac{dv_2}{dt} &= -\tfrac{e}{c}(\dot{z}\mathfrak{H}_x - \dot{x}\mathfrak{H}_z) - e\mathfrak{E}_y, \\ m\tfrac{dv_3}{dt} &= -\tfrac{e}{c}(\dot{x}\mathfrak{H}_y - \dot{y}\mathfrak{H}_x) - e\mathfrak{E}_z, \end{aligned}\right\} \tag{12}$$

$$mc^2\tfrac{dv_0}{dt} = -e(\dot{x}\mathfrak{E}_x + \dot{y}\mathfrak{E}_y + \dot{z}\mathfrak{E}_z). \tag{13}$$

The question arises whether we get correct quantum operators for the four-velocity if we just substitute p_k in (10) and (11) the ordinary operators $\frac{h}{2\pi i}\frac{\partial}{\partial x_k}$. It seems that the answer to this question is affirmative. First, operators (10) have continuous spectra in the interval from $-\infty$ to $+\infty$, and the absolute values of the eigenvalues of the operator for v_0[6] are greater than or equal to unity.[7] These domains of values coincide with the ones for the classical four-velocity. Second, if we apply the equation of motion (4) for $P_1 = mv_1$, $P_2 = mv_2$ and use the expression for the fields

$$\left.\begin{aligned} \mathfrak{E}_x &= -\tfrac{1}{c}\tfrac{\partial A_1}{\partial t} + \tfrac{\partial A_0}{\partial x_1} \quad \text{and so on}, \\ \mathfrak{H}_x &= \tfrac{\partial A_3}{\partial y} - \tfrac{\partial A_2}{\partial z} \quad \text{and so on}, \end{aligned}\right\} \tag{14}$$

[6]Given by formula (16). (*V. Fock*)

[7]The proof is given in Appendix. (*V. Fock*)

we obtain

$$\left.\begin{aligned} \frac{dP_1}{dt} &= -e(\alpha_2\mathfrak{H}_z - \alpha_3\mathfrak{H}_y) - e\mathfrak{E}_x, \\ \frac{dP_2}{dt} &= -e(\alpha_3\mathfrak{H}_x - \alpha_1\mathfrak{H}_z) - e\mathfrak{E}_y, \\ \frac{dP_3}{dt} &= -e(\alpha_1\mathfrak{H}_y - \alpha_2\mathfrak{H}_x) - e\mathfrak{E}_z. \end{aligned}\right\} \tag{15}$$

The operator for the time component can be obtained by eliminating P_0 from (11) using the wave equation. We get

$$\text{the operator for } v_0 = G = \alpha_4 + \frac{1}{mc}(\alpha_1 P_1 + \alpha_2 P_2 + \alpha_3 P_3). \tag{16}$$

Substitution of (16) into (4) gives after a short calculation

$$mc^2 \frac{dG}{dt} = -ec(\alpha_1\mathfrak{E}_x + \alpha_2\mathfrak{E}_y + \alpha_3\mathfrak{E}_z). \tag{17}$$

The quantum-mechanical formulae (15) and (17) are complete analogues of the classical formulae (12) and (13). The former transform into the latter if we let $P_k = mv_k$, $c\alpha_k = \dot{x}_k$, $G = v_0$.

The results obtained seem to show definitely that we have obtained correct quantum-mechanical operators both for three- and four-dimensional velocities.

5. In the above, it was implicitly assumed that operators (7) on one hand and operators (15) and (17) on the other hand correspond to the three- and, respectively, four-dimensional representations of the same physical quantity, namely, the "velocity on the electron." It turns out that this assumption has absolutely no grounds and that one deals here with different physical quantities.

According to classical mechanics the three-dimensional speed can be expressed via the four-dimensional one and vice versa. There is also the relation

$$1 - \frac{1}{c^2}(\dot{x}^2 + \dot{y}^2 + \dot{z}^2) = \frac{1}{v_0^2}\,. \tag{18}$$

On the other hand, there exist the following rules for ordering operators of physical quantities:

a) The square of a quantity corresponds to the operator for this quantity applied twice.

b) The reciprocal quantity corresponds to the inverse of the corresponding operator.

c) If the operators of two quantities commute, the sum of these quantities corresponds to the sum of their operators.

Now we have convinced ourselves that the operator G corresponds to the quantity v_0. Therefore it follows, applying the rules a) and b) to the right-hand side of (18), that

$$\frac{1}{v_0^2} \to G^{-2}. \tag{19}$$

Assume that the operators $c\alpha_k$ correspond to the quantities $\dot{x}_k$. Then their squares are equal to c^2 and are obviously commutative. Therefore from our rules it follows that the operator corresponding to the left-hand side of (18) is –2 (the multiplication by –2):

$$1 - \frac{\dot{x}^2 + \dot{y}^2 + \dot{z}^2}{c^2} \to -2, \tag{20}$$

which is obviously absurd.

We have shown therefore that the operators $c\alpha_k$ and v_k correspond to different quantities. What notions could they be? It follows from the physical sense that the operators v_k can correspond to nothing but the mechanical four-velocity of the electron as a "charged point" and which we would like to interpret as a corpuscular velocity. Concerning the operators $c\alpha_k$, they should correspond to such a triple of quantities for which there would be no sense to sum their squares. One such triple of quantities is given by mutually perpendicular components of the speed of a wave propagating in all directions with the light speed c. Owing to the relation of the operators $c\alpha_k$ to the usual velocity of the electrons (e.g., appearing in the equation of motion) it can be only the de Broglie wave, which is known from the relativistic Dirac equation to propagate with the speed c, and not with the superlight speed $\frac{c^2}{v}$.

We come therefore to the conclusion that it is the dual – corpuscular and wave – nature of the electron which manifests itself in the difference between the electron velocity operators.

6. It can be expected from all described above that one can find the operator corresponding to the ordinary mechanical three-dimensional (corpuscular) velocity. This operator – by the classical analogy – should have a continuous spectrum in the interval from $-c$ to $+c$ and satisfy equation (18) provided the squares of its components mutually commute.

We are going to show now that in the electrostatic case (in the absence of a magnetic field) it is easy to find such an operator.

In classical mechanics the three-dimensional velocity $\dot{x}_k$ can be easily expressed via the four-dimensional one v_k:

$$\dot{x}_k = \frac{v_k}{v_0} \quad (k = 1, 2, 3). \tag{21}$$

In the electrostatic case, the operators for v_k and v_0 are already constructed and we can use the relation (21) in quantum theory offhand. We denote by V_k the operators corresponding to $\dot{x}_k$. So we have

$$V_k = \frac{1}{m} p_k G^{-1}. \tag{22}$$

To obtain the eigenvalues of the operators V_k, we will write down the equation for their eigenfunctions:

$$\frac{1}{m} p_k G^{-1} f = \lambda f. \tag{23}$$

It follows from (23) that

$$\frac{1}{m} p_k f = \lambda G f$$

and after substituting the expression for the operator G we get that

$$c^2 p_k^2 f = \lambda^2 (P_1^2 + P_2^2 + p_3^2 + mc^2) f.$$

For $k = 1$, one can write down this equation explicitly as

$$\left(1 - \frac{c^2}{\lambda^2}\right) \frac{\partial^2 f}{\partial x^2} + \frac{\partial^2 f}{\partial y^2} + \frac{\partial^2 f}{\partial z^2} - \frac{4\pi^2 m^2 c^2}{h^2} f = 0. \tag{24}$$

This differential equation has a finite constant in the whole space solution if and only if the coefficient at $\frac{\partial^2 f}{\partial x^2}$ is negative. It implies that

$$-c < \lambda < c, \tag{25}$$

i.e., that the eigenvalues of the operators V_k form a continuous spectrum in the interval $(-c, +c)$.

One can easily check that the operators V_1, V_2, V_3 satisfy the relation

$$G^2 \left\{1 - \frac{1}{c^2}(V_1^2 + V_2^2 + V_3^2)\right\} = 1\,, \tag{26}$$

which corresponds to the classical relation (18).

We see also that in the electrostatic case the operators V_k have the desired properties and can be considered as representing the three-dimensional corpuscular velocity of the electron. In the generic case the translation of formula (21) into the language of quantum mechanics is not quite clear; however perhaps it is not necessary since the operators V_k have no direct applications in quantum theory; our aim was to show only that in three-dimensional consideration the corpuscular velocity and the wave speed of the electron correspond to different quantum-mechanical operators.

7. However, an analogy exists between the operators $c\alpha_k$ and V_k that we are going to follow now.

The main analogy consists in the behavior of both operators according to the correspondence principle. Assume that the energy operator H does not depend on time and construct the complete system of solutions

$$\psi_k(x, y, z, t; E_n) \qquad (k = 1, 2, 3, 4)$$

of the Dirac equation (2), which are the eigenfunctions of the energy operator. Using this function, we can construct the matrix of the operator corresponding, for instance, to the coordinate x:

$$x_{mn} = \sum_{k=1}^{4} \int \overline{\psi}(E_m)\psi(E_n)dxdydz. \tag{27}$$

According to the correspondence principle, the matrix elements x_{mn} should transform at a certain limiting procedure to the corresponding term of the Fourier decomposition of the coordinate x considered as a function of time. This limiting procedure goes as follows: both quantum numbers[8] should tend to infinity and the Planck constant h should tend to zero in such a way that the difference

$$n - m = s$$

and the quantity

$$nh = J$$

remain finite.[9]

[8]The simplest way here is to assume that the energy depends on only one quantum number. (*V. Fock*)

[9]See, e.g., C. Eckart, *Die korrespondenzmäßige Beziehung zwischen den Matrizen und den Fourierkoeffizienten des Wasserstoffproblems*, Zs. Phys. **48**, 295, 1928. (*V. Fock*)

It follows from equality (6) that

$$\frac{\partial x_{mn}}{dt} = \sum_{k,l=1}^{4} \int \overline{\psi}_k(E_m)(\alpha_1)_{kl}\psi_l(E_n)dxdydz. \tag{28}$$

In the limit, the matrix element of the operator $c\alpha_1$ should coincide with the derivative of the corresponding Fourier decomposition term of the coordinate x, i.e., with the term of the decomposition of the corresponding classical velocity.

Now we would like to demonstrate that the matrix elements of the operator V_1 satisfy this limiting condition. For this purpose, consider the difference

$$B_k = V_k - c\alpha_k. \tag{29}$$

We have

$$B_k = \frac{1}{m}G^{-1}(p_k - mcG\alpha_k) = \frac{1}{2mc}G^{-1}(\alpha_k H - H\alpha_k)$$

or

$$B_k = \frac{ih}{4\pi mc}G^{-1}\dot{\alpha}_k\,, \tag{30}$$

where

$$\dot{\alpha}_k = \frac{2\pi i}{h}(H\alpha_k - \alpha_k H). \tag{31}$$

The matrix elements of the operators $\dot{\alpha}_k$ are the derivatives of the ones of α_k and remain so after taking the limit. Since the operator G^{-1} remains (for all values of h) uniformly bounded, the matrix elements of the operator G^{-1} remain finite. The matrix elements of B_k contain a factor h tending to zero and also tend to zero.

We have demonstrated that in the limit the matrix elements of V_k coincide with the ones of $c\alpha_k$; it also implies that the operator V_k likewise satisfies the corresponding principle.

A further analogy between $c\alpha_k$ and V_k is that the four-velocity v_k can be expressed via these operators in a similar way. On the one hand, we have

$$v_k = \frac{1}{2}(GV_k + V_k G) \tag{32}$$

and on the other hand

$$v_k = \frac{c}{2}(G\alpha_k - \alpha_k G). \tag{33}$$

There is also an equality

$$V_k^2 = \frac{c}{2}(V_k\alpha_k + \alpha_k V_k), \tag{34}$$

which shows that the operator V_k applied twice is a kind of "symmetric" product of the operators V_k and $c\alpha_k$.

8. The results of our investigations can be summarized in the following way.

A single classical quantity – the velocity of the electron – corresponds to two different quantities in the Dirac theory. They can be interpreted as the corpuscular and wave speed of the electron, respectively. The operators of the three-dimensional components of the wave velocity are $c\alpha_k$ ($k = 1, 2, 3$), where α_k are the four-dimensional Dirac matrices. They have the discrete spectrum $\pm c$. The corpuscular velocity writes down better via its four-dimensional components. The operators for these components are related to the operators for the momentum components by the same equations as in the classical theory. These operators have a continuous spectrum that coincides with the domain of the possible values of the corresponding classical quantities. In the electrostatics, it is also possible to write down the corpuscular speed in a three-dimensional form; the operators of the three-dimensional components have the continuous spectrum from $+c$ to $-c$ and satisfy the correspondence principle like the corresponding operators of the wave velocity. In the quantum mechanical equations of motion of the electron, the corpuscular velocity appears in the expression for the acceleration and, therefore, it has the mechanical meaning. The components of the wave velocity appear as factors at the electromagnetic field strength and serve, therefore, to write down the influence of the electromagnetic field on the electron.

Appendix

Eigenvalues spectrum of the four-speed operator

1. Consider now one of the space-like components of the four-velocity, e.g., v_1. The differential equation for the eigenfunctions of the corresponding operator is

$$\frac{1}{m}\left(\frac{h}{2\pi i}\frac{\partial\psi}{\partial x} + \frac{e}{c}A_x \cdot \psi\right) = \lambda\psi. \tag{A1}$$

It is well known that by adding a gradient the vector-potential can be normalized in such a way that its x-component vanishes. Then (A1) transforms into a differential equation with constant coefficients that has all real values of λ in the spectrum. It shows that the operators for the three space-like components of the four-velocity have continuous spectra from $-\infty$ to $+\infty$, which is just what we wanted to demonstrate.

2. Consider now the time-like component v_0 of the four-velocity and the case when there is no magnetic field. It means that one can assume that the space-like components of the four-potential vanish. The corresponding operator is

$$G = \alpha_4 + \frac{1}{mc}(\alpha_1 p_1 + \alpha_2 p_2 + \alpha_3 p_3). \tag{A2}$$

The equation for its eigenvalues reads as

$$G\psi = \lambda\psi. \tag{A3}$$

However, the eigenfunctions also satisfy the differential equation

$$G^2\psi = \lambda^2\psi, \tag{A4}$$

which can be written in the form

$$\psi - \frac{1}{k^2}\Delta\psi = \lambda^2\psi, \tag{A5}$$

where Δ is the usual Laplace operator and

$$k = \frac{2\pi mc}{h} \tag{A6}$$

for brevity.

It is well known that equation (A5) has solutions everywhere finite for $\lambda^2 \geq 1$ only. It implies that in the electrostatic case the operator of the time-like component of the four-velocity of the electron has a continuous spectrum, the eigenvalues being greater than or equal to 1.

It also implies that the inverse operators G^{-1} and G^{-2} are bounded.

It must be mentioned that these inverse operators have (proper) kernels, which can be easily computed.

The kernel of G^{-2} provides the solution for the equation

$$F - \frac{1}{k^2}\Delta F = f\,. \tag{A7}$$

Namely,

$$F(xyz) = \frac{k^2}{4\pi} \int \frac{e^{-kr}}{r} f(\xi\eta\zeta) d\xi d\eta d\zeta, \tag{A8}$$

where

$$r = \sqrt{(x-\xi)^2 + (y-\eta)^2 + (z-\zeta)^2}.$$

The operator G^{-2} has, therefore, the kernel

$$K(xyz, \xi\eta\zeta) = \frac{k^2}{4\pi} \frac{e^{-kr}}{r}, \tag{A9}$$

and G^{-1}, the kernel

$$\frac{k^2}{4\pi} G \frac{e^{-kr}}{r}, \tag{A10}$$

where the differentiations from G act on the variables x, y, z.

3. We would like now to study the spectrum of the operator G for the case of a constant magnetic field parallel to the z-axis. The vector potential for this case is

$$A_x = -\frac{1}{2} H y, \quad A_y = \frac{1}{2} H x; \quad A_z = 0. \tag{A11}$$

The operator G takes the form

$$G = \alpha_4 + \frac{1}{mc} \left[\alpha_1 \left(p_x - \frac{e}{2c} H y \right) + \alpha_2 \left(p_y - \frac{e}{2c} H x \right) + \alpha_3 p_z \right], \tag{A12}$$

$$\alpha_1 = \sigma_1, \quad \alpha_2 = \varrho_3 \sigma_2, \quad \alpha_3 = \sigma_3, \quad \alpha_4 = \varrho_2 \sigma_2, \tag{A13}$$

where ϱ_k and σ_k are the Dirac matrices.[10] Note that the operator

$$L = \varrho_1 \sigma_1 - \frac{p_z}{mc} \varrho_3 \tag{A14}$$

[10] This choice is recommended due to the especially simple and transparent transformation properties of the corresponding ψ-function. Under an arbitrary Lorentz transformation the functions ψ_k transform according to the formula

$$\begin{aligned} \psi_1' &= \alpha\psi_1 + \beta\psi_2, & \psi_3' &= \overline{\alpha}\psi_3 + \overline{\beta}\psi_4, \\ \psi_2' &= \gamma\psi_1 + \delta\psi_2, & \psi_4' &= \overline{\gamma}\psi_3 + \overline{\delta}\psi_4, \end{aligned}$$

where $\alpha, \beta, \gamma, \delta$ are complex parameters satisfying the equation

$$\alpha\delta - \beta\gamma = 1.$$

In the special case of a rotation of the space coordinate system these are the usual Caley–Klein parameters. (*V. Fock*)

commutes with G. Therefore, one can subject the eigenfunctions of G satisfying the equation

$$G\psi = \lambda\psi \tag{A15}$$

by an additional condition

$$L\psi \equiv \left(\varrho_1\sigma_1 - \frac{p_z}{mc}\varrho_3\right)\psi = l\psi. \tag{A16}$$

If we apply the operator L one more time, we obtain

$$L^2\psi = \left(1 + \frac{p_z^2}{m^2c^2}\right)\psi = l^2\psi. \tag{A17}$$

If we take the eigenvalue of the operator $\frac{h}{2\pi i}\frac{\partial}{\partial z}$ for p_z, it follows from (A17) that

$$l^2 = 1 + \frac{p_z^2}{m^2c^2}\,. \tag{A18}$$

The expression $\alpha_4 + \frac{p_z}{mc}\alpha_3$ appearing in G can be rewritten in the form

$$\alpha_4 + \frac{p_z}{mc}\alpha_3 = \varrho_2\sigma_2 + \frac{p_z}{mc}\sigma_3 = -\varrho_3\sigma_3 L\,. \tag{A19}$$

Substituting this expression into G and using (A16) we can obtain from (A15) the following equation:

$$\left[-l\varrho_3\sigma_3 + \frac{1}{mc}(\sigma_1 P_x + \varrho_3\sigma_2 P_y)\right]\psi = \lambda\psi\,, \tag{A20}$$

where for brevity

$$P_x = p_x - \frac{e}{2c}H\cdot y, \quad P_y = p_y + \frac{e}{2c}H\cdot x\,. \tag{A21}$$

The first two equations of the system (A20) contain only the functions ψ_1 and ψ_2 and read as

$$\left.\begin{aligned} -l\psi_1 + \tfrac{1}{mc}(P_x - iP_y)\psi_2 = \lambda\psi_1, \\ l\psi_2 + \tfrac{1}{mc}(P_x + iP_y)\psi_1 = \lambda\psi_2. \end{aligned}\right\} \tag{A22}$$

Eliminating the function ψ_1 from these equations, we get the differential equation on ψ_2:

$$\frac{1}{2m}(P_x^2 + P_y^2)\psi_2 = E\psi_2, \tag{A23}$$

where for brevity

$$E = \frac{mc^2}{2}(\lambda^2 - l^2) + \frac{eh}{4\pi mc}H. \tag{A24}$$

Equation (A23) is the usual Schrödinger equation for a constant magnetic field; its eigenvalues are well known, namely,

$$E_n = (2_n + 1)h\nu, \quad \nu = \left|\frac{eH}{2mc}\right|, \tag{A25}$$

where n is a nonnegative integer. From (A25), (A24) and (A18) we find the final expressions for the eigenvalues of the operator G^2:

$$\lambda^2 = 1 + \frac{p_z^2}{m^2c^2} + \frac{4nh\nu}{mc^2} \qquad (n = 1, 2, \ldots)\,. \tag{A26}$$

The last two terms in this expression for λ^2 are zero or positive and p_z can take any positive real value. Therefore, it is demonstrated that the eigenvalues of G form a continuous spectrum and take all values between $+\infty$ and $-\infty$.

In a generic case,[11] one can easily demonstrate that the eigenvalues of the operator G have absolute values greater than or equal to 1. Denote by P the self-adjoint operator

$$P = \alpha_1 P_1 + \alpha_2 P_2 + \alpha_3 P_3 \tag{A27}$$

with real eigenvalues p. The operator G^2 can be rewritten in the form

$$G^2 = 1 + \frac{1}{m^2c^2}P^2\,. \tag{A28}$$

Its eigenvalues are

$$\lambda^2 = 1 + \frac{p^2}{m^2c^2} \geq 1. \tag{A29}$$

Q.E.D.

Presumably the operators G and P also have a continuous spectrum in a generic case. The proof of this theorem is required; however, it is expected to be rather complicated.

Translated by V. V. Fock

---

[11]This paragraph was added in proof. (*V. Fock*)

29-3
On the Dirac Equations in General Relativity[1]

NOTE[2] BY V. FOCK
PRESENTED BY M. DE BROGLIE

C. R. Acad. Sci. Paris **189**, N1, 25, 1929

1. To take into account that the Einstein form ds^2 is indeterminate, we introduce, following Eisenhart,[3] the numbers $e_1 = e_2 = e_3 = -1$, $e_0 = +1$. The Hermitian Dirac matrices will be denoted by $\alpha_1, \alpha_2, \alpha_3, \alpha_5$ and the unit matrix – by α_0. Indices run through $0, 1, 2, 3$.

Comparing the rules of transport for a semivector

$$\delta\psi = \sum_l e_l C_l ds_l \psi \tag{1}$$

and for a vector $A_i = \overline{\psi}\alpha_i\psi$,

$$\delta A_i = \sum_{kl} e_k e_l \gamma_{ikl} A_k ds_l \,, \tag{2}$$

and noticing that the quantity $A_5 = \overline{\psi}\alpha_5\psi$ is invariant, we obtain

$$C_l^\dagger \alpha_5 + \alpha_5 C_l = 0; \qquad C_l^\dagger \alpha_m + \alpha_m C_l = \sum_k e_k \alpha_k \gamma_{mkl}. \tag{3}$$

One can check that the general solution of these equations is

$$C_l = \frac{1}{4}\sum_{mk} \alpha_m \alpha_k e_k \gamma_{mkl} + \frac{2\pi i e}{hc}\varphi_l, \tag{4}$$

where φ_l is a real vector, which we suppose to be the vector-potential.

[1] See V. Fock and D. Ivanenko, *Linear Quantum Geometry and Parallel Transport*, Comptes Rendus **188**, 1470, 1929.

[2] Session of 24 June 1929.

[3] L. Eisenhart, Riemannian Geometry, Princeton, 1926.

2. To obtain the Dirac equations, we set

$$\mathcal{F}\psi = \frac{h}{2\pi i}\sum_k e_k\alpha_k\left(\frac{\partial\psi}{\partial s_k} - C_k\psi\right) - mc\alpha_5\psi. \tag{5}$$

If the coordinates x^σ and matrices

$$\gamma^\sigma = \sum_k e_k\alpha_k h_k^\sigma; \qquad \Gamma_\sigma = \sum_k e_k\alpha_k h_{\sigma k}C_k \tag{6}$$

are introduced, the operator $\mathcal{F}$ can be written as

$$\mathcal{F}\psi = \frac{h}{2\pi i}\gamma^\sigma\left(\frac{\partial\psi}{\partial x^\sigma} - \Gamma_\sigma\psi\right) - mc\alpha_5\psi. \tag{5*}$$

The following relation can be derived from equation (3) relying on the definition of the Ricci coefficients γ_{mkl} ,

$$\Gamma_\alpha^\dagger\gamma^\sigma + \gamma^\sigma\Gamma_\alpha = -\nabla_\alpha\gamma^\sigma \tag{7}$$

where ∇_α is the symbol of the covariant derivative. Using this relation, one can easily check the identity

$$\overline{\psi}F\psi - \psi\overline{F\psi} = \frac{h}{2\pi i}\frac{1}{\sqrt{g}}\frac{\partial}{\partial x^\sigma}\left(\overline{\psi}\sqrt{g}\gamma^\sigma\psi\right), \tag{8}$$

where g is the absolute value of the determinant of the fundamental tensor $g_{\varrho\sigma}$.

Identity (8) shows that the operator F coincides with its conjugate. This allows us to assert that *in general relativity the Dirac equation has the form*

$$F\psi = 0. \tag{9}$$

Thus, equation (8) shows that *the divergence of the current vector is zero,*

$$\nabla_\sigma S^\sigma = 0, \qquad S^\sigma = \overline{\psi}\gamma^\sigma\psi. \tag{10}$$

3. Setting $D_\sigma = \frac{\partial}{\partial x^\sigma} - \Gamma_\sigma$, one easily finds

$$h_k^\sigma h_l^\varrho\left(D_\sigma D_\varrho - D_\varrho D_\sigma\right) = \frac{1}{4}\sum_{ij}\alpha_i\alpha_j e_j\gamma_{ijkl} + \frac{2\pi ie}{hc}M'_{kl}, \tag{11}$$

where γ_{ijkl} are the n-hedral components of the Riemann tensor and M'_{kl} are those of the electromagnetic field bivector. From formula (11) one gets

$$\gamma^\sigma\left(D_\sigma D_\alpha - D_\alpha D_\sigma\right) = -\frac{1}{2}\gamma^\varrho R_{\varrho\alpha} + \frac{2\pi ie}{hc}\gamma^\varrho M_{\varrho\alpha}, \tag{12}$$

where $R_{\varrho\alpha}$ is the contracted Riemann tensor. If we set $A^{\sigma\cdot}_{\cdot\alpha} = \overline{\psi}\gamma^\sigma D_\alpha\psi$, the divergence of the tensor $A^{\sigma\cdot}_{\cdot\alpha}$ can be checked to be equal to

$$\nabla_\sigma A^{\sigma\cdot}_{\cdot\alpha} = -\frac{1}{2}S^\varrho R_{\varrho\alpha} + \frac{2\pi ie}{hc}S^\varrho M_{\varrho\alpha}. \tag{13}$$

Since the divergence of S^ϱ is zero, one gets the relation $\nabla_\sigma\nabla_\alpha S^\sigma = -S^\varrho R_{\varrho\alpha}$. Comparing this formula with (13), one finds for the divergence of the tensor

$$T^{\sigma\cdot}_{\cdot\alpha} = \frac{ch}{2\pi i}\left[\overline{\psi}\gamma^\sigma\left(\frac{\partial\psi}{\partial x_\alpha} - \Gamma_\alpha\psi\right) - \frac{1}{2}\nabla_\alpha\left(\overline{\psi}\gamma^\sigma\psi\right)\right] \tag{14}$$

the following expression:

$$\Delta_\sigma T^{\sigma\cdot}_{\cdot\alpha} = eS^\varrho M_{\varrho\alpha}. \tag{15}$$

With the help of (9) one can check that the tensor $T^{\sigma\cdot}_{\cdot\alpha}$ is real. *Hence we can consider $T^{\sigma\cdot}_{\cdot\alpha}$ as the matter energy tensor. Equations (15) are thus the equations of motion* in general relativity. The tensor $T_{\sigma\alpha} = g_{\sigma\varrho}T^{\varrho\cdot}_{\cdot\alpha}$ is not symmetric with respect to permutations of indices.

The tensor $T^{\sigma\cdot}_{\cdot\alpha}$ can be used to write down the variational principle equivalent to the Dirac equations (9) in the form

$$\delta\iiiint\left(T^{\sigma\cdot}_{\cdot\alpha} - m\overline{\psi}\alpha_5\psi\right)\sqrt{g}\,dx_0\,dx_1\,dx_2\,dx_3 = 0, \tag{16}$$

demonstrating that *the variation of the mean deviation for the invariant of the matter invariant density tensor T is zero.* Thus, the deviation itself nullifies due to equations (9) in strict accordance with the classical theory.

Translated by V.D. Lyakhovsky

29-4
Dirac Wave Equation and Riemann Geometry[1]

V. Fock

Leningrad University

Le Journal de Physique et de Radium
Série VI, **10**, 392, 1929

1 Lorentz Transformations for a Semi-vector

For an arbitrary Lorentz transformation the four Dirac functions Ψ undergo a definite linear substitution.[2]

The transformation law of a semi-vector differs from that for a vector or a tensor; these functions constitute, therefore, a new geometric quantity which will be called a "semi-vector."

The transformation law of semi-vectors takes a particularly simple form if one chooses for the matrices α entering the Dirac equation[3]

$$\left.\begin{aligned}(\alpha_1 P_1 + \alpha_2 P_2 + \alpha_3 P_3 + mc\alpha_4)\Psi = \alpha_0 P_0 \Psi,\\ P_l = \frac{h}{2\pi i}\frac{\partial}{\partial x_l} - \frac{e}{c}\varphi_l\end{aligned}\right\} \tag{1}$$

in the following form

$$\alpha_0 = 1; \quad \alpha_1 = \sigma_1; \quad \alpha_2 = \varphi_3\sigma_2; \quad \alpha_3 = \sigma_3; \quad \alpha_4 = \varrho_2\sigma_2, \tag{2}$$

where ϱ_i and σ_i $(i = 1, 2, 3)$ are the matrices introduced by Dirac. The matrix α_4 may be replaced by

$$\alpha_5 = \varrho_1\varphi_1 \,. \tag{2*}$$

[1]See 1) V. Fock and D. Ivanenko, *Uber eine mögliche geometrische Deutung der relativitischen Quantentheorie*, Zs. Phys. **54**, 798, 1929; 2) the same authors, *Géomeétrie quantique lineéare et déplacement parallėle*, C. R. **188**, 1470, 1929; 3) V. Fock, *Sur les équations de Dirac dans la teorie de relativité générale*, C. R. **189**, 25, 1929 (see also [29-3], this book); 4) V. Fock, *Geometrisierung der Diracschen Teorie des Elektrons*, Zs. Phys. **57**, 261, 1929. (*V. Fock*)

[2]See F. Möglich, Zs. Phys. **48**, 852, 1928; J. v. Neumann, Zs. Phys. **48**, 868, 1928.

[3]P. Dirac, Proc. Roy. Soc. **A177**, 351, 1928.

To any Lorentz transform, one can put into correspondence the following transformation of the function Ψ:

$$\left.\begin{aligned} \Psi_1'' &= \alpha\Psi_1 + \beta\Psi_2; & \Psi_3'' &= \overline{\alpha}\Psi_3 + \overline{\beta}\Psi_4; \\ \Psi_2'' &= \gamma\Psi_1 + \delta\Psi_2; & \Psi_3'' &= \overline{\gamma}\Psi_3 + \overline{\delta}\Psi_4, \end{aligned}\right\} \tag{3}$$

$$\alpha\delta - \beta\gamma = 1, \tag{4}$$

where $\alpha, \beta, \gamma, \delta$ are the generalized Caley–Klein parameters.

Formulae (3) and their conjugates can be written down in a symbolic way,

$$\Psi'' = S\Psi; \quad \overline{\Psi}'' = \overline{\Psi}S^{\dagger}, \tag{5}$$

where S is the matrix

$$S = \left\{\begin{matrix} \alpha & \beta & 0 & 0 \\ \gamma & \delta & 0 & 0 \\ 0 & 0 & \overline{\alpha} & \overline{\beta} \\ 0 & 0 & \overline{\gamma} & \overline{\delta} \end{matrix}\right\} \tag{6}$$

and $S^{\dagger}$ is the adjoint matrix[4]

$$S^{\dagger} = \left\{\begin{matrix} \overline{\alpha} & \overline{\gamma} & 0 & 0 \\ \overline{\beta} & \overline{\delta} & 0 & 0 \\ 0 & 0 & \alpha & \gamma \\ 0 & 0 & \beta & \delta \end{matrix}\right\}. \tag{6*}$$

Introduce the numbers

$$e_0 = 1; \quad e_1 = e_2 = e_3 = -1 \tag{7}$$

in such a way that the fundamental Minkowski form writes down as

$$\pm s^2 = \sum_{k=0}^{3} e_k x_k^2. \tag{8}$$

The matrices S and $S^{\dagger}$ satisfy the equations

$$S^{\dagger}\alpha_i = \sum_{k=1}^{3} e_k a_{ik} \alpha_k \qquad (i = 0, 1, 2, 3), \tag{9}$$

[4]A transposed and conjugated matrix will be called "adjoint matrix" and denoted by a dagger ($S^{\dagger}$). (*V. Fock*)

$$S^\dagger \alpha_4 = \alpha_4; \qquad S^\dagger \alpha_5 = \alpha_5\,, \tag{9*}$$

where a_{ik} are real numbers satisfying

$$\sum_{k=1}^{3} e_i a_{ik} a_{il} = e_k \delta_{kl}\,, \tag{10}$$

$$\sum_{k=1}^{3} e_i a_{ki} a_{li} = e_k \delta_{kl}\,. \tag{10*}$$

Relations (10) coincide with the ones satisfied by the coefficients a_{ik} of a Lorentz transformation:

$$\left.\begin{aligned} x_i'' &= \sum_{k=1}^{3} e_k a_{ik} x_k, \\ x_i &= \sum_{k=1}^{3} e_k a_{ik} x_k''. \end{aligned}\right\} \tag{11}$$

It is known that these coefficients depend on six arbitrary parameters. On the other hand, the matrix S contains four complex constants restricted by relation (4); it gives six independent real parameters. One deduces that the coefficients a_{ik} of formula (9) correspond to the most general Lorentz transformation.

Assume

$$\mathfrak{A}_i = \overline{\Psi}\alpha_i\Psi \qquad (i = 0,1,2,3,4,5) \tag{12}$$

and apply the transformation S to the semi-vector Ψ.

For the transformed values $\mathfrak{A}_i$ (which we denote by double primes ($\mathfrak{A}_i''$)), one finds using (5) and (9)

$$\mathfrak{A}_i'' = \overline{\Psi}''\alpha_i\Psi'' = \overline{\Psi}S^\dagger\alpha_i S\Psi = \sum_{k=1}^{3} e_k a_{ik}\overline{\Psi}\alpha_k\Psi \qquad (i = 0,1,2,3)$$

$$\overline{\Psi}''\alpha_4\Psi'' = \overline{\Psi}\alpha_4\Psi; \qquad \overline{\Psi}''\alpha_5\Psi'' = \overline{\Psi}\alpha_5\Psi,$$

which means that

$$\mathfrak{A}_i'' = \sum_{k=0}^{3} e_k a_{ik}\mathfrak{A}_k \quad (i = 0,1,2,3), \tag{13}$$

$$\mathfrak{A}_4'' = \mathfrak{A}_4; \quad \mathfrak{A}_5'' = \mathfrak{A}_5. \tag{13*}$$

Comparing (13) and (11) one sees that the first four values $\mathfrak{A}$ $(i = 0, 1, 2, 3)$ are components of a four-dimensional vector and the last two, $\mathfrak{A}_4$ and $\mathfrak{A}_5$, are invariants.

Here are the explicit expressions for $\mathfrak{A}_i$:

$$\begin{aligned}
\mathfrak{A}_0 &= \quad \overline{\Psi_1}\Psi_1 + \overline{\Psi_2}\Psi_2 + \overline{\Psi_3}\Psi_3 + \overline{\Psi_4}\Psi_4, \\
\mathfrak{A}_1 &= \quad \overline{\Psi_1}\Psi_2 + \overline{\Psi_2}\Psi_1 + \overline{\Psi_3}\Psi_4 + \overline{\Psi_4}\Psi_3, \\
\mathfrak{A}_2 &= -i\,\overline{\Psi_1}\Psi_2 + i\,\overline{\Psi_2}\Psi_1 + i\,\overline{\Psi_3}\Psi_4 - i\,\overline{\Psi_4}\Psi_3, \\
\mathfrak{A}_3 &= \quad \overline{\Psi_1}\Psi_1 - \overline{\Psi_2}\Psi_2 + \overline{\Psi_3}\Psi_3 - \overline{\Psi_4}\Psi_4, \\
\mathfrak{A}_4 &= -\overline{\Psi_1}\Psi_4 + \overline{\Psi_2}\Psi_3 + \overline{\Psi_3}\Psi_2 - \overline{\Psi_4}\Psi_1, \\
\mathfrak{A}_5 &= -i\,\overline{\Psi_1}\Psi_4 + i\,\overline{\Psi_2}\Psi_3 - i\,\overline{\Psi_3}\Psi_2 + i\,\overline{\Psi_4}\Psi_1.
\end{aligned}$$

Using these expressions one can easily verify that $\mathfrak{A}_i$ satisfies the relation

$$\mathfrak{A}_1^2 + \mathfrak{A}_2^2 + \mathfrak{A}_3^2 + \mathfrak{A}_4^2 + \mathfrak{A}_5^2 = \mathfrak{A}_0^2 . \tag{14}$$

It gives an important inequality

$$\mathfrak{A}_0^2 - \mathfrak{A}_1^2 - \mathfrak{A}_2^2 - \mathfrak{A}_3^2 \geq 0 , \tag{15}$$

which means that the vector $\mathfrak{A}_i$ belongs to the interior of the characteristic cone (from the same side of the cone as the time axis), or in other words that it has the same four-dimensional character with velocity (the character of time).

2 Parallel Transport of a Semi-Vector in Riemann Geometry

To define a semi-vector in Riemann geometry, introduce, following Ricci and Levi Civita, four orthogonal congruences or equivalently four directions at each point forming an orthogonal "n-hedron." This n-hedron replaces the Cartesian axes of the Minkowski geometry in an infinitesimally small vicinity of a given point. And one does not need to change anything in our formulae of the previous paragraph.

In what follows we are going to consider two kinds of components of a vector and a tensor: the ordinary components (either covariant or contravariant) corresponding to the coordinates and the components corresponding to the direction of the n-hedron. We shall often denote both kinds of components by the same letter indicating the n-hedric components by a prime ($\mathfrak{A}'_k$). The coordinates will be numerated by Greek indices, and the directions of the n-hedron by Latin ones. The indices take values 0, 1, 2, 3. We assume the Einstein convention of omitting the sum sign (Σ) with respect to Greek indices; however, we shall keep it for Latin ones.

We denote the "parameters" of the congruence k by h_k^α and the "momenta" by $h_{k,\alpha}$. Thus, the relation between the ordinary and the n-hedric components is given by

$$\mathfrak{A}'_k = \mathfrak{A}_\sigma h_k^\sigma; \quad \mathfrak{A}_\sigma = \sum_k e_k \mathfrak{A}_k h_{k,\sigma}. \tag{16}$$

Consider the variation of the components of a vector for an infinitesimal shift dx^σ. For the ordinary covariant components one has

$$\delta\mathfrak{A}_\alpha = \Gamma^\beta_{\alpha\sigma}\mathfrak{A}_\beta dx^\sigma; \quad \Gamma^\beta_{\alpha\sigma} = \left\{ \begin{matrix} \alpha\sigma \\ \beta \end{matrix} \right\}. \tag{17}$$

One deduces from (17) the analogous expression for the n-hedric components:

$$\mathfrak{A}'_i = \sum_{kl} e_k e_l \gamma_{ikl} \mathfrak{A}'_k ds_i\,, \tag{18}$$

where ds_i are the n-hedric components of the shift and γ_{ikl} are the Ricci rotation coefficients

$$\gamma_{ikl} = (\nabla_\sigma h_i^\beta) h_{k,\beta} h_l^\sigma = (\nabla_\sigma h_{i,\beta}) h_k^\beta h_l^\sigma\,, \tag{19}$$

where ∇_σ is the symbol of the covariant derivative.

Consider now a semi-vector. The variation of its components must have the form

$$\delta\Psi = \sum_l e_l C_l ds_l \Psi\,, \tag{20}$$

where C_l are certain matrices; the conjugated expression reads as

$$\delta\overline{\Psi} = \overline{\Psi} \sum_l e_l C_l^\dagger ds_l\,. \tag{20*}$$

We have already seen (Section 1) that one can form one vector and two invariants out of the components of a semi-vector. The law (20) for the shift of a semi-vector gives the one for a vector. But the latter should coincide with the law (18); moreover, the two invariants $\mathfrak{A}_4$ and $\mathfrak{A}_5$ should remain unchanged. Therefore, the matrices C_l should satisfy a certain condition allowing one to find them.

One has

$$\delta\mathfrak{A}_i' = \delta(\overline{\Psi}\alpha_i\Psi) = \delta\overline{\Psi}\alpha_i\Psi + \overline{\Psi}\alpha_i\delta\Psi = \overline{\Psi}\sum_l e_l(C_l^\dagger\alpha_i + \alpha_iC_i)ds_l\Psi.$$

On the other hand, $\delta\mathfrak{A}_i'$ is given by formula (18), which can be rewritten as

$$\delta\mathfrak{A}_i' = \overline{\Psi}\sum_{kl} e_ke_l\gamma_{ikl}\alpha_k ds_l\Psi\,.$$

Equating these two expressions for $\delta\mathfrak{A}_i'$, one finds

$$C_l^\dagger\alpha_i + \alpha_iC_l = \sum_k e_k\gamma_{ikl}\alpha_k. \tag{21}$$

It follows from the invariance of $\delta\mathfrak{A}_4$ and $\delta\mathfrak{A}_5$ that

$$C_l^\dagger\alpha_4 + \alpha_4C_l = 0; \quad C_l^\dagger\alpha_5 + \alpha_5C_l = 0. \tag{22}$$

One can check that a particular solution for (21) and (22) is $C_l = g_l{}'$, where

$$g_l{}' = \frac{1}{4}\sum_{mk}\alpha_m\alpha\,. \tag{23}$$

If one substitutes the expression

$$C_l = g_l{}' + i\Phi_l' \tag{24}$$

into (21) and (22), one finds that a general solution is given by the matrices Φ_l', which are Hermitian and commute with all the matrices α_i $(i = 0, 1, 2, 3, 4, 5)$.

It implies the proportionality of φ_l' to the identity, the proportionality factor being a real number.

One can deduce that the solution of equations (21) and (22) is completely determined once its "averaged trace" is given. (By an averaged trace of a matrix we denoted the sum of its diagonal elements divided by its number.) The trace of $g_l{}'$ is equal to zero,

$$\text{trace } g_l{}' = 0\,, \tag{25}$$

and, therefore, the averaged trace of C_l is equal to $i\Phi_l'$.

Once the law (20) of the parallel transport is given, the internal derivative of a semi-vector is written down as

$$\mathfrak{D}_l'\Psi = \frac{\partial\Psi}{\partial s_l} - C_l\Psi. \tag{26}$$

Consider for a while the case of the Minkowski space. Assume that the Cartesian axes coincide with the directions of the n-hedron. Then one has

$$\gamma_{mkl} = 0, \quad g_l{}' = 0, \quad C_l = i\varphi_l'.$$

The internal derivative reduces to

$$\mathfrak{D}_l'\Psi = \frac{\partial \Psi}{\partial s_l} - i\Phi_l'\Psi. \tag{26*}$$

On the other hand, the expression $P_l\Psi$ entering the Dirac equation is proportional to

$$\frac{\partial \Psi}{\partial s_l} - \frac{2\pi i e}{hc}\varphi_l'\Psi\,, \tag{26**}$$

where φ_l' is the vector-potential.

In order that the two last expressions coincide, one should assume that

$$\Phi_l' = \frac{2\pi e}{hc}\varphi_l'\,, \tag{27}$$

which gives us the physical meaning of the vector Φ_l' and formula (24). At the same time we have obtained the geometrical meaning of the operator P_l. As for its physical meaning, it corresponds to the momentum.

In a general Riemann space, we can always assume

$$P_l' = \frac{h}{2\pi i}\mathfrak{D}_l' = \frac{h}{2\pi i}\left(\frac{\partial}{\partial s_l} - C_l\right) \tag{28}$$

and interpret the internal derivative up to the factor

$$\frac{h}{2\pi i}$$

as the momentum operator. This hypothesis will be checked later (Section 5).

If one expresses the matrices C_l in terms of g_l' and the vector potential, formula (20) rewrites as

$$\delta\Psi = \frac{2\pi i e}{hc}\sum_l e_l\varphi_l' ds_l\Psi\,. \tag{29}$$

In the parallel transport law of a semi-vector, one recognizes the Weyl linear differential form.

Consider now the ordinary covariant or contravariant components of vectors. Assume that

$$\gamma^\sigma = \sum_k e_k \alpha_k h_k^\sigma; \quad \gamma_\sigma = \sum_k e_k \alpha_k h_{k,\sigma}, \tag{30}$$

$$\left.\begin{aligned} g_\sigma &= \sum_k e_k h_{k,\sigma} g_k', \\ \Gamma_\sigma &= \sum_k e_k h_{k,\sigma} C_k, \end{aligned}\right\} \tag{31}$$

and rewrite formulae (20) or (29) as

$$\delta\Psi = \Gamma_\sigma dx^\sigma \psi = \left(g_\sigma + \frac{2\pi i e}{hc} \varphi_\sigma \right) dx^\sigma \psi \tag{32}$$

and the internal derivative with respect to x^σ as

$$\mathfrak{D}_\sigma \psi = \frac{\partial \Psi}{\partial x^\sigma} - \Gamma_\sigma \Psi. \tag{33}$$

The components $\mathfrak{A}^\sigma$ and $\mathfrak{A}_\sigma$ can be expressed in terms of $\Psi, \gamma^\sigma, \gamma_\sigma$:

$$\mathfrak{A}^\sigma = \overline{\Psi} \gamma^\sigma \Psi; \quad \mathfrak{A}_\sigma = \overline{\Psi} \gamma_\sigma \Psi. \tag{34}$$

Substituting (32) and (33) into the formula

$$\delta \mathfrak{A}^\sigma + \Gamma^\sigma_{\varrho\tau} \mathfrak{A}^\tau dx^\varrho = 0, \tag{35}$$

one finds that γ^σ and Γ_σ satisfy the equation

$$\Gamma_\varrho^\dagger \gamma^\sigma + \gamma^\sigma \Gamma_\varrho + \nabla_\varrho \gamma^\sigma = 0. \tag{36}$$

One also has

$$\Gamma_\varrho^\dagger \alpha_4 + \alpha_4 \Gamma_\varrho = 0; \quad \Gamma_\varrho^\dagger \alpha_5 + \alpha_5 \Gamma_\varrho = 0. \tag{37}$$

The matrix g_σ (31) satisfies the same equations as Γ_σ does and the auxiliary condition

$$\text{trace } g_\sigma = 0, \tag{38}$$

and it is the unique solution for these equations.

Find now the transformation law for the matrices γ^ϱ, g_σ and Γ_σ for an arbitrary change of the orthogonal congruences. This change corresponds to a rotation of the n-hedron at each point of the space, i.e., to a local Lorentz transformation. One sees that such transformation can be defined by a matrix S with entries depending on time and coordinates.

Recall that the parameters of the congruence are transformed as a vector:

$$\left.\begin{aligned} h''^{\sigma}_{k} &= \sum_i e_i a_{ki} h^{\sigma}_i\,, \\ h^{\sigma}_k &= \sum_i e_i a_{ik} h''^{\sigma}_i\,. \end{aligned}\right\} \tag{39}$$

Multiply the latter formula by $e_k \alpha_k$ and sum it up over k. Taking (9) into account, one gets

$$\sum_k e_k \alpha_k h^{\sigma}_k = S^{\dagger} \left(\sum_i e_i \alpha_i h''^{\sigma}_i \right) S$$

which means that

$$\gamma^{\sigma} = S^{\dagger} \gamma''^{\sigma} S. \tag{40}$$

This relation could be expected since the components (34) of a vector do depend on the choice of the n-hedron.

The transformed matrix g''_{ϱ} can be defined by the equations analogous to (36), (37) and (38). We are going to show that the expression

$$g''_{\varrho} = S g_{\varrho} S^{-1} + \frac{\partial S}{\partial x^{\varrho}} S^{-1} \tag{41}$$

satisfies these equations. In fact, one has

$$g''^{\dagger}_{\varrho} \gamma''^{\sigma} + \gamma''^{\sigma} g''_{\varrho} + \nabla_{\varrho} \gamma''^{\sigma} = (S^{-1})^{\dagger} (g^{\dagger}_{\varrho} \gamma^{\sigma} + \gamma^{\sigma} g_{\varrho} + \nabla_{\varrho} \gamma^{\sigma}) S^{-1}\,, \tag{42}$$

and this expression vanishes as a consequence of the equation satisfied by g. One can check in the same way that the equations analogous to (37) are satisfied by g''_{ϱ} if they are satisfied by g_{ϱ}. Finally one has that identically

$$\text{trace}\, \frac{\partial S}{\partial x^{\varrho}} S^{-1} = 0 \tag{43}$$

and, as a consequence,

$$\text{trace}\; g''_{\varrho} = \text{trace}\; S g_{\varrho} S^{-1} = \text{trace}\; g_{\varrho} = 0\,. \tag{44}$$

All the equations for g''_{ϱ} are, therefore, satisfied, and formula (41) is verified. Since Γ_{ϱ} differs from g_{ϱ} by a multiple of a unity matrix, the law (41) is valid for Γ_{ϱ}:

$$\Gamma''_{\varrho} = S \Gamma_{\varrho} S^{-1} + \frac{\partial S}{\partial x^{\varrho}} S^{-1}\,. \tag{45}$$

For some computations it is convenient to introduce non-Hermitian matrices

$$\gamma^{*\sigma} = \alpha_4 \gamma^\sigma \tag{46}$$

(where α_4 can be replaced by α_5) satisfying the equations

$$\gamma^{*\sigma}\gamma^{*\varrho} + \gamma^{*\varrho}\gamma^{*\sigma} = 2g^{\varrho\sigma}. \tag{47}$$

For these matrices, one has

$$\gamma_\varrho^{*\prime\prime} = S\gamma_\varrho^* S^{-1}. \tag{48}$$

3 Dirac Wave Equation

We are going to impose the following conditions on the Dirac wave equation in general relativity. The wave equation should:

1. be invariant with respect to an arbitrary coordinate change;
2. be invariant with respect to an arbitrary rotation of the n-hedron;
3. coincide with the adjoint equation and also be such that a time-like vector and vector of zero divergency can be defined in order to be interpreted as a current vector;
4. be reduced to the ordinary Dirac equation in the case of the Minkowski space.

We are going to show that all these conditions are satisfied by equation (1) if one replaces P_l by its generalized form (28).

Equation (1) can be written as

$$\mathcal{F}\Psi = 0\,, \tag{49}$$

where

$$\mathcal{F}\Psi = \frac{h}{2\pi i}\sum k e_k \alpha_k \left(\frac{\partial\Psi}{\partial s_k} - C_k\right) - mc\alpha_4\psi. \tag{50}$$

If one introduces the coordinates, the operator $\mathcal{F}$ takes the form

$$\mathcal{F}\Psi = \frac{h}{2\pi i}\gamma^\sigma \left(\frac{\partial\Psi}{\partial x^\sigma} - \Gamma_\sigma\right) - mc\alpha_4\psi. \tag{51}$$

1. The invariance of the operator $\mathcal{F}$ with respect to coordinate changes is obvious from the very form of equation (50) or (51).

2. The invariance under the rotations of the n-hedron can be shown in a straightforward way. Using the formulae derived above

$$\left.\begin{aligned} &\Psi'' = S\Psi; \quad \Gamma''_\sigma = S\Gamma_\sigma S^{-1} + \frac{\partial S}{\partial x^\sigma} S^{-1} \\ &S^\dagger \gamma''^\sigma S = \gamma^\sigma; \quad S^\dagger \alpha_4 S = \alpha_4, \end{aligned}\right\} \tag{52}$$

one finds that

$$\frac{\partial \Psi''}{\partial x^\sigma} - \Gamma''_\sigma \Psi'' = S\left(\frac{\partial \Psi}{\partial x^\sigma} - \Gamma_\sigma \Psi\right),$$

$$S^\dagger \gamma''^\sigma \left(\frac{\partial \Psi''}{\partial x^\sigma} - \Gamma''_\sigma \Psi''\right) = \gamma^\sigma \left(\frac{\partial \Psi}{\partial x^\sigma} - \Gamma_\sigma \Psi\right),$$

$$S^\dagger \alpha_4 \psi'' = \alpha_4 \psi,$$

and, therefore, one has

$$S^\dagger \mathcal{F}'' \psi'' = \mathcal{F}\Psi, \tag{53}$$

where

$$\mathcal{F}'' \psi'' = \frac{h}{2\pi i} \gamma''^\sigma \left(\frac{\partial \Psi''}{\partial x^\sigma} \Gamma''_\sigma \Psi''\right) - mc\alpha_4 \Psi'' \tag{51*}$$

and the operator $\mathcal{F}$ is transformed.

The equation $\mathcal{F}''\Psi'' = 0$ is, therefore, just equivalent to $\mathcal{F}\Psi = 0$.

Note that the expression

$$\overline{\Psi}\mathcal{F}\Psi = \overline{\Psi''}\mathcal{F}''\Psi'' \tag{54}$$

is invariant.

3. Using relation (36) one can verify by a simple computation the identity

$$\overline{\Psi}\mathcal{F}\Psi - (\overline{\mathcal{F}\Psi})\Psi = \frac{h}{2\pi i} \frac{1}{\sqrt{g}} \frac{\partial}{\partial x^\sigma} (\overline{\Psi}\sqrt{g}\gamma^\sigma \Psi) \tag{55}$$

(g is the absolute value of the determinant $|g_{\varrho\sigma}|$), which shows that the operator $\mathcal{F}$ coincides with its adjoint.

Identity (55) allows one to define the vector of current by the formula

$$S^\varrho = \Psi\gamma^\varrho\Psi\,. \tag{56}$$

In fact if Ψ satisfies the equation $\mathcal{F}\Psi = 0$, one has

$$\frac{1}{\sqrt{g}} \frac{\partial}{\partial x^\varrho} (\sqrt{g} S^\varrho) = 0 \tag{57}$$

and, therefore, the divergency of the current vanishes. Moreover, as we have seen from Section 1 [formula (15)], a vector constructed out of the functions Ψ is necessarily a time-like one.

4. For the Cartesian coordinates of the Minkowski space g_l'' vanish and the equation $\mathcal{F}\Psi = 0$ reduces to the ordinary Dirac equation.

Therefore, all the conditions are satisfied.

Each operator commuting with ϱ_3 can be expressed as a linear combination of α_j and $\varrho_3\alpha_j$ $(j = 0, 1, 2, 3)$. Since the operator $\mathcal{F} + mc\alpha_4$ contains only the matrices $\alpha_j\,(j = 0, 1, 2, 3)$ commuting with ϱ_3, it satisfies this property. In order to make the transformation introduce the "quasi-vectors"

$$K_j = -\sum_i e_i\gamma_{ijk} = \frac{1}{\sqrt{g}}\frac{\partial}{\partial x^\sigma}(\sqrt{g}h_j^\sigma)\,, \tag{58}$$

$$f_j = \frac{1}{2}\sum_{ijk} e_ie_je_k\varepsilon_{ijkl}\gamma_{ikl}\,, \tag{59}$$

where $\varepsilon_{ijkl} = 0$ if two of the indices j, i, k, l coincide; $\varepsilon_{ijkl} = 1$ if all of them are different and form an even permutation of $0, 1, 2, 3$; and $\varepsilon_{ijkl} = -1$ for an odd permutation. Note that f_j contain γ_{ijk} with three different indices only. It is known[5] that a necessary condition for a congruence of the n-hedron to be normal to n families of surfaces is just that $\gamma_{ijk} = 0$ for three different indices.

In this case the quasi-vector f_i vanishes.

The transformed operator $\mathcal{F}$ has the form

$$\begin{aligned}\mathcal{F}\Psi = \sum_j e_j\alpha_j\left(\frac{h}{2\pi i}\frac{\partial\Psi}{\partial x_j} - \frac{e}{c}\varphi_j\Psi + \frac{h}{4\pi i}k_j\Psi\right) + \\ + \frac{h}{4\pi}\varrho_3\sum_j e_j\alpha_jf_j\Psi - mc\alpha_4\Psi.\end{aligned} \tag{60}$$

Assume that the Riemann space under consideration is such that there exist n orthogonal surfaces (it is not the case for the generic Riemann space). Then one can introduce an orthogonal coordinate system such that

$$\left.\begin{aligned} ds^2 = \textstyle\sum_i e_iH_i^2dx_i^2; \quad \sqrt{g} = H_0H_1H_2H_3 \\ h_i^i = \tfrac{1}{H_i}; \quad h_{i,i} = e_iH_i; \quad f_i = 0, \end{aligned}\right\} \tag{61}$$

[5] L. Eisenhart, Riemannian Geometry, Princeton, 1926.

and h_i^σ as well as $h_{i,\sigma}$ with different indices vanish. Substituting the values into (60), one gets

$$\mathcal{F}\Psi = \sum_j e_j\alpha_j \frac{1}{H_j}\left[\frac{h}{2\pi i}\frac{\partial\Psi}{\partial x_j} - \frac{e}{c}\varphi_j\Psi + \frac{h}{4\pi i}\frac{\partial}{\partial x_j}\left(\log\frac{\sqrt{g}}{H_j}\right)\Psi\right] - mc\alpha_4. \tag{62}$$

We have obtained a general expression for the Dirac operator for a curvilinear orthogonal coordinate system. This formula is very useful in applications.

4 Energy Tensor

In the classical theory the mixed energy tensor for noncoherent matter has the form

$$W^{\sigma\cdot}_{\cdot\alpha} = \varrho_0 u^\sigma u_\alpha\,, \tag{63}$$

where $u^\sigma = \frac{dx^\sigma}{d\tau}$ is the vector of relativistic velocity and ϱ_0 is the invariant density. This expression can be rewritten as

$$W^{\sigma\cdot}_{\cdot\alpha} = \varrho\frac{dx^\sigma}{dt}\cdot\frac{1}{m}P_\alpha\,, \tag{64}$$

where ϱ is the non-invariant density (mass per unit volume), $\frac{dx^\sigma}{dt} = c\frac{dx^\sigma}{dx^0}$ is the ordinary velocity, P_α is the motion quantity of a particle of mass m.

Find an analogous expression in quantum mechanics. The quantity corresponding to $\frac{\varrho}{m}$ is $\overline{\Psi}\Psi$; we assume that the ordinary velocity operator $c\frac{dx^\sigma}{dx^0}$ is $c\gamma^\sigma$ and the one for the momentum is $P_\alpha = \frac{h}{2\pi i}\cdot\mathfrak{D}_\alpha$ (28). For the special relativity case, these physical meanings of the operators were established by the author in a previous paper;[6] in the general case, it is only a hypothesis to be verified.

We are in a position to consider the expression[7]

$$W^{\sigma\cdot}_{\cdot\alpha} = c\overline{\Psi}\gamma^\sigma P_\alpha\Psi. \tag{65}$$

We have written down the operators γ^σ and P_α in the given order since equation(65) is invariant under the rotation of the n-hedron, though the expression obtained from it by permutation of the factors is not.

[6] V. Fock, Zs. Phys. **55**, 127, 1929. (See [29-2] in this book. (*Editors*))

[7] For the special relativity, this expression reduces to the one proposed by several authors (see H. Tetrode. Zs. Phys. **49**, 858, 1928). In the generalisation proposed by H. Tetrode (Zs. Phys. **50**, 336, 1929) the general relativity can hardly be considered as a correct one since it does not satisfy certain conditions that seem to be necessary for us. (*V. Fock*)

Such tensor $W^{\sigma\cdot}_{\cdot\alpha}$ is complex; since the energy tensor should be real, we may try to identify the latter with the real part of $W^{\sigma\cdot}_{\cdot\alpha}$ and denote it by $T^{\sigma\cdot}_{\cdot\alpha}$ assuming

$$W^{\sigma\cdot}_{\cdot\alpha} = T^{\sigma\cdot}_{\cdot\alpha} + iU^{\sigma\cdot}_{\cdot\alpha}\,. \tag{66}$$

A simple computation allows one to express the imaginary part $U^{\sigma\cdot}_{\cdot\alpha}$ in the form

$$U^{\sigma\cdot}_{\cdot\alpha} = -\frac{h}{4\pi}\nabla_\alpha S^\alpha. \tag{67}$$

Thus, it is proportional to the covariant divergency of the current vector.

Return to the real part $T^{\sigma\cdot}_{\cdot\alpha}$. The classical energy tensor satisfies the equations of motion: its divergency equals the Lorentz force. If it is true for the tensor $T^{\sigma\cdot}_{\cdot\alpha}$, our hypothesis concerning its physical meaning would be verified. We are going to show that this is the case.

In order to find the divergency of $T^{\sigma\cdot}_{\cdot\alpha}$, we compute first the one for $W^{\sigma\cdot}_{\cdot\alpha}$ and then separate the real and imaginary parts of the result. Since the computation is rather long, we shall restrict ourselves to sketches of the main steps.

One finds, taking into account the equation $\mathfrak{F}\Psi = 0$, that

$$\frac{1}{\sqrt{g}}\frac{\partial}{\partial x^\sigma}(\sqrt{g}W^{\sigma\cdot}_{\cdot\alpha}) - \gamma^\varrho_{\alpha\sigma}W^{\sigma\cdot}_{\cdot\varrho} = \frac{hc}{2\pi i}\overline{\Psi}\gamma^\sigma\mathfrak{D}_{\sigma\alpha}\Psi\,, \tag{68}$$

where

$$\mathfrak{D}_{\sigma\alpha} = \mathfrak{D}_\sigma\mathfrak{D}_\alpha - \mathfrak{D}_\alpha\mathfrak{D}_\sigma = \frac{\partial\Gamma_\sigma}{\partial x^\alpha} - \frac{\partial\Gamma_\alpha}{\partial x^\sigma} + \Gamma_\alpha\Gamma_\sigma - \Gamma_\sigma\Gamma_\alpha. \tag{69}$$

The operator $\mathfrak{D}_{\sigma\alpha}$ does not contain differentiation and reduces itself to a matrix that can be written in the form

$$\mathfrak{D}_{\sigma\alpha} = \frac{1}{4}\gamma^{*\mu}\gamma^{*\nu}R_{\mu\nu\sigma\alpha} + \frac{2\pi ie}{hc}\mathfrak{M}_{\sigma\alpha}\,, \tag{70}$$

where $\gamma^{*\nu} = \alpha_4\gamma^\nu$ are the matrices introduced at the end of Section 2,

$R_{\mu\nu\sigma\alpha}$ is the Riemann tensor[8] and

$$\mathfrak{M}_{\sigma\alpha} = \frac{\partial \varphi_\sigma}{\partial x^\alpha} - \frac{\partial \varphi_\alpha}{\partial x^\sigma} \tag{71}$$

is the electric tensor. The expression $\gamma^\sigma \mathfrak{D}_{\sigma\alpha}$ entering (68) is equal to

$$\gamma^\sigma \mathfrak{D}_{\sigma\alpha} \gamma^\varrho \left(-\frac{1}{2} R_{\varrho\alpha} + \frac{2\pi i e}{hc} \mathfrak{M}_{\varrho\alpha} \right), \tag{72}$$

where $R_{\varrho\alpha}$ is the contracted Riemann tensor.

Formula (68) can be thus rewritten as

$$\frac{1}{\sqrt{g}} \frac{\partial}{\partial x^\sigma} (\sqrt{g} W^{\sigma\cdot}_{\cdot\alpha}) - \gamma^\varrho_{\alpha\sigma} W^{\sigma\cdot}_{\cdot\varrho} = S^\varrho \left(e \mathfrak{M}_{\varrho\alpha} - \frac{hc}{4\pi i} R_{\varrho\alpha} \right), \tag{73}$$

where S^ϱ is the current vector (56).

Equating the real and imaginary parts of two terms of the equation obtained, one gets two equations:

$$\nabla_\sigma T^{\sigma\cdot}_{\cdot\alpha} = e S^\varrho \mathfrak{M}_{\varrho\alpha}, \tag{74}$$

$$\nabla_\sigma U^{\sigma\cdot}_{\cdot\alpha} = \frac{4\pi}{hc} S^\varrho R_{\varrho\alpha}, \tag{75}$$

where the second equation is very easy to check starting from expression (67) for $U^{\sigma\cdot}_{\cdot\alpha}$ taking into account that the divergency of the vector S^ϱ vanishes.

The first equation shows that the divergency of the tensor $T^{\sigma\cdot}_{\cdot\alpha}$ is just equal to the Lorentz force. We are therefore able to consider the tensor $T^{\sigma\cdot}_{\cdot\alpha}$ as the energy tensor and equation (74) as an analogue of the classical equation of motion for a continuum medium.

Note that using the tensor $W^{\sigma\cdot}_{\cdot\alpha}$ one can write down a variation principle equivalent to the Dirac equations:

$$\delta \iiiint (W^{\sigma\cdot}_{\cdot\sigma} - mc^2 \overline{\Psi} \alpha_4 \Psi) \sqrt{g}\, dx^0\, dx^1\, dx^2\, dx^3 = 0. \tag{76}$$

[8] Using (70), one can easily derive the Bianchi identities. Introduce

$$\mathfrak{D}_{\varrho\sigma\alpha} = \mathfrak{D}_\varrho \mathfrak{D}_{\sigma\alpha} - \mathfrak{D}_{\sigma\alpha} \mathfrak{D}_\varrho$$

and verify a simple identity

$$B_{\varrho\sigma\alpha} = \mathfrak{D}_{\varrho\sigma\alpha} + \mathfrak{D}_{\sigma\alpha\varrho} + \mathfrak{D}_{\alpha\varrho\sigma} = 0.$$

On the other hand, formula (70) gives

$$B_{\varrho\sigma\alpha} = \frac{1}{4} \gamma^{*\mu} \gamma^{*\nu} (\nabla_\varrho R_{\mu\nu\sigma\alpha} + \nabla_\sigma R_{\mu\nu\alpha\varrho} + \nabla_\alpha R_{\mu\nu\varrho\sigma}).$$

The matrix $B_{\varrho\sigma\alpha}$ is zero if and only if the expression inside the brackets is zero, Q.E.D. (*V. Fock*)

5 Equations of Motion in Quantum Mechanics

It is known[9] that the quantum equations of motion are in fact the relations between entries of certain matrices and their time derivatives. These matrices are constructed using functions $\Psi_1, \Psi_2, \dots \Psi_n, \dots$, etc., satisfying a wave equation. An entry of a matrix of an operator K is defined by the equation

$$K_{mn} = \int \overline{\Psi}_m K \Psi_n dV, \tag{77}$$

where the integration domain is the whole space; K_{mn} is, therefore, a function of only one variable: time t. If the system of functions is complete,[10] the set of entries K_{mn} defines the operator K completely. However, to establish the equations of motion, it is not necessary to have a complete system and it suffices to consider a typical entry (77) constructed using two arbitrary solutions Ψ_m and Ψ_n of the wave equation.

Deriving equation (77) by time, one gets an expression with the derivatives $\frac{\partial \overline{\Psi}_m}{\partial t}$ and $\frac{\partial \Psi_n}{\partial t}$ under the integral. These derivatives can be eliminated using the wave equation and the result can be put into the form

$$\frac{dK_{mn}}{dt} = L_{mn}, \tag{78}$$

where

$$L_{mn} = \int \overline{\Psi}_m L \Psi_n dV, \tag{79}$$

and L is the operator depending on the form of the wave equation and on the operator K. One can write down the system of equations (78) in the symbolic form

$$\frac{dK}{dt} = L, \tag{80}$$

and we say that operator L is the "total derivative" of the operator K by time.

For the Dirac wave equation, the total derivative by x^0 is of the form

$$\frac{dK}{dx^0} = \frac{2\pi i}{h} \left(\mathcal{F}(\gamma^0)^{-1} K (\gamma^0)^{-1} \mathcal{F} \right), \tag{80*}$$

where $\mathcal{F}$ is operator (51) and $(\gamma^0)^{-1}$ is the matrix inverse to γ^0.

---

[9] See, e.g., V. Fock, Zs. Phys **49**, 323, 1928. (See also [28-2] in this book. (*Editors*)).

[10] It means that any function of coordinates of the space can be expanded into a series of the functions Ψ_n. (*V. Fock*)

It is clear that quantum equations (80) are just simple mathematical consequences of the wave equation and one does not need to introduce any extra physical hypothesis to get them.

Equations (80) in general represent an analogue of the classical equations of motion in the Lagrangian form. This analogy is far more complete than the one between the quantum wave equation and the classical Jacobi equation. In the case of the Dirac equation, the latter does not exist anymore; however, as it was shown by the author for the special relativity,[11] one can rewrite equations (80) in a way completely analogous to the Lagrange equations. Let us see if it is also the case for the general relativity.

The classical equations of motion for a particle of mass m and charge e are

$$\frac{d^2x^\sigma}{d\tau^2}+\Gamma^\sigma_{\alpha\varrho}\frac{dx^\alpha}{d\tau}\frac{dx^\varrho}{d\tau}=\frac{e}{mc}\frac{dx^\varrho}{d\tau}g^{\sigma\alpha}\mathcal{M}_{\varrho\alpha}, \tag{81}$$

where τ is the proper time of the particle. Introducing the covariant components of the momentum

$$P_\varrho=mg_{\varrho\nu}\frac{dx^\nu}{d\tau}, \tag{82}$$

one deduces the equations out of (81):

$$\frac{dP_\alpha}{dx^0}-\Gamma^\varrho_{\alpha\sigma}\frac{dx^\sigma}{dx^0}P_\varrho=\frac{e}{c}\frac{dx^\varrho}{dx^0}\mathfrak{M}_{\varrho\alpha}. \tag{83}$$

We are going to look for quantum equations (80) analogous to (83). Denote by Ψ_m and Ψ_n two semi-vectors satisfying the equation $\mathcal{F}\Psi=0$. Note that our formulae (57) and (73) remain valid if one replaces $\overline{\Psi}$ by $\overline{\Psi}_m$ and Ψ by Ψ_n in S^ϱ and $W^{\sigma\cdot}_{\cdot\alpha}$, i.e., by two different solutions of the wave equations.

Represent formula (57) in the form

$$\frac{1}{\sqrt{g}}\frac{\partial}{\partial x^0}(\sqrt{g}\overline{\Psi}_m\gamma^0\Psi_n)+\frac{1}{\sqrt{g}}\sum_{\sigma=1}^{3}\frac{\partial}{\partial x^\sigma}(\sqrt{g}\overline{\Psi}_m\gamma^\sigma\Psi_n)=0. \tag{84}$$

Multiply both terms by

$$\sqrt{g}dx^1dx^2dx^3=dV \tag{85}$$

and integrate them by x^1,x^2,x^3; the sum $\sum_{\sigma=1}^3$ disappears, and we get

$$\frac{d}{dx^0}\int\overline{\Psi}_m\gamma^0\psi_n dV=0, \tag{86}$$

[11] V. Fock, Zs. Phys. **55**, 127, 1929 (*V.F.*), see also [29-2] in this book. (*Editors*)

since, according to (80),

$$\frac{d\gamma^0}{dx^0} = 0. \tag{87}$$

The entries of the matrix of γ^0 are constant, and one can assume that the functions Ψ_m are normalized in such a way that

$$\int \overline{\Psi}_m \gamma^0 \psi_n dV = \delta_{mn}. \tag{88}$$

The matrix γ^0 corresponds, therefore, to the identity operator.

Multiply (84) by $x^\varrho dV$ and integrate it. The result is[12]

$$\frac{d}{dx^0}(\gamma^0 x^\varrho) = \gamma^\varrho. \tag{89}$$

Since γ^0 corresponds to the identity, the classical analogue of γ^ϱ is just the ordinary velocity

$$\frac{dx^\varrho}{dx^0} = \frac{1}{c}\frac{dx^\varrho}{dt}.$$

Consider now equation (73), which we rewrite in the form

$$\frac{1}{\sqrt{g}}\frac{\partial}{\partial x^\sigma}(\sqrt{g}\, c\overline{\Psi}_m \gamma^\sigma P_\alpha \Psi_n) - \Gamma^\varrho_{\alpha\sigma} c\overline{\Psi}_m \gamma^\sigma P_\varrho \Psi_n = = \overline{\Psi}_m \gamma^\varrho \Psi_n \left(e\mathfrak{M}_{\varrho\alpha} - \frac{hc}{4\pi i} R_{\varrho\alpha} \right). \tag{90}$$

Multiply (90) by dV and integrate it. After dividing by c, one gets

$$\frac{d}{dx^0}(\gamma^0 P_\alpha) - \Gamma^\varrho_{\alpha\sigma}\gamma^\sigma P_\varrho = \frac{e}{c}\gamma^\varrho \mathfrak{M}_{\varrho\alpha} - \frac{h}{4\pi i}\gamma^\varrho R_{\varrho\alpha}, \tag{91}$$

and replacing γ^ϱ by its expression (89),

$$\frac{d}{dx^0}(\gamma^0 P_\alpha) - \Gamma^\varrho_{\alpha\sigma}\frac{d}{dx^0}(\gamma^0 x^\sigma) P_\varrho = \frac{e}{c}\frac{d}{dx^0}(\gamma^0 x^\varrho)\mathfrak{M}_{\varrho\alpha} - \frac{h}{4\pi i}\frac{d}{dx^0}(\gamma^0 x^\varrho) R_{\varrho\alpha}. \tag{92}$$

Since γ^0 corresponds to unity, these equations present (except for the last term, which is of the order of magnitude of h) a complete analogue to the classical equations (83).

[12] Equations (87) and (89) are the immediate consequences of (80*). (*V. Fock*)

The matrix γ^0 shows up due to our definition (77) of the matrix. Replacing (77) by

$$K_{mn} = \int \overline{\Psi}_m \gamma^0 K \Psi_n dV, \qquad (77*)$$

one can eliminate the factor γ^0 from formula (92).

Note that in the general Riemann space one can choose coordinates in such a way that $\gamma^0 = 0$, identically.

6 Conclusion

The main aim of our theory is to establish the fact that the Dirac equation is perfectly compatible with the notion of a Riemann space.

The geometric method introduced by Einstein in the macroscopic physics turns out to be applicable to the microscopic physics; a semi-vector and the Dirac wave equation are as fundamental notions as a vector and the d'Alembert equation.

In the described theory, a connection between electromagnetic and gravitational quantities is established via the notion of a semi-vector. The vector-potential finds its place in Riemann geometry and one does not need to generalize it (Weyl, 1918) or to introduce the remote parallelism (Einstein, 1928). From this point of view our theory – developed independently – is compatible with the new theory by H. Weyl described in his memoir "Gravity and electron."[13]

We have left untouched the "Dirac's difficulty" (negative energy, etc). It seems to us that our theory is independent of this difficulty and that it could serve as the basis for a future theory, including maybe the quantization of gravity, which would unify the electricity and the matter theories.

Translated by V.V. Fock

[13]H. Weyl, Proc. Natl. Acad. Soc. **15**, 323, 1929; Zs. Phys. **56**, 330, 1929. The main object of this memoir is the "Dirac's difficulty." The theory suggested by Weyl to solve this difficulty seems to us to be a subject of serious objections; a critique of this theory is given in our article cited at the beginning of Section 4. (*V. Fock*)

30-1
A Comment on the Virial Relation

V. Fock

(Received 30 May 1930)

Zs. Phys. **63**, N11–12, 855, 1930
JRPKhO, **62**, N 4, 379, 1930

The validity of the virial theorem was proved by Born, Heisenberg and Jordan[1] on the ground of matrix calculation and after that by Finkelstein[2] on the ground of the Schrödinger equation. Both proofs essentially reproduced ideas of the classical approach. In the present note, a new and purely quantum mechanical derivation of this relation is given from the variational principle. The derivation is very simple and allows one to get some interesting consequences.

1. The Schrödinger equation for a system of point masses can be obtained from the variational principle

$$\delta J = 0,$$

where the action integral J is of the form

$$J = \int \overline{\psi}(T + U - E)\psi d\tau.$$

Here T stands for the kinetic energy operator

$$T\psi = -\sum_{k=1}^{N} \frac{h^2}{8\pi^2 m_k} \Delta_k \psi,$$

m_k is the mass of a k-th particle, Δ_k is the Laplace operator acting on the coordinates of this particle, $U = U(r_1, r_2, \ldots, r_N)$ is the potential energy and E is the eigenvalue parameter. For the wave function ψ satisfying the normalization condition

$$\int \overline{\psi}\psi d\tau = 1,$$

---

[1]M. Born, W. Heisenberg and P. Jordan, Zs. Phys. **35**, 557, 1926; M. Born, P. Jordan, Elementary Quantum Mechanics, p. 100, Berlin, 1930.

[2]B. Finkelstein, Zs. Phys. **50**, 293, 1928.

the expectation value of the kinetic energy in the state ψ is represented by the integral

$$T_0 = \int \overline{\psi} T \psi d\tau,$$

and similarly that of potential energy is

$$U_0 = \int \overline{\psi} U \psi d\tau.$$

The virial relation can be formulated as follows:
In the state with a function ψ belonging to the point spectrum and for a potential energy U, being a homogeneous degree ϱ function of coordinates, the expectation values T_0 and U_0 of a kinetic and potential energy are connected by the following relation:

$$2T_0 = \varrho U_0.$$

Proof. Let the function $\psi(\mathbf{r}_1, \mathbf{r}_2, \ldots, \mathbf{r}_N)$ solve the variational problem. Consider a set of normalized functions

$$\psi^*(\mathbf{r}_1, \mathbf{r}_2, \ldots, \mathbf{r}_n, \lambda) = \lambda^{3N/2} \psi(\lambda \mathbf{r}_1, \lambda \mathbf{r}_2 \ldots, \lambda \mathbf{r}_n),$$

depending on parameter $\lambda > 0$. Denoting by T_0^* and U_0^* the expectation values of kinetic and potential energy, we obtain

$$\begin{aligned} T_0^* &= \int \overline{\psi}^* T \psi^* d\tau, \\ U_0^* &= \int \overline{\psi}^* U \psi^* d\tau = \\ &= \int \overline{\psi}(\mathbf{r}_1, \mathbf{r}_2 \ldots \mathbf{r}_n) U(\frac{\mathbf{r}_1}{\lambda}, \frac{\mathbf{r}_2}{\lambda}, \ldots, \frac{\mathbf{r}_n}{\lambda}) \psi\, (\mathbf{r}_1, \mathbf{r}_2 \ldots \mathbf{r}_n)\, d\tau \end{aligned}$$

and, due to the homogeneity of U,

$$U_0^* = \lambda^{-\varrho} U_0.$$

Hence, the action integral is equal to

$$J = \lambda^2 T_0 + \lambda^{-\varrho} U_0 U_0 - E,$$

and, putting zero its variation with respect to λ, we get

$$\delta J = (2\lambda T_0 - \varrho \lambda^{-\varrho-1} U_0) \delta\lambda = 0.$$

Solution of the variational problem is obtained at $\lambda = 1$. It gives the required relation

$$2T_0 = \varrho U_0.$$

Generally the potential energy is not necessarily a homogeneous function; hence the following relation holds:

$$2T_0 = -\left(\frac{\partial U_0^*}{\partial \lambda}\right)_{\lambda=1} = \text{E.v.}\left(\sum_{k=1}^{N} r_k \frac{\partial U}{\partial r_k}\right),$$

where E.v. stands for the expectation value.

This proof seems to be very simple, even simpler than that in classical mechanics.

The condition for ψ to belong to the point spectrum is necessary for the convergence of integrals in consideration and it corresponds to the demand of finite values of coordinates in classical mechanics.

2. The given proof admits generalization in two directions.

First, a wave function $\psi(\mathbf{r}_1, \mathbf{r}_2 \ldots \mathbf{r}_n)$ is not necessarily an exact solution of the wave equation. The function ψ is commonly an approximate solution and parameter λ in $\psi^*(\mathbf{r}_1, \mathbf{r}_2 \ldots \mathbf{r}_n)$ should be varied to get the best approximation in the sense of the variational principle. The numerical value of λ is obtained from $\delta J = 0$. This approach can be described as "a stretching of the domain of definition." The condition $J = 0$ gives

$$2T_0^* = \varrho U_0^*.$$

We conclude that *the relation given by the virial theorem would be valid for any approximate solution of the variational problem, when the stretching of the domain of variation is introduced.*

The stretching of the domain of variation was applied for example by Hylleraas in his work on the helium atom.[3] The virial relation is also valid for the solution of the Hartree equations[4] as well as in the case of more general equations by the author.[5]

Second, our method also works for the Dirac equation. As before, we consider the case of a homogeneous potential energy function without magnetic fields. In usual notations, the action integral reads as

$$J = \int \overline{\psi}\left(c(\alpha_1 p_1 + \alpha_2 p_2 + \alpha_3 p_3) + mc^2\alpha_4 + U - E\right)\psi d\tau.$$

[3]E. Hylleraas, Zs. Phys. **54**, 347, 1929.

[4]D.R. Hartree, Proc. Cambridge Phil. Soc. **24**, 89, 111, 1929.

[5]V. Fock, Zs. Phys. **61**, 126, 1930. (See [30-3] in this book. (*Editors*))

We denote by L the operator

$$L\psi = c(\alpha_1 p_1 + \alpha_2 p_2 + \alpha_3 p_3)\psi,$$

which corresponds to the classical quantity

$$mv^2 \frac{1}{\sqrt{1-\frac{v^2}{c^2}}} = \sum_{k=1}^{3} \nu_k p_k, \tag{*}$$

and put further

$$L_0 = \int \overline{\psi} L\psi d\tau, \qquad U_0 = \int \overline{\psi} U\psi d\tau, \qquad A = \int \overline{\psi}\alpha\psi d\tau,$$

L_0 being the expectation value of the quantity (*). Repeating the previous considerations, we introduce the set of functions

$$\psi^*(\mathbf{r};\lambda) = \lambda^{2/3}\psi^*(\mathbf{r};\lambda)$$

and calculate the integral J. We obtain

$$J = \lambda L_0 + \lambda^{\varrho} U_0 + mc^2 A - E.$$

Putting zero its variation with respect to λ, we obtain $\lambda = 1$:

$$L_0 = \varrho U_0$$

and for an arbitrary unnecessary homogeneous potential energy

$$L_0 = \text{E.}v.\left(r\frac{\partial U}{\partial r}\right).$$

For the Coulomb field, $\varrho = -1$ and we have

$$\begin{aligned} 2T_0 + U_0 &= 0 \qquad \text{(Schrödinger)} \\ L_0 + U_0 &= 0 \qquad \text{(Dirac)}. \end{aligned}$$

Due to the wave equation, $J = 0$ and in the Dirac case we arrive at the relation

$$E = mc^2 A = mc^2 \int \overline{\psi}\alpha\psi d\tau,$$

which to our knowledge was not pointed out in the literature before.

Leningrad,
Physical Institute of the University

Translated by A.V. Tulub

30-2
An Approximate Method for Solving the Quantum Many-Body Problem

V.A. Fock

Reported at the Session of the Russian Phys.-Chem. Soc. on 17 December 1929

Zs. Phys. **61**, 126, 1930
TOI **5**, N 51, 1, 1931
UFN **93**, N 2, 342, 1967

1 Introduction

The mathematical formulation of the quantum many-body problem (without relativistic corrections) was given by Schrödinger in one of his pioneer works.[1] Since the wave function sought depends on the great number of variables (namely, there are as many of them as degrees of freedom in the N electron system), the exact solution of this problem encounters the insuperable difficulties and consequently one needs to resort to approximate methods. An extremely ingenious way was proposed by Hartree.[2] However, the derivation of the equations given by Hartree himself was based on the consideration not related to the Schrödinger equation in the configuration space. Another work by Gaunt[3] was devoted to the statement of this link, but the problem was not solved entirely because nothing was mentioned about Hartree's equations being connected with the variational principle. The main point of Hartree's method called by him as that of the "self-consistent field" consists of the following. Hartree preserves the classical notion of the individual electron orbit provided each orbit is described, according to Schrödinger, by the wave function. For each wave function (i.e., for each individual electron) one constructs the Schrödinger equation with the potential energy originating from the interaction first with a nucleus and second with other electrons continuously distributed with the charge density $\varrho = \overline{\psi}\psi$.

[1] E. Schrödinger, *Quantisiering als Eigenwertproblem*, Ann. Phys. (1926).

[2] D.R. Hartree, *The Wave Mechanics of an Atom with a Non-Coulomb Central Field*, Proc. Cambr. Phil. Soc. **24**, 89, 111, 1928.

[3] J.A. Gount, *A Theory of Hartree's Atomic Fields*, Proc. Cambr. Phil. Soc. **24**, 328, 1928.

Because Hartree uses the notion of orbits, his method is more descriptive. However, one can speak about orbits in the context of quantum mechanics only with a certain approximation. At the same time, one can expect that this approximation is good enough, since the preceding Bohr theory, operating with the electronic orbits, gives a quite satisfactory classification of the atomic spectra. The question is whether the Hartree method is able to give the extreme accuracy, combined with a concept of orbits, i.e., with a description of the atom by means of the individual electron wave functions.

In this paper, we shall show that the most extreme accuracy is not achieved in Hartree's method, but can be attained by means of a certain modification. Physically speaking, this corresponds to the so-called exchange energy to be taken into account. In the mathematical sense, it is realized by using the wave functions of the required symmetry in the variational principle.

Our modification of Hartree's equations, combined with the notion of orbits, is the best in the sense of the variational principle.

We take into account the magnetic properties of electrons (so-called spins) insofar as they are required by the Pauli principle. Their presence influences the wave function symmetry, while the correction terms in the energy operator are omitted. Such an approach is quite rightful because correction due to the "spin" is less than that due to the exchange energy.

2 The Idea of the Method

As is well known, the Schrödinger equation can be obtained from the variational principle

$$\delta \int \overline{\Psi}(L-E)\Psi d\tau = 0\,. \tag{1}$$

Here Ψ is the wave function depending on the coordinates of all the N electrons,

$$\Psi = \Psi(x_1y_1z_1, x_2y_2z_2, \ldots, x_Ny_Nz_N)\,,$$

L is the energy operator (its expression will be written later), E is a constant (the atomic energy), and $d\tau$ is an elementary volume of the configuration space,

$$d\tau = dx_1dy_1dz_1 \ldots dx_Ndy_Ndz_N\,.$$

It is possible to show that *in virtu* of the operator L hermiticity, eq. (1) is equivalent to the following:

$$\int \delta\overline{\Psi}(L-E)\Psi d\tau = 0\,.$$

For brevity, we shall denote the three coordinates of a k-th electron by a single letter x_k. Besides, we shall use the so-called atomic system of units, in which for units of length, charge and mass we take correspondingly: $a_H = \frac{h^2}{4\pi^2 me^2} = 0.529 \cdot 10^{-8}$ cm, the radius of the first hydrogen orbit; $e = 4.77 \cdot 10^{-10}$ CGSE, the electron charge; $m = 9.00 \cdot 10^{-28}$ g, the electron mass.

In these units, the Planck constant divided by 2π is equal to one, and the light velocity is equal to 137 (the inverse of Sommerfeld's fine structure constant).

In atomic units, the energy operator for the atom with N electrons assumes the form

$$L = \sum_{k=1}^{N} H_k + \sum_{i,k=1}^{N} \frac{1}{r_{ik}}\,, \tag{2}$$

providing

$$H_k = -\frac{1}{2}\Delta_k - \frac{N}{r_k}\,. \tag{3}$$

Here Δ_k is the Laplace operator applied to the coordinates of the k-th electron, r_k is the distance between the electron and the nucleus, and r_{ik} is the distance between the i-th and k-th electrons.

For the solution of the variational problem (1) we shall use the generalized Ritz method.[4] As it is known, the Ritz method is in the following: one substitutes into the varied functional the expression of the function sought, which depends on some unknown constants; then these constants are determined from the minimum condition of the functional. Generalizing this method, we shall seek Ψ as a sum of a definite number of products of functions $\psi_i(x_k)$, each one depending on the coordinates of a single electron. Thus, we shall have unknown functions instead of unknown coefficients, and this is just our generalization. In the physical sense, the expression of the wave function Ψ by means of a definite number of the one-electron functions $\psi_i(x_k)$ corresponds to the notion of the individual electronic orbits. In fact, we can say that there are definite orbits in the atom that are described by the wave functions $\psi_i(x_k)$ and are occupied by a definite number of electrons.

[4] See, e.g., Walther Ritz, Gesammelte Werke, Gauthier-Villars, Paris, 1911. (*V. Fock*)

Different assumptions about the form of the wave function Ψ lead to different forms of the system of equations for functions $\psi_i(x_k)$. We shall see that the assumption of

$$\Psi = \psi(x_1)\psi(x_2)\ldots\psi(x_N) \tag{4}$$

leads one to the Hartree equations. However, such a form of function does not possess (save a case of the normal state of the He atom) the symmetry required by the group theory. Therefore, Hartree's equations are only a relatively rough approximation that corresponds, as will be seen later, to the neglect of the exchange energy.

We can obtain a better approximation if we seek Ψ in the form of the required symmetry. Then the terms presenting the quantum exchange will appear in the equations for functions $\psi_i(x_k)$.

Generally, the product of two determinants constructed by functions $\psi_i(x_k)$, or a sum of such products, possesses the correct symmetry. In the simplest and commonly occurring case and, therefore, most important partial one (the so-called Heitler's case of complete degeneration of the term system), it is sufficient to take one product of determinants.[5] In our work, we restrict ourself to the consideration of this partial case and infer for it the system of equations, which permit us to determine the functions $\psi_i(x_k)$.

3 Example: A Helium Atom

To clarify the way of calculations, we start from the simplest example, namely, from a helium atom.

In the two-electron problem, the energy operator is of the form

$$L = H_1 + H_2 + \frac{1}{r_{12}}\,. \tag{5}$$

The variational principle is written in the form

$$\iint \delta\overline{\Psi}(L-E)\Psi dx_1 dx_2 = 0\,. \tag{6}$$

Here dx_1 and dx_2 are the elementary volumes; e.g., we write dx_1 instead of $dx_1 dy_1 dz_1$. For the ground state, we take

$$\Psi = \psi(x_1)\psi(x_2)\,. \tag{7}$$

[5] W. Heitler, *Störungsenergie und Austausch beim Mehrkörperproblem*, Zs. Phys. **46**, 47, 1927; J. Waller and D.R. Hartree, *On the Intensity of Total Scattering of X Rays*, Proc. Roy. Soc. London **A124**, 119, 1929.

Substituting this function into equation (6), we obtain

$$\iint \left[\delta\overline{\psi}(x_1)\overline{\psi}(x_2) + \overline{\psi}(x_1)\delta\overline{\psi}(x_2)\right](L - E)\psi(x_1)\psi(x_2)dx_1dx_2 = 0\,.$$

As the consequence of the symmetry of L and Ψ with respect to x_1 and x_2, both terms in square brackets give the same result. Taking this into account and substituting instead of L its expression (5), we shall have

$$\int dx_1\delta\overline{\psi}(x_1)\int\overline{\psi}(x_2)(H_1 + H_2 + \frac{1}{r_{12}})\psi(x_1)\psi(x_2)dx_2 = 0\,. \quad (8)$$

The integral over x_2 is equal to

$$H_1\psi(x_1) + G(x_1)\psi(x_1) - E_0\psi(x_1),$$

where

$$G(x_1) = \int\frac{|\psi(x_2)|^2}{r_{12}}dx_2; \quad (9)$$

$$E_0 = E - \int\overline{\psi}(x_2)H_2\psi(x_2)dx_2\,. \quad (10)$$

Consequently, we have

$$\int dx\delta\overline{\psi}(x)[H + G(x) - E_0]\psi(x) = 0\,. \quad (11)$$

Therefore, for function ψ one obtains the equation:

$$\left[H + G(x) - E_0\right]\psi(x) = 0\,. \quad (12)$$

This equation coincides exactly with that of Hartree. The characteristic number of the energy operator is connected with the parameter E_0 by relations (10), which by means of equation (12) can be written in the form

$$E = 2E_0 - \int G(x)|\psi(x)|^2dx\,. \quad (13)$$

The latter relation coincides with that given by Gaunt (l.c.). Hartree's equation (12) as well as Gaunt's (13) are obtained by us from the variational principle in an absolutely natural way.

Now one can make the following assumption: let the function $\psi(x) = \psi(x, y, z)$ depend only on the distance r from the nucleus. Then the function $G(x, y, z)$ will also depend only on r and one should seek the solution of equation (12), which possesses the spherical symmetry.

For the excited state of a helium atom, one should take

$$\Psi = \psi_1(x_1)\psi_2(x_2) + \psi_1(x_2)\psi_2(x_1)\,, \tag{14}$$

and for parahelium,

$$\Psi = \psi_1(x_1)\psi_2(x_2) - \psi_1(x_2)\psi_2(x_1)\,. \tag{15}$$

In the latter case, one can also assume ψ_1 and ψ_2 to be orthogonal and normalized. If we substitute expression (15) into (6), we obtain the system of equations

$$\begin{aligned} H\psi_1(x) + G_{22}(x)\psi_1(x) - G_{21}(x)\psi_2(x) &= (E - H_{22})\psi_1(x) + H_{21}\psi_2(x)\,, \\ H\psi_2(x) + G_{11}(x)\psi_2(x) - G_{12}(x)\psi_1(x) &= H_{12}\psi_1(x) + (E - H_{11})\psi_2(x)\,, \end{aligned} \tag{16}$$

where for brevity we put

$$G_{ik}(x) = \int \frac{\overline{\psi}_i(x')\psi_k(x')}{r}dx', \qquad H_{ik} = \int \overline{\psi}_i(x)H\psi_k(x)\,, \tag{17}$$

and

$$\begin{aligned} \langle ik \mid sG \mid lm\rangle &= \iint \frac{\overline{\psi}_i(x')\overline{\psi}_k(x)\psi_l(x')\psi_m(x)}{r}dxdx' = \\ &= \int \overline{\psi}_i(x)\psi_l(x)G_{km}(x)dx = \int \overline{\psi}_k(x)\psi_m(x)G_{il}(x)dx\,. \end{aligned} \tag{18}$$

In the system (16), the values $G_{12}(x)$ and $G_{21}(x)$ as well as H_{12} and H_{21} (i.e., the coefficients with different subscripts) present the quantum exchange. In general, the values of $G_{12}(x)$ and $G_{21}(x)$ are less[6] than those of $G_{11}(x)$ and $G_{22}(x)$ but not so much to be neglected.

Multiplying the first of eqs. (16) by $\overline{\psi}(x)$ and integrating, we obtain the following expression for the atomic energy:

$$E = H_{11} + H_{22} + \langle 12 \mid G \mid 12\rangle - \langle 12 \mid G \mid 21\rangle\,. \tag{19}$$

This expression can be presented in another form, which permits a simple interpretation. Putting

$$\varrho(x,x') = \overline{\psi}_1(x)\psi_1(x') + \overline{\psi}_2(x)\psi_2(x')\,, \tag{20}$$

$$\varrho(x) = \varrho(x,x)\,, \tag{21}$$

[6]See Gaunt, the footnote on page 137. (*V. Fock*)

we obtain

$$E = H_{11} + H_{22} + \frac{1}{2} \iint \frac{\varrho(x)\varrho(x') - |\varrho(x,x')|^2}{r} dx dx' \,. \tag{22}$$

Here the first two terms represent the single electron energies, while the integral can be interpreted as the energy of their interaction. It should be noted that in the numerator of the integrand there is not mere $\varrho(x)\varrho(x')$ but $\varrho(x)\varrho(x')$-$|\varrho(x',x)|^2$, the expression which tends to zero when $x = x'$. This situation can be interpreted in the sense that an electron does not interact with itself.

Expression (14) for the wave function of excited parahelium leads to the equations, which are analogous to (16) with a distinction that the terms G_{12}, H_{12} , etc., which characterize the quantum exchange, have a minus, and besides there are some new terms because now the functions ψ_1 and ψ_2 should not be considered to be orthogonal.

4 Hartree's Equations

Turning to the N-electron problem, at first we shall seek Ψ in the form

$$\Psi = \prod_{q=1}^{N} \psi_q(x_q) \,. \tag{23}$$

We have mentioned already that this expression does not possess the required symmetry. In spite of this, we shall perform the calculation to the end to be sure that this expression does lead to Hartree's equation. Due to the Pauli principle there could not be more than two functions among ψ_q that are the same. We have

$$\delta\Psi = \sum_{q=1}^{N} \delta\psi_q(x_q)\Psi^{(q)}, \tag{24}$$

where

$$\Psi^{(q)} = \frac{\Psi}{\psi_q(x_q)} \,. \tag{25}$$

The variational principle has the form

$$\delta I = \int \delta\overline{\Psi}(L - E)\Psi d\tau = \sum_{q=1}^{N} \int \delta\overline{\psi}_q(x_q) A_q dx_q = 0 \,. \tag{26}$$

Here we assume

$$A_q = \int \dots \int \overline{\Psi}^{(q)} \left(\sum_{k=1}^{N} H_k - E + \sum_{i,k=1}^{N} \frac{1}{r_{ik}} \right) \psi_q(x_q) \Psi^{(q)} \cdot dx_1 \dots dx_{q-1} dx_{q+1} \dots dx_N =$$

$$= H_q \psi_q(x_q) + \left[\sum_{i=1}^{N}{}' (G_{ii}(x_q) + H_{ii}) + \sum_{i>k=1}^{N}{}'' W_{ik} - E \right] \psi_q(x_q) , \tag{27}$$

provided that

$$W_{ik} = \langle ik \mid G \mid ik \rangle = \iint \frac{|\psi_i(x)|^2 |\psi_k(x')|^2}{r} dx dx' . \tag{28}$$

Here and in the following, the primed sum means that the terms with $i = q$ and $k = q$ should be omitted.

In expression (26) for δI, one should equal the coefficients under the independent variations $\delta \overline{\psi}_q$. If the function ψ_q is encountered in Ψ only once, then it must be $A_q = 0$; if it is encountered twice, e.g., $\psi_q = \psi_{q+1}$, then, as can be easily seen, the coefficients under the variations $\delta \overline{\psi}_q$ and $\delta \overline{\psi}_{q+1}$ are equal to each other, i.e., $A_q = A_{q+1}$. In both cases, $A_q = 0$.

We shall write these equations in the form

$$H \psi_q(x) + [V(x) - G_{qq}(x)] \psi_q(x) = \lambda_q \psi_q(x) \qquad (q = 1, \dots, N), \tag{29}$$

where

$$V(x) = \sum_{k=1}^{N} G_{kk} , \tag{30}$$

$$\lambda_q = E - \sum_{i=1}^{N}{}' H_{ii} - \sum_{i>k=1}^{N}{}'' W_{ik} . \tag{31}$$

Equations (29) exactly coincide with those inferred by Hartree.

For the energy of the atom, one obtains the expression

$$E = \sum_{i=1}^{N} H_{ii} + \sum_{i>k=1}^{N} W_{ik} . \tag{32}$$

It is easy to verify that this expression is equal to

$$E = \frac{\int \overline{\Psi} L \Psi d\tau}{\int \overline{\Psi} \Psi d\tau}, \tag{33}$$

in spite of the fact that the equality $L\Psi = E\Psi$ does not take place because Ψ is not the exact solution of the Schrödinger equation.

Substitution of expression (32) into (31) gives

$$\lambda_q = H_{qq} + \sum_{i=1}^{N}{}' W_{iq} . \tag{34}$$

If one takes a sum over q, one obtains

$$\sum_{q=1}^{N} \lambda_q = 2E - \sum_{q=1}^{N} H_{qq} . \tag{35}$$

Equations (29) are the variational equations of the problem on the minimum of the integral

$$W = \sum_{i=1}^{N} \int \overline{\psi}_1 H \psi_i dx + \\ + \frac{1}{2} \sum_{i,k=1}^{N} (1 - \delta_{ik}) \iint \frac{\overline{\psi}_k(x')\psi_k(x')\overline{\psi}_i(x)\psi_i(x)}{r} dx dx' \tag{36}$$

under the supplementary conditions

$$\int \overline{\psi}_i(x)\psi_i(x) dx = 1 \qquad (i = 1, \ldots, N) . \tag{37}$$

By comparing equations (36) and (32), one can conclude that the value of the integral is equal to energy E. The coefficients λ_q play the roles of Lagrange's constants.

Equations (29) can be also inferred from another variational principle. The coefficients $V(x)$ and G_{qq} in these equations depend, in turn, on the unknown functions $\psi_i(x)$. But we can consider these coefficients as given and equations (29) as a system of linear differential equations for functions $\psi_q(x)$. The solution of these equations turns the following integral to be minimum

$$W^* = \sum_{i=1}^{N} \int \overline{\psi}_i(x)[H + V(x) - G_{ii}(x)]\psi_i(x)dx \tag{38}$$

provided the above supplementary conditions (37) are fulfilled. This minimum is equal to

$$W^* = 2E - \sum_{q=1}^{N} H_{qq} = \sum_{q=1}^{N} \lambda_q \, . \tag{39}$$

In our equations, the symbol x denotes all three coordinates x, y, z, so that $\psi_q(x)$ stands instead of $\psi_q(x, y, z)$ etc. If one would like to make more partial assumptions regarding $\psi_q(x, y, z)$, one should proceed as follows. One should substitute the function ψ_q of required form into the integral

$$\iiint \delta\overline{\psi}_q(x, y, z)[H + V(x, y, z) - G_{qq}(x, y, z) - \lambda_q]\psi_q(x, y, z)dxdydz \tag{40}$$

and then send to zero the coefficients, standing under the independent variations (of functions or constants, which are needed to be determined). Here we shall present the calculations for the case

$$\psi_q(x, y, z) = f_{nl}(r)Y_l(\vartheta, \varphi) \, ,$$

where Y_l is the spherical function normalized as follows:

$$\iint |Y_l(\vartheta, \varphi)|^2 \sin\vartheta d\vartheta d\varphi = 1 \, . \tag{41*}$$

Here the element to be determined is the function $f_{nl}(r)$, while the spherical function is not varying. In our case

$$\iiint \delta\overline{\psi}_q(x, y, z)\Delta\psi_q(x, y, z)dxdydz = \tag{37}$$

$$= \int \delta\overline{f}_{nl}(r) \left[\frac{\partial}{\partial r}\left(r^2 \frac{\partial f_{nl}}{r}\right) - l(l+1)f_{nl}(r)\right] dr \, .$$

Further

$$\iiint \delta\overline{\psi}_q V(x, y, z)\psi_q dxdydz = \int \delta\overline{f}_{nl}(r)V^0(r)r^2 dr \, ,$$

where we put

$$V^0(r) = \iint V(x,y,z)|Y_l(\vartheta,\varphi)|^2 \sin\vartheta d\vartheta d\varphi\,. \tag{42}$$

The function $G^0_{qq}(r)$ can be constructed by functions $G_{qq}(x,y,z)$ in an analogous way. Then for $f_{nl}(r)$ one can obtain the equation

$$-\frac{1}{2}\left(\frac{d^2 f_{nl}}{dr^2} + \frac{2}{r}\frac{df_{nl}}{\partial r^2} - \frac{l(l+1)}{r^2} f_{nl}\right) + \\ + \left(-\frac{N}{r} + V^0(r) - G^0_{qq}(r)\right) f_{nl} = \lambda_q f_{nl}\,. \tag{43}$$

Our equation (43) does not exactly coincide with that obtained by Hartree's method. According to Hartree, one should average the function $V(x,y,z)$ over a sphere of radius r, i.e., write the expression

$$\frac{1}{4\pi}\iint V(x,y,z)\sin\vartheta d\vartheta d\varphi,$$

while we take the average, in accordance with (42), with a "weight function" $|Y_l(\vartheta,\varphi)|^2$.

5 The Symmetry Properties of the Wave Function. Example: A Lithium Atom

We shall consider the n-electron problem anew, but now seek Ψ in the form, which possesses the required symmetry. The symmetry properties of the wave function corresponding to different terms were investigated in detail by various authors[7] with the help of the group theory, so that we can use here the well-known results.

Generally, some (e.g., s) wave functions

$$\omega_1, \omega_2, \ldots, \omega_s,$$

each depending on the coordinates of all N electrons, correspond to a definite term. If one performs any permutations (say, p^a) of the electron coordinates in the function ω_j, it transforms into a new one, ω'_j. This

[7] See, e.g., Heitler (l.c.), Waller and Hartree (l.c.). (*V. Fock*)

new function can be expressed as a linear combination of the former ones, corresponding to the same term:

$$P^a\omega_j = \omega'_j = \sum_{i=1}^{s} P^a_{ij}\omega_i\,. \tag{44}$$

The set of coefficients gives the irreducible representation of the permutation group. The functions $\omega_1, \omega_2, \ldots, \omega_s$ can be always chosen so that the sum of their square modules

$$|\omega_1|^2 + |\omega_2|^2 + \ldots |\omega_s|^2$$

is symmetric with respect to all the electrons. Then all matrices P^a_{ij} will be unitary.[8]

Now we shall show the following. Let L be a Hermitian operator, which is symmetric with respect to all the electrons. If we construct an arbitrary linear combination Ω of the functions:

$$\Omega = \alpha\omega_1 + \alpha\omega_2 + \ldots + \alpha\omega_s \tag{45}$$

and, by means of it, calculate the expression

$$A = \frac{\int \overline{\Omega} L \Omega d\tau}{\int \overline{\Omega}\Omega d\tau}\,, \tag{46}$$

then the latter does not depend on coefficients $\alpha_1, \ldots, \alpha_s$.

If L is the energy operator, then, equaling the variation of A to zero, one obtains the Schrödinger equation for the function Ω. This statement means that, to infer the Schrödinger equation from the variational principles, one can take any linear combination of functions $\omega_1, \ldots, \omega_s$.

In order to deduce this, we show first the validity of the equalities

$$\int \overline{\omega_j} L \omega_i d\tau = 0 \quad (i \neq j)\,,$$

$$\int \overline{\omega_1} L \omega_1 d\tau = \int \overline{\omega_2} L \omega_2 d\tau = \ldots = \int \overline{\omega_s} L \omega_s d\tau\,, \tag{47}$$

[8]A matrix is unitary if its inverse is obtained by interchanging rows by columns and complex conjugating all its elements, i.e.,

$$(P^a)^{-1}_{ij} = \overline{P}^a_{ji}.$$

The partial case of a unitary matrix is that of orthogonal transformation with real coefficients. (*V. Fock*)

from which it follows that the matrix

$$L_{ji} = \int \overline{\omega_j} L \omega_i d\tau \quad (i, j = 1, 2, \ldots, s) \tag{48}$$

is proportional to the unity matrix.[9]

If L is the symmetric operator, then under permutation P^a the functions

$$\xi_j = L\omega_j$$

expose the same transformation as functions ω_j do themselves, namely,

$$\xi'_i = P^a \xi_i = \sum_{l=1}^{s} P^a_{li} \xi_l \,.$$

If the matrix P^a is unitary, the functions $\eta_j = \overline{\omega}_j$ expose a transformation

$$\eta'_j = \sum_{k=1}^{s} (P^a)^{-1}_{ik} \eta_k \,.$$

Let us construct the product $\eta'_j \xi'_i$ and calculate its average over all $N!$ permutations. In the group theory, it is shown that[10]

$$\frac{1}{N!} \sum_a (P^a)^{-1}_{ik} P^a_{li} = \frac{1}{s} \delta_{ij} \delta_{kl},$$

whence it follows that the average sought is equal to

$$\frac{1}{N!} \sum_\alpha \eta'_j \xi'_i = \frac{1}{s} \delta_{ij} \sum_{k=1}^{s} \eta_k \xi_k \,.$$

Integrating this expression over the configurational space and taking into account that

$$\int \eta'_j \xi'_i d\tau = \int \eta_j \xi_i d\tau = \int \overline{\omega_j} L \omega_i d\tau \,,$$

we obtain

$$\int \overline{\omega_j} L \omega_i d\tau = \frac{1}{s} \delta_{ij} \sum_{k=1}^{s} \int \overline{\omega_k} L \omega_k d\tau, \tag{49}$$

whence equality (47) follows directly.

[9] See, e.g., E. Wigner, Zs. Phys. **43**, 624, 1927, eq. (3). (*V. Fock*)

[10] See Speiser, Theorie der Gruppen von endlicher Ordung, theorem 144. (*V. Fock*)

Now it is easy to verify the validity of our statement. Indeed, if we substitute the expression of Ω into (46), the constants α_i are encountered in the same combination both in the numerator and in the denominator only,

$$|\alpha_1|^2 + |\alpha_2|^2 + \ldots + |\alpha_s|^2,$$

and this expression in the fraction cancels.

Now our challenge consists of the following. We must construct such an expression by the one-electron functions $\psi_i(x_k)$, which ought to have the same symmetric properties with respect to any permutations of electrons as a linear combination of functions $\omega_1, \ldots, \omega_s$, belonging to the same term. Thus, the expression obtained should be substituted into the variational principle.

As it has been shown in the cited papers by Heitler, Waller and Hartree, in one important partial case one can write the corresponding expression as the product of two determinants. By Heitler's terminology, this partial case corresponds to the term, which belongs to "the quite degenerate system of terms" (vollständig ausgeartetes Termsystem).

To clarify the above treatment, let us consider a lithium atom as an example. Suppose that two of three electrons are on the same orbit. Since each orbit is described by the respective function, we shall have two functions $\psi_1(x)$ and $\psi_2(x)$. By means of these functions, one can construct three products

$$\psi_1(x_2)\psi_1(x_3)\psi_2(x_1),$$

$$\psi_1(x_3)\psi_1(x_1)\psi_2(x_2),$$

$$\psi_1(x_1)\psi_1(x_2)\psi_2(x_3);$$

and then construct three linear combinations

$$\omega = \psi_1(x_2)\psi_1(x_3)\psi_2(x_1) + \psi_1(x_3)\psi_1(x_1)\psi_2(x_2) + \psi_1(x_1)\psi_1(x_2)\psi_2(x_3),$$

$$\omega_1 = -\psi_1(x_3) \begin{vmatrix} \psi_1(x_1) & \psi_2(x_1) \\ \psi_1(x_2) & \psi_2(x_2) \end{vmatrix},$$

$$\omega_2 = \frac{1}{\sqrt{3}}\psi_1(x_2) \begin{vmatrix} \psi_1(x_3) & \psi_2(x_3) \\ \psi_1(x_1) & \psi_2(x_1) \end{vmatrix} + \frac{1}{\sqrt{3}}\psi_1(x_1) \begin{vmatrix} \psi_1(x_3) & \psi_2(x_3) \\ \psi_1(x_2) & \psi_2(x_2) \end{vmatrix};$$

the function ω is symmetric of all three electrons and the corresponding term is forbidden by the Pauli principle; the functions ω_1 and ω_2 belong to just the same term. Their substitutions under permutations of the electrons give the following irreducible orthogonal representation of the permutation group:

$$P^{(23)} = \begin{Bmatrix} \frac{1}{2} & \frac{\sqrt{3}}{2} \\ \frac{\sqrt{3}}{2} & -\frac{1}{2} \end{Bmatrix}, \quad P^{(31)} = \begin{Bmatrix} \frac{1}{2} & -\frac{\sqrt{3}}{2} \\ -\frac{\sqrt{3}}{2} & -\frac{1}{2} \end{Bmatrix}, \quad P^{(12)} = \begin{Bmatrix} -1 & 0 \\ 0 & 1 \end{Bmatrix},$$

$$P^{(123)} = \begin{Bmatrix} -\frac{1}{2} & \frac{\sqrt{3}}{2} \\ -\frac{\sqrt{3}}{2} & -\frac{1}{2} \end{Bmatrix}, \quad P^{(213)} = \begin{Bmatrix} -\frac{1}{2} & -\frac{\sqrt{3}}{2} \\ \frac{\sqrt{3}}{2} & -\frac{1}{2} \end{Bmatrix}, \quad E = \begin{Bmatrix} 1 & 0 \\ 0 & 1 \end{Bmatrix}.$$

From the general theory, it follows:

$$\int \overline{\omega_1} L \omega_2 d\tau = 0, \qquad \int \overline{\omega_1} L \omega_1 d\tau = \int \overline{\omega_2} L \omega_2 d\tau \,,$$

which can be verified directly in the simple case considered.

As is seen, though several functions (in our case two) correspond to a single term, one can choose a single combination of these functions for the variational principle. For a lithium atom, one can choose the product of determinants, for example, $\Psi = -\omega_1$:

$$\Psi = \psi_1(x_3) \begin{vmatrix} \psi_1(x_1) & \psi_2(x_1) \\ \psi_1(x_2) & \psi_2(x_2) \end{vmatrix}$$

(here $\psi_1(x_3)$ can be treated as the determinant of the first order).

6 The Derivation of Generalized Hartree's Equations

We present here the derivation of equations for the one-electron wave functions that describe individual orbits. We restrict ourselves to the case when a multi-electron function in the configurational space can be expressed in the form of the product of two determinants:

$$\Psi = \Psi_1 \Psi_2 \,, \tag{50}$$

$$\Psi_1 = \begin{vmatrix} \psi_1(x_1) & \psi_2(x_1) & \dots & \psi_q(x_1) \\ \psi_1(x_2) & \psi_2(x_2) & \dots & \psi_q(x_2) \\ \dots & \dots & \dots & \dots \\ \psi_1(x_q) & \psi_2(x_q) & \dots & \psi_q(x_q) \end{vmatrix}, \tag{51}$$

$$\Psi_2 = \begin{vmatrix} \psi_1(x_{q+1}) & \psi_2(x_{q+1}) & \dots & \psi_q(x_{q+1}) \\ \psi_1(x_{q+2}) & \psi_2(x_{q+2}) & \dots & \psi_q(x_{q+2}) \\ \dots & \dots & \dots & \dots \\ \psi_1(x_{q+p}) & \psi_2(x_{q+p}) & \dots & \psi_q(x_{q+p}) \end{vmatrix}. \tag{51*}$$

Here $q + p = N$ is the number of electrons. For definiteness, we put $q > p$. The function Ψ describes an atomic state, in which there are q different orbits, each of p being occupied by two electrons.

Without the loss of generality, we can suppose that the functions

$$\psi_1(x), \psi_2(x), \ldots, \psi_q(x)$$

are normalized and orthogonal to each other. Indeed, if one replaces the first p functions by their linear combinations, the determinants will be multiplied by constants only, but we can choose these linear combinations to be normalized and mutually orthogonal. Further leaving these new p functions unchanged, we can replace other functions $\psi_{p+1}, \ldots, \psi_q$ by such linear combinations of all the functions $\psi_1, \ldots, \psi_q$ that are normalized and orthogonal both with each other and with the first p functions $\psi_1, \ldots, \psi_p$. As a result, all the functions will be normalized and orthogonal. This set of functions is determined up to an arbitrary unitary transformation of the first p and the last $q - p$ functions separately.

To facilitate manipulations with determinants, we introduce the coefficients

$$\varepsilon_{(\alpha)} = \varepsilon_{\alpha_1 \alpha_2 \ldots \alpha_q} , \tag{52}$$

which are equal to $+1$, when all the subscripts $\alpha_1 \alpha_2 \ldots \alpha_q$ are different and present an even permutation of numbers $1, 2, \ldots, q$; equal to -1 for an odd permutation; and equal to zero, when among subscripts there are the same ones. Then the determinant Ψ_1 can be written in the form

$$\Psi_1 = \sum_{(\alpha)} \varepsilon_{\alpha_1 \alpha_2 \ldots \alpha_q} \psi_{\alpha_1}(x_1) \psi_{\alpha_2}(x_2) \ldots \psi_{\alpha_q}(x_q) . \tag{53}$$

The energy operator L has been already written down [see (3)]. We split it into three parts. The first one, L_1, contains terms depending on coordinates $x_1, \ldots, x_q$ only; the second one, L_2, depends on $x_{q+1}, \ldots, x_{q+p}$; the last, L_{12}, contains all the other terms. We have

$$L = L_1 + L_2 + L_{12} , \tag{54}$$

$$L_1 = \sum_{k=1}^{q} H_k + \sum_{i>k=1}^{q} \frac{1}{r_{ik}} , \tag{55}$$

$$L_2 = \sum_{k=q+1}^{q+p} H_k + \sum_{i>k=q+1}^{q+p} \frac{1}{r_{ik}} , \tag{56}$$

$$L_{12} = \sum_{k=i}^{q} \sum_{k=q+1}^{q+p} \frac{1}{r_{ik}}. \tag{57}$$

The variation of the integral

$$\delta I = \int \delta\overline{\Psi}(L - E)\Psi d\tau$$

will be

$$\delta I = \int\int \delta(\overline{\Psi}_1\overline{\Psi}_2)(L_1 + L_2 + L_{12} - E)\Psi_1\Psi_2 d\tau_1 d\tau_2, \tag{58}$$

where

$$d\tau_1 = dx_1 dx_2 \ldots dx_q\,,$$
$$d\tau_2 = dx_{q+1} dx_{q+2} \ldots dx_{q+p}\,.$$

For brevity, we put

$$\begin{aligned} A_1 &= \int \overline{\Psi}_1\Psi_1 d\tau_1, & A_2 &= \int \overline{\Psi}_2\Psi_2 d\tau_2, \\ B_1 &= \int \overline{\Psi}_1 L_1\Psi_1 d\tau_1, & B_2 &= \int \overline{\Psi}_2 L_2\Psi_2 d\tau_2, \\ F_1 &= \int \overline{\Psi}_1 L_{12}\Psi_1 d\tau_1, & F_2 &= \int \overline{\Psi}_2 L_{12}\Psi_2 d\tau_2\,. \end{aligned} \tag{59}$$

The values A_i and B_i are constants, while F_i are functions of coordinates. By means of these notations, the variation can be written as

$$\delta I = \int \delta(\overline{\Psi}_1[A_2(L_1 - E) + B_2 + F_2]\Psi_1 d\tau_1 +$$
$$+ \int \delta(\overline{\Psi}_2\,[A_1(L_2 - E) + B_1 + F_1]\,\Psi_2 d\tau_2\,. \tag{60}$$

Now we must calculate the integrals involved. We start with the following integral:

$$\int (\overline{\Psi}_1\Psi_1) dx_3 dx_4 \ldots dx_q\,.$$

It will be equal to

$$\int (\overline{\Psi}_1\Psi_1) dx_3 dx_4 \ldots dx_q =$$
$$= \sum_{(\alpha)} \sum_{\alpha'} \varepsilon_{(\alpha)}\varepsilon_{(\alpha')}\overline{\psi}_{\alpha_1}(x_1)\overline{\psi}_{\alpha_2}(x_2)\psi_{\alpha_1}(x_1)\psi_{\alpha_2}(x_2)\delta_{\alpha_3\alpha'_3} \ldots \delta_{\alpha_q\alpha'_q}$$

because ψ_i are normalized and orthogonal. Summation over α' gives

$$\int (\overline{\Psi}_1 \Psi_1) dx_3 dx_4 \ldots dx_q = \sum_{\alpha} \varepsilon_{(\alpha)} \overline{\psi}_{\alpha_1}(x_1) \psi_{\alpha_2}(x_2) \begin{vmatrix} \psi_1(x_1) & \ldots & \psi_q(x_1) \\ \psi_1(x_2) & \ldots & \psi_q(x_2) \\ \delta_{\alpha_3 1} & \ldots & \delta_{\alpha_3 q} \\ \ldots & \ldots & \ldots \\ \delta_{\alpha_q 1} & \ldots & \delta_{\alpha_q q} \end{vmatrix} .$$

The expression under the summation sign is invariant with regard to any permutations of subscripts $\alpha_3, ..., \alpha_q$. Therefore, the summation over these subscripts gives $(q-2)!$ of identical terms. Further,

$$\varepsilon_{(\alpha)} \begin{vmatrix} \psi_1(x_1) & \ldots & \psi_q(x_1) \\ \psi_1(x_2) & \ldots & \psi_q(x_2) \\ \delta_{\alpha_3 1} & \ldots & \delta_{\alpha_3 q} \\ \ldots & \ldots & \ldots \\ \delta_{\alpha_q 1} & \ldots & \delta_{\alpha_q q} \end{vmatrix} = \begin{vmatrix} \delta_{11} & \ldots & \delta \\ \ldots & \ldots & \ldots \\ \psi_1(x_1) & \ldots & \psi_q(x_1) \\ \ldots & \ldots & \ldots \\ \psi_1(x_2) & \ldots & \psi_q(x_2) \\ \delta_{\alpha_3 1} & \ldots & \delta_{\alpha_3 q} \\ \ldots & \ldots & \ldots \\ \delta_{q1} & \ldots & \delta_{qq} \end{vmatrix} .$$

Here $\psi_1(x_1) \ldots \psi_q(x_1)$ form a row number α_1 and $\psi_1(x_2) \ldots \psi_q(x_2)$ forms a row number α_2. This determinant is equal to

$$\begin{vmatrix} \psi_{\alpha_1}(x_1) & \psi_{\alpha_2}(x_1) \\ \psi_{\alpha_1}(x_2) & \psi_{\alpha_2}(x_2) \end{vmatrix} .$$

If we denote the summation subscripts by i and k instead of α_1 and α_2, the integral under consideration can be written in the form

$$\int (\overline{\Psi}_1 \Psi_1) dx_3 dx_4 \ldots dx_q =$$
$$= (q-2)! \textstyle\sum_{i,k=1}^{q} \overline{\psi}_i(x_1) \overline{\psi}_k(x_2) [\psi_i(x_1)\psi_k(x_2) - \psi_k(x_1)\psi_i(x_2)] . \quad (61)$$

We put

$$\varrho_1(x_1, x_2) = \sum_{i=1}^{q} \overline{\psi}_i(x_1) \psi_i(x_2) = \overline{\varrho_1(x_1, x_2)}, \qquad \varrho_1(x) = \varrho_1(x, x) . \quad (62)$$

Then

$$\int (\overline{\Psi}_1 \Psi_1) dx_3 dx_4 \ldots dx_q = (q-2)! [\varrho_1(x_1)\varrho_1(x_2) - |\varrho_1(x_1, x_2)|^2] . \quad (63)$$

Now, integrals (59) can be easily calculated by means of (63). We have

$$\int \varrho_1(x)dx = q, \qquad \int |\varrho_1(x_1, x_2)|^2 dx_2 = \varrho_1(x_1), \tag{64}$$

whence

$$\int (\overline{\Psi}_1 \Psi_1) dx_2 \dots dx_q = (q-1)! \varrho_1(x_1), \tag{65}$$

$$A_1 = \int \overline{\Psi}_1 \Psi_1 dx_1 \dots dx_q = q! \,. \tag{66}$$

The integral B_1 is equal to

$$B_1 = \int \overline{\Psi}_1 L_1 \Psi_1 d\tau = \sum_{k=1}^{q} \int \overline{\Psi}_1 H_k \Psi_1 d\tau_1 + \sum_{i>k=1}^{q} \int \overline{\Psi}_1 \frac{1}{r_{ik}} \Psi_1 d\tau_1 \,.$$

Since the expression of $\overline{\Psi}_1 \Psi_1$ is symmetric with respect to all coordinates $x_1, x_2, \dots, x_q$, all $\frac{1}{2}q(q-1)$ terms of the double sum are equal to one another; likewise, all q terms of the ordinary sum are also equal to one another. Thus,

$$B_1 = q \int \overline{\Psi}_1 H_1 \Psi_1 d\tau_1 + \frac{1}{2} q(q-1) \int \overline{\Psi}_1 \frac{1}{r_{12}} \Psi_1 d\tau_1 \,.$$

Integration over $x_2 \dots x_q$ in the first integral and integration over $x_3 \dots x_q$ in the second integral are performed by means of the above expressions. As a result, we obtain

$$B_1 = q! E_1, \tag{67}$$

where

$$E_1 = \sum_{k=1}^{q} H_{kk} + \frac{1}{2} \iint \frac{\varrho_1(x_1)\varrho_1(x_2) - |\varrho_1(x_1, x_2)|^2}{r_{12}} dx_1 dx_2 \,; \tag{68}$$

here, in accordance with (17),

$$H_{ik} = \int \overline{\psi}_i(x) H \psi_k(x) dx.$$

The integral F_1 can be written in the form

$$F_1 = q! \sum_{k=q+1}^{q+p} V^{(1)}(x_k) \,, \tag{69}$$

where

$$V^{(1)}(x_k(= \frac{1}{q!}\sum_{i=1}^{q}\int \frac{\overline{\Psi}_1\Psi_1}{r_{ik}}d\tau_1 . \tag{70}$$

In the latter sum, all terms are equal to one another. Applying relation (65), we obtain

$$V^{(1)}(x_k) = \int \frac{\varrho_1(x_1)}{r_{ik}}dx_1 = \sum_{i=1}^{q} G_{ii}(x_k) , \tag{71}$$

where

$$G_{ik}(x) = \int \frac{\overline{\psi}_i(x')\psi_k(x')}{r}dx' . \tag{72}$$

The values of A_2, B_2 and F_2 can be obtained from A_1, B_1 and F_1 without calculations. Analogously to (62), (68) and (71), we put

$$\varrho_2(x_1, x_2) = \sum_{i=1}^{p}\overline{\psi}_i(x_1)\psi_i(x_2) = \overline{\varrho_2(x_2, x_1)} ,$$

$$\varrho_2(x) = \varrho_2(x, x) , \tag{62*}$$

$$E_2 = \sum_{k=1}^{p} H_{kk} + \frac{1}{2}\iint \frac{\varrho_2(x_1)\varrho_2(x_2) - |\varrho_2(x_1, x_2)|^2}{r_{12}}dx_1dx_2 , \tag{68*}$$

$$V^{(2)}(x_k) = \int \frac{\varrho_2(x_1)}{r_{ik}}dx_1 = \sum_{i=1}^{p} G_{ii}(x_k), \tag{71*}$$

and then obtain

$$A_2 = p!, \qquad B_2 = p!E_2, \qquad F_2 = p!\sum_{k=1}^{q} V^{(2)}(x_k) . \tag{73}$$

Substitution of the values of the sought integrals (59) into expression (60) leads to

$$\delta I = p!\int \delta\overline{\psi}_1 \left[L_1 - E + E_2 + \sum_{k=1}^{q} V^{(2)}(x_k)\right] \psi_1 d\tau_1 +$$

$$+ q!\int \delta\overline{\psi}_2 \left[L_2 - E + E_1 + \sum_{k=1}^{q} V^{(1)}(x_k)\right] \psi_2 d\tau_2 . \tag{74}$$

The calculations of the integrals in this expression for δI can be performed analogously to the above calculations by expanding the determinants in accordance with (53). Here we present only the results. We put

$$E_{12} = \iint \frac{\varrho_1(x_1)\varrho_2(x_2)}{r_{12}} dx_1 dx_2 \tag{75}$$

and introduce the values

$$V_{ki}^{(s)} = \int \overline{\psi}_k(x) V^{(s)}(x) \psi_i(x) dx = \iint \overline{\psi}_k(x_1)\psi_i(x_1) \frac{\varrho_s(x_2)}{r_{12}} dx_1 dx_2,$$

$$T_{ki}^{(s)} = \iint \overline{\psi}_k(x_2)\psi_i(x_1) \frac{\varrho_s(x_1, x_2)}{r_{12}} dx_1 dx_2 \,, \tag{76}$$

$$U_{ki}^{(s)} = V_{ki}^{(s)} - T_{ki}^{(s)}, \qquad s = 1, 2 \,.$$

The values $V_{ki}^{(s)}$, $T_{ki}^{(s)}$ and $U_{ki}^{(s)}$ are the definite (positive) Hermitian matrices. The values $V_{ki}^{(s)}$ and $T_{ki}^{(s)}$ can be expressed by the matrix elements $(ik|G|lm)$, which have been introduced in Section 3 [see (18)], in the following manner:

$$V_{ki}^{(1)} = \sum_{l=1}^{q} \langle kl \mid G \mid il \rangle, \quad V_{ki}^{(2)} = \sum_{l=1}^{p} \langle kl \mid G \mid il \rangle,$$

$$T_{ki}^{(1)} = \sum_{l=1}^{q} \langle kl \mid G \mid li \rangle, \quad T_{ki}^{(2)} = \sum_{l=1}^{q} \langle kl \mid G \mid li \rangle \,. \tag{77}$$

In turn, the values E_1, E_2 and E_{12} are expressed by $V_{ki}^{(s)}$ and $U_{ki}^{(s)}$:

$$E_1 = \sum_{k=1}^{q} \left(H_{kk} + \frac{1}{2} U_{kk}^{(1)} \right),$$

$$E_2 = \sum_{k=1}^{q} \left(H_{kk} + \frac{1}{2} U_{kk}^{(2)} \right), \tag{78}$$

$$E_{12} = \sum_{k=1}^{q} V_{kk}^{(2)} = \sum_{k=1}^{p} V_{kk}^{(1)} \,.$$

Now the integrals in (74) can be written in the form

$$\int \delta \overline{\Psi}_1 \Psi_1 \partial \tau_1 = q! \sum_{i=1}^{q} \int \delta \overline{\psi}_i(x) \psi_i(x) dx \,, \tag{79}$$

$$\int \delta\overline{\Psi}_1 \sum_{k=1}^{q} (H_k + V^{(2)}(x_k)\Psi_1 d\tau_1 =$$

$$= q! \sum_{i=1}^{q} \int \delta\overline{\psi}_1 \left[H_{kk} + V^{(2)}(x) \sum_{k=1}^{q} H_{kk} + E_{12} \right] \psi_i(x)dx -$$

$$-q! \sum_{i=1}^{q} \int \delta\overline{\psi}_i(x) \sum_{k=1}^{q} (H_{ki} + V_{ki}^{(2)})\psi_k(x)dx\,, \tag{80}$$

$$\sum_{i>k+1}^{q} \int \delta\overline{\Psi}_1 \frac{1}{r_{ik}} \Psi_1 d\tau_1 =$$

$$= q! \sum_{i=1}^{q} \int \delta\overline{\psi}_1 \left[V^{(1)}(x) + E_1 - \sum_{k=1}^{q} H_{kk} \right] \psi_i(x)dx -$$

$$-q! \sum_{i=1}^{q} \int \delta\overline{\psi}_i(x) \sum_{k=1}^{q} \left[G_{ki}(x) + U_{ki}^{(1)} \right] \psi_k(x)dx\,. \tag{81}$$

Substituting these expressions into (74) gives

$$\frac{1}{p!q!}\delta I =$$

$$= \sum_{i=1}^{q} \int \delta\overline{\psi}_i(x) \Big[H + V^{(1)}(x) + V^{(2)}(x) + E_1 + E_2 + E_{12} - E \Big] \psi_i(x)dx$$

$$- \sum_{i=1}^{q} \int \delta\overline{\psi}_i(x) \sum_{k=1}^{q} \left[G_{ki}(x) + H_{ki} + V_{ki}^{(2)} + U_{ki}^{(1)} \right] \psi_k(x)dx$$

$$+ \sum_{i=1}^{p} \int \delta\overline{\psi}_i(x) \Big[H + V^{(1)}(x) + V^{(2)}(x) + E_1 + E_2 + E_{12} - E \Big] \psi_k(x)dx$$

$$- \sum_{i=1}^{p} \int \delta\overline{\psi}_i(x) \sum_{k=1}^{p} \left[G_{ki}(x) + H_{ki} + V_{ki}^{(1)} + U_{ki}^{(2)} \right] \psi_k(x)dx\,. \tag{82}$$

Here one needs to put zero the coefficients at independent variations $\delta\overline{\psi}_i(x)$, keeping in mind that in the expression of δI for $i = 1, 2, \ldots, p$ the values of $\delta\overline{\psi}_i(x)$ are encountered twice, while for $i = p+1, \ldots, q$ only once.

For $i = 1, 2, \ldots, p$ we get

$$2[H + V(x) + E_1 + E_2 + E_{12} - E]\psi_i(x) -$$

$$-\sum_{k=1}^{p}[2G_{ik}(x)+2H_{ki}+U_{ki}+V_{ki}]\psi_k(x)-$$

$$-\sum_{k=p+1}^{q}[G_{ik}(x)+H_{ki}+U_{ki}^{(1)}+V_{ki}^{(2)}]\psi_k(x)=0\,, \tag{83}$$

while for $i=p+1,\ldots,q$

$$[H+V(x)+E_1+E_2+E_{12}-E]\psi_i(x)-$$

$$-\sum_{k=1}^{q}[G_{ki}(x)+H_{ki}+U_{ki}^{(1)}+V_{ki}^{(2)}]\psi_k(x)=0\,. \tag{84}$$

Here for brevity we assume

$$V(x)=V^{(1)}(x)+V^{(2)}(x)\,, \tag{85}$$

$$V_{ki}=V_{ki}^{(1)}+V_{ki}^{(2)} \qquad U_{ki}=U_{ki}^{(1)}+U_{ki}^{(2)}\,. \tag{86}$$

Let us multiply equality (83) by $\overline{\psi}_l(x)$ and integrate it over x. If we take into account (70) and (77) as well as the orthogonality of functions $\psi_i(x)$, then for $l\neq i$ we obtain an identity, while for $l=i$ we get the relation

$$E=E_1+E_2+E_{12}. \tag{87}$$

We should get the same result if we multiply equality (83) by $\overline{\psi}_l(x)$ $(l=p+1,\ldots,q)$ and then integrate it.

However if we multiply (83) by one of the functions $\overline{\psi}_{p+1}(x)$, ..., $\overline{\psi}_q(x)$ or (84) by $\overline{\psi}_1(x)$, ..., $\overline{\psi}_p(x)$, we obtain

$$H_{ki}+V_{ki}^{(1)}+U_{ki}^{(2)}. \tag{88}$$

It can be verified that now as well as for the case of Hartree's equation the relation

$$E=\frac{\int\overline{\Psi}L\Psi d\tau}{\int\overline{\Psi}\Psi d\tau} \tag{89}$$

is fulfilled.

7 Analysis of the Equations and a New Formulation of the Variational Principle

Let us consider the system of equations (83) and (84). We have

$$\begin{aligned} \lambda_{ki} &= 2H_{ki} + U_{ki} + V_{ki} \qquad (i \quad \text{or} \quad k = \quad 1,2,\ldots,p), \\ \lambda_{ki} &= H_{ki} + U_{ki}^{(1)} + V_{ki}^{(2)} \qquad (i \quad \text{or} \quad k = p+1,2,\ldots,q). \end{aligned} \tag{90}$$

$$\begin{aligned} \varepsilon_k &= 2 \qquad (k = 1,2,\ldots,p), \\ \varepsilon_k &= 1 \qquad (k = p+1,\ldots,q). \end{aligned} \tag{91}$$

(For $i = 1,2,\ldots p$ and $k = p+1,\ldots q$, both determinations of λ_{ki} coincide due to relation (88).)

Then equations (83) and (84) can be written in the form

$$\begin{aligned} 2[H + V(x)]\psi_i(x) - \sum_{k=1}^{q} \varepsilon_k G_{ki}(x)\psi_k(x) &= \sum_{k+1}^{q} \lambda_{ki}\psi_k(x) \quad i = 1,\ldots,p, \\ [H + V(x)]\,\psi_i(x) - \sum_{k=1}^{q} G_{ki}(x)\psi_k(x) &= \sum_{k+1}^{q} \lambda_{ki}\psi_k(x) \quad i = p+1,\ldots,q. \end{aligned} \tag{92}$$

If one omits all terms for which $i \neq k$, then one obtains the Hartree equations (29). Thus the terms with different subscripts represent the quantum exchange, which is not taken into account in the Hartree equations. Generally speaking, these non-diagonal terms ($i \neq k$) are small as compared with the diagonal ones ($i = k$).

The characteristic feature of Hartree's theory, that the effect of an electron on itself $G_{ii}(x)$ is subtracted from the total potential energy $V(x)$, is also reproduced here.

Like (29) our equations (92) can be derived from the three-dimensional variational principle. The varied integral is of the form

$$\begin{aligned} W = \sum_{i=1}^{q} \int \overline{\psi}_i H\psi_i + \frac{1}{2} \iint \frac{\varrho_1(x)\varrho_1(x') - |\varrho_1(x,x')|^2}{r} dx dx' + \\ + \sum_{i=1}^{p} \int \overline{\psi}_i H\psi_i + \frac{1}{2} \iint \frac{\varrho_2(x)\varrho_2(x') - |\varrho_2(x,x')|^2}{r} dx dx' + \\ + \iint \frac{\varrho_1(x)\varrho_2(x')}{r} dx dx'\,. \end{aligned} \tag{93}$$

Let us take a variation of this integral under additional conditions

$$\int \overline{\psi}_i(x)\psi_k(x) = \delta_{ik}\,. \tag{94}$$

The coefficients λ_{ik} forming a Hermitian matrix play the role of the Lagrange arbitrary multipliers. Expression (93) represents the atomic energy because it reads

$$W = E_1 + E_2 + E_{12} = E\,. \tag{95}$$

Just as we have proceeded above in Section 4, we can treat coefficients in equations (92) as given. Then these equations will represent a self-adjoint linear system. This system can be obtained by a variation of the integral

$$W^* = \int \left[\sum_{i=1}^{q} \overline{\psi}_i (H+V)\psi_i - \sum_{i,k=1}^{q} G_{ki}\overline{\psi}_i\psi_k \right] dx +$$

$$+ \int \left[\sum_{i=1}^{p} \overline{\psi}_i (H+V)\psi_i - \sum_{i,k=1}^{p} G_{ki}\overline{\psi}_i\psi_k \right] dx \tag{96}$$

under the previous additional conditions (94).

Now the value of the varied integral is not equal to an energy, but is equal to

$$W^* = 2E - \sum_{i=1}^{p} H_{ii} - \sum_{i=1}^{q} H_{ii} = \sum_{i=1}^{q} \lambda_{ii}\,. \tag{97}$$

Our results, in particular, expression (93) for the energy, can be interpreted as follows. The atomic electrons are divided into two groups, having q and p electrons, correspondingly ($q + p = N$). The electrons of the same group obey the Pauli principle in the narrow sense (i.e., without taking a spin into account). In expression (93) for the energy, the first double integral over volume represents the interaction energy of electrons of the first group; the second integral is that for the second group; the last double integral represents the mutual potential energy of both groups. Note that in the first two integrals the integrands tend to zero when $x = x'$. Our expressions for the energy are much similar to those obtained by Jordan by means of the second quantization method.[11]

[11] See, e.g., a review by Jordan in Phys. Zs. **30**, 700, (1929). (*V. Fock*)

The numerical solution of our system of equations can be done by the method of successive approximations.

If we assume the spherical symmetry, the equations can be considerably simplified.[12] The solution of our equations for the case of the spherical symmetry is, apparently, not more complicated than that of the corresponding Hartree's equations, but its result has to be much more precise.

8 The Spectral Line Intensities

Having obtained the wave functions ψ_i, one can calculate the frequencies and intensity of the spectral lines. The energy levels can be calculated by means of formula (93). It should be taken into account that this formula gives not the term but the total atomic energy. To obtain the term, one has to calculate the difference between the atomic and ion energy.

To obtain the intensity, one needs to calculate the integral

$$\langle E \mid f \mid E' \rangle = \frac{1}{p!q!} \int \overline{\Psi} \sum_{i=k}^{N} f(x_k) \Psi' d\tau \,, \tag{98}$$

where

$$f(x_k) = f(x_k \,, y_k \,, z_k) \,.$$

We denote the basic functions ψ_i for the level E in more detail as $\psi_i(x, E)$. The total wave function for the entire atom in state E is denoted as before by Ψ and the same function for state E' by Ψ'. If, for brevity, we put

$$\begin{aligned} a_{ik} &= \int \overline{\psi}_i(x, E) \psi_k(x, E') dx, \\ f_{ik} &= \int \overline{\psi}_i(x, E) f(x) \psi_k(x, E') dx \,, \end{aligned} \tag{99}$$

the value of $(E|f|E')$ will be equal to

$$\langle E \mid f \mid E' \rangle = \begin{vmatrix} a_{11} & \dots & a_{1p} \\ a_{21} & \dots & a_{2p} \\ \dots & \dots & \dots \\ a_{p1} & \dots & a_{pp} \end{vmatrix} \cdot \sum_{k=1}^{q} \begin{vmatrix} a_{11} & \dots & a_{1q} \\ \dots & \dots & \dots \\ f_{k1} & \dots & f_{kq} \\ \dots & \dots & \dots \\ a_{q1} & \dots & a_{qq} \end{vmatrix} +$$

[12] The detailed derivation of the equations with the spherical symmetry for the sodium atom ($N = 11$) will be given in a separate paper. (*V. Fock*) (See [30-3] in this book. (*Editors*))

$$+\begin{vmatrix} a_{11} & \dots & a_{1q} \\ a_{21} & \dots & a_{2q} \\ \dots & \dots & \dots \\ a_{q1} & \dots & a_{qq} \end{vmatrix} \cdot \sum_{k=1}^{p} \begin{vmatrix} a_{11} & \dots & a_{1p} \\ \dots & \dots & \dots \\ f_{k1} & \dots & f_{kp} \\ \dots & \dots & \dots \\ a_{p1} & \dots & a_{pp} \end{vmatrix} . \tag{100}$$

Here the entries $f_{k1}, f_{k2} \dots$ form the k-th row. We shall show that in the case of a single valence electron this expression for the matrix element is approximately equal to an ordinary expression obtained by solution of the one-electron Schrödinger equation.

In this case, one needs to put $q = p + 1$. The functions $\psi_i(x, E)$ for subscripts i equal to $1, 2, \dots, p$ differ only slightly from the corresponding functions $\psi_i(x, E')$ because the transition of the valence electron to another orbit affects the internal electron relatively weakly. Whence it follows that for $i, k = 1, 2, \dots, p$ the coefficients are close to δ_{ik}. The wave function $\psi_q(x, E)$ of the valence electron will be approximately orthogonal not only to $\psi_i(x, E')$ for $i = 1, 2, \dots, p$, but to the function $\psi_q(x, E')$ of the valence electron in state E'. In fact, if one supposes the field to be unchanged by the core electrons, then $\psi_q(x, E)$ and $\psi_q(x, E')$ will be approximately equal to the eigenfunctions of the Schrödinger energy operator for the valence electron with different eigenvalues. Whence it follows that the coefficients a_{qq}, as well as $a_{1q}, \dots, a_{q-1,q}$ and $a_{q,1}, \dots, a_{q,q-1}$, will be small. The single term in the expression for $\langle E \mid f \mid E' \rangle$, not containing small factors, is equal to

$$f_{qq} = \int \overline{\psi}_q(x, E) f(x) \psi_q(x, E') dx \, , \tag{101}$$

i.e., to usual expression for the matrix entry.

In some cases the deviation of $\langle E | f | E' \rangle$ from its approximate value (101) can be noticeable enough. This deviation can be attributed to the rearrangement of the core electrons.

Translated by E.D. Trifonov

30-3
Application of the Generalized Hartree Method to the Sodium Atom

V.A. Fock

Zs. Phys. **62**, N11–12, 795, 1930
TOI **5**, N51, 29, 1931

1 A Form of Wave Functions

The number of electrons in a sodium atom is equal to 11; one of them is a valence electron. Ten inner electrons are distributed in pairs at five orbitals. Thus, we have six orbitals, and, hence, six different wave functions. In the equations of the previous article, it should be taken

$$N = 11, \ p = 5, \ q = 6.$$

Let us denote a wave function of a valence electron by ψ_6, and wave functions of inner electrons by $\psi_1, \ldots, \psi_5$. In the case of spherical symmetry, these functions should have the form

$$\left.\begin{aligned}
\psi_1 &= \frac{1}{r} f_1(r) \frac{1}{\sqrt{4\pi}}, \\
\psi_2 &= \frac{1}{r} f_2(r) \frac{1}{\sqrt{4\pi}}, \\
\psi_3 &= \frac{1}{r} f_3(r) \frac{\sqrt{3}}{\sqrt{4\pi}} \cos\vartheta, \\
\psi_4 &= \frac{1}{r} f_3(r) \frac{\sqrt{3}}{\sqrt{4\pi}} \sin\vartheta \cos\varphi, \\
\psi_5 &= \frac{1}{r} f_3(r) \frac{\sqrt{3}}{\sqrt{4\pi}} \sin\vartheta \sin\varphi, \\
\psi_6 &= \frac{1}{r} f_4(r) \frac{1}{\sqrt{4\pi}} Y_l(\vartheta, \varphi).
\end{aligned}\right\} \tag{1}$$

The spherical function $Y_l(\vartheta, \varphi)$ is normalized by the following condition:

$$\frac{1}{4\pi} \iint | Y_l(\vartheta, \varphi) |^2 \sin\vartheta \, d\vartheta \, d\varphi = 1 \, . \tag{2}$$

The normalized condition for $f_i(r)$ is

$$\int_0^\infty [f_i(r)]^2\,dr = 1 \qquad (i = 1, 2, 3, 4). \tag{3}$$

In order for the functions $\psi_1, \ldots, \psi_5$ to be orthogonal, $f_i(r)$ should obey the following conditions:

$$\left.\begin{aligned} &\int_0^\infty f_1(r) f_2(r) dr = 0; \qquad &\delta_{l0}\cdot\int_0^\infty f_1(r) f_4(r) dr = 0;\\ \delta_{l0}\cdot&\int_0^\infty f_2(r) f_4(r) dr = 0; \qquad &\delta_{l1}\cdot\int_0^\infty f_3(r) f_4(r) dr = 0. \end{aligned}\right\} \tag{4}$$

The functions f_1, f_2, f_3, f_4 correspond to different electronic shells. For simplicity we denoted them by one subscript only; more complete notation would be $f_{10}, f_{20}, f_{21}, f_{nl}$, where two subscripts are nothing else but the quantum numbers of the electron shell under consideration.

2 Expressions for the Energy of an Atom

To obtain the equations, we will vary the expression for the energy of an atom derived in the previous article [formula (93)]. In this formula we should carry out integration over the variables ϑ and φ. Let us denote a sum of simple volume integrals by W_1 and a sum of double volume integrals by W_2, so that

$$W = W_1 + W_2. \tag{5}$$

Evaluation of W_1 is quite trivial; as the result, one gets

$$\begin{aligned} W_1 = \int_0^\infty \Bigg[&\left(\frac{df_1(r)}{dr}\right)^2 + \left(\frac{df_2(r)}{dr}\right)^2 + 3\left(\frac{df_3(r)}{dr}\right)^2 + \frac{1}{2}\left(\frac{df_4(r)}{dr}\right)^2 + \\ &+ \frac{6}{r^2} f_3^2 + \frac{l(l+1)}{2r^2} f_4^2 \Bigg] dr - \int_0^\infty \left(2f_1^2 + 2f_2^2 + 6f_3^2 + f_4^2\right) \frac{11}{r} dr. \end{aligned} \tag{6}$$

In this expression the first integral represents an average kinetic energy, and the second integral gives the average potential energy of electrons with respect to the nucleus (without the energy of interaction).

Evaluation of the interaction energy W_2 is more complicated; it can be carried out as follows. Let us suppose in formulae (62) and (62*) of the previous article

$$\left.\begin{aligned} \varrho_2(x,x') &= \varrho(\mathbf{r},\mathbf{r}'); \qquad & \varrho_1(x,x') &= \varrho(\mathbf{r},\mathbf{r}') + \sigma(\mathbf{r},\mathbf{r}'), \\ \varrho(\mathbf{r},\mathbf{r}) &= \varrho(\mathbf{r}); & \sigma(\mathbf{r},\mathbf{r}) &= \sigma(\mathbf{r}), \end{aligned}\right\} \tag{7}$$

where

$$\varrho(\mathbf{r},\mathbf{r}') \text{ and } \sigma(\mathbf{r},\mathbf{r}')$$

have the following values:

$$\left.\begin{aligned} \varrho(\mathbf{r},\mathbf{r}') &= \frac{1}{4\pi r r'}(f_1 f_1' + f_2 f_2' + 3 f_3 f_3' \cos\gamma), \\ \sigma(\mathbf{r},\mathbf{r}') &= \frac{1}{4\pi r r'} f_4 f_4' \,\overline{Y_l(\vartheta,\varphi)}\, Y_l(\vartheta',\varphi'). \end{aligned}\right\} \tag{8}$$

Here we assumed for brevity that

$$f_i = f_i(r); \qquad f_i' = f_i(r')$$

and

$$\cos\gamma = \cos\vartheta\cos\vartheta' + \sin\vartheta\sin\vartheta'\cos(\varphi-\varphi'). \tag{9}$$

We also suppose that

$$d\omega = \sin\vartheta\, d\vartheta\, d\varphi; \qquad d\omega' = \sin\vartheta'\, d\vartheta'\, d\varphi'. \tag{10}$$

The interaction energy W_2 will have the form

$$\begin{aligned} W_2 = \iint \big[2\varrho(\mathbf{r})\,\varrho(\mathbf{r}') - |\,\varrho(\mathbf{r},\mathbf{r}')\,|^2 + 2\varrho(\mathbf{r})\,\sigma(\mathbf{r}') - \\ - \varrho(\mathbf{r},\mathbf{r}')\,\sigma(\mathbf{r}',\mathbf{r})\big] \frac{r^2\, dr\, d\omega\, r'^2\, dr'\, d\omega'}{\sqrt{r^2 + r'^2 - 2rr'\cos\gamma}}. \end{aligned} \tag{11}$$

On account of

$$\sigma(\mathbf{r})\sigma(\mathbf{r}') - |\,\sigma(\mathbf{r},\mathbf{r}')\,|^2 = 0$$

the integrand does not contain terms quadratic with respect to σ, which greatly simplifies the calculations. Evaluating the integral, we will use the known expansion

$$\frac{1}{\sqrt{r^2 + r'^2 - 2rr'\cos\gamma}} = \sum_{n=0}^{\infty}(2n+1)K_n(r,r')P_n(\cos\gamma), \tag{12}$$

which assumed that

$$\left.\begin{aligned} K_n(r,r') &= \frac{1}{2n+1}\frac{r'^n}{r^{n+1}} \qquad \text{for } r' \le r, \\ K_n(r,r') &= \frac{1}{2n+1}\frac{r^n}{r'^{n+1}} \qquad \text{for } r' \ge r, \end{aligned}\right\} \tag{13}$$

as well as the integral property of the spherical functions

$$\frac{2n+1}{4\pi}\int Y_l(\vartheta',\varphi')\,P_n(\cos\gamma)\,d\omega' = \delta_{nl}\,Y_n(\vartheta,\varphi)\,. \tag{14}$$

Here we will write down the evaluation for only one term of expression (11) for W_2, namely, for

$$J = \iint \varrho(\mathbf{r},\mathbf{r}')\sigma(\mathbf{r}',\mathbf{r})\frac{r^2\,dr\,d\omega\,r'^2\,dr'\,d\omega'}{\sqrt{r^2+r'^2-2rr'\cos\gamma}}\,.$$

Let us write this integral in more detail:

$$J = \frac{1}{(4\pi)^2}\iint f_4f_4'\,dr\,dr'\cdot\iint[f_1f_1'+f_2f_2'+3f_3f_3'\cos\gamma]\cdot$$

$$\cdot\overline{Y_l(\vartheta,\varphi)}\,Y_l(\vartheta',\varphi')\frac{d\omega\,d\omega'}{\sqrt{r^2+r'^2-2rr'\cos\gamma}}\,.$$

Using the known relation between the spherical functions

$$(2n{+}1)\;x\;P_n(x) = (n{+}1)P_{n+1}(x) + n\,P_{n-1}(x)\,,$$

one gets

$$\frac{f_1f_1'+f_2f_2'+3f_3f_3'\cos\gamma}{\sqrt{r^2+r'^2-2rr'\cos\gamma}} = \sum_{n=0}^{\infty}\Big\{(2n{+}1)(f_1f_1'+f_2f_2')K_n+$$

$$+3f_3f_3'\,[nK_{n-1}+(n{+}1)K_{n+1}]\Big\}\cdot P_n(\cos\gamma)\,.$$

Integration over $d\omega'$ and $d\omega$ gives

$$J = \iint f_4f_4'\Big\{(f_1f_1'+f_2f_2')K_l+$$

$$+3f_3f_3'\left[\frac{l}{2l{+}1}K_{l-1}+\frac{l{+}1}{2l{+}1}K_{l+1}\right]\Big\}dr\,dr'.$$

In order to carry out integration over r', let us suppose

$$F_l^{ik}(r) = \int_0^\infty f_i(r')\, f_k(r')\, K_l(r,r')\, dr' ; \tag{15}$$

then we will finally get

$$J = \int_0^\infty \left\{ f_1 f_4\, F_l^{14} + f_2 f_4\, F_l^{24} + 3 f_3 f_4 \left[\frac{l}{2l+1} F_{l-1}^{34} + \frac{l+1}{2l+1} F_{l+1}^{34} \right] \right\} dr \, .$$

The other integrals in expression (11) for W_2 can be evaluated in the same manner. As the result, one has

$$\begin{aligned} W_2 = \int_0^\infty \Big\{ & \frac{1}{2}(2f_1^2 + 2f_2^2 + 6f_3^2 + f_4^2)(2F_0^{11} + 2F_0^{22} + 6F_0^{33} + F_0^{44}) - \\ & - f_1^2 F_0^{11} - f_2^2 F_0^{22} - 3f_3^2(F_0^{33} + 2F_2^{33}) - \frac{1}{2} f_4^2 F_0^{44} - \\ & - 2f_1 f_2 F_0^{12} - 6 f_1 f_2 F_1^{13} - f_1 f_4 F_l^{14} - 6 f_2 f_3 F_1^{23} - \\ & - f_2 f_4 F_l^{24} - 3 f_3 f_4 \left[\frac{l}{2l+1} F_{l-1}^{34} + \frac{l+1}{2l+1} F_{l+1}^{34} \right] \Big\} dr \, . \end{aligned} \tag{16}$$

In this formula the terms containing products of different functions $f_i(r)$ (the last two lines) represent the energy of quantum exchange.[1]

3 Variational Equations

To derive variational equations, we should vary the energy $W = W_1 + W_2$ under additional conditions (3) and (4) ensued from orthogonalization and normalization of the wave functions. Constructing the variation of W_2, we should take into account that the coefficients $F_l^{ik}(r)$ also depend on the varied functions. We can distinguish between two kinds of W_2 variation: total variation δW_2, when both $f_i(r)$ and $F_l^{ik}(r)$ are varied, and partial variation $\delta^* W_2$, when only $f_i(r)$ is varied. It is not difficult to see that total and partial variations are related as follows:

$$\delta W_2 = \delta^* W_2 \, . \tag{17}$$

[1] According to preliminary calculations these terms are about 3% of the total value W_2 for the ground state of sodium. (*V. Fock*)

This remark simplifies constructing of the equations because a partial variation is evaluated easier than the total one.

The variational equations have the form:

$$-\frac{d^2f_1}{dr^2}+2\left(-\frac{11}{r}+F_0^{11}+2F_0^{22}+6F_0^{33}+F_0^{44}\right)f_1- \\ -2F_0^{21}f_2-6F_1^{31}f_3-F_l^{41}f_4=\lambda_{11}f_1+\lambda_{21}f_2+\lambda_{41}\delta_{l0}f_4\,, \tag{18}$$

$$-\frac{d^2f_2}{dr^2}+2\left(-\frac{11}{r}+2F_0^{11}+F_0^{22}+6F_0^{33}+F_0^{44}\right)f_2- \\ -2F_0^{12}f_1-6F_1^{32}f_3-F_l^{42}f_4=\lambda_{12}f_1+\lambda_{22}f_2+\lambda_{42}\delta_{l0}f_4\,, \tag{19}$$

$$-3\frac{d^2f_3}{dr^2}+6\left(\frac{1}{r^2}-\frac{11}{r}+2F_0^{11}+2F_0^{22}+5F_0^{33}-2F_2^{33}+F_0^{44}\right)f_3 \\ -6F_1^{13}f_1-6F_1^{23}f_2-3\left(\frac{l+1}{2l+1}F_{l+1}^{34}+\frac{l}{2l+1}F_{l-1}^{34}\right)f_4= \\ =\lambda_{33}f_3+\lambda_{43}\delta_{l1}f_4\,, \tag{20}$$

$$-\frac{1}{2}\frac{d^2f_4}{dr^2}+\left[\frac{l(l+1)}{2r^2}-\frac{11}{r}+2F_0^{11}+2F_0^{22}+6F_0^{33}\right]f_4- \\ -F_l^{14}f_1-F_l^{24}f_2-3\left[\frac{l+1}{2l+1}F_{l+1}^{34}+\frac{l}{2l+1}F_{l-1}^{34}\right]f_3= \\ =\lambda_{14}\delta_{l0}f_1+\lambda_{24}\delta_{l0}f_2+\lambda_{34}\delta_{l1}f_3+\lambda_{44}f_4\,. \tag{21}$$

The last of these equations is the wave equation for a valence electron, and the first three are the ones for inner electrons. Constants $\lambda_{ik}=\lambda_{ki}$ are nothing else but the Lagrangian factors. Off-diagonal terms (for example, the terms with f_1, f_2, f_3 in the last equation) describe the influence of quantum exchange. Equations (18), (19), (20) are invariant with respect to substitution f_1, f_2, f_3, f_4, keeping the quadratic form $2f_1^2+2f_2^2+6f_3^2+f_4^2$ invariant. We can assume this substitution to be chosen so that

$$\lambda_{ik}=\lambda_i\delta_{ik}\,.$$

4 Properties of the Coefficients of the Equations

The coefficients $F_l^{ik}(r)$ of our system of equations defined by formula (15) satisfy the differential equation

$$r^2 \frac{d^2 F_l^{ik}}{dr^2} + 2r \frac{dF_l^{ik}}{dr} - l(l+1) F_l^{ik} = -f_i(r)\, f_k(r)\,. \tag{22}$$

The function $F_l^{ik}(r)$ can be determined as a solution of this equation, which remains finite at $r = 0$ and tends to zero at infinity. The value $K_l(r, r')$ is the Green's function of the self-conjugate differential operator in the left-hand side of (22). Numerical integration of the differential equation (22) by the Adams–Störmer method gives a convenient way for calculating the functions $F_l^{ik}(r)$ when $f_i(r)$ are known.

Let us write down expression (15) for $F_l^{ik}(r)$ in more detail:

$$F_l^{ik}(r) = \frac{1}{2l+1} \left(\frac{1}{r^{l+1}} \int_0^r f_i(r') f_k(r') r'^l \, dr' + r^l \int_r^\infty f_i(r') f_k(r') \, \frac{dr'}{r'^{l+1}} \right). \tag{23}$$

As we treat only those functions $f_i(r)$, which belong to a discrete spectrum and rapidly decrease at infinity, we can transform this expression in the following manner:

$$F_l^{ik}(r) = \frac{C_l^{ik}}{r^{l+1}} - R_l^{ik}(r)\,, \tag{24}$$

where it is supposed that

$$C_l^{ik} = \frac{1}{2l+1} \int_0^\infty f_i(r)\, f_k(r)\, r^l dr \tag{25}$$

and

$$R_l^{ik}(r) = \frac{1}{2l+1} \int_r^\infty \left(\frac{r'^l}{r^{l+1}} - \frac{r^l}{r'^{l+1}} \right) f_i(r') f_k(r') dr'\,. \tag{26}$$

This formula allows one to obtain an approximate expression for $F_l^{ik}(r)$ at large values of r. Let us suppose that for sufficiently large values of r (in any case, larger than the largest root of $f_i(r)$) the function $f_i(r)$ approximately equals[2]

$$f_i(r) = M_i \, r^{\alpha_i} \, e^{-\beta r} \left[1 + O\left(\frac{1}{r}\right) \right], \tag{27}$$

[2]On account of the fact that the functions $f_i(r)$ are bound by the system of equations, they have the equal coefficients β in indices. (*V. Fock*)

where the symbol $O\left(\frac{1}{r}\right)$ denotes a value of the order of $\frac{1}{r}$. In this case formula (26) gives the following approximate expression for $R_l^{ik}(r)$:

$$R_l^{ik}(r) = \frac{M_i M_k}{4\beta^2}\, r^{\alpha_i+\alpha_k-2}\, e^{-2\beta r}\left[1 + O\left(\frac{1}{r}\right)\right]. \qquad (28)$$

We see that this expression is notably small as compared with the first term of formula (24) and does not depend on the subscript l in the approximation considered.

Thus, formula (24) can be treated as the asymptotic expression for $F_l^{ik}(r)$, such that the first term gives an approximate value of the function, and $R_l^{ik}(r)$ does the remainder. When $i = k$ and $l = 0$, the constant value C_0^{kk} is equal to unity due to the normalization of $f_k(r)$, so that the functions $F_0^{kk}(r)$ are asymptotically equal to $\frac{1}{r}$, as should be expected, because they are the potential of a unit charge, the density of which decreases rather quickly with the separation from the origin.

5 Calculation of Terms

Numerical solution of the system of equations (18)–(21) can be carried out by means of consequent approximations. After getting the functions $f_i(r)$, as well as $F_l^{ik}(r)$, the energy of an atom $W = W_1 + W_2$ can be found by means of formulae (6) and (16). Herewith, in order to control the calculations it is possible to use the following relation:

$$W_1 + \frac{1}{2}W_2 = \lambda_{11} + \lambda_{22} + \lambda_{33} + \lambda_{44}\,. \qquad (29)$$

Besides, the calculations can be controlled by means of the virial theorem, according to which the double kinetic energy must be equal to the absolute value of the potential energy (including the energy of the electron interaction). As was shown by the author,[3] this relation takes place not only for the exact solution of the Schrödinger equation, but also for an approximate solution obtained by the method described in the present article.

It is necessary to take into account that the total energy W of an atom does not coincide with the value of the term; the term is equal to the difference between the energies of an atom in the present state and

[3]V.A. Fock, *Comment on the Virial Relation*, JRPKhO **62**, N4, 379, 1930. (See [30-1] in this book. (*Editors*))

in the ionization state. In order to get the value of the energy in the ionization state, it is necessary to solve a new system of equations that is derived from the present one, if all functions having the symbol 4 are assumed to equal zero. However, the first-hand calculation of the term as the energy difference is disadvantageous in the sense that the term is obtained as a small difference between two large quantities. In view of this, it is more expedient to do as follows. Let us denote the solutions of the equation system for an ionized atom by $f_i^0(r)$ $(i = 1, 2, 3)$ and for an atom in the present state by $f_i(r)$, and construct the differences (for the first three functions)

$$\delta f_i = f_i(r) - f_i^0(r) \qquad (i = 1, 2, 3)\,. \tag{30}$$

For these differences, it is possible to develop a system of equations that allows one to calculate them directly (i.e., without knowing $f_i(r)$). If $f_i(r)$, as well as $f_4(r)$, are known with sufficient accuracy, it is possible to get the value of the energy difference, i.e., the value of a term, also with the same accuracy.

6 Intensities

Finally we need to obtain the formulae for intensities. For the general case, these formulae have been derived in our first article (formulae (99) and (100)).[4] In them we will make the simplifications that follow from the assumption of a spherical symmetry.

Let us calculate a matrix element for the coordinate $z = r\cos\vartheta$. The matrix entries for $x = r\sin\vartheta\cos\varphi$ and for $y = r\sin\vartheta\sin\varphi$ can be written by analogy.

For convenience, we will present formula (100) of the previous article replacing f_{ik} by z_{ik} in it and assuming $p = 5$, $q = 6$:

$$\langle E \mid z \mid E' \rangle = \begin{vmatrix} a_{11} & \cdots & a_{15} \\ \cdots & \cdots & \cdots \\ \cdots & \cdots & \cdots \\ a_{51} & \cdots & a_{55} \end{vmatrix} \cdot \sum_{k=1}^{6} \begin{vmatrix} a_{11} & \cdots & a_{16} \\ \cdots & \cdots & \cdots \\ z_{k1} & \cdots & z_{k6} \\ \cdots & \cdots & \cdots \\ a_{61} & \cdots & a_{66} \end{vmatrix} +$$

[4]See [30-2] in this book. (*Editors*)

$$+\begin{vmatrix} a_{11} \dots a_{16} \\ \dots\dots\dots \\ \dots\dots\dots \\ a_{61} \dots a_{66} \end{vmatrix} \cdot \sum_{k=1}^{5} \begin{vmatrix} a_{11} \dots a_{15} \\ \dots\dots\dots \\ z_{k1} \dots z_{k5} \\ \dots\dots\dots \\ a_{51} \dots a_{55} \end{vmatrix} . \tag{31}$$

The values a_{ik} and $z_{ik} = f_{ik}$ are written in formula (99) of the previous article. They are equal to

$$a_{ik} = \int \overline{\psi_i}(r,E)\, \psi_k(r,E')\, d\tau\,; \qquad z_{ik} = \int \overline{\psi_i}(r,E)\, z\, \psi_k(r,E')\, d\tau\,. \tag{32}$$

Let us make a table of values a_{ik} and z_{ik} for the case of spherical symmetry under consideration. Because of the orthogonality of the spherical functions, many of these values will be equal to zero, and we will get

$$((a_{ik})) = \begin{Bmatrix} a_{11} & a_{12} & 0 & 0 & 0 & a_{16} \\ a_{21} & a_{22} & 0 & 0 & 0 & a_{26} \\ 0 & 0 & a_{33} & 0 & 0 & a_{36} \\ 0 & 0 & 0 & a_{44} & 0 & a_{46} \\ 0 & 0 & 0 & 0 & a_{55} & a_{56} \\ a_{61} & a_{62} & a_{63} & a_{64} & a_{65} & a_{66} \end{Bmatrix},$$

$$((z_{ik})) = \begin{Bmatrix} 0 & 0 & z_{13} & 0 & 0 & z_{16} \\ 0 & 0 & z_{23} & 0 & 0 & z_{26} \\ z_{31} & z_{32} & 0 & 0 & 0 & z_{36} \\ 0 & 0 & 0 & 0 & 0 & z_{46} \\ 0 & 0 & 0 & 0 & 0 & z_{56} \\ z_{61} & z_{62} & z_{63} & z_{64} & z_{65} & z_{66} \end{Bmatrix}. \tag{33}$$

It is not difficult to see that all determinants in the second sum of formula (31) are equal to zero. In order to calculate the first sum, as well as the factor in front of it, we will introduce the values

$$\beta_{ik} = \int_0^\infty f_i(r,E)\, f_k(r,E')\, dr\,; \qquad \gamma_{ik} = \int_0^\infty f_i(r,E)\, r\, f_k(r,E')\, dr \tag{34}$$

and denote by b_{ik} and c_{ik} the matrix elements

$$((a_{ik})) = \begin{Bmatrix} \beta_{11} & \beta_{12} & 0 & \delta_{0l'}\beta_{14} \\ \beta_{21} & \beta_{22} & 0 & \delta_{0l'}\beta_{24} \\ 0 & 0 & \beta_{33} & \delta_{1l'}\beta_{34} \\ \delta_{l0}\beta_{41} & \delta_{l0}\beta_{42} & \delta_{l1}\beta_{43} & \beta_{44} \end{Bmatrix}, \tag{35}$$

$$((c_{ik})) = \left\{ \begin{matrix} 0 & 0 & \gamma_{13} & \gamma_{14} \\ 0 & 0 & \gamma_{23} & \gamma_{24} \\ \gamma_{31} & \gamma_{32} & \gamma_{33} & \gamma_{34} \\ \gamma_{41} & \gamma_{42} & \gamma_{43} & \gamma_{44} \end{matrix} \right\}. \tag{36}$$

Then the matrix element $\langle E|z|E'\rangle$ corresponding to the transition from the level E to the level E' will be equal to

$$\langle E \mid z \mid E'\rangle = C \cdot \frac{1}{4\pi} \iint \overline{Y_l}\, Y_{l'} \cos\vartheta \, \sin\vartheta \, d\vartheta \, d\varphi \,, \tag{37}$$

where

$$C = (b_{33})^5 \cdot \begin{vmatrix} b_{11} & b_{12} \\ b_{21} & b_{22} \end{vmatrix} \cdot \sum_{k=1}^{4} \begin{vmatrix} b_{11} & \dots & b_{14} \\ \dots & \dots & \dots \\ c_{k1} & \dots & c_{k4} \\ \dots & \dots & \dots \\ b_{41} & \dots & b_{44} \end{vmatrix} . \tag{38}$$

The matrix elements for the coordinates x and y are expressed absolutely analogously. Thus, we will have

$$\left.\begin{aligned} \langle E|x|E'\rangle &= C \cdot \frac{1}{4\pi} \int \overline{Y_l}\, Y_{l'} \, \sin\vartheta \, \cos\varphi \, d\omega, \\ \langle E|y|E'\rangle &= C \cdot \frac{1}{4\pi} \int \overline{Y_l}\, Y_{l'} \, \sin\vartheta \, \sin\varphi \, d\omega, \\ \langle E|z|E'\rangle &= C \cdot \frac{1}{4\pi} \int \overline{Y_l}\, Y_{l'} \, \cos\vartheta \, d\omega. \end{aligned}\right\} \tag{39}$$

These expressions have the same form, as in the usual theory where only a valence electron is treated. The selection rule remains valid without any changes. Here the distinction is only in the factor C, which has a slightly different meaning than in the usual theory, when it is equal to

$$C = c_{44} = \int_0^\infty f_4(r, E) \; r \; f_4(r, E') \; dr \,. \tag{40}$$

In our theory the equality $C = c_{44}$ is only approximate. The exact calculation of the factor C by means of formula (38) has no problems, as many elements in the determinants in this formula are equal to zero.

Translated by A.K. Belyaev

30-4
New Uncertainty Properties of the Electromagnetic Field

P. JORDAN AND V. FOCK

At present in Kharkov

(Received 15 October 1930)

Zs. Phys. **66**, 206, 1930

If we assume that the most precise measurements of electromagnetic fields using *electrons or protons* as "test bodies" have been carried out, then by reason of the quantum mechanical uncertainty principle for the coordinate and momentum of the bodies, we get certain bounds on the *measurability* of the field intensities, namely, the same bounds for measurements both with electrons and protons. In the sense of the quantum mechanical concepts, it means that these fundamental restrictions on the *measurability* of the electromagnetic fields suggest a reasonable limitation on the possibility *to determine* precisely the state of the classical field.

Electric Field Intensities

In order to measure an electrical field, varying in space and time, at a point x, y, z at time t, we place an electron which we represent as a wave packet localized at the point x, y, z with zero velocity[1] at this time t. Then we determine the corresponding *acceleration* of the electron caused by the field. *Being averaged over a small time-interval* $t, t + \delta t$ for the x-component $\mathfrak{E}_x$ it is given by the equation

$$-e\mathfrak{E}_x = m\frac{\delta v_x}{\delta t}$$

($-e$ and m stand for the charge and mass of the electron, respectively), where δv_x is the increment of the velocity in the x-direction of the elec-

[1] In order to exclude a simultaneous deflection of the electron due to the magnetic field. (*Authors*)

tron within this time interval. Now we can attribute a certain velocity component v_x to time t and to time $t+\delta t$ only with a finite accuracy Δv_x, which is connected with the uncertainty of the electron x-coordinate

$$m\Delta v_x \Delta x \geq \hbar.$$

Consequently, the measured value of $\mathfrak{E}_x$ acquires an indefiniteness that has the following estimate:

$$\Delta\mathfrak{E}_x = \frac{m\Delta v_x}{e\delta t} \geq \frac{\hbar}{e}\frac{1}{\Delta x \Delta t}.$$

Hence, we can measure only an *average* of $\mathfrak{E}_x$, which refers to a space interval of the length Δx and to a time interval of the duration $\Delta t = \delta t$. The uncertainty $\Delta\mathfrak{E}_x$ of the value $\mathfrak{E}_x$ is inversely proportional to the product $\Delta x \Delta t$ of the corresponding values x and t, on which the measured quantity of $\mathfrak{E}$ depends.

As a complementation of the corresponding equations for $\mathfrak{E}_x$ and $\mathfrak{E}_y$, we obtain

$$\left.\begin{aligned} \Delta\mathfrak{E}_x \Delta x \Delta t &\geq \frac{\hbar}{e}, \\ \Delta\mathfrak{E}_y \Delta y \Delta t &\geq \frac{\hbar}{e}, \\ \Delta\mathfrak{E}_x \Delta z \Delta t &\geq \frac{\hbar}{e}. \end{aligned}\right\} \qquad (1)$$

Magnetic Field Intensities

The same considerations are also valid for magnetic field intensities. Below we investigate two cases that are distinguished by the curvature of the electron trajectory.

a) Trajectories of sharp curvature. Consider an electron having a helical trajectory in a magnetic field which is directed along the x-axis. Then we obtain

$$\mathfrak{H}_x = \frac{mv}{a}\frac{c}{e} = \frac{p}{a}\frac{c}{e},$$

where a is the radius of the orbit. The uncertainty of $\mathfrak{H}_x$ is

$$\Delta\mathfrak{H}_x = \frac{\Delta p}{a}\frac{c}{e}.$$

Now a is the uncertainty of the coordinate in the direction of the radius and Δp, of the momentum in the perpendicular direction; we can set, e.g.,

$$a = \Delta r; \qquad \Delta p = \Delta p_y.$$

Owing to

$$\Delta p_y \geq \frac{\hbar}{\Delta y},$$

we obtain

$$\mathfrak{H}_x \geq \frac{\hbar c}{e} \frac{1}{\Delta y \Delta z}.$$

b) Trajectories of small curvature. We denote by v_I the unit vector in the direction of electron velocity and by $\mathfrak{H}_\perp$ the component of $\mathfrak{H}$, which is perpendicular to v_I. Then we obtain

$$\mathfrak{H}_\perp = \frac{c}{e} v_I \times \frac{m\delta v}{\delta s}.$$

We suppose that the electron moves in the y-direction, therefore,

$$v_{Ix} = 0, \quad v_{Iy} = 1, \quad v_{Iz} = 0, \quad \delta s = \delta y$$

and the x-component of $\mathfrak{H}$ is

$$\mathfrak{H}_x = \frac{c}{e} \frac{m\delta v_z}{\delta y} = \frac{c}{e} \frac{\delta p_z}{\delta y}.$$

We hold the value of y and measure the corresponding shift δp_z of the momentum p_z. By the order of magnitude, the uncertainty δp_z is equal to Δp_z, which gives for the uncertainty of $\mathfrak{H}_x$ the relation

$$\Delta \mathfrak{H}_x = \frac{c}{e} \frac{\Delta p_z}{\delta y}.$$

Let further δy be the uncertainty of the y-value, to which the measured value of $\mathfrak{H}_x$ corresponds, i.e., $\delta y = \Delta y$. Because of the relation $\Delta p_z \geq \frac{\hbar}{\Delta z}$, we obtain, as before,

$$\Delta \mathfrak{H}_x \geq \frac{\hbar c}{e} \frac{1}{\Delta y \Delta z}.$$

This relation and those that are obtained from it by the cyclic permutations of indices are written in the following form:

$$\left.\begin{aligned} \Delta\mathfrak{H}_x \Delta y \Delta z &\geq \frac{\hbar c}{e}, \\ \Delta\mathfrak{H}_y \Delta z \Delta x &\geq \frac{\hbar c}{e}, \\ \Delta\mathfrak{H}_z \Delta x \Delta y &\geq \frac{\hbar c}{e}. \end{aligned}\right\} \qquad (2)$$

Another Form of Relations (1) and (2)

Obviously, equations (1), (2) have the correct relativistic symmetry. It is also possible to formulate their contents as follows: let us consider an integral over an arbitrary two-dimensional surface in the four-dimensional coordinate–time space

$$J = \int F_{\mu\nu} dS^{\mu\nu},$$

where $F_{\mu\nu}$ are the electromagnetic field intensity components and $dS^{\mu\nu}$ is an infinitely small two-dimensional surface element. Thus, J is defined with the uncertainty[2]

$$J = \frac{\hbar c}{e}.$$

The same reasoning is also valid for the integral

$$J = \oint \Phi_\mu dx^\mu,$$

which is taken along the closed contour in the coordinate–time space where Φ_μ denotes a 4-potential.

Concluding Remarks

The contemporary quantum theory of electromagnetic fields fails to give well-justified deductive derivation of equations (1) and (2). It states that a possibility to measure any field component, e.g., $\mathfrak{E}_x$, with an arbitrary precision at an exactly defined space–time point x, y, z, t is excluded by

[2]We were informed about this formulation by D. Ivanenko. See also J.Q. Stewart, Phys. Rev. **34**, 1290, 1929. (*Authors*)

our relations. Furthermore, according to Heisenberg,[3] the theory permits a simultaneous measurement of the field intensities $\mathfrak{E}_x$ and $\mathfrak{H}_y$ within the space volume $(\Delta l)^3$ with the uncertainty

$$\Delta\mathfrak{E}_x\Delta\mathfrak{H}_y \geq \frac{\hbar c}{(\Delta l)^4},$$

while, conversely, our theory gives the uncertainty for separately measured intensities of $\mathfrak{E}_x$ and $\mathfrak{H}_y$ the value that is obtained by the multiplication of the quantities mentioned above,

$$\Delta\mathfrak{E}_x\Delta\mathfrak{H}_y \geq \frac{\hbar^2 c^2}{e^2(\Delta I)^4},$$

with the minimal value, which is greater by a factor of $\frac{\hbar c}{e^2}$. The derivation of equations (1) and (2) on the basis of a proper multiplication law for field operators representing field quantities is expected to be given only by a future theory that calculates quantities with the accuracy of the fine structure constant.

Kharkov,
Physical-Technical Institute

Translated by A.V. Tulub

[3]W. Heisenberg, The Physical Principles of Quantum Theory, Leipzig 1930.

30-5
The Mechanics of Photons

NOTE BY V. FOCK
PRESENTED[1] BY M. DE BROGLIE

Comptes Rendus **187**, 1280, 1928

We propose to treat Maxwell equations as a wave equation of motion for a photon (a quantum of light) and to develop the photon mechanics on this basis.

Consider vectors of electric and magnetic fields $\mathcal{E}$ and $\mathcal{H}$ as a single object (a bivector) $\mathcal{F}$, which corresponds to the function ψ in the wave theory of the electron. Multiplying the vacuum Maxwell equations

$$-\operatorname{rot}\mathcal{H}+\frac{1}{c}\frac{\partial\mathcal{E}}{\partial t}=0,\qquad \operatorname{rot}\mathcal{E}+\frac{1}{c}\frac{\partial\mathcal{H}}{\partial t}=0 \tag{1}$$

by $\dfrac{hc}{2\pi i}$, it is possible to rewrite them in the form

$$\mathbf{H}\mathcal{F}+\frac{h}{2\pi i}\frac{\partial\mathcal{F}}{\partial t}=0, \tag{2}$$

where $\mathbf{H}$ is the operator

$$\mathbf{H}=\frac{hc}{2\pi i}\begin{Bmatrix} 0 & -\operatorname{rot}\\ \operatorname{rot} & 0\end{Bmatrix}; \tag{3}$$

its meaning follows from the comparison of equation (2) with (1). *Equation (2) can be interpreted as a wave equation for a photon.*

Let us consider the volume V delimited by conceptual planes. The boundary conditions $\mathcal{E}\times\mathbf{n}=0,\quad \mathcal{H}\cdot\mathbf{n}=0$ being imposed, the operator $\mathbf{H}$ is self-adjoint since for two bivectors $\mathcal{F}(\mathcal{E},\mathcal{H})$ and $\mathcal{F}'=(\mathcal{E}',\mathcal{H}')$ one has

$$\int_V\left(\overline{\mathcal{F}'}\mathbf{H}\mathcal{F}-\overline{\mathbf{H}\mathcal{F}'}\mathcal{F}\right)d\tau=\frac{hc}{2\pi i}\int\left[(\overline{\mathcal{E}'}\times\mathcal{H})+(\overline{\mathcal{H}'}\times\mathcal{E})\right]\mathbf{n}d\mathbf{S}=0. \tag{4}$$

Having established this, one can form the self-adjoint equation $\mathbf{H}\mathcal{F}=\lambda\mathcal{F}$. It can be checked that *the eigenvalues λ_n are just the quanta of*

[1]Session of 16 June 1930.

energy $\lambda_n = h\nu_n$, ν_n being the proper oscillation frequencies in the volume considered. The basic functions can be normalized, so that one would have

$$\int_V \left(\overline{\mathcal{E}_m}\mathcal{E}_n + \overline{\mathcal{H}_m}\mathcal{H}_n\right) d\tau = \delta_{mn}. \tag{5}$$

These functions can be used to construct matrices for different operators and to establish the equations of motion for a photon.

The matrix for the velocity component $v_x = \frac{dx}{dt}$ has the elements

$$(v_x)_{mn} = c \int \left[\left(\mathcal{E}_n \times \overline{\mathcal{H}_m}\right)_x + \left(\overline{\mathcal{E}_m} \times \mathcal{H}_n\right)_x\right] d\tau. \tag{6}$$

Thus, the operator for v_x is

$$v_x = \frac{dx}{dt} = \frac{2\pi i}{h}(\mathbf{H}x - x\mathbf{H}) = c\left\{\begin{matrix} 0 & -\mathbf{i}x \\ \mathbf{i}x & 0 \end{matrix}\right\}, \tag{7}$$

where $\mathbf{i}$ is the unit vector in the positive direction of the axis x. The eigenvalues of this operator are $\lambda = 0$, $\lambda = \pm c$. For $\lambda = 0$, the radius vector is zero; one can say that the photon is in the state of zero energy. For $\lambda = \pm c$ the photon moves with the velocity of light. The operators for v_y and v_z have the same eigenvalues.

The operator $p_x = \frac{hc}{2\pi i}\frac{\partial}{\partial x}$ in the electron mechanics has no analog in the mechanics of photons; here we have only one operator for the velocity instead of two[2] in the case of an electron.

One can also construct the acceleration operator. Its matrix elements can be expressed in terms of the Maxwell tensor and can be written in the form of a surface integral. One can say that the acceleration of a photon results from its reflections by surfaces.

The field corresponding to a photon in some state can be presented by the series

$$\mathcal{E} = \Sigma_n c_n \mathcal{E}_n; \qquad \mathcal{H} = \Sigma_n c_n \mathcal{H}_n,$$

where c_n are scalar coefficients (the same for both series). In view of relations (5), the energy can be written as

$$\mathrm{E} = \int \overline{\mathcal{F}}\mathbf{H}\mathcal{F} d\tau = \sum_n \overline{c_n} c_n h\nu_n. \tag{8}$$

Up to this point, we considered a single photon; to evolve the statistics for an ensemble of photons, one must apply the latest method of

[2]V. Fock, Zs. Phys. **55**, 127, 1929. (See [29-2] in this book. (*Editors*))

quantization developed by Dirac. The coefficients c_n must be treated as matrices (b_n) and $\overline{c_n}$ as the conjugate matrices $(b_n^\dagger)$; these matrices must satisfy the relation $b_n b_n^\dagger - b_n^\dagger b_n = 1$. Then the eigenvalues of $b_n^\dagger b_n$ are integers $0, 1, 2$, etc. This fact can be interpreted by regarding the energy of the frequency ν_n as a multiple integer (including zero) of $h\nu_n$. We see that the difficulty of the old theory (the quantization of an ensemble of oscillators) where the eigenvalues were $1/2, 3/2$, etc., leading to the existence of infinite energy in the zero state (Nullpunktsenergie), does not appear in our theory.

It must be possible to obtain the Einstein law for energy fluctuations from our theory. Our interpretation of Maxwell equations could be also used in the difficult problem of matter and light interaction; so one must consider the bivector $\mathcal{F}$ as an operator acting on the wave function ψ of matter.

Translated by V.D. Lyakhovsky

32-1
A Comment on the Virial Relation in Classical Mechanics

V. Fock and G. Krutkow

Received 10 May 1932

Phys. Zs. Sowjetunion **1**, N 6, 756, 1932

As is known, the conservation laws in the classical mechanics of point masses can be deduced directly from the Hamilton principle without appealing to the equations of motion. On the other hand, it was pointed out that in quantum mechanics it is possible to derive the virial theorem from the variational principle.[1] The purpose of this note is to demonstrate that this relation can also be easily obtained from the Hamilton principle in classical mechanics.

1. In the variational equation

$$\delta A = \delta \int_{t_0}^{t} L(q_k, \dot{q}_k)\, dt = 0, \tag{1}$$

every q_k is replaced by λq_k and every $\dot{q}_k$ by $\lambda\dot{q}_k + \dot{\lambda} q_k$, considering λ as the variational time-dependent function. After carrying out its variation, λ must be put equal to 1. The variation $\delta\lambda$ is taken constant in the whole time interval $(t - t_0)$ excluding the domains near the end points. In the intervals $(t_0, t_0 + \Delta t_0)$ and $(t - \Delta t, t)$ $\delta\lambda$ goes to zero and satisfies the equations

$$\int_{t_0}^{t_0+\Delta t_0} \delta\dot{\lambda}\, dt = \delta\lambda,$$

$$\int_{t-\Delta t}^{t} \delta\dot{\lambda}\, dt = -\delta\lambda, \tag{2}$$

[1] V. Fock, Zs. Phys. **63**, 855, 1930; JRPKhO **62**, 379, 1930. (See [30-1] in this book. (*Editors*))

where $\delta\lambda$ denotes the constant value of this quantity in the interval $(t_0 + \delta t_0, t - \Delta t)$. Hence, the variation δA splits into two terms:

$$\delta A = \delta_1 A + \delta_2 A,$$

where $\delta_1 A$ means the term resulting from $\delta\lambda = \text{const}$ in the whole interval $(t - t_0)$ and $\delta_2 A$ appears due to the non-zero value of $\delta\dot{\lambda}$ at the end points. We have

$$\delta_1 A = \delta\lambda \int_{t_0}^{t} \sum_k \left(\frac{\partial L}{\partial q_k} q_k + \frac{\partial L}{\partial \dot{q}_k} \dot{q}_k \right) dt,$$

$$\delta_2 A = \int_{t_0}^{t} \delta\dot{\lambda} \sum_k \frac{\partial L}{\partial \dot{q}_k} q_k \, dt. \tag{3}$$

2. First let us consider $\delta_2 A$. According to (2), $\delta_2 A$ can be written as

$$\delta_2 A = -\delta\lambda \sum_k \frac{\partial L}{\partial \dot{q}_k} q_k \Bigg|_{t_0}^{t} = -\delta\lambda\, \Lambda. \tag{3*}$$

For a periodic system, when $(t - t_0)$ is taken equal to the period, Λ disappears. In the general case of a finite motion, it holds

$$\lim_{t\to\infty} \frac{\Lambda}{t - t_0} = 0. \tag{4}$$

Here $\delta_2 A$ disappears due to the virial relation.

3. Assuming the validity of (4) when $(t - t_0)$ tends to infinity, we arrive at the equation

$$\lim_{t\to\infty} \frac{\delta_1 A}{t - t_0} = 0\,, \tag{5}$$

representing the virial relation in its general form. In rectangular coordinates, which are commonly used in the proof and application of the virial relations, we obtain

$$L = T - U = \frac{1}{2} \sum_k m_k (\dot{x}_k^2 + \dot{y}_k^2 + \dot{z}_k^2) - U(x_k, y_k, z_k), \tag{6}$$

$$\delta_1 A = \delta\lambda \int_{t_0}^{t} \left[2T - \sum_k \left(\frac{\partial U}{\partial x_k} x_k + \frac{\partial U}{\partial y_k} y_k + \frac{\partial U}{\partial z_k} z_k \right) \right] dt \tag{7}$$

and equation (5) takes the usual form of the virial relation:

$$2\overline{T} - \overline{\sum_k \left(\frac{\partial U}{\partial x_k} x_k + \frac{\partial U}{\partial y_k} y_k + \frac{\partial U}{\partial z_k} z_k \right)} = 0 \tag{8}$$

or, for a homogenous degree ϱ function,

$$2\overline{T} - \varrho\overline{U} = 0. \tag{8*}$$

Equation (8*) also follows directly from (7), namely,

$$\delta_1 A = (t - t_0)\delta(\lambda^2\overline{T} - \lambda^\varrho\overline{U}) \tag{9}$$

and

$$\lim \left. \frac{\delta_1 A}{t - t_0} \right|_{\lambda=1} = 2\overline{T} - \varrho\overline{U}. \tag{10}$$

The virial relation in an arbitrary coordinate system follows immediately from (3).

Leningrad, 7 May 1932

Translated by A.V. Tulub

32-2*
Configuration Space and Second Quantization[1]

V. Fock

Zs. Phys. **75**, 662, 1932
Fock57, pp. 25–52

In principle, it is known that the method of the quantized wave function is equivalent to the method of the usual wave function in the configuration space. However, the close connection between both methods was not observed in a proper way. In the proposed paper, this connection is traced in detail. It appears that it is close to such a degree that on each step the calculation with the quantized wave function admits the direct transition to the configuration space.

The paper consists of two parts. The first one is of introductory character and contains the derivation and comparison of the known results. Here the transition from the configuration space to the second quantization one is considered both for Bose and Fermi statistics, the uniqueness of the definition of the order of noncommutative factors being especially stressed. The starting point of the second part is the commutation relations between the quantum wave functions (operators Ψ). It is demonstrated that these relations are satisfied with some operators acting on the sequences of the usual wave functions of $1, 2, \ldots, n$ particles. Thereby, the representation of operators Ψ in the configuration space is obtained (more exactly, in sequences of configuration spaces). Further, the dependence of the operators Ψ on time is considered and the form of the operator $\dot{\Psi} = \partial\Psi/\partial t$ is defined. With the help of the representation obtained, it is shown that the Schrödinger equation for the operator Ψ containing the time derivative can be written as a set of usual Schrödinger equations for $1, 2, \ldots, n$ particles. As an application of the representation obtained, the simple derivation of the Hartree equation with exchange is given.

---

[1]This paper was reported to a theoretical seminar at Leningrad State University in January 1931.

I Transition From the Configuration Space to Second Quantization[2]

Let us denote as x_r the set of variables of an r-th particle (for example, the coordinates and spin of an electron, $x_r = (x_r, y_r, z_r, \sigma_r)$) and consider the wave function

$$\psi(x_1 x_2 \ldots x_n; t), \tag{1}$$

which describes the set of n identical particles in the configuration space. For convenience, let us pass by canonical transformation from the initial variable x to a new variable E, taking only discrete values

$$E = E^{(1)}, E^{(2)}, \ldots E^{(r)}, \ldots . \tag{2}$$

By quantities (2), one can mean the eigenvalues of the operators with a discrete spectrum. If one denotes the corresponding eigenfunctions as

$$\psi_r(x) = \psi_r(E^{(r)}, x), \tag{3}$$

then the transformed wave function

$$c\,(E_1, E_2, \ldots, E_n; t) \tag{4}$$

is related to initial wave function (1) as

$$\psi(x_1 x_2, \ldots, x_n; t) =$$

$$= \sum_{E_1, \ldots, E_n} c\,(E_1, E_2, \ldots, E_n; t)\psi(E_1; x_1) \ldots \psi(E_n; x_n), \tag{5}$$

where each summation variable $E_1, E_2, \ldots, E_n$ runs all the values (2).

The Schrödinger equation in the configuration space will be written as follows:[3]

$$H\psi(x_1 \ldots x_n; t) - i\hbar \frac{\partial \psi}{\partial t} = 0. \tag{6}$$

Let the energy operator H have the form

$$H = \sum_{k=1}^{n} H(x_k) + \sum_{k<l=1}^{n} G(x_k; x_l). \tag{7}$$

[2]The reader familiar with the theory of second quantization can skip the first part and begin reading from the second part at once. (*V. Fock*)

[3]Here $\hbar$ denotes the Planck constant divided by 2π. (*V. Fock*)

Here the usual sum gives the energy of separate particles and the double sum gives their interaction energy. For Coulomb forces

$$G(x, x') = \frac{e^2}{|\mathbf{r} - \mathbf{r}'|}. \tag{8}$$

In order to get the Schrödinger equation for transformed wave function (4) one should substitute series (5) into equation (6), decompose the result over products

$$\psi(E_1; x_1) \dots \psi(E_n; x_n)$$

of functions (3) and put the coefficients of each product zero. In this way, we obtain

$$\begin{aligned} &\sum_{k=1}^{n}\sum_{W} \langle E_k \mid H \mid W\rangle c\,(E_1 \dots E_{k-1} W E_{k+1} \dots E_n; t) + \\ &+ \sum_{k \le l = 1}^{n} \sum_{WW'} \langle E_k E_l \mid G \mid WW'\rangle c\,(E_1 \dots E_{k-1} W E_{k+1} \dots \\ &\dots E_{l-1} W' E_{l+1} \dots E_n; t) - i\hbar \frac{\partial}{\partial t} c\,(E_1 E_2 \dots E_n; t) = 0 \end{aligned} \tag{9}$$

where the following notations are introduced for the matrix elements

$$\langle E \mid H \mid W\rangle = \int \overline{\psi}(E; x) H(x) \psi(W; x) dx, \tag{10}$$

$$\begin{aligned} &\langle EE' \mid G \mid WW'\rangle = \\ &= \iint \overline{\psi}(E; x)\overline{\psi}(E'; x') G(x; x') \psi(W; x) \psi(W'; x')\, dx\, dx'. \end{aligned} \tag{10*}$$

Let the arguments

$$E_1, E_2, \dots, E_k, \dots E_n$$

of the wave function c in equation (9) be equal, respectively, to the eigenvalues

$$E^{(r_1)}, E^{(r_2)}, \dots E^{(r_k)}, \dots E^{(r_n)}.$$

If we write for brevity

$\langle r \mid H \mid s\rangle$	instead of	$\langle E^{(r)} \mid H \mid E^{(s)}\rangle$,
$\langle rt \mid H \mid su\rangle$	instead of	$\langle E^{(r)} E^{(t)} \mid H \mid E^{(s)} E^{(u)}\rangle$,
$c\,(r_1, r_2 \dots r_n; t)$	instead of	$c\,(E^{(r_1)} E^{(r_2)} \dots E^{(r_n)}; t)$,

then wave equation (9) will take the form

$$\sum_r \sum_{k=1}^{n} \langle r_k \mid H \mid r\rangle c\,(r_1 \ldots r_{k-1} r r_{k+1} \ldots r_n; t) +$$

$$+ \sum_{rs} \sum_{k<l=1}^{n} \langle r_k r_l \mid G \mid rs\rangle c\,(r_1 \ldots r_{k-1} r r_{k+1} \ldots r_{l-1} s r_{l+1} \ldots r_n; t) -$$

$$-i\hbar \frac{\partial}{\partial t} c\,(r_1 \ldots r_n; t) = 0\,. \tag{9*}$$

Until now we did not take into account the symmetry properties of the wave function and, therefore, the kind of statistics. But the wave function (both ψ and c) is either symmetric (Bose statistics) or antisymmetric (Fermi statistics). In the case of the symmetric wave function, the value $c\,(r_1, r_2, \ldots r_n; t)$ is defined by the set of numbers

$$n_1, n_2, \ldots n_r, \tag{11}$$

which indicate how many times the corresponding arguments $1, 2, \ldots, r$ or $E^{(1)}, E^{(2)}, \ldots E^{(r)}$ occur in c. Therefore, we can put

$$c\,(r_1 r_2 \ldots r_n; t) = c^*(n_1 n_2 \ldots; t). \tag{12}$$

Now, the set of values $r_1 r_2 \ldots r_n$ (defined independently of their order) corresponds to each series of numbers (11). For example, for $n = 3$ we have $c\,(4,4,5) = c\,(4,5,4) = c\,(5,4,4) = c^*(0,0,0,2,1,0,0\ldots)$.

In the normalization condition

$$\sum_{r_1,\ldots r_n} |c\,(r_1 r_2 \ldots r_n; t)|^2 = 1 \tag{13}$$

one can make the first summation over all permutations of a given set of values $r_1, r_2 \ldots r_n$ and then over different sets

$$\sum_{(r_1,\ldots r_n)} \sum_{\text{Perm}} |c\,(r_1 r_2 \ldots r_n; t)|^2 = 1.$$

The sum $\sum_{\text{Perm}}$ contains $\frac{n!}{n_1! n_2! \ldots}$ equal terms; consequently, we have

$$\sum_{(r_1,\ldots r_n)} \frac{n!}{n_1! n_2! \ldots} |c\,(r_1 r_2 \ldots r_n; t)|^2 = 1$$

or, introducing quantities n_r as variables according to (12),

$$\sum_{n_1 n_2 \dots} \frac{n!}{n_1! n_2! \dots} |c^*(n_1 n_2 \dots; t)|^2 = 1. \tag{14}$$

In normalization condition (14), one can bring the weight function $\frac{n!}{n_1! n_2! \dots}$ to unity with the substitution

$$c^*(n_1 n_2 \dots; t) = \sqrt{\frac{n_1! n_2! \dots}{n!}} f(n_1 n_2 \dots; t). \tag{15}$$

For a new wave function f, the normalization condition takes the form

$$\sum_{n_1 n_2 \dots} |f(n_1 n_2 \dots; t)|^2 = 1. \tag{16}$$

In the case of Fermi statistics, it is yet not enough to fix numbers n_r for the unique definition of $c(r_1 r_2 \dots r_n; t)$ since the quantity c is defined by them up to a sign. However, we can also keep equations (12) and (15) for Fermi statistics if we impose an additional condition that in this case the arguments in $c(r_1 r_2 \dots r_n; t)$ form the "natural" sequence, for instance:

$$r_1 < r_2 < r_3 \dots < r_n.$$

If the sequence of arguments is obtained from the natural one by even permutation then equation (12) remains unchanged. For odd permutation, its sign should be changed, for example:

$$c(1,4,5) = -c(4,1,5) = c^*(1,0,0,1,1,0,0 \dots).$$

In what follows, Bose and Fermi statistics will be treated separately.

Bose Statistics

In the case of Bose statistics, few equal arguments can occur in the wave function $c(r_1 r_2 \dots r_n; t)$, for example:

$$c = c(u,u,u,v,v,w,\dots).$$

Therefore, in the first sum in expression (9*), functions can appear that differ by the order of arguments only, namely, n_u items can occur, whose argument r stays on the place u, n_v items with r on the place v and so on. If we collect the identical items, then for the first sum in (9*) we obtain the expression

$$\sum_r \langle u \mid H \mid r\rangle n_u c(r,u,u,v,v,w,\dots)+$$

$$+\sum_r \langle v \mid H \mid r\rangle n_v c(u,u,u,r,v,w,\dots)+\dots.$$

Introducing quantities n_k as variables according to (12), we obtain

$$\sum_r \langle u \mid H \mid r\rangle n_u c^*(\dots n_u-1,\dots n_r+1,\dots)+$$

$$+\sum_r \langle v \mid H \mid r\rangle n_v c^*(\dots n_v-1,\dots n_r+1,\dots)+\dots$$

or simpler

$$\sum_p \sum_r \langle p \mid H \mid r\rangle n_p c^*(\dots n_p-1,\dots n_r+1,\dots), \tag{17}$$

where the index p can run now all the values (but not only values $p = u,v,w,\dots$), since extra terms vanish due to factor n_p. Herewith, for $r=p$ the quantity $c^*(\dots n_p-1,\dots n_r+1,\dots)$ should be understood simply as $c^*(\dots n_r \dots)$.

Similarly, one can also transform the second sum in expression (9*). Taking into account the number of identical items, we obtain:

$$\sum_{r,s}\{\langle uu \mid G \mid rs\rangle \frac{1}{2} n_u(n_u-1)c(r,s,u,v,v,w,\dots)+$$

$$+\langle uv \mid G \mid rs\rangle n_u n_v c(r,u,u,s,v,w,\dots)+$$

$$\langle vv \mid G \mid rs\rangle \frac{1}{2} n_v(n_v-1)c(u,u,u,r,s,w,\dots)+\dots\}.$$

Introducing the quantities $c^*(n_1 n_2 \dots)$, we can write

$$\sum_{r,s}\{\langle uu \mid G \mid rs\rangle \frac{1}{2} n_u(n_u-1)c^*(\dots n_u-2,\dots n_r+1,\dots n_s+1,\dots)+$$

$$+\langle uv \mid G \mid rs\rangle n_u n_v c^*(\dots n_n-1,\dots n_v-1,\dots n_r+1,\dots n_s+1,\dots)+$$

$$+\langle vv \mid G \mid rs\rangle \frac{1}{2} n_v(n_v-1)c^*(\dots n_v-2,\dots n_r+1,\dots n_s+1,\dots)+\dots\}$$

or simpler

$$\frac{1}{2}\sum_{p,q}\sum_{r,s}\langle pq \mid G \mid rs\rangle n_p(n_q - \delta_{pq})\cdot$$

$$\cdot c^*(\ldots n_p - 1, \ldots n_q - 1, \ldots n_r + 1, \ldots n_s + 1, \ldots\ ). \qquad (18)$$

Here the summation indices p and q can also run all the values without exception (but not only $p, q = u, v, w$). The factor $\frac{1}{2}$ should stay with *all* the items, since the combination $p = u, q = v$ is met in (18), e.g., as well as the combination $p = v, q = u$. The meaning of the terms in (18), in which two or more numbers p, q, r, s coincide, is clear by itself.

With the help of (17) and (18), equation (9*) can be written as

$$\begin{aligned}&\sum_p\sum_r\langle p \mid H \mid r\rangle n_p c^*(\ldots n_p - 1, \ldots n_r + 1, \ldots\ ) +\\ &+\frac{1}{2}\sum_{p,q}\sum_{r,s}\langle pq \mid G \mid rs\rangle n_p(n_q - \delta_{pq})\cdot\\ &\cdot c^*(\ldots n_p - 1, \ldots n_q - 1, \ldots n_r + 1, \ldots n_s + 1, \ldots\ ) -\\ &-i\hbar\frac{\partial c^*(n_1 n_2 \ldots ; t)}{\partial t} = 0.\end{aligned} \qquad (19)$$

Henceforth, it is expedient to introduce the operator U_r which transforms the function $f(n_1, n_2, \ldots\ )$ into the function

$$U_r f(n_1 n_2 \ldots n_r, \ldots\ ) = f(n_1 n_2 \ldots n_r + 1, \ldots\ ). \qquad (20)$$

The quantity U_r considered as a matrix with respect to the variable n_r and its conjugate matrix $U_r^\dagger$ are of the form

$$U_r = \begin{pmatrix} 0 & 1 & 0 & 0 & \ldots \\ 0 & 0 & 1 & 0 & \ldots \\ 0 & 0 & 0 & 1 & \ldots \\ 0 & 0 & 0 & 0 & \ldots \\ \ldots & \ldots & \ldots & \ldots & \ldots \end{pmatrix}, \quad U_r^\dagger = \begin{pmatrix} 0 & 0 & 0 & 0 & \ldots \\ 1 & 0 & 0 & 0 & \ldots \\ 0 & 1 & 0 & 0 & \ldots \\ 0 & 0 & 1 & 0 & \ldots \\ \ldots & \ldots & \ldots & \ldots & \ldots \end{pmatrix}. \qquad (21)$$

Consequently, the conjugate operator $U^\dagger$ transforms the function $f(n_1 n_2 \ldots n_r \ldots\ )$ into f'', where $f''(n_1 n_2 \ldots n_r \ldots\ ) = f(n_1 n_2 \ldots n_r - 1 \ldots\ )$ in the case of $n_r \neq 0$ and $f'' = 0$ at $n_r = 0$. Thus,

$$U_r^\dagger f(n_1 n_2 \ldots n_r \ldots\ ) = \begin{cases} f(n_1 n_2 \ldots n_r - 1, \ldots\ ) & \text{for} \quad n_r \neq 0, \\ 0 & \text{for} \quad n_r = 0. \end{cases} \qquad (22)$$

It follows from the definition U_r that

$$U_r U_r^\dagger = 1\,. \tag{23}$$

On the contrary, $U_r^\dagger U_r \neq 1$. We have:

$$U_r^\dagger U_r = \begin{pmatrix} 0 & 0 & 0 & 0 & \dots \\ 0 & 1 & 0 & 0 & \dots \\ 0 & 0 & 1 & 0 & \dots \\ 0 & 0 & 0 & 1 & \dots \\ \dots & \dots & \dots & \dots & \dots \end{pmatrix}. \tag{23*}$$

Thus, the operator U_r is non-unitary.

At $p \neq r$ the operators U_r and $U_r^\dagger$ commute with U_p and $U_p^\dagger$. With the help of operators U_p one can write down functions c^* appearing in formula (19) as:

$$c^*(\dots n_p - 1, \dots n_r + 1 \dots\ ) = U_r U_p^\dagger c^*(\dots n_p \dots n_r \dots\ )\,,$$

$$c^*(\dots n_p - 1, \dots n_q - 1, \dots n_r + 1, \dots n_s + 1 \dots\ ) =$$
$$= U_r U_s U_p^\dagger U_q^\dagger c^*(\dots n_p, \dots n_q, \dots n_r, \dots n_s \dots\ )\,.$$

The order of factors, i.e., $U^\dagger$ stand to the right of U, follows uniquely from the definition c^* for $p = r$ in connection with (23) and (23*).[4] These expressions hold for all values p, q, r, s (and also for coincident ones). If we introduce them in (19), we obtain:

$$\sum_p \sum_r \langle p \mid H \mid r \rangle n_p U_r U_p^\dagger c^*\,(n_1 n_2 \dots\ ) +$$
$$+ \frac{1}{2} \sum_{p,q} \sum_{r,s} \langle pq \mid G \mid rs \rangle n_p (n_q - \delta_{pq}) U_r U_s U_p^\dagger U_q^\dagger c^*(n_1 n_2 \dots\ )$$
$$- i\hbar \frac{\partial}{\partial t} c^*(n_1 n_2 \dots\ ) = 0\,. \tag{19*}$$

But if we take into account that

$$n_p U_r U_p^\dagger = n_p U_p U_r^\dagger \tag{23**}$$

and

$$n_p (n_q - \delta_{pq}) U_r U_s U_p^\dagger U_q^\dagger = n_p (n_q - \delta_{pq}) U_p^\dagger U_q^\dagger U_s U_r\,, \tag{23***}$$

[4]However, the order of factors U and $U^\dagger$ becomes inessential after multiplying them by n_p and $(n_q - \delta_{pq})$; see below equations (23**) and (23***). (*V. Fock*)

we can write equation (19*) in the form

$$\sum_p \sum_r \langle p \mid H \mid r\rangle n_p U_p^\dagger U_r c^*(n_1 n_2 \dots) +$$

$$+\frac{1}{2}\sum_{p,q}\sum_{r,s}\langle pq \mid G \mid rs\rangle n_p(n_q - \delta_{pq})U_p^\dagger U_q^\dagger U_s U_r c^*(n_1 n_2 \dots) -$$

$$-i\hbar\frac{\partial}{\partial t}c^*(n_1 n_2 \dots) = 0. \tag{19**}$$

Here we should express quantity $c^*(n_1, n_2, \dots)$ in terms of $f(n_1, n_2, \dots)$ in accordance with (15). The operator n for the total number of particles and consequently for $n!$ obviously commutes with the products $U_p^\dagger U_r$ and $U_p^\dagger U_q^\dagger U_r U_s$; apart from that, we have

$$\frac{1}{\sqrt{n_1!n_2!\dots}}U_r\sqrt{n_1!n_2!\dots} = \sqrt{n_r+1}U_r = U_r\sqrt{n_r}, \tag{24}$$

$$\frac{1}{\sqrt{n_1!n_2!\dots}}U_r^\dagger\sqrt{n_1!n_2!\dots} = \frac{1}{\sqrt{n_r}}U_r^\dagger. \tag{24*}$$

Therefore, the term $n_p U_p^\dagger U_r c^*(n_1 n_2 \dots)$ of the first sum in (19**) multiplied by $\frac{\sqrt{n!}}{\sqrt{n_1!n_2!\dots}}$ is equal to

$$\frac{\sqrt{n!}}{\sqrt{n_1!n_2!\dots}}n_p U_p^\dagger U_r\frac{\sqrt{n_1!n_2!\dots}}{\sqrt{n!}}f(n_1 n_2 \dots) = \sqrt{n_p}U_p^\dagger U_r\sqrt{n_r}f(n_1 n_2 \dots).$$

Similarly, with the help of formulae (24) and the relation

$$(n_q - \delta_{pq})U_p^\dagger = U_p^\dagger n_q$$

for the terms of the second sum in (19**), we obtain the expression

$$\frac{\sqrt{n!}}{\sqrt{n_1!n_2!\dots}}n_p(n_q-\delta_{pq})U_p^\dagger U_q^\dagger U_r U_s\frac{\sqrt{n_1!n_2!\dots}}{\sqrt{n!}}f(n_1 n_2 \dots) =$$

$$= \sqrt{n_p}U_p^\dagger\sqrt{n_q}U_q^\dagger U_r\sqrt{n_r}U_s\sqrt{n_s}f(n_1 n_2 \dots).$$

Inserting these expressions in (19**), we obtain for $f(n_1 n_2 \dots)$ the wave equation in the form

$$\mathbf{H}f(n_1 n_2 \dots) - i\hbar\frac{\partial f}{\partial t} = 0, \tag{25}$$

where $\mathbf{H}$ means the transformed energy operator

$$\mathbf{H} = \sum_{pr} \langle p \mid H \mid r \rangle \sqrt{n_p} U_p^\dagger U_r \sqrt{n_r} + \\ + \frac{1}{2} \sum_{p,q} \sum_{r,s} \langle pq \mid G \mid rs \rangle \sqrt{n_p} U_p^\dagger \sqrt{n_q} U_q^\dagger U_r \sqrt{n_r} U_s \sqrt{n_s}. \tag{26}$$

Here the operators U_r and n_r enter in combinations

$$b_r = U_r \sqrt{n_r}, \ b_r^\dagger = \sqrt{n_r} U_r^\dagger \tag{27}$$

only. If one inserts expressions (27) in (26), then the operator $\mathbf{H}$ takes the form

$$\mathbf{H} = \sum_{pr} b_p^\dagger \langle p \mid H \mid r \rangle b_r + \frac{1}{2} \sum_{pr} \sum_{qs} b_p^\dagger b_q^\dagger \langle pq \mid G \mid rs \rangle b_r b_s. \tag{28}$$

As it follows from definitions (20) and (22) of the operators U_r and $U_r^\dagger$, the operators b_r just introduced satisfy the relations

$$b_r^\dagger b_r = n_r, \qquad b_r b_r^\dagger = n_r + 1 \tag{29}$$

and, moreover, since for $r \neq s$ the operators $b_r^\dagger$ and $b_s^\dagger$ commute with b_r and b_s, the well-known commutation relations take place:

$$b_r^\dagger b_s - b_s b_r^\dagger = \delta_{rs}\,, \tag{30}$$

$$b_r b^s - b^s b_r = 0\,. \tag{30*}$$

Forming with the help of b_r the quantized wave function

$$\Psi(x) = \sum_r b_r \psi_r(x) \tag{31}$$

and its conjugate wave function

$$\Psi^\dagger(x) = \sum_r b_r^\dagger \overline{\psi}_r(x)\,, \tag{31*}$$

we can represent the energy operator as

$$\mathbf{H} = \int \Psi^\dagger(x) H(x) \Psi(x)\, dx + \\ + \frac{1}{2} \iint \Psi^\dagger(x) \Psi^\dagger(x') G(xx') \Psi(x') \Psi(x)\, dx\, dx'\,. \tag{32}$$

From relations (30) and (30*), taking into account the equality

$$\sum_r \overline{\psi}_r(x)\psi_r(x') = \delta(x - x'),$$

it is easy to obtain the commutation relations for the quantized wave functions (operators Ψ).

In this way, we obtain

$$\Psi(x')\Psi^\dagger(x) - \Psi^\dagger(x)\Psi(x') = \delta(x - x'), \tag{33}$$

$$\Psi(x')\Psi(x) - \Psi(x)\Psi(x') = 0. \tag{33*}$$

Fermi Statistics

Let us turn again to wave equation (9*). We assume that the numbers $r_1 r_2 r_3 r_4, \ldots$ are arranged in the natural order

$$r_1 < r_2 < r_3 < \ldots < r_n, \tag{34}$$

so that, according to (12), we have:

$$c(r_1 r_2 \ldots r_n; t) = c^*(n_1 n_2 \ldots; t). \tag{35}$$

In the natural order, the number r_k stays in the place k, where

$$k = n_1 + n_2 + \ldots + n_{r_k}. \tag{36}$$

In the k-th item of the first sum in (9*) r_k was replaced by r, so that the arguments in c are arranged in the order

$$r_1 r_2 \ldots r_{k-1}\, r\, r_{k+1} \ldots r_n, \tag{*}$$

which is not natural, because r stays in the place k, while it should stay in the place

$$k' = n_1' + n_2' + \ldots + n_r'.$$

(The primed quantities are those new variables n_s, which correspond to arguments (*) in c.) Therefore,

$$c(r_1 r_2 \ldots r_{k-1}\, r\, r_{k+1} \ldots r_n; t) = (-1)^{k+k'} c^*(\ldots n_{r_k} - 1, \ldots n_r + 1, \ldots).$$

Now, let us introduce new operators $\alpha_r^\dagger$ and α_r assuming

$$\alpha_r f(n_1 \ldots n_r \ldots) = \begin{cases} f(n_1 \ldots n_r + 1 \ldots) & \text{for } n_r = 0, \\ 0 & \text{for } n_r = 1. \end{cases} \tag{37}$$

$$\alpha_r^\dagger f(n_1 \dots n_r \dots) = \begin{cases} f(n_1 \dots n_r - 1 \dots) & \text{for } n_r = 1, \\ 0 & \text{for } n_r = 0. \end{cases} \tag{37*}$$

It follows from this definition that for $r \neq s$ the operators α_r and $\alpha_r^\dagger$ commute with α_s and $\alpha_s^\dagger$ (since they act on the different variables), while for $r = s$ the following equalities hold

$$\alpha_r^\dagger \alpha_r = n_r, \qquad \alpha_r \alpha_r^\dagger = 1 - n_r \,. \tag{38}$$

It is easy to prove the equality

$$\alpha_r(1 - 2n_r) = -(1 - 2n_r)\alpha_r \,. \tag{39}$$

Using the operators α_r, one can write

$$c^*(\dots n_{r_k} - 1, \dots n_r + 1, \dots) = \alpha_{r_k}^\dagger \alpha_r c^*(n_1 n_2 \dots) \,.$$

The order of factors $\alpha_r^\dagger$ and α_r is defined here uniquely, since for $r = r_k$ and $n_{r_k} = 1$ the factor in front of c^* in the right-hand side is reduced to unity, as it should be. We have

$$(-1)^k = (-1)^{n_1 + \dots + n_{r_k}},$$

but since for $n = 0$ and $n = 1$ the quantity $(-1)^n$ coincides with $(1-2n)$; instead of this, one can write

$$(-1)^k = \prod_{p=1}^{r_k} (1 - 2n_p) = \nu_{r_k} \,,$$

where

$$\nu_s = \prod_{p=1}^{s} (1 - 2n_p) \tag{40}$$

denotes the sign-function of Wigner. Similarly, $(-1)^{k'} = \nu_r'$, where ν_r' is constructed from the numbers n_r'. Thus, we have

$$c\,(r_1 \dots r_{k-1}\, r\, r_{k+1} \dots ; t) = \nu_{r_k} \nu_r' \alpha_{r_k}^\dagger \alpha_r c^*(n_1 n_2 \dots) \,.$$

On the basis of

$$\nu_r' \alpha_{r_k}^\dagger \alpha_r = \alpha_{r_k}^\dagger \alpha_r \nu_r,$$

we can also write

$$c\,(r_1 r_2 \dots r_{k-1}\, r\, r_{k+1} \dots ; t) = \nu_{r_k} \alpha_{r_k}^\dagger \alpha_r \nu_r c^*(n_1, n_2 \dots) \,.$$

Then the first sum in (9*) is equal to

$$\sum_r \sum_{k-1}^{n} \langle r_k \mid H \mid r\rangle \nu_{r_k} \alpha_{r_k}^{\dagger} \alpha_r \nu_r c^*(n_1 n_2 \dots) .$$

In summing over k the index r_k runs the values $r_1 r_2 \dots r_n$. Instead of this, one can suppose that r_k runs *all* the values since the superfluous items vanish due to the properties of the operator $\alpha^{\dagger}$. Thus, for the sum considered, we obtain the expression

$$\sum_p \sum_r \langle p \mid H \mid r\rangle \nu_p \alpha_p^{\dagger} \alpha_r \nu_r c^*(n_1, n_2 \dots) . \tag{41}$$

Now let us transform the second sum in (9*). To do this, first of all one needs to define the sign in the equality

$$\pm c\,(r_1 \dots r_{k-1}\; r\; r_{k+1} \dots r_{l-1}\; s\; r_{l+1} \dots r_n; t) =$$

$$= c^*\,(\dots n_{r_k} - 1 \dots n_{r_l} - 1 \dots n_r + 1, \dots n_s + 1, \dots) = c^*(n_1', n_2' \dots) .$$

In the function c the argument r stays in the k-th place. First, we shift it to the first place; then c acquires the factor $(-1)^k = -\nu_{r_k}$ and we obtain

$$c\,(r_1 \dots r_{k-1}\; r\; r_{k+1} \dots r_{l-1}\; s\; r_{l+1} \dots r_n; t) =$$

$$= -\nu_{r_k} c\,(r\; r_1 \dots r_{k-1} r_{k+1} \dots r_{l-1}\; s\; r_{l+1} \dots r_n; t) .$$

When $r_l > r_k$, the argument remains in the place number $l+1$, where

$$l = n_1 + n_2 \dots n_{r_l} .$$

In the case of $r_l < r_k$, the argument s is shifted by one step to the right and, thus, appears in the place number $l+1$. Now if we move s to the second place, we obtain

$$c\,(\dots r_{k-1}\; r\; r_{k+1} \dots r_{l-1}\; s\; r_{l+1} \dots) =$$

$$= \begin{cases} -\nu_{r_k} \nu_{r_l} c\,(r\; s\; r_1\; r_2 \dots) & \text{for } r_l > r_k, \\ +\nu_{r_k} \nu_{r_l} c\,(r\; s\; r_1\; r_2 \dots) & \text{for } r_l < r_k. \end{cases}$$

On the other hand, if we denote the places r and s in the natural sequence as k' and l', where

$$k' = n_1' + n_2' + \dots + n_r',$$

$$l' = n_1' + n_2' + \dots + n_s',$$

then, applying absolutely similar considerations, we obtain

$$c(\overbrace{r_1 \ldots r \ldots s \ldots}^{\text{natural sequence}}) = c^*(n_1' \; n_2' \; \ldots \;) =$$

$$= \begin{cases} -\nu_r' \; \nu_s' \; c(r \; s \; r_1 \ldots \;) & \text{for } s > r, \\ +\nu_r' \; \nu_s' \; c(r \; s \; r_1 \ldots \;) & \text{for } s < r. \end{cases}$$

Together with the previous equalities, it gives

$$c(\ldots r_{k-1} \; r \; r_{k+1} \ldots r_{l-1} \; s \; r_{l+1} \ldots \;) =$$

$$= \begin{cases} +\nu_{r_k}\nu_{r_l} \; \nu_r' \; \nu_s' c(n_1' \; n_2' \ldots \;) & \text{in case I,} \\ -\nu_{r_k}\nu_{r_l} \; \nu_r' \; \nu_s' c(n_1' \; n_2' \ldots \;) & \text{in case II,} \end{cases}$$

where cases I and II are characterized by the inequalities

$$\left. \begin{array}{ll} r_l > r_k & \text{and } s > r \\ r_l < r_k & \text{and } s < r \end{array} \right\} \quad \text{case I,}$$

or

$$\left. \begin{array}{ll} r_l > r_k & \text{and } s < r \\ r_l < r_k & \text{and } s > r \end{array} \right\} \quad \text{case II.}$$

The change of arguments $n_1 n_2 \ldots$ to $n_1' n_2' \ldots$ in the function c^* is produced by means of the operator $\alpha_{r_k}^\dagger \; \alpha_{r_l}^\dagger \; \alpha_s \; \alpha_r$:

$$c^*(n_1' \; n_2' \ldots \;) = \alpha_{r_k}^\dagger \; \alpha_{r_l}^\dagger \; \alpha_s \; \alpha_r \; c^*(n_1 \; n_2 \ldots \;).$$

We make sure that the order of factors $\alpha^\dagger$ and α (so far as it matters) was chosen correctly considering particular cases $r = r_k, s = r_l$ and $r = r_l, s = r_k$. If we also take into account the equality

$$\nu_r' \; \nu_s' \; \alpha_{r_k}^\dagger \; \alpha_{r_l}^\dagger \; \alpha_s \; \alpha_r = \alpha_{r_k}^\dagger \; \alpha_{r_l}^\dagger \; \alpha_s \; \alpha_r \; \nu_r \; \nu_s,$$

we obtain

$$c\,(\ldots r_{k-1} \; r \; r_{k+1} \; \ldots r_{l-1} \; s \; r_{l+1} \ldots \;) =$$

$$= \begin{cases} +\nu_{r_k} \; \nu_{r_l} \; \alpha_{r_k}^\dagger \; \alpha_{r_l}^\dagger \; \alpha_s \; \alpha_r \; \nu_s \; \nu_r \; c^* \; (n_1 \; n_2 \ldots \;) & \text{in case I.} \\ -\nu_{r_k} \; \nu_{r_l} \; \alpha_{r_k}^\dagger \; \alpha_{r_l}^\dagger \; \alpha_s \; \alpha_r \; \nu_s \; \nu_r \; c^* \; (n_1 \; n_2 \ldots \;) & \text{in case II.} \end{cases}$$

However, it follows from formula (39) and from definition (40) of the quantity ν_s that

$$\begin{aligned} \alpha_r \nu_s &= \nu_s \alpha_r && \text{for } r > s, \\ \alpha_r \nu_s &= -\nu_s \alpha_r && \text{for } r \leq s. \end{aligned} \tag{42}$$

Therefore, in case I it will be either simultaneously

$$\alpha_r \nu_s = \nu_s \alpha_r \text{ and } \alpha^\dagger_{r_k} \nu_{r_l} = \nu_{r_l} \alpha^\dagger_{r_k} \quad (\text{for } r > s \text{ and } r_k > r_l),$$

or simultaneously

$$\alpha_r \nu_s = -\nu_s \alpha_r \text{ and } \alpha^\dagger_{r_k} \nu_{r_l} = -\nu_{r_l} \alpha^\dagger_{r_k} \quad (\text{for } r \leq s \text{ and } r_k \leq r_l).$$

Therefore, in case I the operator acting on $c^*(n_1 n_2 \dots)$ is equal to

$$\nu_{r_k} \alpha^\dagger_{r_k} \nu_{r_l} \alpha^\dagger_{r_l} \alpha_s \nu_s \alpha_r \nu_r \, .$$

But this operator has the same sign also in case II, since then we have either simultaneously

$$\alpha_r \nu_s = \nu_s \alpha_r \text{ and } \alpha^\dagger_{r_k} \nu_{r_l} = -\nu_{r_l} \alpha^\dagger_{r_k} \quad (\text{for } r > s \text{ and } r_k < r_l),$$

or simultaneously

$$\alpha_r \nu_s = -\nu_s \alpha_r \text{ and } \alpha^\dagger_{r_k} \nu_{r_l} = \nu_{r_l} \alpha^\dagger_{r_k} \quad (\text{for } r < s \text{ and } r_k > r_l).$$

Thus, always

$$c\,(\dots r_{k-1}\; r\; r_{k+1} \dots r_{l-1}\; s\; r_{l+1} \dots\,) =$$
$$= \nu_{r_k} \alpha^\dagger_{r_k}\; \nu_{r_l} \alpha^\dagger_{r_l}\; \alpha_s\; \nu_s\; \alpha_r\; \nu_r\; c^*\; (n_1\; n_2\; \dots\;).$$

Now, we should substitute this expression in the second sum in formula (9*). This sum will be equal to

$$\sum_{rs} \sum_{k<l}^{n} \langle r_k r_l \mid G \mid rs \rangle \nu_{r_k} \alpha^\dagger_{r_k} \nu_{r_l} \alpha^\dagger_{r_l} \alpha_s \nu_s \alpha_r \nu_r c^*(n_1 n_2 \dots).$$

If we discard the restriction $k < l$, then the sum will be doubled and we should add the factor $\frac{1}{2}$. Then, we obtain

$$\frac{1}{2} \sum_{rs} \sum_{pq} \langle pq \mid G \mid rs \rangle \nu_p \alpha^\dagger_p \nu_q \alpha^\dagger_q \alpha_s \nu_s \alpha_r \nu_r c^*(n_1 n_2 \dots). \tag{43}$$

Here, according to the initial definition, p and q run only the values $r_1, r_2, \dots r_n$ (hereby $p \neq q$). However, one can admit for them all the values without exceptions if we take into account that superfluous items vanish.

Substitution of expressions (35), (41) and (43) into formula (9*) gives the wave equation for the wave function $c^*\,(n_1\; n_2\; \dots;\, t)$. But for Fermi statistics this wave function differs from the wave function $f(n_1 n_2 \dots ; t)$ of equation (15) only by the factor (namely, $\sqrt{n!}$), which commutes with separate items of the energy operator. Therefore, the wave equation for $f(n_1 n_2 \dots\ )$ has the same form as for $c^*(n_1 n_2 \dots\ )$, namely,

$$\mathbf{H} f(n_1 n_2 \dots ; t) - i\hbar \frac{\partial f}{\partial t} = 0,$$

where, according to (41) and (43), the energy operator $\mathbf{H}$ is

$$\mathbf{H} = \sum_{pr} \langle p \mid H \mid r \rangle\ \nu_p\ \alpha_p^\dagger\ \alpha_r\ \nu_r + \tag{44}$$

$$+ \frac{1}{2} \sum_{pqrs} \langle pq \mid G \mid rs \rangle \nu_p\ \alpha_p^\dagger\ \nu_q\ \alpha_q^\dagger\ \alpha_s\ \nu_s\ \alpha_r\ \nu_r.$$

The operators α_r and ν_r enter the energy operator only in combinations

$$a_r = \alpha_r \nu_r, \qquad a_r^\dagger = \nu_r \alpha_r^\dagger. \tag{45}$$

Indeed, we have

$$\mathbf{H} = \sum_{pr} a_p^\dagger \langle p \mid H \mid r \rangle a_r + \frac{1}{2} \sum_{pqrs} a_p^\dagger a_q^\dagger \langle pq \mid G \mid rs \rangle a_r^\dagger a_s^\dagger . \tag{44*}$$

As it can be easily proven with the help of (38) and (42), for the quantum amplitudes a_r there are the equalities

$$a_r^\dagger a_r = n_r, \qquad a_r a_r^\dagger = 1 - n_r \tag{46}$$

and the commutation relations

$$a_r a_s^\dagger + a_s^\dagger a_r = \delta_{rs}, \tag{47}$$

$$a_r a_s + a_s a_r = 0. \tag{47*}$$

Using "amplitudes" a_r, one can construct the quantized wave functions a_r

$$\Psi(x) = \sum_r a_r \psi_r(x), \tag{48}$$

$$\Psi^\dagger(x) = \sum_r a_r^\dagger \overline{\psi}_r(x), \tag{48*}$$

which satisfy the commutation relations

$$\Psi(x')\Psi^\dagger(x) + \Psi^\dagger(x)\Psi(x') = \delta(x - x'), \tag{49}$$

$$\Psi(x')\Psi(x) + \Psi(x)\Psi(x') = 0\,. \tag{49*}$$

As in the case of Bose statistics, the energy operator can be written in the form

$$\mathbf{H} = \int \Psi(x)^{\dagger} H(x)\Psi(x)\, dx + \tag{50}$$
$$+\frac{1}{2}\int\!\!\int \Psi^{\dagger}(x)\Psi^{\dagger}(x')G(xx')\Psi(x)\Psi(x')\, dx\, dx'.$$

Transition from the amplitudes a_r (or b_r in the case of Bose statistics) to quantized wave functions $\Psi(x)$ presents a unitary canonical transformation of single particle variables (transition from variables $E^{(r)}$ of formula (2) to variables x). One can consider the amplitudes a_r (or b_r) and also $\Psi(x)$ as quantized wave functions and formulae such as, e.g., the commutation relations or the expression for the energy operator written in terms of a_r (or b_r) are essentially interchangeable with those written in terms of $\Psi(x)$.

Note also that all other operators in the configuration space are transformed in the pattern of the energy operator and can be presented by means of the quantum wave functions. Herewith, as in the energy operator the order of the non-commuting factors is obtained in a completely unique way.

II Representation of Operators Ψ in Configuration Space

Total number of the particles does not enter explicitly the formulae of the second quantization; these formulae hold for arbitrary and even indefinite n. To the number n, one can put in correspondence the operator

$$\mathbf{n} = \int \Psi^{\dagger}(x)\Psi(x)dx \tag{1}$$

having the eigenvalues $n = 0, 1, 2\ldots.$

With respect to operator $\mathbf{n}$, one can split all the operators into two classes: the operators commuting with $\mathbf{n}$ belong to the first class, and the non-commuting ones belong to the second class.

Here we will deal with the representation in the configuration space of general operators non-commutating with $\mathbf{n}$ and, first of all, with the representation of operators $\Psi(x)$. It goes without saying that the results will be applicable also to operators commuting with $\mathbf{n}$ since they can be expressed in terms of $\Psi(x)$ and $\Psi^{\dagger}(x)$.

To unify both kinds of statistics, let us write the commutation relations for the quantized wave function in the form:

$$\Psi(x')\Psi^\dagger(x) - \varepsilon\Psi^\dagger(x)\Psi(x') = \delta(x - x'), \tag{2}$$

$$\Psi(x')\Psi(x) - \varepsilon\Psi(x)\Psi(x') = 0, \tag{2*}$$

where it should set $\varepsilon = 1$ for Bose statistics and $\varepsilon = -1$ for Fermi statistics. It follows from the definition (1) of operator $\mathbf{n}$ and from commutation relations (2) for both kinds of statistics that

$$\mathbf{n}\Psi - \Psi(\mathbf{n} - 1) = 0\,. \tag{3}$$

For $\Psi(x)$ we choose the representation in which $\mathbf{n}$ has a diagonal form. If we denote the matrix elements of $\Psi(x)$ in this representation as $\langle n \mid \Psi \mid n' \rangle$, then "selection rules" follow from relation (3):

$$(n - n' + 1)\langle n \mid \Psi \mid n' \rangle = 0\,, \tag{3*}$$

according to which only the matrix elements of the form $\langle n \mid \Psi \mid n' + 1 \rangle$ differ from zero. Hence, the matrix $\Psi(x)$ takes the form

$$\Psi(x) = \begin{pmatrix} 0 & \langle 0 \mid \Psi \mid 1 \rangle & 0 & 0 & \dots \\ 0 & 0 & \langle 1 \mid \Psi \mid 2 \rangle & 0 & \dots \\ 0 & 0 & 0 & \langle 2 \mid \Psi \mid 3 \rangle & \dots \\ \dots & \dots & \dots & \dots & \dots \end{pmatrix}. \tag{4}$$

A separate matrix element $\langle n - 1 \mid \Psi \mid n \rangle$ can be treated as an operator acting on the function of n variables[5] $x_1 x_2 \dots x_n$ and transforming this function into the function of $n-1$ variables $x_1 x_2 \dots x_{n-1}$ and parameter x. Thus, the operator $\Psi(x)$ acts on the sequence of functions

$$\begin{pmatrix} \text{const} \\ \psi(x_1) \\ \psi(x_1 x_2) \\ \psi(x_1 x_2 x_3) \\ \dots\dots\dots\dots \end{pmatrix} \tag{5}$$

of $0, 1, 2, 3, \dots$ variables and transforms it into a similar sequence; functions (5) can be interpreted as usual Schrödinger wave functions in the

[5]Each variable x_r is actually the set of variables, for example, x_r, y_r, z_r, σ_r describing the r-th particle. (*V. Fock*)

configuration space.[6] We shall say that $\psi(x_1 x_2 \ldots x_n)$ is the wave function in the n-th space. Let us suppose that the action of operator $\langle n-1 | \Psi(x) | n\rangle$ is defined as

$$\langle n-1 \mid \Psi(x) \mid n\rangle \, \psi(x_1 x_2 \ldots x_n) = \sqrt{n}\, \psi(x\; x_1\; x_2\; \ldots x_{n-1}) \qquad (6)$$

and show that for the proper definition of the conjugate to $\Psi(x)$ operator $\Psi^\dagger(x)$ commutation relations (2) will be implemented. Due to (4), the matrix $\Psi^\dagger(x)$ has the form

$$\Psi^\dagger(x) = \begin{pmatrix} 0 & 0 & 0 \ldots \\ \langle 1 \mid \Psi^\dagger \mid 0\rangle & 0 & 0 \ldots \\ 0 & \langle 2 \mid \Psi^\dagger \mid 1\rangle & 0 \ldots \\ \ldots & \ldots & \ldots\ldots \end{pmatrix}, \qquad (4^*)$$

where $\langle n \mid \Psi^\dagger \mid n-1\rangle$ is the operator conjugated to $\langle n-1 \mid \Psi \mid n\rangle$. This operator sends the function of $n-1$ variables $x_1 x_2 \ldots x_{n-1}$ into the function of n variables $x_1 x_2 \ldots x_n$ and parameter x. Herewith, one needs to take into account that the operator $\langle n \mid \Psi^\dagger \mid n-1\rangle$ should not change the symmetry properties of the wave function; it should transform the symmetric function into the symmetric one and the antisymmetric function into the antisymmetric one. Let us find the operator $\langle n|\Psi^\dagger|n-1\rangle$ defining its kernel; we can make it if we construct the kernel $\langle n-1| \, \Psi \mid n\rangle$ and pass to the conjugate kernel.

Because the wave function is either symmetric or antisymmetric, instead of (6) one can also write

$$\langle n-1 \mid \Psi \mid n\rangle \, \psi(x_1\, x_2 \ldots\, x_n) = \frac{1}{\sqrt{n}}[\psi(x\; x_1 \ldots\; x_{n-1}) +$$

$$+\varepsilon\, \psi(x_1\; x\; x_2 \ldots\; x_{n-1}) + \ldots + \varepsilon^{n-1}\psi(x_1 \ldots\; x_{n-1}\; x)]. \qquad (6^*)$$

The kernel of the operator defined by (6*) is

$$\langle n-1; x_1\; x_2 \ldots\; x_{n-1} \mid \Psi(x) \mid n; \xi_1\; \xi_2 \ldots\; \xi_n\rangle =$$

$$= \frac{1}{\sqrt{n}}[\delta(\xi_1 - x)\delta(\xi_2 - x_1)\ldots\delta(\xi_n - x_{n-1}) + \ldots$$

$$+\varepsilon^{k-1}\delta(\xi_1 - x_1)\ldots\delta(\xi_{k-1} - x_{k-1})\delta(\xi_k - x)\delta(\xi_{k+1} - x_k)\ldots\delta(\xi_n - x_{n-1}) +$$

$$\ldots + \varepsilon^{n-1}\; \delta(\xi_1 - x_1)\; \ldots\; \delta(\xi_{n-1} - x_{n-1})\delta(\xi_n - x)]. \qquad (7)$$

[6]For the first time such sequences of functions were considered by L. Landau and R. Peierls (Zs. Phys. **62** 188 (1930)). (*V. Fock*)

The kernel of the conjugate operator $\langle n \mid \Psi^\dagger(x) \mid n-1\rangle$ will be obtained if one replaces the variables $\xi_1, \xi_2, \ldots, \xi_n$ in formula (7) to $x_1, x_2, \ldots, x_n$ and the variables $x_1, x_2, \ldots, x_{n-1}$ to $\xi_1, \xi_2, \ldots, \xi_{n-1}$. Then the operator $\langle n \mid \Psi^\dagger(x) \mid n-1\rangle$ acts on the wave function $\psi(x_1, x_2, \ldots, x_n)$ as

$$\langle n \mid \Psi^\dagger \mid n-1\rangle\psi(x_1\, x_2\, \ldots\, x_{n-1}) =$$

$$= \frac{1}{\sqrt{n}}[\delta(x_1 - x)\psi(x_2\, x_3\, \ldots\, x_n) + \varepsilon\delta(x_2 - x)\psi(x_1\, x_3\, \ldots\, x_n) + \ldots$$
$$+\varepsilon^{k-1}\delta(x_k - x)\psi(x_1\, x_2\, \ldots\, x_{k-1}\, x_{k+1}\, \ldots\, x_n) + \ldots$$
$$+\varepsilon^{n-1}\delta(x_n - x)\psi(x_1\, x_2\, \ldots\, x_{n-1})]. \tag{8}$$

The operator $\langle n \mid \Psi^\dagger \mid n-1\rangle$ defined with this equality satisfies the requirement that it leaves the symmetry properties of the wave function unchanged; for this purpose we passed from relation (6) to relation (6*).

Once $\Psi(x)$ and $\Psi^\dagger(x)$ have been defined, we can turn to the proof of commutation relations (2) and (2*). Let us construct the operators $\Psi^\dagger(x)\Psi(x')$ and $\Psi(x')\Psi^\dagger(x)$. These operators commute with **n**; therefore, they have the diagonal form with respect to **n**. We have

$$\Psi^\dagger(x)\Psi(x') = \begin{pmatrix} 0 & 0 & 0 & \ldots \\ 0 & A_1 & 0 & \ldots \\ 0 & 0 & A_2 & \ldots \\ \ldots & \ldots & \ldots & \ldots \end{pmatrix} \tag{9}$$

and

$$\Psi(x')\Psi^\dagger(x) = \begin{pmatrix} B_0 & 0 & 0 & \ldots \\ 0 & B_1 & 0 & \ldots \\ 0 & 0 & B_2 & \ldots \\ \ldots & \ldots & \ldots & \ldots \end{pmatrix}, \tag{9*}$$

where A_n and B_n are the operators;

$$A_n = \langle n \mid \Psi^\dagger(x)\Psi(x') \mid n\rangle = \langle n \mid \Psi^\dagger(x) \mid n-1\rangle\langle n-1 \mid \Psi(x') \mid n\rangle, \tag{10}$$

$$B_n = \langle n \mid \Psi(x')\Psi^\dagger(x) \mid n\rangle = \langle n \mid \Psi(x') \mid n+1\rangle\langle n+1 \mid \Psi^\dagger(x) \mid n\rangle, \tag{10*}$$

which act in the n-th subspace. Applying initially (6) and then (8), we find

$$A_n\psi(x_1\, x_2\, \ldots\, x_n) = \delta(x_1 - x)\psi(x'\, x_2\, \ldots\, x_n) + \ldots$$
$$+\varepsilon^{k-1}\delta(x_k - x)\psi(x'\, x_1\, \ldots\, x_{k-1}\, x_{k+1}\, \ldots\, x_n) + \ldots$$
$$+\varepsilon^{n-1}\delta(x_n - x)\psi(x'\, x_1\, \ldots\, x_{n-1}) \tag{11}$$

or taking into account the symmetry properties of the wave function

$$\langle n \mid \Psi^\dagger(x)\Psi(x') \mid n\rangle\psi(x_1\, x_2\, \dots\, x_n) = \delta(x_1 - x)\psi(x'\, x_2\, \dots x_n) + \dots$$
$$+\delta(x_k - x)\psi(x_1\, \dots\, x_{k-1}\, x'\, x_{k+1}\, \dots\, x_n) + \dots$$
$$+\delta(x_n - x)\psi(x_1\, \dots\, x_{n-1}\, x'). \tag{11*}$$

If one applies first (8) and then (6), after changing n to $n+1$ we find

$$B_n\psi(x_1\, \dots\, x_n) = \langle n \mid \Psi(x')\Psi^\dagger(x) \mid n\rangle\psi(x_1\, \dots\, x_n) =$$
$$= \delta(x' - x)\psi(x_1\, x_2\, \dots\, x_n) + \varepsilon\delta(x_1 - x)\psi(x'\, x_2\, \dots\, x_n) + \dots$$
$$+\varepsilon^k\delta(x_k - x)\psi(x'\, x_1\, \dots\, x_{k-1}\, x_{k+1}\, \dots\, x_n) + \dots$$
$$+\varepsilon^n\delta(x_n - x)\psi(x'\, x_1\, \dots\, x_{n-1}). \tag{12}$$

Comparison of (11) and (12) shows that

$$B_n\psi(x_1x_2\dots x_n) - \varepsilon A_n\psi(x_1x_2\dots x_n) =$$
$$= \delta(x - x')\psi(x_1x_2\dots x_n)\,. \tag{13}$$

In virtue of (9) and (9*) it means that there are the commutation relations

$$\Psi(x')\Psi^\dagger(x) - \varepsilon\Psi^\dagger(x)\Psi(x') = \delta(x - x'), \tag{2}$$

where the unit matrix in the right-hand side (with respect to $\mathbf{n}$) is implied.

Relation (2*) is proven even simpler. By formula (6) the operator $\Psi(x)$ transforms the sequence of functions (5) into the sequence

$$\begin{pmatrix} \psi(x) \\ \sqrt{2}\psi(xx_1) \\ \sqrt{3}\psi(xx_1x_2) \\ \dots\dots\dots\dots \end{pmatrix},$$

i.e.,

$$\Psi(x)\begin{pmatrix} \text{const} \\ \psi(x_1) \\ \psi(x_1x_2) \\ \psi(x_1x_2x_3) \\ \dots\dots\dots\dots \end{pmatrix} = \begin{pmatrix} \psi(x) \\ \sqrt{2}\psi(xx_1) \\ \sqrt{3}\psi(xx_1x_2) \\ \dots\dots\dots\dots \\ \dots\dots\dots\dots \end{pmatrix}. \tag{14}$$

Applying the operator $\Psi(x')$ to (14), we shall have

$$\Psi(x')\Psi(x)\begin{pmatrix} \text{const} \\ \psi(x_1) \\ \psi(x_1x_2) \\ \psi(x_1x_2x_3) \\ \dots\dots\dots \end{pmatrix} = \begin{pmatrix} \sqrt{2\cdot 1}\psi(xx') \\ \sqrt{3\cdot 2}\psi(xx'x_1) \\ \sqrt{4\cdot 3}\psi(xx'x_1x_2) \\ \dots\dots\dots\dots\dots \\ \dots\dots\dots\dots\dots \end{pmatrix}. \qquad (15)$$

From this with substitution of x and x' we obtain

$$\Psi(x)\Psi(x')\begin{pmatrix} \text{const} \\ \psi(x_1) \\ \psi(x_1x_2) \\ \psi(x_1x_2x_3) \\ \dots\dots\dots \end{pmatrix} = \begin{pmatrix} \sqrt{2\cdot 1}\psi(x'x) \\ \sqrt{3\cdot 2}\psi(x'xx_1) \\ \sqrt{4\cdot 3}\psi(x'xx_1x_2) \\ \dots\dots\dots\dots\dots \\ \dots\dots\dots\dots\dots \end{pmatrix}. \qquad (15^*)$$

However, expressions (15) and (15*) are either equal ($\varepsilon = +1$ for symmetric functions) or equal in quantity but opposite in sign ($\varepsilon = -1$ for antisymmetric functions); hence, the commutation relation (2*) is proven.

By means of the formulae obtained, all the operators of the second quantization can be constructed in the configuration space. Those that do not commute with **n** act on the sequence of functions of the form (5) and cannot be represented in the configuration space of the definite number of particles. For commuting with **n** operators, it is sufficient to consider a diagonal (with respect to **n**) element of the matrix that can be interpreted as an operator in the n-th subspace, i.e., in the configuration space of a given number of n particles. For instance, the energy operator commutes with **n** (formula (50), Section 1), and as a result of its construction in the configuration space, we arrive back at the usual Schrödinger energy operator for n particles. Let us consider some more examples of commuting with **n** operators.

The operator $\Psi^+(x)\Psi(x)$ of the particle density in the n-th subspace has a representation

$$\langle n \mid \Psi^\dagger(x)\Psi(x) \mid n\rangle\psi(x_1x_2\dots x_n) =$$

$$= [\delta(x_1 - x) + \delta(x_2 - x) + \dots + \delta(x_n - x)]\psi(x_1\ x_2\ \dots\ x_n). \qquad (16)$$

This formula is a particular case of formula (11) that is obtained if one sets in (11) $x = x'$ and uses the relation $\delta(x_k - x)f(x) = \delta(x_k - x)f(x_k)$ valid for any continuous function.

Multiplying expression (16) by the volume element dx and integrating over some volume V, we can conclude that the operator

$$\mathbf{n}_V = \int_V \Psi^\dagger(x)\Psi(x)dx \tag{17}$$

has the following representation in the n-th subspace:

$$\langle n \mid \mathbf{n}_V \mid n\rangle\psi(x_1 \,\dots\, x_n) = n'_V\,(x_1 \,\dots\, x_n)\psi(x_1 \,\dots\, x_n)\,. \tag{18}$$

The value of the function $n'_V(x_1 \dots\, x_n)$ in (18) is equal to the number of those arguments $x_1 \dots\, x_n$ that belong to the volume V. Therefore, as was expected, the operator $\mathbf{n}_V$ has integer eigenvalues.

As another example, let us consider the operator of the Coulomb potential

$$\mathbf{V}(\mathbf{r}) = e^2 \int \frac{\Psi^\dagger(x')\Psi(x')}{|\mathbf{r}-\mathbf{r}'|}dx'\,. \tag{19}$$

To construct the matrix element $\langle n \mid \mathbf{V}(\mathbf{r}) \mid n\rangle$ in (16), we replace x by x', multiply it by $\frac{e^2}{|\mathbf{r}-\mathbf{r}'|}$ and integrate over x' . We obtain:

$$\langle n \mid \mathbf{V}(\mathbf{r}) \mid n\rangle\psi(x_1 \,\dots\, x_n) = \sum_{k=1}^{n} \frac{e^2}{|\mathbf{r}-\mathbf{r}_k|}\psi(x_1 \,\dots\, x_n)\,. \tag{20}$$

Hence, the operator $\mathbf{V}(\mathbf{r})$ in the n-th space means "multiplication by $\sum_{k=1}^{n} \frac{e^2}{|\mathbf{r}-\mathbf{r}_k|}$."

Time Dependence of Ψ Operators and Quantized Wave Equation

As is known, the evolution in time of the state of a physical system manifests itself as either the time dependence of the wave function or the time dependence of the operator. Following Dirac,[7] we shall call the Schrödinger representation the representation of the operators, in which the time dependence is transferred to the wave function. The representation, in which the time dependence is transferred to the operators (matrices), shall be called the Heisenberg representation.

Let ψ be the wave function of the system and $S(t)$ the unitary operator that transforms the initial wave function $\psi(\cdot, 0)$ into wave function

[7] P.A.M. Dirac, The Principles of Quantum Mechanics, Oxford (1930).

$\psi(\cdot, t)$ corresponding to time t (here, the variables of the system are denoted by a point). We have

$$\psi(\cdot, t) = S(t)\psi(\cdot, 0)\,. \tag{21}$$

Differentiating this equation with respect to time and replacing $\psi(\cdot, 0)$ with

$$\psi(\cdot, 0) = S^\dagger(t)\psi(\cdot, t), \tag{21*}$$

we obtain

$$\frac{\partial \psi}{\partial t} = \dot{S}(t)\psi(\cdot, 0) = \dot{S}(t)S^\dagger(t)\psi(\cdot, t). \tag{22}$$

We denote the operator $i\dot{S}(t)S^\dagger(t)$, which will be Hermitean in virtue $SS^\dagger = 1$ as

$$i\dot{S}(t)S^\dagger(t) = -iS(t)\dot{S}^\dagger(t) = \frac{1}{\hbar}H\,. \tag{23}$$

Then H is the Hamilton operator of the system. Denoting by L some operator in the Schrödinger representation and by $L'(t)$ the same operator in the Heisenberg representation, we shall have:

$$L'(t) = S^\dagger(t)LS(t)\,. \tag{24}$$

From (24) and (21), we obtain

$$L'(t)\psi(\cdot, 0) = S^\dagger(t)L\psi(\cdot, t). \tag{25}$$

The time derivative of this expression is equal to

$$\frac{dL'(t)}{dt}\psi(\cdot, 0) = \frac{d}{dt}[S^\dagger(t)L\psi(\cdot, t)]. \tag{26}$$

Here, the operator $\frac{dL'(t)}{dt}$ on the left is in the Heisenberg representation. Denoting the same operator in the Schrödinger representation as $\frac{dL}{dt}$, we obtain, similar to (24) and (25),

$$\frac{dL'(t)}{dt} = S^\dagger(t)\frac{dL}{dt}S(t) \tag{27}$$

and

$$\frac{dL'(t)}{dt}\psi(\cdot, 0) = S^\dagger(t)\frac{dL}{dt}\psi(\cdot, t)\,. \tag{28}$$

Comparison of (26) and (28) gives

$$\frac{dL}{dt}\psi(\cdot, t) = S(t)\frac{d}{dt}[S^\dagger L\psi(\cdot, t)]\,, \tag{29}$$

and after differentiation

$$\frac{dL}{dt}\psi(\cdot,t) = S(t)\dot{S}^{\dagger}L\psi(\cdot,t) + \frac{d}{dt}[L\psi(\cdot,t)]\,. \tag{30}$$

In accordance with (21), the operator $S(t)$ gives the time evolution of the wave function for time t in positive direction.

Analogously, the operator $S^{\dagger}(t)$ gives the time evolution of the wave function for time t in negative direction. Taking this into account one can formulate the meaning of equation (29) in the following way.

The result of the action of the operator $\frac{dL}{dt}$ on the wave function $\psi(\cdot,t)$ in the Schrödinger representation is obtained by means of the following operations:

1. Application of the operator L.
2. Continuation in time by the quantity t in negative direction.
3. Differentiation with respect to t.
4. Continuation in time by the quantity t in positive direction.

This formulation has the advantage that it does not use the Hamilton operator, at least in an explicit way.

Now let us apply this rule to the definition of the operator $\frac{d\Psi}{dt}$, which enters the quantized wave equation. In this case, the operator is the quantum wave function $\Psi(x,t)$ and $\psi(\cdot,t)$ is the sequence of functions (5). We restrict ourselves to the case when the number of particles does not change with time, thus excluding photons from consideration. For lucidity, we consider the quantum Schrödinger equation

$$[H^0(x) + \mathbf{V}(x)]\Psi(x) = i\hbar\frac{\partial\Psi}{\partial t}, \tag{31}$$

where H^0 denotes the usual Schrödinger operator for the one-body problem

$$H^0(x) = -\frac{\hbar^2}{2m}\left(\frac{\partial^2}{\partial x^2} + \frac{\partial^2}{\partial y^2} + \frac{\partial^2}{\partial z^2}\right) + U(x,y,z) \tag{32}$$

and $\mathbf{V}(x) = \mathbf{V}(\mathbf{r})$ is the Coulomb potential operator defined in (19).

In our case, the operator $S(t)$ produces in the configuration space just the continuation of the separate wave functions of sequence (5) in time:

$$S(t)\begin{pmatrix} \text{const} \\ \psi(x_1;0) \\ \psi(x_1\ x_2;0) \\ \cdots\cdots\cdots \end{pmatrix} = \begin{pmatrix} \text{const} \\ \psi(x_1;t) \\ \psi(x_1\ x_2;t) \\ \cdots\cdots\cdots \end{pmatrix}. \tag{33}$$

Hence, the operator $S(t)$ is of the diagonal form

$$S(t) = \begin{pmatrix} S_0 & 0 & 0 & \cdots \\ 0 & S_1 & 0 & \cdots \\ 0 & 0 & S_3 & \cdots \\ \cdot & \cdot & \cdot & \cdots \end{pmatrix}, \tag{34}$$

where the operator $S_n = S_n(t)$ continues the wave function in the n-th subspace in time:

$$S_n(t)\psi(x_1\; x_2\; \ldots\; x_n; 0) = \psi(x_1\; x_2\; \ldots\; x_n; t)\,. \tag{35}$$

Furthermore, by (23) we have

$$-i\hbar S_n(t)\dot{S}_n^\dagger(t) = H(x_1 \ldots x_n), \tag{36}$$

where $H(x_1 \ldots x_n)$ denotes the Hamilton operator in the n-th subspace. The operator $-i\hbar S(t)\dot{S}^\dagger(t)$ will, therefore, be also diagonal and its diagonal entries will be operators (36).

Now, let us form the operator $\dot{\Psi}(xt) = \frac{\partial\Psi}{\partial t}$. By (29) or (30), we have

$$\dot{\Psi}(x,t)\begin{pmatrix} \text{const} \\ \psi(x_1\; t) \\ \psi(x_1\; x_2\; t) \\ \cdots\cdots\cdots \end{pmatrix} = S(t)\dot{S}^\dagger(t)\begin{pmatrix} \psi(x\; t) \\ \sqrt{2}\psi(x\; x_1\; t) \\ \sqrt{3}\psi(x\; x_1\; x_2\; t) \\ \cdots\cdots\cdots\cdots\cdots \end{pmatrix} +$$

$$+\begin{pmatrix} \dot{\psi}(x\; t) \\ \sqrt{2}\dot{\psi}(x\; x_1\; t) \\ \sqrt{3}\dot{\psi}(x\; x_1\; x_2\; t) \\ \cdots\cdots\cdots\cdots\cdots \end{pmatrix} \tag{37}$$

or, taking into account (36),

$$i\hbar\dot{\Psi}(x\; t)\begin{pmatrix} \text{const} \\ \psi(x_1\; t) \\ \psi(x_1\; x_2\; t) \\ \cdots\cdots\cdots \end{pmatrix} =$$

$$\begin{pmatrix} 0 \\ -\sqrt{2}H(x_1)\quad \psi(x\; x_1\; t) \\ -\sqrt{3}H(x_1\; x_2)\psi(x\; x_1\; x_2\; t) \\ \cdots\cdots\cdots\cdots\cdots\cdots\cdots\cdots \end{pmatrix} + i\hbar\begin{pmatrix} \dot{\psi}(x\; t) \\ \sqrt{2}\dot{\psi}(x\; x_1\; t) \\ \sqrt{3}\dot{\psi}(x\; x_1\; x_2\; t) \\ \cdots\cdots\cdots\cdots\cdots \end{pmatrix}. \tag{38}$$

Taking into consideration (20), we obtain the following expression for the operator in the left-hand side of the quantized wave equation (30):

$$[H_0(x) + \mathbf{V}(x)]\Psi \begin{pmatrix} \text{const} \\ \psi(x_1\ t) \\ \psi(x_1\ x_2\ t) \\ \cdots\cdots\cdots \end{pmatrix} =$$

$$= [H^0(x) + \mathbf{V}(x)] \begin{pmatrix} \psi(xt) \\ \sqrt{2}\psi(x\ x_1\ t) \\ \sqrt{3}\psi(x\ x_1\ x_2\ t) \\ \cdots\cdots\cdots\cdots \end{pmatrix} =$$

$$= \begin{pmatrix} H^0(x)\psi(x\ t) \\ \sqrt{2}\left[H^0(x) + \dfrac{e^2}{|\mathbf{r}-\mathbf{r}_1|}\right]\psi(x\ x_1\ t) \\ \sqrt{3}\left[H^0(x) + \dfrac{e^2}{|\mathbf{r}-\mathbf{r}_1|} + \dfrac{e^2}{|\mathbf{r}-\mathbf{r}_2|}\right]\psi(x\ x_1\ x_2\ t) \\ \cdots\cdots\cdots\cdots\cdots\cdots\cdots\cdots\cdots\cdots \end{pmatrix}. \tag{39}$$

Equating (38) and (39), we obtain (after cancellations of $\sqrt{2}$, $\sqrt{3}$, etc.) the chain of equations

$$H^0(x)\psi(x\ t) = i\hbar\frac{\partial\psi(x\ t)}{\partial t}, \tag{40_1}$$

$$\left[H^0(x) + \frac{e^2}{|\mathbf{r}-\mathbf{r}_1|} + H(x_1)\right]\psi(x\ x_1\ t) = i\hbar\frac{\partial\psi(x\ x_1\ t)}{\partial t}, \tag{40_2}$$

$$\left[H^0(x) + \frac{e^2}{|\mathbf{r}-\mathbf{r}_1|} + \frac{e^2}{|\mathbf{r}-\mathbf{r}_2|} + H(x_1\ x_2)\right]\psi(x\ x_1\ x_2\ t) =$$
$$= i\hbar\frac{\partial\psi(x\ x_1\ x_2\ t)}{\partial t}, \tag{40_3}$$

. . . .

$$\left[H^0(x) + \sum_{k=1}^{n}\frac{e^2}{|\mathbf{r}-\mathbf{r_k}|} + H(x_1\ x_2\ \ldots\ x_n)\right]\psi(x\ x_1\ \ldots\ x_n\ t) =$$
$$= i\hbar\frac{\partial\psi(x\ x_1\ \ldots\ x_n\ t)}{\partial t}. \tag{40_{n+1}}$$

From equation (40_1), we conclude that

$$H(x) = H^0(x).$$

Then equation (40_2) gives

$$H(x\ x_1) = H^0(x) + H^0(x_1) + \frac{e^2}{|\mathbf{r} - \mathbf{r}_1|}$$

and so on. Generally, the $(n+1)$-th equation gives the recurrent relation between the Schrödinger operators for n and $n+1$ particles, namely,

$$H(x\ x_1\ x_2\ \ldots\ x_n) = H^0(x) + \sum_{k=1}^{n} \frac{e^2}{|\mathbf{r} - \mathbf{r}_k|} + H(x_1\ x_2\ \ldots\ x_n)\,. \quad (41)$$

Now, expressing $H(x_1\ x_2\ \ldots\ x_n)$ directly through H^0, we obtain for the Hamilton operator of the n-body problem the usual Schrödinger's expression

$$H(x_1 x_2 \ldots x_n) = \sum_{k=1}^{n} H^0(x_k) + \sum_{k>l=1}^{n} \frac{e^2}{|\mathbf{r}_k - \mathbf{r}_l|}. \quad (42)$$

Thus, in the configuration space the quantized wave equation is decomposed into a set of the usual Schrödinger's equations of the form

$$H(x_1\ x_2\ \ldots\ x_n)\psi(x_1\ x_2\ \ldots\ x_n\ t) = i\hbar\frac{\partial\psi}{\partial t}. \quad (43)$$

It is seen from this example that the reasoning with quantum wave function admits the direct transfer to the usual configuration space at any stage.

Derivation of the Hartree Equation by the Second Quantization Method

As a simple application of the results obtained, we derive the Hartree equations[8] extended by account of the exchange.

As is known the equations for the eigenfunction of the energy operator (and the Hartree equations, too) can be derived from the variation principle

$$\delta W = 0\,, \quad (44)$$

where W denotes the energy of an atom in the stationary state considered. Therefore, it is sufficient to find the expression for the energy. But the quantity W is equal to the diagonal element

$$W = \langle Wn \mid H \mid Wn \rangle \quad (45)$$

[8] V. Fock, Zs. Phys. **61**, 126 (1930). (See [30-2] in this book (*Editors*)).

of the matrix of the quantum energy operator **H** (formula (50) of the first part)

$$\mathbf{H} = \int \Psi^\dagger(x)H(x)\Psi(x)dx + \frac{e^2}{2}\int \frac{\Psi^\dagger(x)\Psi^\dagger(x')\Psi(x')\Psi(x)dxdx'}{|\mathbf{r}-\mathbf{r}'|}. \quad (46)$$

To define W, we should calculate the matrix elements of integrand operators. We have

$$\begin{aligned} &\langle Wn \mid \Psi^\dagger(x)H(x)\Psi(x) \mid Wn\rangle = \\ &= H(x')\langle Wn \mid \Psi^\dagger(x)\Psi(x') \mid Wn\rangle, \qquad x = x'. \end{aligned} \quad (47)$$

Therefore, for the calculation of the matrix element of the integrand operator of the first integral, it is sufficient to define the quantity

$$\varrho(x\,x') = \langle Wn \mid \Psi^\dagger(x)\Psi(x') \mid Wn\rangle\,. \quad (48)$$

The expression for the operator $\Psi^\dagger(x)\Psi(x')$ in the n-th subspace has been already found in (11*). Let $\psi_W(x_1x_2\ldots x_n)$ be the eigenfunction belonging to the eigenvalue W of the energy operator in the n-th subspace. Then, by taking into account the symmetry properties of the wave function, formula (48) gives

$$\varrho(x\,x') = n\int\cdots\int \overline{\psi}_W(x\,x_2\,\ldots\,x_n)\psi_W(x'x_2\,\ldots\,x_n)dx_2\,\ldots\,dx_n. \quad (49)$$

To derive the Hartree equations, we should replace the wave function in the exact expression by the determinant approximation

$$\psi_W(x_1x_2\ldots x_n) = \frac{1}{\sqrt{n!}} \parallel \varphi_i(x_k) \parallel \qquad i,k = 1,2,\ldots n\,, \quad (50)$$

where $\varphi_i(x)$ are assumed to be orthogonal and normalized

$$\int \overline{\varphi}_i(x)\varphi_k(x)dx = \delta_{ik}\,. \quad (51)$$

Then, we obtain

$$\varrho(x\,x') = \sum_{i=1}^{n} \overline{\varphi}_i(x)\varphi_i(x') \quad (49^*)$$

and formulae (47) and (48) give

$$\langle Wn \mid \Psi^\dagger(x)H\Psi(x) \mid Wn\rangle = \sum_{i=1} \overline{\varphi}_i(x)H(x)\varphi_i(x). \quad (52)$$

Now, let us calculate the matrix element in the double integral (46). By means of relations (14), (16) and (8), it is easy to find that in the n-th subspace this operator has the form

$$\langle n \mid \Psi^\dagger(x)\Psi^\dagger(x')\Psi(x')\Psi(x) \mid n\rangle\psi(x_1x_2\ldots x_n) =$$

$$= \sum_{k,\, l=1,\, k\neq l}^{n} \delta(x_k - x)\delta(x_l - x')\psi(x_1x_2\ldots x_n). \tag{53}$$

Hence, its matrix element is equal to

$$\langle Wn \mid \Psi^\dagger(x)\Psi^\dagger(x')\Psi(x')\Psi(x) \mid Wn\rangle =$$

$$= n(n-1)\int\cdots\int |\psi(x\, x'x_3\, \ldots\, x_n|^2\, dx_3\, \ldots\, dx_n\,. \tag{54}$$

After substitution ψ_n in the form of determinant (50), we obtain from (54) an approximative formula

$$\langle Wn \mid \Psi^\dagger(x)\Psi^\dagger(x')\Psi(x')\Psi(x) \mid Wn\rangle =$$

$$= \varrho(x\, x)\varrho(x'\, x') - |\varrho(x\, x')|^2. \tag{55}$$

Now, substituting (52) and (55) into formula (46) for the matrix element H, i.e., for the energy W, we get the expression:

$$W = \int \sum_{r=1}^{n} \overline{\varphi}_r H(x)\varphi_r(x)dx + \frac{e^2}{2}\int\int \frac{\varrho(x\, x)\varrho(x'\, x') - |\varrho(x\, x')|^2}{|\mathbf{r} - \mathbf{r}'|} dx dx'\,. \tag{56}$$

This differs from the result obtained in our quoted paper only in that here we suppose the spin coordinate to be included in the variable x and, hence, we can operate with purely antisymmetric wave functions.

Translated by Yu.M. Pis'mak

32-3*

On Dirac's Quantum Electrodynamics

V. Fock and B. Podolsky

Phys. Zs. Sowjetunion **1**, 798, 1932 (in English)
Fock57, pp. 52–54

In his new paper,[1] Dirac suggested an original combination of the quantum electrodynamics of vacuum with the wave equation for matter. For a one-dimensional example, he demonstrated how the Coulomb interaction can appear in some approximation.

Since the one-dimensional case cannot make physical sense, it is natural to try to make calculations for a three-dimensional case. This will be done here but very briefly; a more detailed exposition will be presented in another article. We denote the components of four-dimensional potential in Heaviside units as $A_0 = V$; A_1, A_2, A_3 and write the Fourier expansion for these quantities as follows:

$$A_l(\mathbf{r},t) = \frac{1}{(2\pi)^{3/2}} \int a_l(\mathbf{k}) e^{-ic|k|t+i\mathbf{k}\cdot\mathbf{r}} (d\mathbf{k}) +$$

$$+ \frac{1}{(2\pi)^{3/2}} \int a_l^\dagger(\mathbf{k}) e^{ic|k|t-i\mathbf{k}\cdot\mathbf{r}} (d\mathbf{k}),$$

where

$$\mathbf{r} = (x_1, x_2, x_3); \qquad \mathbf{k} = (k_1, k_2, k_3).$$

Applying properly the known quantization rules, we find the following commutation relations for the amplitudes $a_l(\mathbf{k})$:

$$a_l(\mathbf{k}')a_m^\dagger(\mathbf{k}) - a_m^\dagger(\mathbf{k})a_l(\mathbf{k}') = -\frac{\hbar c}{2|k|} e_l \delta_{lm} \delta(\mathbf{k} - \mathbf{k}')$$

(here $e_0 = 1$; $e_1 = e_2 = e_3 = -1$; $l, m = 0, 1, 2, 3$; $2\pi\hbar$ is the Planck constant). As we will neglect the relativistic effects, we restrict our consideration only to the case of the scalar potential. According to Dirac, the wave equation is

$$W\psi - i\hbar \frac{\partial \psi}{\partial t} = -(\varepsilon_1 V(\mathbf{r}_1, t) + \varepsilon_2 V(\mathbf{r}_2, t))\psi,$$

[1] P.A.M. Dirac, Proc. Roy. Soc. **A136**, 453, 1932.

where

$$W = \frac{1}{2m_1}\mathbf{p}_1^2 + \frac{1}{2m_2}\mathbf{p}_2^2.$$

We shall look for the wave function in the form

$$\psi = \psi_0 + \psi_1 + \psi_2 + \dots ,$$

where ψ_n is of the n-th degree in charges ε_1 and ε_2. We transform the equation to the momentum space and write down the expansion

$$\varphi = \varphi_0 + \varphi_1 + \varphi_2 + \dots$$

for the corresponding wave function. We put

$$\psi_0 = \frac{1}{(2\pi h)^3} e^{\frac{i}{\hbar}(\mathbf{p}_1^0\cdot\mathbf{r}_1 + \mathbf{p}_2^0\cdot\mathbf{r}_2 - W_0 t)} \cdot \delta_{j0}.$$

Then

$$\varphi_0 = \delta(\mathbf{p}_1 - \mathbf{p}_1^0)\delta(\mathbf{p}_2 - \mathbf{p}_2^0)e^{-\frac{i}{\hbar}W_0 t} \cdot \delta_{j0}.$$

For the function φ_1, we find the expression linear in operators $a_0(\mathbf{k}')$ and $a_0^\dagger(\mathbf{k})$ that contains four terms. The right-hand side of the equation for φ_2 will be quadratic in $a_0(\mathbf{k}')$ and $a_0^\dagger(\mathbf{k})$. We should replace the terms of the type $a_0(\mathbf{k}')a_0(\mathbf{k})$; $a_0^\dagger(\mathbf{k}')a_0^\dagger(\mathbf{k})$; $a_0(\mathbf{k}')a_0^\dagger(\mathbf{k})$ by zero and in the term $a_0^+ a_0$ make the substitution[2]

$$a_0^\dagger(\mathbf{k})a_0(\mathbf{k}') \rightarrow \frac{\hbar c}{2|k|}\delta(\mathbf{k} - \mathbf{k}').$$

Then, we obtain

$$W\varphi_2 - i\hbar\frac{\partial\varphi_2}{\partial t} =$$

$$= K\varphi_0 - \frac{\varepsilon_1\varepsilon_2}{(2\pi)^3\hbar}\frac{1}{|\mathbf{p}_1 - \mathbf{p}_1^0|^2}\delta(\mathbf{p}_1 + \mathbf{p}_2 - \mathbf{p}_1^0 - \mathbf{p}_2^0)e^{-\frac{i}{\hbar}W_0 t} \cdot \delta_{j0},$$

where K is a constant (in fact, an infinite one). Transforming this equation back to the coordinate space, we find

$$W\psi_2 - i\hbar\frac{\partial\psi_2}{\partial t} = \left(K - \frac{\varepsilon_1\varepsilon_2}{4\pi|\mathbf{r}_1 - \mathbf{r}_2|}\right)\psi_0 = -U\psi_0.$$

[2]We are looking for that part of the wave functional φ (or ψ) that corresponds to the zero-quantum state. (This is denoted symbolically by the factor δ_{j0}, where j is the number of light quanta.) Only the retained operator $a_0^\dagger a_0$ sends this zero-quantum part again to a zero-quantum one, whereas the removed operators $a_0 a_0^\dagger$, $a_0^\dagger a_0^\dagger$ and $a_0 a_0$ either cancel it or transform it into a two-quantum state. (*V. Fock*)

From this equation, one sees that the interaction energy is equal to

$$U = \frac{\varepsilon_1\varepsilon_2}{4\pi|\mathbf{r}_1 - \mathbf{r}_2|}) - K.$$

But this is just the Coulomb energy (in Heaviside units) and, moreover, with the correct sign. It seems very likely that if one will not neglect the vector potential then Breit's retardation correction will appear in the corresponding approximation.

Typeset by V.V. Vechernin

32-4*
On Quantization of Electro-magnetic Waves and Interaction of Charges in Dirac Theory[1]

V.A. Fock and B. Podolsky

Received 2 July 1932

Phys. Zs. Sowijetunion **1**, 798, 1932 (in English)
Fock1957, pp. 55–69.

In the previous note [1], we have stated the result of the application of Dirac's new ideas to calculation of the electro-static interaction of two charges. In this article, we give the detailed account of this result and also give the application of Dirac's ideas to a more exact account of the interaction with the help of Dirac's linear Hamilton function. The work is divided into three parts. In the first part, we consider the problem of quantization of the electro-magnetic field in an empty space, which is of new interest in view of Dirac's ideas. In the second part, an application is made to the calculation of the interaction resulting from the scalar potential in the Schrödinger equation, and the Coulomb force is obtained. This part, in an abbreviated form, was the subject of the previous note. In the third part, an application to Dirac's Hamilton operator is made.

I Quantization of Electro-Magnetic Fields

§1. In applying here Dirac's theory only the fields satisfying electro-magnetic equations for an empty space need to be considered. Thus, each component of the electric and magnetic fields as well as each component of the vector and scalar potentials must satisfy the D'Alembert wave equation. Applying the general theory of quantization of fields developed by Heisenberg and Pauli [2], we consider the D'Alembert equations as fundamental and Maxwell's equations as secondary ones. Therefore, we

[1]This paper includes the correction published separately as [7]. Up to formula (61) the text coincides with article [6] and after formula (61) with the text of §2 of article [7]. (*V. Fock, 1957*)

chose a Lagrangian function to obtain the D'Alembert equation as the equation of motion. In this way, the difficulty of vanishing impulse P_0 is avoided and Maxwell's equations appear as a result of an additional condition $P_0 = 0$.

We identify four coordinates of the field Q_0, Q_1, Q_2, Q_3 with Φ, A_x, A_y, A_z, respectively, Φ being the scalar and $\mathbf{A} = (A_x, A_y, A_z)$ the vector potentials. As usual, we put

$$\mathbf{E} = -\operatorname{grad}\Phi - \frac{1}{c}\dot{\mathbf{A}} \tag{1}$$

and

$$\mathbf{H} = \operatorname{curl}\mathbf{A}. \tag{2}$$

Heaviside units are used throughout this article.

The Lagrangian function is assumed to be

$$L = \frac{1}{2}(\mathbf{E}^2 - \mathbf{H}^2) - \frac{1}{2}\left(\operatorname{div}\mathbf{A} + \frac{1}{c}\dot{\Phi}\right)^2, \tag{3}$$

which is a four-dimensional invariant. In coordinates of the field and their derivatives, the Lagrangian function becomes

$$L = \frac{1}{2}\sum_{l=1}^{3}\left(\frac{1}{c}\dot{Q}_l + \frac{\partial Q_0}{\partial x_l}\right)^2 - \frac{1}{2}\sum_{l>m}\left(\frac{\partial Q_l}{\partial x_m} - \frac{\partial Q_m}{\partial x_l}\right)^2 -$$

$$-\frac{1}{2}\left(\frac{1}{c}\dot{Q}_0 + \sum_{l=1}^{3}\frac{\partial Q_l}{\partial x_l}\right)^2. \tag{4}$$

It follows for $l = 1, 2, 3$

$$P_l = \frac{\partial L}{\partial \dot{Q}_l} = \frac{1}{c}\left(\frac{\partial Q_0}{\partial x_l} + \frac{1}{c}\dot{Q}_l\right) \tag{5}$$

and for $l = 0$

$$P_0 = \frac{\partial L}{\partial \dot{Q}_0} = -\frac{1}{c}\left(\frac{1}{c}\dot{Q}_0 + \sum_{l=1}^{3}\frac{\partial Q_l}{\partial x_l}\right) = -\frac{1}{c}\left(\operatorname{div}\mathbf{A} + \frac{1}{c}\dot{\Phi}\right). \tag{6}$$

The Hamilton function is formed in the usual way and reads as

$$H = \frac{1}{2}(\mathbf{E}^2 + \mathbf{H}^2) - c\mathbf{P}\cdot\operatorname{grad}\Phi - cP_0\operatorname{div}\mathbf{A} - \frac{1}{2}c^2P_0^2 \tag{7}$$

or, in terms of momenta,

$$H = \frac{c^2}{2}(\mathbf{P}^2 - P_0^2) + \frac{1}{2}\sum_{l>m}\left(\frac{\partial Q_l}{\partial x_m} - \frac{\partial Q_m}{\partial x_l}\right)^2 -$$

$$-cP_0\sum_{l=1}^{3}\frac{\partial Q_l}{\partial x_l} - c\mathbf{P}\cdot \operatorname{grad} Q_0, \tag{8}$$

where $\mathbf{P}$ is the vector $\mathbf{P} = (P_1, P_2, P_3)$. It is to be noted that this expression for H does not contain non-commuting factors and can, therefore, be directly accepted in quantum mechanics.

According to Heisenberg and Pauli [2] (formula (10)) equations of motion follow from the Hamilton function by the rule

$$\dot{Q}_\alpha = \frac{\partial H}{\partial P_\alpha}; \qquad \dot{P}_\alpha = -\frac{\partial H}{\partial Q_\alpha} + \sum_{l=1}^{3}\frac{\partial}{\partial x_l}\frac{\partial H}{\partial\left(\dfrac{\partial Q_\alpha}{\partial x_l}\right)}. \tag{9}$$

Application of this rule to our case gives

$$\left.\begin{aligned}
&\dot{\mathbf{A}} = c^2\mathbf{P} - c\operatorname{grad}\Phi,\\
&\dot{\Phi} = -c^2P_0 - c\operatorname{div}\mathbf{A},\\
&\dot{\mathbf{P}} = \Delta\mathbf{A} - \operatorname{grad}\operatorname{div}\mathbf{A} - c\operatorname{grad}P_0,\\
&\dot{P}_0 = -c\operatorname{div}\mathbf{P}.
\end{aligned}\right\} \tag{10}$$

Eliminating $\mathbf{P}$ and P_0 from these equations, we obtain the desirable equations

$$\Delta\mathbf{A} - \frac{1}{c^2}\ddot{\mathbf{A}} = 0, \qquad \Delta\Phi - \frac{1}{c^2}\ddot{\Phi} = 0. \tag{11}$$

If we add the condition $P_0 = 0$, that according to equation (6) reads as

$$\operatorname{div}\mathbf{A} + \frac{1}{c}\dot{\Phi} = 0, \tag{12}$$

then formulae (1), (2), (11) and (12) lead to Maxwell's equations for an empty space as it should be.

§2. To each field variable $F = F(x, y, z, t)$, we put in correspondence the amplitudes $F(\mathbf{k})$ and $F^\dagger(\mathbf{k})$ defined by decomposition of F into plane monochromatic waves. It reads

$$F = \frac{1}{(2\pi)^{3/2}}\int\left\{F(\mathbf{k})e^{-ic|k|t+i\mathbf{k}\cdot\mathbf{r}} + F^\dagger(\mathbf{k})e^{ic|k|t-i\mathbf{k}\cdot\mathbf{r}}\right\}(d\mathbf{k}), \tag{13}$$

where $\mathbf{r} = (x, y, z)$ is the position vector and $\mathbf{k} = (k_x, k_y, k_z)$ is the wave vector, whose direction is the direction of the propagation of the wave and the magnitude is equal to $|k| = \frac{2\pi}{\lambda}$, λ being the wave length. Further, $(d\mathbf{k}) = dk_x dk_y dk_z$, the integration being performed over all values of each component of $\mathbf{k}$ from $-\infty$ to ∞.

Obviously, if F is a vector, then each vector component has its own amplitude, so $F(\mathbf{k})$ is also a vector.[2] A relation between the amplitudes corresponds to each relation between the field quantities. We shall denote these relations by the same numbers with an asterisk. Thus, corresponding to equation (10), we have:

$$\begin{aligned} -ic|k|\mathbf{A}(\mathbf{k}) &= c^2\mathbf{P}(\mathbf{k}) - ic\mathbf{k}\Phi(\mathbf{k}), \\ -ic|k|\Phi(\mathbf{k}) &= -c^2 P_0(\mathbf{k}) - ic\mathbf{k}\cdot\mathbf{A}(\mathbf{k}), \\ -ic|k|\mathbf{P}(\mathbf{k}) &= -|k|^2\mathbf{A}(\mathbf{k}) + \mathbf{k}(\mathbf{k}\cdot\mathbf{A}(\mathbf{k})) - ic\mathbf{k}P_0(\mathbf{k}), \\ -ic|k|P_0(\mathbf{k}) &= -ic\mathbf{k}\cdot\mathbf{P}(\mathbf{k}). \end{aligned} \tag{10*}$$

It is easy to see that in equation (10*) the last two equations are algebraic consequences of the first two equations. These equations define the amplitudes of momenta in terms of the amplitudes of potentials

$$\mathbf{P}(\mathbf{k}) = \frac{i}{c}[\mathbf{k}\Phi(\mathbf{k}) - |k|\mathbf{A}(\mathbf{k})] = -\frac{1}{c}\mathbf{E}(\mathbf{k}), \tag{5*}$$

$$P_0(\mathbf{k}) = \frac{\mathbf{k}\cdot\mathbf{P}(\mathbf{k})}{|k|} = \frac{i}{c}[|k|\Phi(\mathbf{k}) - \mathbf{k}\cdot\mathbf{A}(\mathbf{k})]. \tag{6*}$$

Recall that each of these equations implies its conjugate, which in the classical theory is an ordinary complex conjugate. It is suggested that the adjoint equation holds. In terms of amplitudes, the D'Alembert equation becomes an algebraic identity.

Due to (5*) and (6*), the additional condition $P_0 = 0$ becomes

$$i\mathbf{k}\cdot\mathbf{E}(\mathbf{k}) = \operatorname{div}\mathbf{E}(\mathbf{k}) = 0, \tag{14}$$

so, when Maxwell's equations hold, $\mathbf{E}(\mathbf{k}) \perp \mathbf{k}$ and $\operatorname{div}\mathbf{E} = 0$.

§3. The volume integral of a product of two field variables M and N can be expressed in terms of the amplitudes as

$$\int MN d\tau = \int \Big\{ M(\mathbf{k})N(-\mathbf{k})e^{-2ic|k|t} + M^\dagger(\mathbf{k})N^\dagger(-\mathbf{k})e^{2ic|k|t} +$$

[2] We mean a three-dimensional vector. In a four-dimensional case, it can he shown that not $F(\mathbf{k})$ but $|k|F(\mathbf{k})$ will be a four-vector, if it is $F(x, y, z, t)$. (*V. Fock, 1957*)

$$+M(\mathbf{k})N^\dagger(\mathbf{k}) + M^\dagger(\mathbf{k})N(\mathbf{k})\} \, (d\mathbf{k}). \tag{15}$$

Here we used the relation

$$\int e^{i(\mathbf{k}-\mathbf{k}')\cdot\mathbf{r}} d\tau = (2\pi)^3 \delta(\mathbf{k}-\mathbf{k}').$$

Applying equation (15) to the calculation of the volume integral of the Hamilton function H, we get

$$\overline{H} = \int H d\tau = \int \{[\mathbf{A}(\mathbf{k})\cdot\mathbf{A}^\dagger(\mathbf{k}) + \mathbf{A}^\dagger(\mathbf{k})\cdot\mathbf{A}(\mathbf{k})] -$$

$$-[\Phi(\mathbf{k})\Phi^\dagger(\mathbf{k}) + \Phi^\dagger(\mathbf{k})\Phi(\mathbf{k})]\} \, |k|^2 (d\mathbf{k}),$$

or, rearranging the order of factors[3] and going over to the components,

$$\overline{H} = 2\int H d\tau = \int \left\{A_1^\dagger(\mathbf{k})A_1(\mathbf{k}) + A_2^\dagger(\mathbf{k})A_2(\mathbf{k}) + \right.$$

$$\left. + A_3^\dagger(\mathbf{k})A_3(\mathbf{k}) - \Phi(\mathbf{k})\Phi^\dagger(\mathbf{k})\right\} |k|^2 (d\mathbf{k}). \tag{16}$$

It is to be noted that the expression for H does not contain any time-dependent terms, although equation (15) contains them. It confirms the correct choice of the Lagrangian function. Without the last term in equation (8) the time-dependent terms would not cancel.

§4. Now passing from classical to quantum equations we must first establish the commutation rules for the amplitudes $\mathbf{A}(\mathbf{k})$, $\Phi(\mathbf{k})$ and their conjugates $\mathbf{A}^\dagger(\mathbf{k})$ and $\Phi^\dagger(\mathbf{k})$. This can be done in one of two ways: either by the direct application of the Heisenberg–Pauli commutation rules ([2], equation (11)),

$$\begin{gathered} [Q_\alpha, Q'_\beta] = 0; \qquad [P_\alpha, P'_\beta] = 0; \\ [P_\alpha, Q'_\beta] = [P'_\alpha, Q_\beta] = \tfrac{\hbar}{i}\delta_{\alpha\beta}\delta(\mathbf{r}-\mathbf{r}'), \end{gathered} \tag{17}$$

or using the equation of motion (ibid., equation (21))

$$\dot{F} = \frac{i}{\hbar}[\overline{H}, F]. \tag{18}$$

[3]Here the rearranging of the order of factors is made in a slightly different way than in the original paper [6]. Namely, we write $\Phi\Phi^\dagger$ instead of $\Phi^\dagger\Phi$. In the same way, equations (18*) and (21) are corrected. These corrections are made in correspondence with [8] (*V. Fock, 1957*). (See [34-3*] in this book. (*Editors*))

We will apply the latter method. Expressing F through the amplitudes, substituting this expression into equation (18) and comparing the coefficients, we obtain

$$\int |k|^2 \left\{[\mathbf{A}^\dagger(\mathbf{k})\cdot\mathbf{A}(\mathbf{k}) - \Phi(\mathbf{k})\Phi^\dagger(\mathbf{k})]F(\mathbf{k}')-\right.$$

$$\left.-F(\mathbf{k}')[\mathbf{A}^\dagger(\mathbf{k})\cdot\mathbf{A}(\mathbf{k}) - \Phi(\mathbf{k})\Phi^\dagger(\mathbf{k})]\right\}(d\mathbf{k}) = -\frac{1}{2}c\hbar|k'|F(\mathbf{k}'). \qquad (18^*)$$

Substituting sequentially $F = \Phi$ and $F = A_l$, and assuming that for any F and G all expressions of the type

$$F(\mathbf{k})G(\mathbf{k}') - G(\mathbf{k}')F(\mathbf{k}), \qquad F(\mathbf{k})G^\dagger(\mathbf{k}') - G^\dagger(\mathbf{k}')F(\mathbf{k})$$

etc. are proportional to $\delta(\mathbf{k}-\mathbf{k}')$, we readily obtain:

$$\begin{aligned} \Phi^\dagger(\mathbf{k})\Phi(\mathbf{k}') - \Phi(\mathbf{k}')\Phi^\dagger(\mathbf{k}) &= \frac{c\hbar}{2|k|}\delta(\mathbf{k}-\mathbf{k}'); \\ A_l^\dagger(\mathbf{k})A_m(\mathbf{k}') - A_m(\mathbf{k}')A_l^\dagger(\mathbf{k}) &= -\frac{c\hbar}{2|k|}\delta_{lm}\delta(\mathbf{k}-\mathbf{k}'); \end{aligned} \qquad (17^*)$$

$$\begin{aligned} [\Phi(\mathbf{k}), \Phi(\mathbf{k}')] = 0; &\quad [\Phi^\dagger(\mathbf{k}), \Phi^\dagger(\mathbf{k}')] = 0; \\ [A_l(\mathbf{k}), A_m(\mathbf{k}')] = 0; &\quad [A_l^\dagger(\mathbf{k}), A_m^\dagger(\mathbf{k}')] = 0; \\ [A_l(\mathbf{k}), \Phi(\mathbf{k}')] = 0; &\quad [A_l^\dagger(\mathbf{k}), \Phi(\mathbf{k}')] = 0. \end{aligned} \qquad (17^{**})$$

The same results follow by the direct use of the commutation rules (17).

Using relations (1) and (2), we immediately obtain

$$\mathbf{E}(\mathbf{k}) = i[|k|\mathbf{A}(\mathbf{k}) - \mathbf{k}\Phi(\mathbf{k})]; \qquad (1^*)$$

$$\mathbf{H}(\mathbf{k}) = i\mathbf{k}\times\mathbf{A}(\mathbf{k}). \qquad (2^*)$$

Thus, the commutation relations for the amplitudes of the field variables are as follows:

$$\left.\begin{aligned} E_l(\mathbf{k})E_m^\dagger(\mathbf{k}') - E_m^\dagger(\mathbf{k}')E_l(\mathbf{k}) &= \frac{c\hbar}{2|k|}[k^2\delta_{lm} - k_lk_m]\delta(\mathbf{k}-\mathbf{k}'), \\ H_l(\mathbf{k})H_m^\dagger(\mathbf{k}') - H_m^\dagger(\mathbf{k}')H_l(\mathbf{k}) &= \frac{c\hbar}{2|k|}[k^2\delta_{lm} - k_lk_m]\delta(\mathbf{k}-\mathbf{k}'), \\ E_s(\mathbf{k})H_m^\dagger(\mathbf{k}') - H_m^\dagger(\mathbf{k}')E_s(\mathbf{k}) &= \frac{c\hbar}{2}(k_n\delta_{sl} - k_l\delta_{sn})\delta(\mathbf{k}-\mathbf{k}'), \end{aligned}\right\} \qquad (19)$$

$$\left.\begin{aligned}
&[E_l(\mathbf{k}), E_m(\mathbf{k}')] = 0; \qquad &&[E_l^\dagger(\mathbf{k}), E_m^\dagger(\mathbf{k}')] = 0,\\
&[H_l(\mathbf{k}), H_m(\mathbf{k}')] = 0; \qquad &&[H_l^\dagger(\mathbf{k}), H_m^\dagger(\mathbf{k}')] = 0,\\
&[E_l(\mathbf{k}), H_m(\mathbf{k}')] = 0; \qquad &&[E_l^\dagger(\mathbf{k}), H_m^\dagger(\mathbf{k}')] = 0,
\end{aligned}\right\} \qquad (19^*)$$

where in the last equation l, m, n denote any right-handed system of three mutually perpendicular directions.

§5. Finally, we consider the eigenvalues of the field energy $\overline{H}$. It consists of four terms commuting with each other. Therefore, we may consider each term separately and then take the sum. Integrating over a volume (Δk), so small that k throughout it may be regarded as constant, we find

$$\text{eigenvalue of} \quad 2\int_{\mathbf{k}}^{\mathbf{k}+\Delta\mathbf{k}} \mathbf{A}^\dagger(\mathbf{k})\mathbf{A}(\mathbf{k})|k|^2(d\mathbf{k}) = c|k|\hbar(n_1+n_2+n_3), \qquad (20)$$

$$\text{eigenvalue of} \quad 2\int_{\mathbf{k}}^{\mathbf{k}+\Delta\mathbf{k}} \Phi(\mathbf{k})\Phi^\dagger(\mathbf{k})|k|^2(d\mathbf{k}) = c|k|\hbar n_0, \qquad (21)$$

where each of the quantum numbers n_0, n_1, n_2, n_3 run 0, 1, 2, 3,.... Hence,[4]

$$\text{eigenvalue of} \quad \overline{H}(\mathbf{k}, \Delta\mathbf{k}) = c|k|\hbar(n_1+n_2+n_3-n_0). \qquad (22)$$

However, this consideration suffers from the defect that it does not take into account the additional condition (12) and, therefore, the sign of the field energy remains unfixed. This defect is easy to correct, using the relation

$$k^2(\mathbf{A}^\dagger\mathbf{A} - \Phi\Phi^\dagger) = [\mathbf{k}\times\mathbf{A}^\dagger]\cdot[\mathbf{k}\times\mathbf{A}] +$$
$$+(\mathbf{k}\cdot\mathbf{A}^\dagger)\{(\mathbf{k}\cdot\mathbf{A}) - |k|\Phi\} + |k|\Phi\{(\mathbf{k}\cdot\mathbf{A}^\dagger) - |k|\Phi^\dagger\}. \qquad (23)$$

By virtue of the additional condition (12) and its conjugate the last two terms in (23) give zero, whereas the first term is positive. This first term contains not all three components of the vector potential but only two of them, which are transverse to the wave vector. The corresponding eigenvalues differ from (20) only by that they have the sum not of three but only of two non-negative integers as a factor. It is also a non-negative integer. So, finally, we have:

$$\text{eigenvalue of} \quad \overline{H}(\mathbf{k}, \Delta\mathbf{k}) = c|k|\hbar n \qquad (n = 0, 1, 2, \ldots). \qquad (24)$$

[4]Consideration at the end of §5 (after equation (21)) was changed in correspondence with the paper [8] (*V. Fock, 1957*). (See [34-3*] in this book. (*Editors*))

II Application to the Schrödinger Equation

§6. According to Dirac's ideas [3], one must solve the system of two equations:

$$\left.\begin{aligned}\left(\tfrac{1}{2m_1}p_1^2+\varepsilon_1\Phi(\mathbf{r}_1)\right)\psi+\frac{\hbar}{i}\frac{\partial\psi}{\partial t_1}=0\\ \left(\tfrac{1}{2m_2}p_2^2+\varepsilon_2\Phi(\mathbf{r}_2)\right)\psi+\frac{\hbar}{i}\frac{\partial\psi}{\partial t_2}=0\end{aligned}\right\}. \tag{25}$$

Putting $t=t_1=t_2$, and observing that

$$\frac{\partial\psi}{\partial t}=\frac{\partial\psi}{\partial t_1}+\frac{\partial\psi}{\partial t_2}, \tag{26}$$

we conclude that the solution of equations (25), in which one puts $t=t_1=t_2$, must satisfy the equation

$$\left(W-i\hbar\frac{\partial}{\partial t}\right)\psi=-(\varepsilon_1\Phi(\mathbf{r}_1)+\varepsilon_2\Phi(\mathbf{r}_2))\psi, \tag{27}$$

where

$$W=\frac{1}{2m_1}\mathbf{p}_1^2+\frac{1}{2m_2}\mathbf{p}_2^2. \tag{28}$$

One must remember that these are three-dimensional equations, and that the operator $\mathbf{p}_1^2$, e.g., when written in full, is

$$\mathbf{p}_1^2=\left(\frac{\hbar}{i}\right)^2\left(\frac{\partial^2}{\partial x_1^2}+\frac{\partial^2}{\partial y_1^2}+\frac{\partial^2}{\partial z_1^2}\right). \tag{29}$$

Let

$$\psi=\psi_0+\psi_1+\psi_2+\dots,$$

where ψ_n is of the n-th degree in charges ε_1 and ε_2, then:

$$\left(W-i\hbar\frac{\partial}{\partial t}\right)\psi_0=0; \tag{30}$$

$$\left(W-i\hbar\frac{\partial}{\partial t}\right)\psi_1=-(\varepsilon_1\Phi(\mathbf{r}_1)+\varepsilon_2\Phi(\mathbf{r}_2))\psi_0; \tag{31}$$

$$\left(W-i\hbar\frac{\partial}{\partial t}\right)\psi_2=-(\varepsilon_1\Phi(\mathbf{r}_1)+\varepsilon_2\Phi(\mathbf{r}_2))\psi_1, \tag{32}$$

etc. According to the general theory of Part I, we also have

$$\Phi(\mathbf{r}_1) = \frac{1}{(2\pi)^{3/2}} \int \left\{ \Phi(\mathbf{k}) e^{-ic|k|t + i\mathbf{k}\cdot\mathbf{r}_1} + \Phi^\dagger(\mathbf{k}) e^{ic|k|t - i\mathbf{k}\cdot\mathbf{r}_1} \right\} (d\mathbf{k}) \quad (33)$$

and a similar equation for $\Phi(\mathbf{r}_2)$.

§7. It is convenient to solve equations (30)–(32) by passing to the momentum space by

$$\psi_n(\mathbf{r}_1, \mathbf{r}_2) = \frac{1}{(2\pi\hbar)^3} \iint \varphi_n(\mathbf{p}_1, \mathbf{p}_2) e^{\frac{i}{\hbar}(\mathbf{p}_1\cdot\mathbf{r}_1 + \mathbf{p}_2\cdot\mathbf{r}_2)} (dp_1)(dp_2). \quad (34)$$

Here $\mathbf{p}_1$ and $\mathbf{p}_2$ are vectors; $(dp_1) = dp_{1x}dp_{1y}dp_{1z}$ etc.; now p_{1x}, p_{1y}, ..., p_{2z} are not operators but c-numbers. Integration was from $-\infty$ to ∞ in all the variables.

As a solution of (30), we take[5]

$$\psi_0 = \frac{1}{(2\pi\hbar)^3} e^{\frac{i}{\hbar}(\mathbf{p}_1^0\cdot\mathbf{r}_1 + \mathbf{p}_2^0\cdot\mathbf{r}_2 - W^0 t)} \delta_{j0}, \quad (35)$$

which, together with relation (34), gives

$$\varphi_0 = \delta(\mathbf{p}_1 - \mathbf{p}_1^0)\delta(\mathbf{p}_2 - \mathbf{p}_2^0) e^{-\frac{i}{\hbar}W^0 t} \delta_{j0}. \quad (36)$$

Substitution of relation (34) into (31) and (32) gives

$$\begin{aligned} -\left(W - i\hbar\frac{\partial}{\partial t}\right)\varphi_n(\mathbf{p}_1, \mathbf{p}_2) = \\ = \frac{\varepsilon_1}{(2\pi)^{3/2}} \int \Phi(\mathbf{k})\varphi_{n-1}(\mathbf{p}_1 - \hbar\mathbf{k}, \mathbf{p}_2) e^{-ic|k|t} (d\mathbf{k}) + \\ + \frac{\varepsilon_1}{(2\pi)^{3/2}} \int \Phi^\dagger(\mathbf{k})\varphi_{n-1}(\mathbf{p}_1 + \hbar\mathbf{k}, \mathbf{p}_2) e^{ic|k|t} (d\mathbf{k}) + \\ + \frac{\varepsilon_2}{(2\pi)^{3/2}} \int \Phi(\mathbf{k})\varphi_{n-1}(\mathbf{p}_1, \mathbf{p}_2 - \hbar\mathbf{k}) e^{-ic|k|t} (d\mathbf{k}) + \\ + \frac{\varepsilon_2}{(2\pi)^{3/2}} \int \Phi^\dagger(\mathbf{k})\varphi_{n-1}(\mathbf{p}_1, \mathbf{p}_2 + \hbar\mathbf{k}) e^{ic|k|t} (d\mathbf{k}), \end{aligned} \quad (37)$$

where W still has the form (28) and now $\mathbf{p}_1$ and $\mathbf{p}_2$ are not operators but c-numbers.

[5]The symbol δ_{j0} means that we take the zero-quantum state (j is the number of light quanta). See [3]. (*V. Fock, 1957*)

§8. Letting $n = 1$ and taking relation (36) for φ_0, we have an expression for $\left(W - i\hbar\frac{\partial}{\partial t}\right)\varphi_1$ that can be solved for φ_1. Repeating the process with $n = 2$, we can obtain φ_2. The result will contain integrals involving the products

$$\Phi(\mathbf{k})\Phi(\mathbf{k}'), \quad \Phi(\mathbf{k})\Phi^\dagger(\mathbf{k}'), \quad \Phi^\dagger(\mathbf{k})\Phi(\mathbf{k}'), \quad \Phi^\dagger(\mathbf{k})\Phi^\dagger(\mathbf{k}').$$

The commutation rules for $\Phi(\mathbf{k})$ and $\Phi^\dagger(\mathbf{k})$ are up to a constant factor similar to the commutation rules for the operators expressing the creation and annihilation of one quantum in state $\mathbf{k}$. The products $\Phi^\dagger(\mathbf{k})\Phi(\mathbf{k})$ and $\Phi(\mathbf{k})\Phi^\dagger(\mathbf{k})$ are proportional to $n_k + 1$ and n_k, where n_k is the number of quanta in the k. Thus, if we start with an unexcited field, i.e., when $n_k = 0$, we must put, in accordance with the commutation rules (17),[6]

$$\Phi(\mathbf{k})\Phi^\dagger(\mathbf{k}') \sim 0,$$

$$\Phi^\dagger(\mathbf{k})\Phi(\mathbf{k}') \sim \frac{c\hbar}{2|k|}\delta(\mathbf{k} - \mathbf{k}'). \tag{38}$$

We are interested in that part of ψ, which corresponds to the initial and final states without the field. This part may be obtained by putting $\Phi(\mathbf{k})\Phi(\mathbf{k}')$ and $\Phi^\dagger(\mathbf{k})\Phi^\dagger(\mathbf{k}')$ each equal to zero, as they correspond to the creation and annihilation of two quanta.

Keeping this in mind, we may write:

$$-\left(W - i\hbar\frac{\partial}{\partial t}\right)\varphi_2 \sim \frac{\varepsilon_1}{(2\pi)^{3/2}}\int \Phi^\dagger(()\mathbf{k})\varphi_1(\mathbf{p}_1 + \hbar\mathbf{k}, \mathbf{p}_2)e^{ic|k|t}(d\mathbf{k}) +$$

$$+\frac{\varepsilon_2}{(2\pi)^{3/2}}\int \Phi^\dagger(()\mathbf{k})\varphi_1(\mathbf{p}_1, \mathbf{p}_2 + \hbar\mathbf{k})e^{ic|k|t}(d\mathbf{k}) \tag{39}$$

and

$$-\left(W - i\hbar\frac{\partial}{\partial t}\right)\varphi_1 \sim \frac{\varepsilon_1}{(2\pi)^{3/2}}\int \Phi(\mathbf{k})\varphi_0(\mathbf{p}_1 - \hbar\mathbf{k}, \mathbf{p}_2)e^{-ic|k|t}(d\mathbf{k}) +$$

$$+\frac{\varepsilon_2}{(2\pi)^{3/2}}\int \Phi(\mathbf{k})\varphi_0(\mathbf{p}_1, \mathbf{p}_2 - \hbar\mathbf{k})e^{-ic|k|t}(d\mathbf{k}). \tag{40}$$

Solving equation (40) for φ_1, we obtain

[6]See, e.g., [4], §§41, 67, 68 and especially (11) and (12). (*V. Fock, 1957*)

$$\varphi_1 \sim$$

$$-\frac{\varepsilon_1}{(2\pi)^{3/2}}\frac{1}{\hbar^3}\Phi\left(\frac{\mathbf{p}_1-\mathbf{p}_1^0}{\hbar}\right)\frac{\delta(\mathbf{p}_2-\mathbf{p}_2^0)\delta_{j0}}{W-W^0-c|\mathbf{p}_1-\mathbf{p}_1^0|}e^{-\frac{i}{\hbar}(W^0+c|\mathbf{p}_1-\mathbf{p}_1^0|)t}-$$

$$-\frac{\varepsilon_2}{(2\pi)^{3/2}}\frac{1}{\hbar^3}\Phi\left(\frac{\mathbf{p}_2-\mathbf{p}_2^0}{\hbar}\right)\frac{\delta(\mathbf{p}_1-\mathbf{p}_1^0)\delta_{j0}}{W-W^0-c|\mathbf{p}_2-\mathbf{p}_2^0|}e^{-\frac{i}{\hbar}(W^0+c|\mathbf{p}_2-\mathbf{p}_2^0|)t}. \tag{41}$$

Substituting expression (41) into equation (39) and using (38), we have

$$\left(W-i\hbar\frac{\partial}{\partial t}\right)\varphi_2 = K\varphi_0-$$
$$-\frac{\varepsilon_1\varepsilon_2}{(2\pi)^3\hbar}\frac{\delta(\mathbf{p}_1-\mathbf{p}_1^0+\mathbf{p}_2-\mathbf{p}_2^0)\delta_{j0}}{|\mathbf{p}_1-\mathbf{p}_1^0|^2}e^{-\frac{i}{\hbar}W^0 t}. \tag{42}$$

Here we neglected $W-W^0$ compared to $c|\mathbf{p}_1-\mathbf{p}_1^0|$, $K=\varepsilon_1^2K_1+\varepsilon_2^2K_2$, where K_1 and K_2 are infinite constants. These terms are analogous to an infinite action of an electron on itself, the difficulty remaining in this theory.

§9. Passing back to the coordinate space by means of relation (34), we get the result

$$\left(W-i\hbar\frac{\partial}{\partial t}\right)\psi_2 \sim \left(K-\frac{\varepsilon_1\varepsilon_2}{4\pi|\mathbf{r}_1-\mathbf{r}_2|}\right)\psi_0. \tag{43}$$

This equation gives the term of ψ_2, which corresponds to the state without quanta. Function ψ_1 does not contain such terms. Thus, the second approximation of ψ gives the first correction to ψ_0 and is the first approximation for the part of ψ we are interested in. This part could, evidently, be obtained by solving the equation

$$\left(W-i\hbar\frac{\partial}{\partial t}+\frac{\varepsilon_1\varepsilon_2}{4\pi|\mathbf{r}_1-\mathbf{r}_2|}-K\right)\psi=0. \tag{44}$$

It corresponds to an interaction energy $\frac{\varepsilon_1\varepsilon_2}{4\pi|\mathbf{r}_1-\mathbf{r}_2|}$, i.e., to the Coulomb interaction.

III Application to the Dirac Equation

§10. Here we abandon Dirac's suggestion to use the quadratic Hamilton operator; the Dirac linear operator would be used. Instead of equation (27), we now have

$$\left(D_1 + D_2 - i\hbar\frac{\partial}{\partial t}\right)\psi = (\varepsilon_1 V_1 + \varepsilon_2 V_2)\psi, \tag{45}$$

where

$$\left.\begin{aligned} D_1 &= c\boldsymbol{\alpha}_1 \cdot \mathbf{p}_1 + c^2 m_1 \beta_1, \\ D_2 &= c\boldsymbol{\alpha}_2 \cdot \mathbf{p}_2 + c^2 m_2 \beta_2, \end{aligned}\right\} \tag{46}$$

$$\left.\begin{aligned} V_1 &= \boldsymbol{\alpha}_1 \cdot \mathbf{A}(\mathbf{r}_1) - \Phi(\mathbf{r}_1), \\ V_2 &= \boldsymbol{\alpha}_2 \cdot \mathbf{A}(\mathbf{r}_2) - \Phi(\mathbf{r}_2). \end{aligned}\right\} \tag{47}$$

The quantities $\boldsymbol{\alpha}_1 = (\alpha_{1x}, \alpha_{1y}, \alpha_{1z})$, β_1; $\boldsymbol{\alpha}_2 = (\alpha_{2x}, \alpha_{2y}, \alpha_{2z})$, β_2 are the Dirac matrices referring to the internal variables of the first and second charge, respectively. Regarding the expansions of V_1 and V_2 into plane waves, according to decomposition (13), we set

$$\left.\begin{aligned} V_1(\mathbf{k}) &= \boldsymbol{\alpha}_1 \cdot \mathbf{A}(\mathbf{k}) - \Phi(\mathbf{k}), \\ V_2(\mathbf{k}) &= \boldsymbol{\alpha}_2 \cdot \mathbf{A}(\mathbf{k}) - \Phi(\mathbf{k}). \end{aligned}\right\} \tag{48}$$

As before, we suppose $\psi = \psi_0 + \psi_1 + \psi_2 + \ldots$ and obtain equations quite analogous to equations (30)–(32). We can still use the transformation to the momentum space by formula (34). Now instead of (37), we have

$$\begin{aligned} &\left(D_1 + D_2 - i\hbar\frac{\partial}{\partial t}\right)\varphi_n(\mathbf{p}_1, \mathbf{p}_2) = \\ &= \frac{\varepsilon_1}{(2\pi)^{3/2}} \int V_1(\mathbf{k})\varphi_{n-1}(\mathbf{p}_1 - \hbar\mathbf{k}, \mathbf{p}_2) e^{-ic|k|t}(d\mathbf{k}) + \\ &+ \frac{\varepsilon_1}{(2\pi)^{3/2}} \int V_1^\dagger(\mathbf{k})\varphi_{n-1}(\mathbf{p}_1 + \hbar\mathbf{k}, \mathbf{p}_2) e^{ic|k|t}(d\mathbf{k}) + \\ &+ \frac{\varepsilon_2}{(2\pi)^{3/2}} \int V_2(\mathbf{k})\varphi_{n-1}(\mathbf{p}_1, \mathbf{p}_2 - \hbar\mathbf{k}) e^{-ic|k|t}(d\mathbf{k}) + \\ &+ \frac{\varepsilon_2}{(2\pi)^{3/2}} \int V_2^\dagger(\mathbf{k})\varphi_{n-1}(\mathbf{p}_1, \mathbf{p}_2 + \hbar\mathbf{k}) e^{ic|k|t}(d\mathbf{k}). \end{aligned} \tag{49}$$

§11. We take ψ_0 as

$$\psi_0 = \frac{1}{(2\pi\hbar)^3}\psi_{00}e^{\frac{i}{\hbar}(\mathbf{p}_1^0\cdot\mathbf{r}_1+\mathbf{p}_2^0\cdot\mathbf{r}_2-W^0t)} \tag{50}$$

(the factor δ_{j0} defining the zero-quantum state is omitted for brevity). Here ψ_{00} is a function of momenta and internal variables satisfying the equations

$$D_1^0\psi_{00} = W_1^0\psi_{00}; \qquad D_2^0\psi_{00} = W_2^0\psi_{00}. \tag{51}$$

D_1^0 and D_2^0 are given by relation (46) with $\mathbf{p}_1^0$ and $\mathbf{p}_2^0$, respectively, being substituted by $\mathbf{p}_1$ and $\mathbf{p}_2$; W_1^0 and W_2^0 are the energy of free electrons corresponding to momenta $\mathbf{p}_1^0$ and $\mathbf{p}_2^0$, and

$$W^0 = W_1^0 + W_2^0. \tag{52}$$

Four solutions of the Dirac equation correspond to the given value of the momentum $\mathbf{p}$. We will distinguish them by means of the subscript s, running four values ($s = 1, 2, 3, 4$), when necessary. In these notations, the wave function in the momentum space is equal to

$$\varphi_0 = \psi_{00}\delta(\mathbf{p}_1 - \mathbf{p}_1^0)\delta(\mathbf{p}_2 - \mathbf{p}_2^0)e^{-\frac{i}{\hbar}W^0t}, \tag{53}$$

where

$$\psi_{00} = \psi_{00}(\mathbf{p}_1^0, \mathbf{p}_2^0, s_1^0, s_2^0).$$

The substitution of expression (53) into (49) gives:

$$\left(D_1 + D_2 - i\hbar\frac{\partial}{\partial t}\right)\varphi_1 =$$

$$= \frac{1}{(2\pi)^{3/2}\hbar^3}\left\{\varepsilon_1\xi_1 e^{-\frac{i}{\hbar}(W^0+c|\mathbf{p}_1-\mathbf{p}_1^0|)t} + \varepsilon_1\eta_1 e^{-\frac{i}{\hbar}(W^0-c|\mathbf{p}_1-\mathbf{p}_1^0|)t} + \right.$$
$$\left. + \varepsilon_2\xi_2 e^{-\frac{i}{\hbar}(W^0+c|\mathbf{p}_2-\mathbf{p}_2^0|)t} + \varepsilon_2\eta_2 e^{-\frac{i}{\hbar}(W^0-c|\mathbf{p}_2-\mathbf{p}_2^0|)t}\right\}, \tag{54}$$

where

$$\left.\begin{aligned}
\xi_1 &= V_1\left(\tfrac{\mathbf{p}_1-\mathbf{p}_1^0}{\hbar}\right)\delta(\mathbf{p}_2 - \mathbf{p}_2^0)\psi_{00},\\
\eta_1 &= V_1^\dagger\left(-\tfrac{\mathbf{p}_1-\mathbf{p}_1^0}{\hbar}\right)\delta(\mathbf{p}_2 - \mathbf{p}_2^0)\psi_{00},\\
\xi_2 &= V_2\left(\tfrac{\mathbf{p}_2-\mathbf{p}_2^0}{\hbar}\right)\delta(\mathbf{p}_1 - \mathbf{p}_1^0)\psi_{00},\\
\eta_2 &= V_2^\dagger\left(-\tfrac{\mathbf{p}_2-\mathbf{p}_2^0}{\hbar}\right)\delta(\mathbf{p}_1 - \mathbf{p}_1^0)\psi_{00}.
\end{aligned}\right\} \tag{55}$$

From this it is clear that φ_1 must have the form of the right-hand side of (54) with ξ_s and η_s replaced by

$$\xi'_s = (D_1 + D_2 - W^0 - c|\mathbf{p}_s - \mathbf{p}_s^0|)^{-1}\xi_s = \Theta_s\xi_s \tag{56}$$

and

$$\eta'_s = (D_1 + D_2 - W^0 - c|\mathbf{p}_s - \mathbf{p}_s^0|)^{-1}\eta_s = \Omega_s\eta_s\,,$$

respectively. Now, ξ_1 and η_1 are the eigenfunctions of the operator D_2 with the eigenvalue W_2^0. Since all the other terms in the operators Θ_1 and Ω_1 defined by (56) commute with D_2, we can replace the operator D_2 in Θ_1 and Ω_1 by the number W_2^0. Using relation (52), we write down Θ_s and Ω_s in the form

$$\left.\begin{aligned} \Theta_s &= (D_s - W_s^0 - c|\mathbf{p}_s - \mathbf{p}_s^0|)^{-1}, \\ \Omega_s &= (D_s - W_s^0 + c|\mathbf{p}_s - \mathbf{p}_s^0|)^{-1}. \end{aligned}\right\} \tag{57}$$

Hence

$$\varphi_1(\mathbf{p}_1, \mathbf{p}_2) =$$

$$= \frac{1}{(2\pi)^{3/2}\hbar^3}\Big\{\varepsilon_1\Theta_1\xi_1 e^{-\frac{i}{\hbar}(W^0+c|\mathbf{p}_1-\mathbf{p}_1^0|)t} + \varepsilon_1\Omega_1\eta_1 e^{-\frac{i}{\hbar}(W^0-c|\mathbf{p}_1-\mathbf{p}_1^0|)t} +$$

$$+\varepsilon_2\Theta_2\xi_2 e^{-\frac{i}{\hbar}(W^0+c|\mathbf{p}_2-\mathbf{p}_2^0|)t} + \varepsilon_2\Omega_2\eta_2 e^{-\frac{i}{\hbar}(W^0-c|\mathbf{p}_2-\mathbf{p}_2^0|)t}\Big\}, \tag{58}$$

where Θ_s and Ω_s are defined by (57).

Substitution of (56) and (58) into formula (49) for $n = 2$ gives sixteen terms, quadratic in the V and $V^\dagger$, of which we reject eight involving VV and $V^\dagger V^\dagger$ as corresponding to two-quantum transitions. We are left with the eight terms involving only products $VV^\dagger$ and $V^\dagger V$. Of these, two contain ε_1^2, two ε_2^2, and four $\varepsilon_1\varepsilon_2$.

We omit terms in ε_1^2 and ε_2^2 similar to Part II.

Terms in the expression for $\left(D_1 + D_2 - i\hbar\frac{\partial}{\partial t}\right)\varphi_2$ containing $\varepsilon_1\varepsilon_2$ are

$$\frac{\varepsilon_1\varepsilon_2}{(2\pi)^3\hbar}\int(d\mathbf{k})\Bigg\{\Bigg[V_1(\mathbf{k})e^{\frac{i}{\hbar}c|\mathbf{p}_2-\mathbf{p}_2^0|t}\Omega_2 V_2^\dagger\left(-\tfrac{\mathbf{p}_2-\mathbf{p}_2^0}{\hbar}\right)\delta(\mathbf{p}_1 - \mathbf{p}_1^0 - \hbar\mathbf{k})+$$

$$+V_2(\mathbf{k})e^{\frac{i}{\hbar}c|\mathbf{p}_1-\mathbf{p}_1^0|t}\Omega_1 V_1^\dagger\left(-\tfrac{\mathbf{p}_1-\mathbf{p}_1^0}{\hbar}\right)\delta(\mathbf{p}_2 - \mathbf{p}_2^0 - \hbar\mathbf{k})\Bigg]e^{-\frac{i}{\hbar}(W^0+c\hbar|k|)t} +$$

$$+\left[V_1^\dagger(\mathbf{k})e^{-\frac{i}{\hbar}c|\mathbf{p}_2-\mathbf{p}_2^0|t}\Theta_2 V_2\left(\tfrac{\mathbf{p}_2-\mathbf{p}_2^0}{\hbar}\right)\delta(\mathbf{p}_1-\mathbf{p}_1^0+\hbar\mathbf{k})+\right.$$

$$\left.+V_2^\dagger(\mathbf{k})e^{-\frac{i}{\hbar}c|\mathbf{p}_1-\mathbf{p}_1^0|t}\Theta_1 V_1\left(\tfrac{\mathbf{p}_1-\mathbf{p}_1^0}{\hbar}\right)\delta(\mathbf{p}_2-\mathbf{p}_2^0+\hbar\mathbf{k})\right]\times$$

$$\left.\times e^{-\frac{i}{\hbar}(W^0-c\hbar|k|)t}\right\}\psi_{00}. \tag{59}$$

Here, analogously to (38) and precisely by the same reasons, we must put

$$\left.\begin{aligned}\Phi(\mathbf{k})\Phi^\dagger(\mathbf{k}')\sim 0,\qquad & A_l^\dagger(\mathbf{k})A_m(\mathbf{k}')\sim 0,\\ A_l(\mathbf{k})\Phi^\dagger(\mathbf{k}') &= \Phi^\dagger(\mathbf{k}')A_l(\mathbf{k})\sim 0,\\ \Phi^\dagger(\mathbf{k})\Phi(\mathbf{k}') &\sim \tfrac{c\hbar}{2|k|}\delta(\mathbf{k}-\mathbf{k}'),\\ A_l(\mathbf{k})A_m^\dagger(\mathbf{k}') &\sim \tfrac{c\hbar}{2|k|}\delta_{lm}\delta(\mathbf{k}-\mathbf{k}').\end{aligned}\right\} \tag{60}$$

Doing integration we obtain an expression that at substitution in the equation for φ_2 gives

$$\left(D_1+D_2-i\hbar\frac{\partial}{\partial t}\right)\varphi_2\sim K\varphi_0+U\varphi_0, \tag{61}$$

where

$$U\varphi_0=-\frac{c\varepsilon_1\varepsilon_2}{2\hbar(2\pi)^3}\frac{\delta(\mathbf{p}_1-\mathbf{p}_1^0+\mathbf{p}_2-\mathbf{p}_2^0)}{|\mathbf{p}_1-\mathbf{p}_1^0|}\times$$

$$\times\{\Theta_1+\Theta_2+\boldsymbol{\alpha}_1\cdot\Omega_2\boldsymbol{\alpha}_2+\boldsymbol{\alpha}_2\cdot\Omega_1\boldsymbol{\alpha}_1\}\psi_{00}e^{-\frac{i}{\hbar}W^0t}. \tag{62}$$

Here the terms containing Θ and Ω are due to the scalar and vector potentials, respectively.

Let us find the matrix elements of U. We have

$$\langle\mathbf{p}_1,\mathbf{p}_2,s_1,s_2|U|\mathbf{p}_1^0,\mathbf{p}_2^0,s_1^0,s_2^0\rangle=$$

$$=\iint(d\mathbf{p}_1')(d\mathbf{p}_2')\widetilde{\varphi}_0(\mathbf{p}_1',\mathbf{p}_2';\mathbf{p}_1,\mathbf{p}_2,s_1,s_2)U\varphi_0(\mathbf{p}_1',\mathbf{p}_2';\mathbf{p}_1^0,\mathbf{p}_2^0,s_1^0,s_2^0)=$$

$$=e^{\frac{i}{\hbar}Wt}\,\widetilde{\psi}_{00}(\mathbf{p}_1,\mathbf{p}_2,s_1,s_2)U\varphi_0(\mathbf{p}_1,\mathbf{p}_2;\mathbf{p}_1^0,\mathbf{p}_2^0,s_1^0,s_2^0), \tag{63}$$

where formula (53) was used for φ_0. Substituting here expression (62) for $U\varphi_0$, we find

$$\langle \mathbf{p}_1, \mathbf{p}_2, s_1, s_2 | U | \mathbf{p}_1^0, \mathbf{p}_2^0, s_1^0, s_2^0 \rangle =$$

$$= -\frac{c\varepsilon_1\varepsilon_2}{2\hbar(2\pi)^3} \frac{\delta(\mathbf{p}_1 + \mathbf{p}_2 - \mathbf{p}_1^0 - \mathbf{p}_2^0)}{|\mathbf{p}_1 - \mathbf{p}_1^0|} \times$$

$$\times \widetilde{\psi}_{00}(\mathbf{p}_1, \mathbf{p}_2, s_1, s_2) B \psi_{00}(\mathbf{p}_1^0, \mathbf{p}_2^0, s_1^0, s_2^0) e^{\frac{i}{\hbar}(W - W^0)t}, \tag{64}$$

where

$$B = \Theta_1 + \Theta_2 + \boldsymbol{\alpha}_1 \cdot \Omega_2 \boldsymbol{\alpha}_2 + \boldsymbol{\alpha}_2 \cdot \Omega_1 \boldsymbol{\alpha}_1. \tag{65}$$

Since $\widetilde{\psi}_{00}$ on the left of B is an eigenfunction of D_1 and D_2, we can simplify relation (64) replacing the operators D_1 and D_2 entering B by the numbers W_1 and W_2. It gives

$$B = \frac{1}{W_1 - W_1^0 - c|\mathbf{p}_1 - \mathbf{p}_1^0|} + \frac{1}{W_2 - W_2^0 - c|\mathbf{p}_2 - \mathbf{p}_2^0|} +$$

$$+ \left(\frac{1}{W_1 - W_1^0 - c|\mathbf{p}_1 - \mathbf{p}_1^0|} + \frac{1}{W_2 - W_2^0 - c|\mathbf{p}_2 - \mathbf{p}_2^0|} \right) \boldsymbol{\alpha}_1 \cdot \boldsymbol{\alpha}_2. \tag{66}$$

Further, we have

$$\mathbf{p}_1 + \mathbf{p}_2 = \mathbf{p}_1^0 + \mathbf{p}_2^0, \tag{67}$$

and, as we are interested only in the matrix elements for which the conservation of energy holds,

$$W_1 + W_2 = W_1^0 + W_2^0. \tag{68}$$

Using (67) and (68), formula (66) can be reduced to

$$B = -\frac{2c|\mathbf{p}_1 - \mathbf{p}_1^0|}{c^2|\mathbf{p}_1 - \mathbf{p}_1^0|^2 - (W_1 - W_1^0)^2}(1 - \boldsymbol{\alpha}_1 \cdot \boldsymbol{\alpha}_2). \tag{69}$$

With this value of B, formula (64) takes the form

$$\langle \mathbf{p}_1, \mathbf{p}_2, s_1, s_2 | U | \mathbf{p}_1^0, \mathbf{p}_2^0, s_1^0, s_2^0 \rangle =$$

$$= -\frac{\varepsilon_1\varepsilon_2}{\hbar(2\pi)^3} \delta(\mathbf{p}_1 + \mathbf{p}_2 - \mathbf{p}_1^0 - \mathbf{p}_2^0) \times$$

$$\times \frac{\widetilde{\psi}_{00}(\mathbf{p}_1, \mathbf{p}_2, s_1, s_2)(1 - \boldsymbol{\alpha}_1 \cdot \boldsymbol{\alpha}_2)\psi_{00}(\mathbf{p}_1^0, \mathbf{p}_2^0, s_1^0, s_2^0)}{|\mathbf{p}_1 - \mathbf{p}_1^0|^2 - \frac{1}{c^2}(W_1 - W_1^0)^2}, \tag{70}$$

which is Möller's result [5] in slightly different notations.

References

1. V.A. Fock and B. Podolsky, Sow. Phys. **1**, 798–800, 1932.
2. W. Heisenberg and W. Pauli, Zs. Phys. **56**, 1, 1929.
3. P.A.M. Dirac, Proc. Roy. Soc. **A136**, 453, 1932.
4. P.A.M. Dirac, The Principles of Quantum Mechanics, Oxford, 1930.
5. C. Möller, Zs. Phys. **70**, 786, 1931.
6. V.A. Fock and B. Podolsky, Sow. Phys. **1**, 801–817, 1932.
7. V.A. Fock and B. Podolsky, Sow. Phys. **2**, 275–277, 1934.
8. V.A. Fock, Sow. Phys. **6**, 425–469, 1934.

Typeset by V.V. Vechernin

32-5*
On Quantum Electrodynamics

P.A.M. Dirac, V.A. Fock and Boris Podolsky

Received 25 October 1932

Phys. Zs. Sowjetunion **2**, 468, 1932
Fock57, pp.70–82

In the first part of this paper, the equivalence of the new form of relativistic quantum mechanics [1] to that of Heisenberg and Pauli [2] is proved in a new way that has the advantage of showing its physical relation and serves to suggest further development considered in the second part.

Part I Equivalence of Dirac's and Heisenberg–Pauli's Theories

§1. Recently, Rosenfeld showed [3] that the new form of relativistic quantum mechanics [1] is equivalent to that of Heisenberg and Pauli [2]. Rosenfeld's proof is, however, obscure and does not bring out some features of the relation of two theories. To assist in the further development of the theory, we give here a simplified proof of the equivalence.

Consider a system, with Hamiltonian H, consisting of two parts A and B with their respective Hamiltonians H_a and H_b and the interaction V. We have

$$H = H_a + H_b + V\,, \tag{1}$$

where

$$H_a = H_a(p_a,\, q_a,\, T)\,, \qquad H_b = H_b(p_b,\, q_b,\, T)\,,$$
$$V = V(p_a,\, q_a,\, p_b,\, q_b,\, T)\,,$$

and T is the time for the entire system. The wave function for the entire system will satisfy the equation[1]

$$\left(H - i\hbar\frac{\partial}{\partial T}\right)\psi(q_a,\, q_b,\, T) = 0 \tag{2}$$

[1] $\hbar$ is the Planck constant divided by 2π. (*Authors*)

and will be a function of the variables indicated.

Now, upon performing the canonical transformation[2]

$$\psi^* = e^{\frac{i}{\hbar}H_bT}\psi\,, \tag{3}$$

by which dynamic variables, say, F, transform as

$$F^* = e^{\frac{i}{\hbar}H_bT}Fe^{-\frac{i}{\hbar}H_bT}\,, \tag{4}$$

equation (2) takes the form

$$\left(H_a^* + V^* - i\hbar\frac{\partial}{\partial T}\right)\psi^* = 0\,. \tag{5}$$

Since H_a commutes with H_b, $H_a^* = H_a$. On the other hand, since the functional relation between the variables is not disturbed by the canonical transformation (3), V^* is the same function of the transformed variables p^*, q^* as V is of p, q. But p_a and q_a commute with H_b so that

$$p_a^* = p_a; \qquad q_a^* = q_a\,.$$

Therefore,

$$V^* = V(p_a,\, q_a,\, p_b^*,\, q_b^*)\,, \tag{6}$$

where

$$\left.\begin{aligned} q_b^* &= e^{\frac{i}{\hbar}H_bT}q_be^{-\frac{i}{\hbar}H_bT}\,,\\ p_b^* &= e^{\frac{i}{\hbar}H_bT}p_be^{-\frac{i}{\hbar}H_bT}\,. \end{aligned}\right\} \tag{7}$$

It will be shown in **§7**, after a suitable notation is developed, that equations (7) are equivalent to

$$\left.\begin{aligned} \frac{\partial q_b^*}{\partial t} &= \frac{i}{\hbar}(H_bq_b^* - q_b^*H_b),\\ \frac{\partial p_b^*}{\partial t} &= \frac{i}{\hbar}(H_bp_b^* - p_b^*H_b), \end{aligned}\right\} \tag{8}$$

where t is the separate time of part B.

These, however, are just the equations of motion for part B alone, unperturbed by the presence of part A.

---

[2]Strictly speaking, the canonical transformation (3) should be written in the form $\psi^* = S\psi$, where S is the unitary operator, representing the solution of the equation $-i\hbar S^{-1}\frac{\partial S}{\partial T} = H_b$. Here, the relation $SH_b - H_bS = 0$ used in the text remains valid. The operator S has the form $S = e^{i/\hbar H_bT}$ only in the case when H_b does not depend on T explicitly. (*V. Fock, 1957*)

§2. Now let part B correspond to the field and part A to the particles present. Equations (8) must then be equivalent to Maxwell's equations for empty space. Equation (2) is then the wave equation of Heisenberg–Pauli theory, while (5), in which the perturbation is expressed in terms of potentials corresponding to empty space, is the wave equation of the new theory. Thus, this theory corresponds to treating separately part of the system, which is more convenient in some problems.[3]

Now, H_a can be represented as a sum of Hamiltonians for separate particles. The interaction between the particles is not included in H_a, for this is taken to be the result of interaction between the particles and the field. Similarly, V is the sum of interactions between the field and the particles. Thus, we may write

$$H_a = \sum_{s=1}^{n} (c\,\boldsymbol{\alpha}_s \cdot \mathbf{p}_s + m_s c^2 \beta_s) = \sum_{s=1}^{n} H_s\,,$$

$$V^* = \sum_{s=1}^{n} \varepsilon_s [\Phi(\mathbf{r}_s,\, T) - \boldsymbol{\alpha} \cdot \mathbf{A}(\mathbf{r}_s,\, t)] = \sum_{s=1}^{n} V_s^*\,, \tag{9}$$

where $\mathbf{r}_s$ are coordinates of the s-th particle and n is the number of particles. Equation (5) takes the form

$$\left[\sum_{s=1}^{n} (H_s + V_s^*) - i\hbar \frac{\partial}{\partial T}\right] \psi^*(\mathbf{r}_s,\, J,\, T)\,, \tag{10}$$

where J stands for the variables describing the field. Besides the common time T and the field time t an individual time $t_s = t_1, t_2, \ldots t_n$ is introduced for each particle. Equation (10) is satisfied by the common solution of the set of equations

$$\left(R_s - i\hbar \frac{\partial}{\partial t_s}\right) \psi^* = 0\,, \tag{11}$$

where

$$R_s = c\boldsymbol{\alpha}_s \cdot \mathbf{p}_s + m_s c^2 \beta_s + \varepsilon_s [\Phi(\mathbf{r}_s,\, t_s) - \boldsymbol{\alpha} \cdot \mathbf{A}(\mathbf{r}_s,\, t_s)] \tag{11*}$$

and $\psi^* = \psi^*(\mathbf{r}_1,\, \mathbf{r}_2,\, \ldots\, \mathbf{r}_n;\, t_1,\, t_2,\, \ldots\, t_n;\, J)$ when all t's are put equal to the common time T.

[3]This is somewhat analogous to Frenkel's method of treating incomplete systems; see [4]. (*Authors*)

Now, (11) are the equations of Dirac's theory. They are obviously relativistically invariant and form a generalization of (10). This obvious relativistic invariance is achieved by the introduction of separate time for each particle.

§3. For further development, we shall need some formulae of quantization of electromagnetic fields and shall use for this purpose some formulae obtained by Fock and Podolsky.[4] Starting with the Lagrangian function[5]

$$L = \frac{1}{2}(\mathbf{E}^2 - \mathbf{H}^2) - \frac{1}{2}\left(\operatorname{div}\mathbf{A} + \frac{1}{c}\dot{\Phi}\right)^2, \tag{12}$$

taking the potentials (Φ, A_1, A_2, A_3) as coordinates (Q_0, Q_1, Q_2, Q_3) and retaining the usual relations

$$\mathbf{E} = -\operatorname{grad}\Phi - \frac{1}{c}\dot{\mathbf{A}}; \qquad \mathbf{H} = \operatorname{curl}\mathbf{A}, \tag{13}$$

one obtains

$$\left.\begin{aligned} (P_1,\, P_2,\, P_3) &= \mathbf{P} = -\tfrac{1}{c}\mathbf{E}\,; \\ P_0 &= -\tfrac{1}{c}\left(\operatorname{div}\mathbf{A} + \tfrac{1}{c}\dot{\Phi}\right); \end{aligned}\right\} \tag{14}$$

and the Hamiltonian

$$\begin{aligned} H = \frac{c^2}{2}(\mathbf{P}^2 - P_0^2) + \frac{1}{2}\sum_{l>m}\left(\frac{\partial Q_l}{\partial x_m} - \frac{\partial Q_m}{\partial x_l}\right)^2 - \\ -cP_0\sum_{l=1}^{3}\frac{\partial Q_l}{\partial x_l} - c\mathbf{P}\cdot\operatorname{grad}Q_c\,. \end{aligned} \tag{15}$$

The equations of motion are[6]

$$\left.\begin{aligned} \dot{\mathbf{A}} &= c^2\mathbf{P} - c\operatorname{grad}\Phi\,, \\ \dot{\Phi} &= -c^2P_0 - c\operatorname{div}\mathbf{A}\,, \\ \dot{\mathbf{P}} &= \Delta\mathbf{A} - \operatorname{grad}\operatorname{div}\mathbf{A} - c\operatorname{grad}P_0\,, \\ \dot{P}_0 &= -c\operatorname{div}\mathbf{P}\,. \end{aligned}\right\} \tag{16}$$

---

[4] [5], later quoted as l.c. For other treatments see [6] or [7]. The Lagrangian (12) differs from that of Fermi [7] only by a four-dimensional divergence. (*Authors*)

[5] In the English version the authors use gothic script for electric $\mathfrak{E}$ and magnetic $\mathfrak{H}$ fields. Here we follow the notations from the Russian edition 1957. (*Editors*)

[6] A dot over a field quantity will be used to designate a derivative with respect to the field time t. (*Authors*)

On elimination of P and P_0, equations (16) give the D'Alembert equations for the potential Φ and $\mathbf{A}$. To obtain Maxwell's equation for empty space, one must set $P_0 = 0$.

The quantization rules are expressed in terms of the amplitudes of the Fourier's integral. Thus, for every $F = F(x, y, z, t)$, amplitudes $F(\mathbf{k})$ and $F^\dagger(\mathbf{k})$ are introduced by the equation

$$F = \frac{1}{(2\pi)^{3/2}} \int \{F(\mathbf{k})e^{-ic|k|t+i\mathbf{k}\cdot\mathbf{r}} + F^\dagger(\mathbf{k})e^{ic|k|t-i\mathbf{k}\cdot\mathbf{r}}\}(d\mathbf{k}), \qquad (17)$$

where $\mathbf{r} = (x, y, z)$ is the position vector, $\mathbf{k} = (k_x, k_y, k_z)$ is the wave vector having the magnitude $|k| = \frac{2\pi}{\lambda}$, $(dk) = dk_x dk_y dk_z$, the integration being performed for each component k from $-\infty$ to $+\infty$.

In terms of the amplitudes, equations of motion can be written as

$$\left.\begin{aligned} \mathbf{P}(\mathbf{k}) &= \frac{i}{c}[\mathbf{k}\Phi(\mathbf{k}) - |k|\mathbf{A}(\mathbf{k})] = -\frac{1}{c}\mathbf{E}(\mathbf{k}), \\ P_0(\mathbf{k}) &= \frac{1}{c}[|k|\Phi(\mathbf{k}) - \mathbf{k}\cdot\mathbf{A}(\mathbf{k})], \end{aligned}\right\} \qquad (18)$$

the other two equations being the algebraic consequences of these equations.

The commutation rules for the potentials are

$$\left.\begin{aligned} \Phi^\dagger(\mathbf{k})\Phi(\mathbf{k}') - \Phi(\mathbf{k}')\Phi^\dagger(\mathbf{k}) &= \frac{c\hbar}{2|k|}\delta(\mathbf{k} - \mathbf{k}'), \\ A_l^\dagger(\mathbf{k})A_m(\mathbf{k}') - A_m(\mathbf{k}')A_l^\dagger(\mathbf{k}) &= -\frac{c\hbar}{2|k|}\delta_{lm}\delta(\mathbf{k} - \mathbf{k}'). \end{aligned}\right\} \qquad (19)$$

All other combinations of amplitudes are commuting.

Part II The Maxwellian Case

§4. For the Maxwellian case, the following additional considerations are necessary. In obtaining the field variables, besides the regular equations of motion of the electromagnetic field, one must use the additional condition $P_0 = 0$ or div $\mathbf{A} + \frac{1}{c}\dot{\Phi} = 0$. This condition cannot be regarded as a quantum-mechanical equation, but rather as a condition on permissible ψ functions.

This can be seen, e.g., from the fact that, when regarded as a quantum-mechanical equation, div $\mathbf{A} + \frac{1}{c}\dot{\Phi} = 0$ contradicts the commutation rules. Thus, only those ψ's that satisfy the condition

$$-cP_0\psi = \left(\mathrm{div}\mathbf{A} + \frac{1}{c}\dot{\Phi}\right)\psi = 0 \tag{20}$$

should be regarded as physically permissible. Condition (20), expressed in terms of amplitudes by the use of (18), takes the form

$$i[\mathbf{k}\cdot\mathbf{A}(\mathbf{k}) - |k|\Phi(\mathbf{k})]\psi = 0\,,$$

$$-i[\mathbf{k}\cdot\mathbf{A}^\dagger(\mathbf{k}) - |k|\Phi^\dagger(\mathbf{k})]\psi = 0\,. \tag{20*}$$

To these, we must, of course, add the wave equation

$$\left(H_b - i\hbar\frac{\partial}{\partial t}\right)\psi = 0\,, \tag{21}$$

where H_b is the Hamiltonian for the field

$$H_b = 2\int\{\mathbf{A}^\dagger(\mathbf{k})\mathbf{A}(\mathbf{k}) - \Phi(\mathbf{k})\Phi^\dagger(\mathbf{k})\}|k|^2(d\mathbf{k})\,, \tag{22}$$

as in l.c.

If a number of equations $A\psi = 0$, $B\psi = 0$, etc. are simultaneously satisfied, then $AB\psi = 0$, $BA\psi = 0$ etc., and therefore $(AB - BA)\psi = 0$ etc. All such new equations must be the consequences of the old ones, i.e., must not give any new conditions on ψ. This may be regarded as a test of consistency of the original equations.

Applying this to our equations (20*) and (21), we have

$$\begin{aligned}
&P_0(\mathbf{k})P_0^\dagger(\mathbf{k}') - P_0^\dagger(\mathbf{k}')P_0(\mathbf{k}) = \\
&= c^2[(\mathbf{k}\cdot\mathbf{A}^\dagger(\mathbf{k}))(\mathbf{k}'\cdot\mathbf{A}^\dagger(\mathbf{k}')) - (\mathbf{k}'\cdot\mathbf{A}^\dagger(\mathbf{k}'))(\mathbf{k}\cdot\mathbf{A}^\dagger(\mathbf{k}))] + \\
&+ c^2|k|\cdot|k'|[\Phi(\mathbf{k})\Phi^\dagger(\mathbf{k}') - \Phi^\dagger(\mathbf{k}')\Phi(\mathbf{k})]\,,
\end{aligned} \tag{23}$$

since all A_l and $A_l^\dagger$ commute with Φ and $\Phi^\dagger$. Applying now the commutation rules of equations (19), we obtain

$$\begin{aligned}
&P_0(\mathbf{k})P_0^\dagger(\mathbf{k}') - P_0^\dagger(\mathbf{k}')P_0(\mathbf{k}) = \\
&= \frac{c^3\hbar}{2|k|}\Big(\sum_{l,m} k_l k_m \delta_{lm} - |k|^2\Big)\delta(\mathbf{k}-\mathbf{k}') = 0\,.
\end{aligned} \tag{24}$$

Equation (24) is satisfied as an operator equality; therefore,

$$\left[P_0(\mathbf{k})P_0^\dagger(\mathbf{k}') - P_0^\dagger(\mathbf{k}')P_0(\mathbf{k})\right]\psi = 0$$

is not a condition on ψ. Thus, conditions (20*) are consistent.

Since $P_0(\mathbf{k}) = 0$ and $P_0^\dagger(\mathbf{k}) = 0$ commute with $\partial/\partial t$, to test the consistence of condition (20) with (21), one must test the condition

$$(H_b P_0 - P_0 H_b)\psi = 0\,. \tag{25}$$

Since $\dot{P}_0 = \frac{i}{\hbar}(H_b P_0 - P_0 H_b)$, equation (25) takes the form $\dot{P}_0\psi = 0$, or in Fourier's components

$$\dot{P}_0(\mathbf{k})\psi = -ic|k|P_0(\mathbf{k})\psi = 0$$

and

$$\dot{P}_0^\dagger(\mathbf{k})\psi = ic|k|P_0^\dagger(\mathbf{k})\psi = 0\,.$$

But these are just the conditions (20*). Thus, conditions (20) and (21) are consistent.

§5. The extra condition of equation (20) is not an equation of motion, but a "constraint" imposed on the initial coordinates and velocities that the equations of motion then preserve for all time. The existence of this constraint for the Maxwellian case is the reason for the additional considerations, mentioned at the beginning of §4.

It turns out that we must modify this constraint when particles are present, in order to get the new constraint, which the equations of motion will preserve for all time.

Conditions (20*), as they stand, when applied to ψ are not consistent with equations (11). It is, however, not difficult to see that they can be replaced by a somewhat different set of conditions,[7]

$$C(\mathbf{k})\psi = 0 \quad \text{and} \quad C^\dagger(\mathbf{k})\psi = 0\,, \tag{26*}$$

where

$$C(\mathbf{k}) = i[(\mathbf{k}\cdot\mathbf{A}(\mathbf{k})) - |k|\Phi(\mathbf{k})] + \frac{i}{2(2\pi)^{3/2}|k|}\sum_{s=1}^{n}\varepsilon_s e^{ic|k|t_s - i\mathbf{k}\cdot\mathbf{r}_s}\,. \tag{27*}$$

Terms in $C(\mathbf{k})$ not contained in $-cP_0(\mathbf{k})$ are functions of the coordinates and the time for the particles. They commute with $H_b - i\hbar\frac{\partial}{\partial t}$,

[7]We shall drop the asterisk and in the following use ψ instead of ψ^*. (*Authors*)

with $P_0(\mathbf{k})$ and with each other. Therefore, equations (26*) are consistent with each other and with (21). It remains to show that equations (26*) are consistent with (11). In fact, $C(\mathbf{k})$ and $C^\dagger(\mathbf{k})$ commute with $R_s - i\hbar\frac{\partial}{\partial t_s}$. We shall show this for $C(\mathbf{k})$. Designating, in the usual way, $AB - BA$ as $[A, B]$, we see that it is sufficient to show that

$$\left[C(\mathbf{k}),\, \mathbf{p}_s - \frac{\varepsilon_s}{c}\mathbf{A}(\mathbf{r}_s,\, t_s)\right] = 0 \tag{28}$$

and

$$\left[C(\mathbf{k}),\, i\hbar\frac{\partial}{\partial t_s} - \varepsilon_s\Phi(\mathbf{r}_s,\, t_s)\right] = 0\,. \tag{29}$$

By considering the form of $C(\mathbf{k})$, this becomes, respectively,

$$[(\mathbf{k}\cdot\mathbf{A}(\mathbf{k})),\, \mathbf{A}(\mathbf{r}_s,\, t_s)] - \frac{ce^{ic|k|t_s}}{2(2\pi)^{3/2}|k|}\left[e^{-i\mathbf{k}\mathbf{r}_s},\, \mathbf{p}_s\right] = 0 \tag{30}$$

and

$$[|k|\Phi(\mathbf{k}),\, \Phi(\mathbf{r}_s,\, t_s)] + \frac{e^{-i\mathbf{k}\mathbf{r}_s}}{2(2\pi)^{3/2}|k|}\left[e^{ic|k|t_s},\, i\hbar\frac{\partial}{\partial t_s}\right] = 0\,. \tag{31}$$

Now

$$[(\mathbf{k}\cdot\mathbf{A}(\mathbf{k})),\, \mathbf{A}(\mathbf{r}_s,\, t_s)] =$$

$$= \frac{1}{(2\pi)^{3/2}|k|}\int[(\mathbf{k}\cdot\mathbf{A}(\mathbf{k})),\, \mathbf{A}^\dagger(\mathbf{k}')]e^{ic|k'|t_s - i\mathbf{k}'\mathbf{r}_s}(dk')\,,$$

by equality (17) and because $A(\mathbf{k})$ commutes with $A(\mathbf{k}')$. Using the commutation formulae and performing the integration, it becomes

$$[(\mathbf{k}\cdot\mathbf{A}(\mathbf{k})),\, \mathbf{A}(\mathbf{r}_s,\, t_s)] = \frac{c\hbar\mathbf{k}}{2(2\pi)^{3/2}|k|}e^{ic|k|t_s - i\mathbf{k}\cdot\mathbf{r}_s}\,. \tag{32}$$

On the other hand,

$$\left[e^{-i\mathbf{k}\cdot\mathbf{r}_s},\, \mathbf{p}_s\right] = i\hbar\,\mathrm{grad}\, e^{-i\mathbf{k}\cdot\mathbf{r}_s} = h\mathbf{k}e^{-i\mathbf{k}\mathbf{r}_s}\,. \tag{33}$$

Inserting these expressions into equation (30), we find that it is satisfied. Similarly, equation (31) is satisfied because

$$[|k|\Phi(\mathbf{k}),\, \Phi(\mathbf{r}_s,\, t_s)] = -\frac{c\hbar}{2(2\pi)^{3/2}}e^{ic|k|t_s - i\mathbf{k}\cdot\mathbf{r}_s} \tag{34}$$

and

$$\left[e^{ic|k|t_s},\, i\hbar\frac{\partial}{\partial t_s}\right] = c\hbar|k|e^{ic|k|t_s}\,. \tag{35}$$

Thus, conditions (26*) satisfy all the requirements of consistency. It can be shown that these requirements determine $C(\mathbf{k})$ uniquely up to an additive constant if it is taken to have the form

$$C(\mathbf{k}) = i[(\mathbf{k}\cdot\mathbf{A}(\mathbf{k})) - |k|\Phi(\mathbf{k})] + \sum_s f(\mathbf{r}_s,\, t_s)\,.$$

§6. We shall now show that the introduction of separate time for the field and for each particle allows the use of the entire vacuum electrodynamics of **§3** and l.c., except for the change discussed in **§5**. In fact, we shall show that Maxwell's equations of electrodynamics, in which current or charge densities enter, become *conditions* on the ψ function.

For convenience, we collect together our fundamental equations.

The equations of vacuum electrodynamics are

$$\mathbf{E} = -\operatorname{grad}\Phi - \frac{1}{c}\dot{\mathbf{A}}; \quad \mathbf{H} = \operatorname{curl}\mathbf{A}\,, \tag{13}$$

$$\Delta\Phi - \frac{1}{c^2}\ddot{\Phi} = 0; \quad \Delta\mathbf{A} - \frac{1}{c^2}\ddot{\mathbf{A}} = 0\,. \tag{36}$$

The wave equations are

$$\left(R_s - i\hbar\frac{\partial}{\partial t_s}\right)\psi = 0\,, \tag{11}$$

where

$$R_s = c\boldsymbol{\alpha}_s\cdot\mathbf{p}_s + m_s c^2\beta_s - \varepsilon_s\boldsymbol{\alpha}\cdot\mathbf{A}(\mathbf{r}_{s,\,t_s}) + \Phi(\mathbf{r}_s,\, t_s)\,. \tag{11*}$$

The additional conditions on the ψ function are

$$C(\mathbf{k})\psi = 0; \qquad C^\dagger(\mathbf{k})\psi = 0\,, \tag{26*}$$

where

$$C(\mathbf{k}) = i[(\mathbf{k}\cdot\mathbf{A}(\mathbf{k})) - |k|\Phi(\mathbf{k})] + \frac{i}{2(2\pi)^{3/2}|k|}\sum_{s=1}^{n}\varepsilon_s e^{ic|k|t_s - i\mathbf{k}\cdot\mathbf{r}_s}\,. \tag{27*}$$

We transform the last two equations by passing from the amplitudes $C(\mathbf{k})$ and $C^\dagger(\mathbf{k})$ to the field function $C(\mathbf{r},t)$ by means of relation (17). Thus, we obtain

$$C(\mathbf{r},t)\psi = 0 \tag{26}$$

and

$$C(\mathbf{r},t) = \operatorname{div}\mathbf{A} + \frac{1}{c}\frac{\partial\Phi}{\partial t} - \sum_{s=1}^{n}\frac{\varepsilon_s}{4\pi}\Delta(X - X_s)\,, \tag{27}$$

where X and X_s are four-dimensional vectors $X = (x, y, z, t)$, $X_s = (x_s, y_s, z_s, t_s)$ and Δ is the so called invariant delta function[8]

$$\Delta(X) = \frac{1}{r}[\delta(r + ct) - \delta(r - ct)]\,. \tag{37}$$

From equations (13), it follows immediately that

$$\operatorname{div}\mathbf{H} = 0; \qquad \operatorname{curl}\mathbf{E} + \frac{1}{c}\frac{\partial \mathbf{H}}{\partial t} = 0\,, \tag{38}$$

so that these remain as quantum-mechanical operator equations.

Using equations (13) and (36) and condition (26), we obtain by direct calculation

$$\left(\operatorname{curl}\mathbf{H} - \frac{1}{c}\frac{\partial \mathbf{E}}{\partial t}\right)\psi = \operatorname{grad}\sum_{s=1}^{n}\frac{\varepsilon_s}{4\pi}\Delta(X - X_s)\psi \tag{39}$$

and

$$(\operatorname{div}\mathbf{E})\psi = -\frac{1}{c}\left(\frac{\partial}{\partial t}\sum_{s=1}^{n}\frac{\varepsilon_s}{4\pi}\Delta(X - X_s)\right)\psi\,. \tag{40}$$

Now, let us consider what becomes of these equations when we put $t = t_1 = t_2 = \ldots = t_n = T$, which is implied in Maxwell's equations and which we shall write for brevity as $t_s = T$.

For any quantity $f(t, t_1, t_2, \ldots t_n)$

$$\frac{\partial f(T, T, \ldots T)}{\partial T} = \left[\frac{\partial f}{\partial t} + \frac{\partial f}{\partial t_1} + \ldots + \frac{\partial f}{\partial t_n}\right]_{t_s = T}\,, \tag{41}$$

and for each of the n derivatives $\frac{\partial}{\partial_s}$ we have an equation of motion

$$\frac{\partial t}{\partial t_s} = \frac{i}{\hbar}(R_s f - f R_s)\,. \tag{42}$$

If we put $f = \mathbf{A}(\mathbf{r}, t)$ or $f = \Phi(\mathbf{r}, t)$, then, since both commute with R_s, $\frac{\partial f}{\partial t_s} = 0$, and we get

$$\frac{\partial \mathbf{A}}{\partial t} = \frac{\partial \mathbf{A}}{\partial T}; \qquad \frac{\partial \Phi}{\partial t} = \frac{\partial \Phi}{\partial T}\,. \tag{43}$$

It follows that

$$\mathbf{E} = -\frac{1}{c}\frac{\partial \mathbf{A}}{\partial T} - \operatorname{grad}\Phi, \qquad \mathbf{H} = \operatorname{curl}\mathbf{A}\,, \tag{44}$$

[8]See [6]. (*Authors*)

so that the form of the connection between the field and the potentials is preserved. Remembering that for $t = t_s$ we have $\Delta(X - X_s) = 0$ and, hence, $\operatorname{grad}\Delta(X - X_s) = 0$, and using (26),(39) and (40), we obtain:

$$\left(\operatorname{div}\mathbf{A} + \frac{1}{c}\frac{\partial \Phi}{\partial T}\right)_{t_s=T} \psi = 0\,, \tag{45}$$

$$\left(\operatorname{curl}\mathbf{H} - \frac{1}{c}\frac{\partial \mathbf{E}}{\partial t}\right)_{t_s=T} \psi = 0 \tag{46}$$

and

$$(\operatorname{div}\mathbf{E})\psi = -\sum_{s=1}^{n} \frac{\varepsilon_s}{4\pi}\left[\frac{1}{c}\frac{\partial}{\partial t}\Delta(X - X_s)\right]_{t=t_s} \psi\,. \tag{47}$$

For further reduction of (46), we must use equations (41) and (42), from which it follows that

$$\left(\frac{1}{c}\frac{\partial \mathbf{E}}{\partial t}\right)_{t_s=T} = \frac{1}{c}\frac{\partial \mathbf{E}}{\partial t} - \frac{i}{c\hbar}\sum_{s=1}^{n}[R_s, \mathbf{E}]\,, \tag{48}$$

and $[R_s, \mathbf{E}]$ is easily calculated, because the only term in R_s which does not commute with $\mathbf{E}$ is $-\varepsilon_s \boldsymbol{\alpha}_s \cdot \mathbf{A}(\mathbf{r}_s, t_s)$, and $-\frac{1}{c}\mathbf{E}$ is the momentum conjugate to $\mathbf{A}$. In this way, we obtain

$$[R_s, \mathbf{E}] = ic\hbar\varepsilon_s\boldsymbol{\alpha}_s\delta(\mathbf{r} - \mathbf{r}_s)\,. \tag{49}$$

For the reduction of (47), we need only to remember [9] that

$$\left(\frac{1}{c}\frac{\partial}{\partial t}\Delta(X)\right)_{t=0} = -4\pi\delta(\mathbf{r})\,. \tag{50}$$

Thus, equations (46) and (47) become

$$\left(\operatorname{curl}\mathbf{H} - \frac{1}{c}\frac{\partial \mathbf{E}}{\partial t}\right)\psi = \sum_{s=1}^{n}\varepsilon_s\boldsymbol{\alpha}_s\delta(\mathbf{r} - \mathbf{r}_s)\psi \tag{51}$$

and

$$(\operatorname{div}\mathbf{E})\,\psi = \sum_{s=1}^{n}\varepsilon_s\delta(\mathbf{r} - \mathbf{r}_s)\psi\,, \tag{52}$$

which are just the remaining Maxwell's equations appearing as conditions on ψ. Equation (52) is an additional condition of the Heisenberg–Pauli's theory.

§7. We shall now derive equation (8) of **§1**. For this we need to recall that transformation (7) is a canonical transformation that preserves the form of the algebraic relations between the variables, as well as equations of motion.[9]

These will be, in exact notation now developed,

$$\left.\begin{aligned}\frac{\partial q_b^*}{\partial T} &= \frac{i}{\hbar}[H^*,\, q_b^*]_{t_s=T}\\ \frac{\partial p_b^*}{\partial T} &= \frac{i}{\hbar}[H^*,\, p_b^*]_{t_s=T}\end{aligned}\right\}. \tag{53}$$

As we have seen in the discussion following equation (5),

$$H^* = H_a + H_b + V^*, \tag{54}$$

and since q_b and p_b commute with H_a, q_b^* and p_b^* commute with H_a^* and, hence, with H_a. Therefore, equations (53) become

$$\left.\begin{aligned}\frac{\partial q_b^*}{\partial T} &= \frac{i}{\hbar}\{[H_b,\, q_b^*] + [V^*,\, q_b^*]\}_{t_s=T}\\ \frac{\partial p_b^*}{\partial T} &= \frac{i}{\hbar}\{[H_b,\, p_b^*] + [V^*,\, p_b^*]\}_{t_s=T}\end{aligned}\right\}. \tag{55}$$

On the other hand, we have from (41) and (42)

$$\left.\begin{aligned}\frac{\partial q_b^*}{\partial T} &= \left\{\frac{\partial q_b^*}{\partial t} + \frac{i}{\hbar}\sum_{s=1}^{n}[R_s,\, q_b^*]\right\}_{t_s=T}\\ \frac{\partial p_b^*}{\partial T} &= \left\{\frac{\partial p_b^*}{\partial t} + \frac{i}{\hbar}\sum_{s=1}^{n}[R_s,\, p_b^*]\right\}_{t_s=T}\end{aligned}\right\}. \tag{56}$$

Now the only term in R_s that does not commute with p_b^* and q_b^* is V_s^*. Thus,

$$[R_s,\, q_b^*] = [V_s^*,\, q_b^*] \quad \text{and} \quad [R_s,\, p_b^*] = [V_s^*,\, p_b^*]. \tag{57}$$

[9]Symbols $\frac{\partial q}{\partial T}$, $\frac{\partial q}{\partial t}$, $\frac{\partial q}{\partial t_s}$ etc. written with the round ∂ mean here partial derivatives in the sense that these derivatives are being taken in different variables T, t or t_s but not in the sense of the *explicit* dependence of the given operator q on the variable considered. If we write the derivatives in the latter meaning with the letter δ, then the expression usually called the "complete" derivative of an operator q in time will be written as

$$\frac{\partial q}{\partial t} = \frac{\delta q}{\delta t} + \frac{i}{\hbar}(Hq - qH).$$

(Note that in this expression d is usually written d instead of ∂ and ∂ instead of δ.) (*V. Fock, 1957*)

Since $\sum_s V_s^* = V^*$, equations (56) become

$$\left.\begin{aligned} \frac{\partial q_b^*}{\partial T} &= \left\{ \frac{\partial q_b^*}{\partial t} + \frac{i}{\hbar}[V^*,\, q_b^*] \right\}_{t_s=T}, \\ \frac{\partial p_b^*}{\partial T} &= \left\{ \frac{\partial p_b^*}{\partial t} + \frac{i}{\hbar}[V^*,\, p_b^*] \right\}_{t_s=T}. \end{aligned}\right\} \tag{58}$$

Comparison of (55) with (58) finally gives

$$\left.\begin{aligned} \left(\frac{\partial q_b^*}{\partial t}\right)_{t=T} &= \frac{i}{\hbar}[H_b,\, q_b^*]_{t=T}\,, \\ \left(\frac{\partial p_b^*}{\partial t}\right)_{t=T} &= \frac{i}{\hbar}[H_b,\, p_b^*]_{t=T}\,, \end{aligned}\right\} \tag{59}$$

which are, in the more exact notation, just equations (8).

References

1. P.A.M. Dirac, Proc. Roy. Soc. **A 136**, 453, 1932.
2. W. Heisenberg and W. Pauli, Zs. Phys. **56**, 1, 1929; **59**, 168, 1930.
3. L. Rosenfeld, Zs. Phys. **76**, 729, 1932.
4. J. Frenkel, Sow. Phys. **1**, 99, 1932.
5. V.A. Fock and B. Podolsky, Phys. Zs. Sowijetunion **1**, 798, 1932; Fock57, pp. 55–56. (See [32-4*] in this book. (*Editors*))
6. P. Jordan and W. Pauli, Zs. Phys. **47**, 151, 1928.
7. E. Fermi, Rend. Lincei **9**, 881, 1929.
8. P. Jordan and W. Pauli, Zs. Phys. **47**, 159, 1928.
9. W. Heisenberg and W. Pauli, Zs. Phys. **56**, 34, 1929.

Cambridge, Leningrad and Kharkov

Typeset by A.A. Bolokhov

33-1*
On the Theory of Positrons

V.A. Fock

(Received 21 November 1933)

DAN N6, 265, 1933
Fock57, pp. 83–87

1 Definition of States with Negative Energy

To formulate the Dirac theory, one first of all must start with definitions of positive and negative states of the electron. If the operator of electron kinetic energy is denoted as T and the operator of its modulus as $|T|$, the operator of the kinetic energy sign is

$$\varepsilon = \frac{T}{|T|} \,. \tag{1}$$

The eigenfunction ψ_+ of the operator ε, corresponding to the eigenvalue $\varepsilon = +1$, describes the "positive" state, while the eigenfunction ψ_-, corresponding to the eigenvalue $\varepsilon = -1$, describes the "negative" state of an electron. An arbitrary wave function ψ might be represented as a sum $\psi = \psi_+ + \psi_-$. The set of matrix elements of an operator corresponding to transitions from ψ_+ to ψ_- represents, according to Schrödinger, the odd part of the operator considered, while all the other matrix elements represent its even part.

When formulating the Dirac theory of positrons, one is forced to use the Schrödinger decomposition of operators into odd and even parts. By this, it becomes rather difficult to satisfy the condition of invariance in respect to the addition of gradients to potentials (Gauge invariance). Indeed, first, the Schrödinger decomposition of operators (at least in its original form) does not obey the property of this invariance, and, second, the operator of the electron total energy itself is not invariant. Therefore, the Dirac theory of positrons also is not invariant in the indicated sense.

2 Quantized Wave Function

Let us assume that the form of the energy operator as well as the decomposition of states into even and odd ones are fixed. Then for the purpose of the further development of the Dirac theory we may use the results by Heisenberg that were obtained in his work on "holes" in the electron shells of atoms.[1]

We shall consider the representation of operators in which the operator $\varepsilon(1)$ is brought to the diagonal form. The other independent variables (for example, the spin and the momentum) will be denoted by q. The quantized wave function satisfies the commutation relation

$$\left.\begin{aligned} \psi(q,\varepsilon)\psi^\dagger(q',\varepsilon') + \psi^\dagger(q',\varepsilon')\psi(q,\varepsilon) &= \delta(q-q')\,\delta_{\varepsilon\varepsilon'}, \\ \psi(q,\varepsilon)\psi(q',\varepsilon') + \psi(q',\varepsilon')\psi(q,\varepsilon) &= 0. \end{aligned}\right\} \tag{2}$$

To generalize the corresponding Heisenberg formula, let us introduce instead of ψ the function φ according to the equations

$$\psi(q,+1) = \varphi(q,1), \qquad \psi^\dagger(q,-1) = \varphi(q,2). \tag{3}$$

Then the commutation relations for $\varphi(q,\alpha)$ $(\alpha = 1, 2)$ will have the same form as for $\psi(q,\varepsilon)$:

$$\left.\begin{aligned} \varphi(q,\alpha)\varphi^\dagger(q',\alpha') + \varphi^\dagger(q',\alpha')\varphi(q,\alpha) &= \delta(q-q')\,\delta_{\alpha\alpha'}, \\ \varphi(q,\alpha)\varphi(q',\alpha') + \varphi(q',\alpha')\varphi(q,\alpha) &= 0. \end{aligned}\right\} \tag{4}$$

The assumption about filling all negative states made by Dirac can be now formulated in the following way. It is necessary to express $\psi(q,\varepsilon)$ and $\psi^\dagger(q,\varepsilon)$ in terms of $\varphi(q,\alpha)$ and $\varphi^\dagger(q,\alpha)$ in the secondary quantized operators and to consider $\varphi(q,\alpha)$ as the quantized wave function of a particle that might be either an electron $(\alpha = 1)$ or a positron $(\alpha = 2)$.

3 Transformation of the Quantized Operator

We apply the above-formulated rule to the operator, which in its initial form (i.e., being expressed in terms of ψ and $\psi^\dagger$) acts on one particle.

[1] W. Heisenberg, *Zum Paulischen Auschliessungsprinzip*, Ann. Phys. **5**, N 10, 888, 1931.

For the purpose of definiteness, we shall consider the energy operator H. Let the matrix of this operator in the representation adopted be

$$\langle q,\, \varepsilon \mid H \mid q',\, \varepsilon' \rangle\,.$$

The matrix elements for $\varepsilon = \varepsilon'$ constitute the even part of the operator, while the ones for $\varepsilon \neq \varepsilon'$ constitute the odd part. The former will be denoted as H_g and the latter as H_u. The quantized operator H will have the form

$$H = \sum_{\varepsilon\varepsilon'} \iint dq dq' \psi^\dagger(q,\, \varepsilon) \langle q,\, \varepsilon \mid H \mid q',\, \varepsilon' \rangle \psi(q',\, \varepsilon')\,. \tag{5}$$

First, we consider the terms with $\varepsilon = \varepsilon'$, constituting the even part H_g, and introduce the notations

$$\begin{aligned} \langle q, +1 \mid H \mid q', +1 \rangle &= \langle q \mid H_1 \mid q' \rangle\,, \\ \langle q', -1 \mid H \mid q, -1 \rangle = \overline{\langle q, -1 \mid H \mid q', -1 \rangle} &= -\langle q \mid H_2 \mid q' \rangle\,. \end{aligned} \tag{6}$$

According to (3), we substitute φ for ψ in the expression for H_g and transform the result in such a way that $\varphi^\dagger$ always appears to the left of φ. On the grounds of commutation rules (4) we obtain, using the notation (6), the expression

$$H_g = \sum_{\alpha=1}^{2} \iint dq dq' \varphi^\dagger(q,\, \alpha) \langle q \mid H_\alpha \mid q' \rangle \varphi(q',\, \alpha) - \int dq \langle q \mid H_2 \mid q \rangle\,. \tag{7}$$

The last term in (7) represents an infinite negative constant (the infinite energy of electrons in the negative states); this term must be rejected.

If H is assumed to be the operator T of the kinetic energy of a particle, H_1 and H_2 represent the kinetic energies of an electron and a positron. Both these operators have only positive eigenvalues; this was the goal of the hypothesis introduced by Dirac.

The odd part H_u of the operator H with the help of notations

$$\left.\begin{aligned} \langle q, +1 \mid H \mid q', -1 \rangle &= \langle q \mid U \mid q' \rangle, \\ \langle q, -1 \mid H \mid q', +1 \rangle &= \langle q \mid U^\dagger \mid q' \rangle \end{aligned}\right\} \tag{8}$$

might be rewritten without significant modifications. Then we get the expression for H_u:

$$\begin{aligned} H_u = &\iint dq dq' \varphi^\dagger(q,\, 1) \varphi^\dagger(q',\, 2) \langle q \mid U^\dagger \mid q' \rangle + \\ &+ \iint dq dq' \langle q \mid U^\dagger \mid q' \rangle \varphi(q,\, 2) \varphi(q',\, 1)\,. \end{aligned} \tag{9}$$

The operator H_u will commute with the operator L for the total charge

$$L = (-e)\int dq[\varphi^\dagger(q,\,1)\varphi(q,\,1) - \varphi^\dagger(q,\,2)\varphi(q,\,2)]\,, \tag{10}$$

but it will not commute with the operator N,

$$N = \int dq[\varphi^\dagger(q,\,1)\varphi(q,\,1) + \varphi^\dagger(q,\,2)\varphi(q,\,2)]\,, \tag{11}$$

for the total number of particles (electrons and positrons). Therefore, if there is an operator like H_u in a quantized wave equation, then it causes the change of the particle number. This change is, thus, provided by the odd part of the energy operator in the Dirac theory.

4 The System of Equations in the Configuration Space

Let us assume that there is no interaction between particles, so the energy operator has the form $H = H_g + H_u$. This operator acts on the set of functions

$$\varphi_1(q_1,\,\alpha_1);\;\varphi_2(q_1,\,\alpha_1;\;q_2,\,\alpha_2);\;\;\ldots\varphi_n(q_1,\,\alpha_1;\ldots q_n,\,\alpha_n)\,, \tag{12}$$

all of which are antisymmetric in variables $q_i,\,\alpha_i$. For these functions φ_n, the following set of equations holds:

$$[H_{\alpha_1}(q_1) + \ldots + H_{\alpha_n}(q_n)]\varphi_n(q_1,\,\alpha_1;\ldots q_n,\,\alpha_n) - i\hbar\frac{\partial\varphi_n}{\partial t} =$$

$$= -\sqrt{n(n-1)}\,\{\langle q_1 \mid U \mid q_2\rangle\,\delta_{1_{\alpha_1}}\delta_{2_{\alpha_2}}\varphi_{n-2}(q_3,\,\alpha_3;\,\ldots q_n,\,\alpha_n)\}_{\mathrm{ant}} -$$

$$-\sqrt{n(n+1)}\iint dqdq'\langle q' \mid U^\dagger \mid q\rangle\;\varphi_{n+2}(q,\,1;\,q',\,2;\,q_1,\,\alpha_1;\,\ldots q_n,\,\alpha_n) \tag{13}$$

(the subscript "ant" at the first term of the right-hand side means that the corresponding expression must be antisymmetrized in variables $q_1,\,\alpha_1;\ldots q_n,\,\alpha_n$). Equations (13) relate the wave functions corresponding only to the even or odd number of particles. By physics, this is evident since due to the charge conservation, new particles can be created in pairs only (an electron and a positron). The probability of the existence of exactly n particles is expressed by the formula

$$W_n = \sum_{\alpha_1 \dots \alpha_n} \int dq_1 \dots dq_n |\varphi_n(q_1, \alpha_1; \dots q_n, \alpha_n)|^2 . \tag{14}$$

The sum of probabilities $\left(\sum_n W_n\right)$ does not depend on time as it should be.

If we should take into account the particle interaction in the course of developing the wave equation, then equations (13) would be modified by adding the "even–even" part of the interaction energy to the operators H_α and the "odd–even" part — to the operator U. Besides, new terms originating from the "odd–odd" part of this operator and corresponding to the simultaneous creation of two pairs of particles would appear.

It should be expected from the physical considerations that the terms of equations (13) containing operators U and $U^\dagger$ might be considered as perturbations resulting in some (generally speaking, small) probability of changing the number of particles. Unfortunately, such an interpretation of these terms meets an obstacle in the mathematical side because these terms lead to infinities in the solution of equations (for example, of the kind Spur $UU^\dagger$).[2]

Translated by A.A. Bolokhov

[2]We omit the concluding remarks (§5). (*V. Fock, 1957*)

33-2
On Quantum Exchange Energy

V.A. Fock

Received 3 November 1933

Zs. Phys. **81**, 195, 1933
JETP **4**, N 1, 1–16, 1934
TOI **10**, N 92, 1–16, 1934

1 One-body Problem with Quantum Exchange

As is known,[1] the Hartree equations with the quantum exchange can be derived from the variation principle $\delta W = 0$, where W is the energy of an atom, which has the form

$$W = \int \sum_{i=1}^{n+1} \overline{\varphi_i}(x)\, H(x)\, \varphi_i(x)\; dx + \\ + \frac{e^2}{2} \iint \frac{\varrho(xx)\, \varrho(x'x') - \mid \varrho(xx') \mid^2}{\mid \mathbf{r} - \mathbf{r}' \mid} dx dx'. \quad (1)$$

In this formula, $n+1$ is the number of electrons in the atom and x is the set of the variables corresponding to a separate electron, i.e.,

$$x = (x, y, z, \sigma) \quad \text{and} \quad \int f(x)\, dx = \sum_{\sigma} \int f(x)\, d\tau,$$

$H(x)$ is the energy operator of a separate electron (it does not include the interaction energy with remaining electrons), $\varphi_i(x)$ $(i = 1, 2, \ldots, n+1)$ are the one-electron wave functions of atomic electrons that are supposed to be mutually orthogonal and normalized, so that

$$\int \overline{\varphi_i}(x)\, \varphi_j(x)\, dx = \delta_{ij}, \quad (2)$$

and, finally, $\varrho(x, x')$ denotes the "mixed" density, i.e.,

$$\varrho(x, x') = \sum_{i=1}^{n+1} \overline{\varphi_i}(x)\, \varphi_i(x'). \quad (3)$$

---

[1] V. Fock, Zs. Phys. **61**, 126, 1930; **75**, 622, 1932; P.A.M. Dirac, Proc. Cambridge Phil. Soc. **27**, 240, 1930. (See also [30-1], [30-2] in this book. (*Editors*))

When varying expression (1), one has to take into account the additional conditions (2).

We suppose now that the motion of the $(n+1)$-th electron (the valence electron) does not make a noticeable influence on the remaining n electrons (the atomic core). In this case we can proceed in the following way. In expression (1), we can vary not all the functions simultaneously but first we determine the wave functions of the internal electrons $(i = 1, 2, \ldots, n)$, equating to zero a variation of

$$W_0 = \int \sum_{i=1}^{n} \overline{\varphi_i}(x)\, H(x)\, \varphi_i(x)\, dx + \\ + \frac{e^2}{2} \iint \frac{\varrho_0(xx)\,\varrho_0(x'x') - |\,\varrho_0(xx')\,|^2}{|\,\mathbf{r} - \mathbf{r}'\,|} dx dx', \tag{1*}$$

where

$$\varrho_0(xx') = \sum_{i=1}^{n} \overline{\varphi_i}(x)\, \varphi_i(x') \tag{3*}$$

denotes a mixed density of the internal electrons. In the additional conditions (2), the indices i and j run over the values $i, j = 1, 2, \ldots, n$. Then, keeping fixed the functions $\varphi_i(x)$ $(i = 1, 2, \ldots, n)$, we will vary expression (1) for W with respect to the function $\varphi_{n+1}(x)$ of the valence electron, which we denote as $\psi(x)$:

$$\varphi_{n+1}(x) = \psi(x)\,. \tag{4}$$

The additional conditions will now be written as

$$\int \overline{\varphi_i}(x)\, \psi(x)\, dx = 0 \qquad i = 1, 2, \ldots, n\,. \tag{5}$$

One can expect that the solutions obtained in this way will only slightly differ from the solutions of the more complete initial problem (besides, it is also an approximation).

We separate the energy of the internal electrons W_0 from the total energy W and put

$$W = W_0 + W'. \tag{6}$$

Since the energy W is constant, the wave function $\psi(x)$ can be determined from the condition $\delta W' = 0$. Further, for the same reason, as the difference of the energy of two terms we can take, instead of

$W_1 - W_2$, the value $W_1{}' - W_2{}'$. Even if we would vary all the functions $\varphi_i(x)$ $(i = 1, 2, \ldots, n+1)$ simultaneously, a more exact expression for W_0 differs from our W_0 by small quantities of the second order with respect to the difference $\varphi_i(x)_{\text{exact}} - \varphi_i(x)$, since we determine W_0 from the minimum condition.

Due to this, we could not consider W_0 at all and treat W' as the energy of the valence electron.[2] The quantity W' is the difference between the total energy of the atom and its energy in the ionized state; therefore it gives directly the optical term. From our formulae (1)–(6) the expression for W' is as follows:

$$W' = \int \overline{\psi}(x) H(x) \psi(x)\, dx + \int \overline{\psi}(x) V(\mathbf{r}) \psi(x)\, dx - \\ -e^2 \int\int \overline{\psi}(x) \frac{\overline{\varrho_0}(x, x')}{|\mathbf{r} - \mathbf{r}'|}) \psi(x')\, dx dx', \tag{7}$$

where the quantity

$$V(\mathbf{r}) = e^2 \int \frac{\varrho_0(x', x')}{|\mathbf{r} - \mathbf{r}'|}) dx' \tag{8}$$

is the part of the potential energy[3] of the valence electron that originates from the field of the internal electrons. The last term in (7) is important for us because it can be treated as the quantum exchange energy[4] of the valence electron.

Bearing in mind condition (5), from $\delta W' = 0$ the next equation follows:

$$H(x)\psi(x) + V(\mathbf{r})\psi(x) - e^2 \int \frac{\overline{\varrho_0}(x, x')}{|\mathbf{r} - \mathbf{r}'|} \psi(x')\, dx' = \\ = E\psi(x) + \sum_{i=1}^{n} \lambda_i \varphi_i(x), \tag{9}$$

where λ_i are these Lagrange multipliers. It is easy to show that as the equations are satisfied by $\varphi_i(x)$, all these Lagrange multipliers are equal to zero. Therefore, equation (9) takes the following simple form:

$$H(x)\psi(x) + V(\mathbf{r})\psi(x) - e^2 \int \frac{\overline{\varrho_0}(x, x')}{|\mathbf{r} - \mathbf{r}'|} \psi(x')\, dx' = E\psi(x). \tag{9*}$$

---

[2] In the same way we could also single out several valence electrons but here we restrict ourselves to the simplest case of one valence electron. (*V. Fock*)

[3] As an operator, it is a multiplication operator on the function $V(\mathbf{r})$. (*V. Fock*)

[4] As a mathematical expectation value for the state $\psi(x)$. (*V. Fock*)

Thus, the wave function $\psi(x)$ of the valence electron satisfies the integro-differential equation (9*). It is an eigenfunction of the linear operator in (9*) and the corresponding eigenvalue E is the optical term of the atom. Further it is possible to show that the wave functions $\varphi_i(x)$ of the internal electrons are the eigenfunctions of the same operator; the corresponding eigenvalues can be treated as the Röntgen terms. Hence, the whole spectrum of the atom is described by the operator in equation (9*).

Since $\psi(x)$ and $\varphi_i(x)$ are eigenfunctions of the same operator, the orthogonality conditions are satisfied automatically. Therefore, it is possible not to introduce them and the corresponding Lagrange multipliers λ_i explicitly.

We have already emphasized the linearity of the operator in equation (9*). *Thus, a linear operator can be associated with the energy of quantum exchange.* If we denote this operator as $-A$, the operator A will have the kernel

$$\langle x \mid A \mid x' \rangle = e^2 \frac{\overline{\varrho_0}\,(x\, x')}{\mid \mathbf{r} - \mathbf{r}' \mid}. \tag{10}$$

An important conclusion follows from our considerations. The quantum mechanical system of "an electron in a given external[5] field" will be totally determined only when the quantum exchange field is introduced, besides the electrostatic (or electromagnetic) field. This idea is completely outlying to the classical theory and we hope that it might be fruitful in different areas of the quantum mechanics. Addition of the quantum exchange to the external field is to a certain extent exhaustive, because the most precise description of the electron in the framework of the one-body problem is given by the generalized Hartree equations.

2 Schrödinger Equation

Now in our formulae we introduce the spatial coordinates $\mathbf{r} = (x, y, z)$ and the variable $\sigma = \pm 1$, which describes the spin. We designate the Schrödinger functions (without the spin) for internal electrons as $\psi_i(\mathbf{r})$ $(i = 1, 2, \ldots, \frac{n}{2})$ and set

$$\varrho\,(\mathbf{r}\,\mathbf{r}') = \sum_{i=n}^{\frac{n}{2}} \overline{\psi_i}\,(\mathbf{r})\,\psi_i\,(\mathbf{r}')\,. \tag{11}$$

[5]Because of the assumption that the valence electron does not affect the internal electrons, their field should be considered external to the valence electron. (*V.F.*)

(Here the symbol ϱ is used in a slightly different sense than before in equation (2).) For a closed shell,[6] we get

$$\varrho_0 (x\, x') = \varrho_0 (\mathbf{r}\sigma; \mathbf{r}'\sigma') = \varrho (\mathbf{r}\, \mathbf{r}')\, \delta_{\sigma\sigma'}. \tag{12}$$

Further, $\varrho(\mathbf{rr}')$ is invariant at the spatial rotations of the coordinate system. Therefore, if we introduce the polar coordinates (r, θ, φ), (r', θ', φ'), we will have

$$\varrho (\mathbf{r}\, \mathbf{r}') = \varrho (r, r' \cos\gamma), \tag{13}$$

where

$$\cos\gamma = \cos\theta \cos\theta' + \sin\theta \sin\theta' \cos(\varphi - \varphi'). \tag{14}$$

Thus, the part of the potential energy that originates from the internal electrons will get the following form:[7]

$$V = V(r) = 2e^2 \int \frac{\varrho(\mathbf{r}', \mathbf{r}')}{|\, \mathbf{r} - \mathbf{r}' \,|} d\tau' = \\ = 4\pi e^2 \int_0^\infty \int_0^\pi \frac{\varrho(\mathbf{r}', \mathbf{r}', 1)}{\sqrt{r^2 + r'^2 - 2rr'\cos\gamma}} r'^2 dr' \sin\gamma d\gamma, \tag{15}$$

so that it will depend only on r. If we denote the new wave function of the valence electron as $\psi(x)$ or, in detail, as $\psi(r\sigma)$, the action of the quantum exchange operator on the function ψ results in

$$A\psi(\mathbf{r}\sigma) = e^2 \int \frac{\overline{\varrho}(\mathbf{r}\, \mathbf{r}')}{|\, \mathbf{r} - \mathbf{r}' \,|} \psi(\mathbf{r}'\sigma)\, d\tau'. \tag{16}$$

The quantity $H(x)\psi$ for a non-perturbed atom is[8]

$$H\psi = -\frac{\hbar^2}{2m}\Delta\psi - \frac{(n+1)e^2}{r}\psi. \tag{17}$$

[6] As is known, the electron shell of an atom consists of a set of "closed shells" where each shell includes $4l + 2$ electrons having definite principal and azimuth quantum numbers and all possible values of other quantum numbers, namely, the magnetic quantum number $m = -l, -l+1, \ldots, l-1, l$ and the number $\sigma = -1, +1$, which describes the spin. (*V. Fock*)

[7] Integration over x' also includes summation over σ', which leads to factor 2. (*V. Fock*)

[8] Here $\hbar$ denotes the Planck constant h divided by 2π. (*V. Fock*)

The Schrödinger equation for the valence electron is written as

$$-\frac{\hbar^2}{2m}\Delta\psi + \left(V(r) - \frac{(n+1)e^2}{r}\right)\psi - A\psi = E\psi, \tag{18}$$

where A is defined by equation (16). The kernel of the operator A is

$$\langle \mathbf{r}\sigma \mid A \mid \mathbf{r}'\sigma' \rangle = \langle \mathbf{r} \mid A \mid \mathbf{r}' \rangle \, \delta_{\sigma\sigma'}, \tag{19}$$

where

$$\langle \mathbf{r} \mid A \mid \mathbf{r}' \rangle = \frac{e^2 \overline{\varrho}(r, r', \cos\gamma)}{\sqrt{r^2 + r'^2 - 2rr'\cos\gamma}}. \tag{20}$$

Since this kernel depends on the angles $\theta, \varphi, \theta', \varphi'$ only in the combination $\cos\gamma$ (equation (14)) the operator A will commute with the components of the angular momentum operator

$$m_x = yp_z - zp_y, \quad m_y = zp_x - xp_z, \quad m_z = xp_y - yp_x. \tag{21}$$

Therefore, in the ordinary way we can single out the factor in the Schrödinger wave function ψ, depending on the angles, and set

$$\psi = \frac{1}{\sqrt{4\pi}} \frac{f(r)}{r} Y_l(\theta, \varphi), \tag{22}$$

where Y_l is the spherical harmonics normalized by the formula

$$\int \overline{Y_l} Y_l \, d\omega = 4\pi \qquad (d\omega = \sin\theta \, d\theta \, d\varphi). \tag{23}$$

The normalization condition for $f(r)$ reads as

$$\int_0^\infty \overline{f(r)} f(r)\, dr = 1. \tag{24}$$

In order to integrate over θ' and φ' in $A\psi$, we expand $\langle \mathbf{r} \mid A \mid \mathbf{r}' \rangle$ over the Legendre polynomials $P_l(\cos\gamma)$ and set

$$\langle \mathbf{r} \mid A \mid \mathbf{r}' \rangle = \frac{1}{4\pi rr'} \sum_{l=0}^{\infty} (2l+1)\, G_l(rr')\, P_l(\cos\gamma), \tag{25}$$

so that

$$G_l(r, r') = 2\pi rr' e^2 \int_0^\pi \frac{\overline{\varrho}(r, r', \cos\gamma)}{\sqrt{r^2 + r'^2 - 2rr'\cos\gamma}} P_l(\cos\gamma) \sin\gamma \, d\gamma. \tag{26}$$

If we use the property of the spherical harmonics

$$\frac{2n+1}{4\pi}\int P_n(\cos\gamma)\,Y_l(\theta'\varphi')\,d\omega' = \delta_{nl}Y_l(\theta,\varphi)\,, \tag{27}$$

then from equation (18) we obtain the following linear integro-differential equation for $f(r)$:

$$-\frac{\hbar^2}{2m}\left[\frac{d^2f}{dr^2}-\frac{l(l+1)}{r^2}f\right]+\left[V(r)-\frac{(n+1)e^2}{r}\right]f-$$
$$-\int\limits_0^\infty G_l(r,r')\,f(r')\,dr' = Ef. \tag{28}$$

This equation can be also obtained by the direct transformation of expression (7) for W'. In this way, we would get first

$$W' = \int\limits_0^\infty\left\{\frac{\hbar^2}{2m}\left(\frac{df}{dr}\right)^2+\left[V(r)-\frac{(n+1)e^2}{r}+\frac{\hbar^2 l(l+1)}{2mr^2}\right]f^2(r)\right\}-$$
$$-\int\limits_0^\infty\int\limits_0^\infty f(r)\,G_l(r,r')\,f(r')\,dr dr', \tag{29}$$

and herefrom our equation (28) follows directly.

For the sodium atom, the quantity $G_l(r,r')$ in the atomic units is equal to

$$G_l(r,r') = [f_1(r)\,f_1(r')+f_2(r)\,f_2(r')]\,K_l(r,r') -$$
$$+3f_3(r)\,f_3(r')\left[\frac{l}{2l+1}K_{l-1}(r,r')+\frac{l+1}{2l+1}K_{l+1}(r,r')\right], \tag{30}$$

where

$$\left.\begin{aligned} K_l(r,r') &= \frac{1}{2l+1}\frac{r'^l}{r^{l+1}} \quad \text{for } r'\le r,\\ K_l(r,r') &= \frac{1}{2l+1}\frac{r^l}{r'^{l+1}} \quad \text{for } r'\ge r, \end{aligned}\right\} \tag{31}$$

and $f_1(r), f_2(r), f_3(r)$ are the wave functions of the internal electron shells defined similarly to (22). Equation (28) for the sodium atom coincides with the equation given in our previous paper.[9] Numerical integration of this equation is now in progress in the Optical Institute. Equation (28) gives a good approximation for values of the terms of the sodium atom (the error is less than 2%); moreover, the contribution of the quantum exchange plays a rather significant role (up to 20%).[10]

3 Classical Analog of Quantum Exchange Energy

Let us try to find a classical quantity that corresponds to our operator A. For this, we transform the operator in the following way. According to (11) and (16), we have

$$A\psi = e^2 \sum_{i=1}^{\frac{n}{2}} \psi_i(\mathbf{r}) \int \frac{\overline{\psi_i}(\mathbf{r}')}{\mid \mathbf{r} - \mathbf{r}' \mid} \psi(\mathbf{r}')\, d\tau'. \tag{32}$$

But $1/\mid \mathbf{r} - \mathbf{r}' \mid$ is the kernel of the operator $-4\pi/\Delta$ (the Poisson equation in the potential theory). Thus, we have

$$A = -4\pi e^2 \sum_{i=1}^{\frac{n}{2}} \psi_i(\mathbf{r}) \frac{1}{\Delta} \overline{\psi_i}(\mathbf{r}), \tag{33}$$

or, if we introduce the operator $\mathbf{p}^2$ for the square of the momentum,

$$\mathbf{p}^2 = -\hbar^2 \Delta, \tag{34}$$

we obtain

$$A = 4\pi e^2 \hbar^2 \sum_{i=1}^{\frac{n}{2}} \psi_i(\mathbf{r}) \frac{1}{p^2} \overline{\psi_i}(\mathbf{r}). \tag{35}$$

Herewith, it is seen that in the classical limit ($\hbar \to 0$) the quantity A turns to be zero, as it should be. But we can define in the corresponding classical quantity the terms proportional to $\hbar^2$. To do this, we can

---

[9] V.A. Fock, TOI **5**, N 51, 1, 1931, equation (26). (*V. Fock*). (See [30-2] in this book. (*Editors*))

[10] The ground-state term of the sodium atom ($n = 3, l = 0$) is obtained as 0.1862 atomic units (the experimental value 0.1888); the calculated value for the next term ($n = 3, l = 1$) is 0.1093 (the experimental value 0.1115). The calculations were carried out by M.I. Petrashen and A.R. Krichagina for the State Optical Institute. The calculation method will be published in a separate paper. (*V. Fock*). (See [34-1] in this book. (*Editors*))

approximately suppose that p^2 is permuting with x, y, z and treat the quantity

$$N(\mathbf{r}) = 2\sum_{i=1}^{\frac{n}{2}} \psi_i(\mathbf{r})\,\overline{\psi_i}(\mathbf{r}) \tag{36}$$

as a number of electrons in the unit volume. Then we get

$$A = 2\pi e^2\hbar^2 \frac{N(\mathbf{r})}{p^2}. \tag{37}$$

This formula can be used to estimate the value of the quantum exchange energy in the framework of the classical model. To do this, one can substitute, for example, instead of $N(\mathbf{r})$, the corresponding value from the Fermi statistical model, express p^2 as a function of the coordinate with the help of the energy conservation law and then take the time average of expression (37) over the classical electron orbit.

The classical Hamilton function with the quantum exchange according to (37) will be

$$H_{\text{class}} = \frac{1}{2m}p^2 + U - 2\pi e^2\hbar^2 \frac{N(\mathbf{r})}{p^2}, \tag{38}$$

where $U = U(\mathbf{r})$ denotes the potential energy and $N(\mathbf{r})$ is connected with $U(\mathbf{r})$ by the relation

$$\Delta N(\mathbf{r}) = -4\pi e^2 N(\mathbf{r}). \tag{39}$$

4 Dirac Equation

Here we will try to apply our idea about the necessity of accounting the quantum exchange energy in the one-body problem for the Dirac equation. The main problem here is caused by the fact that there is no reliable relativistic formulation of the many-body problem. Besides, here it is not clear at all if there is any sense to consider relativistic corrections to the values of the terms when the error from the replacement of the many-body problem by the one-body problem, generally speaking, is much more than all these corrections.[11] But one can still hope that the difference of the corrections for two terms of a doublet gives a good

[11] This doubt is more legitimate in theory, which neglects completely the energy of the quantum exchange. (*V. Fock*)

estimate for the term splitting. In any case, we have to remember that here we leave the solid ground of the Schrödinger equation and have to look for a verification of the theory not only in the theoretical considerations but also in a possible experimental test of the resulting formulae for the doublet splitting, which is based on our assumptions.

We derive first some consequences from the Breit formulation of the quantum many-body problem. If one takes into account the Breit corrections for the retardation, the interaction energy of two electrons will be

$$H(x,x') = \frac{e^2}{|\mathbf{r}-\mathbf{r}'|} - \frac{e^2}{2}\frac{\boldsymbol{\alpha}\boldsymbol{\alpha}'}{|\mathbf{r}-\mathbf{r}'|} - \frac{e^2}{2}\frac{[\boldsymbol{\alpha}(\mathbf{r}-\mathbf{r}')][\boldsymbol{\alpha}'(\mathbf{r}-\mathbf{r}')]}{|\mathbf{r}-\mathbf{r}'|^3}, \tag{40}$$

where $\boldsymbol{\alpha} = (\alpha_x, \alpha_y, \alpha_z)$ is the vector composed of the first three Dirac matrices. In order to derive an expression for the energy of the $(n+1)$-th electron, which follows from (40) (cf. equation (1)), by analogy with (2) we introduce the "mixed density"

$$\varrho(x,x') = \sum_{i=1}^{n+1} \overline{\varphi_i}(x)\,\varphi_i(x') = \sum_{i=1}^{n+1} \overline{\varphi_i}\,\varphi_i', \tag{41}$$

and also the "mixed current" $\mathbf{s}$ with the z-component given by

$$\mathbf{s}_z(x\,x') = \sum_{i=1}^{n+1} \overline{\varphi_i}\,(\alpha_z\,\varphi_i)'. \tag{42}$$

Here x denotes the set of variables $\mathbf{r}, \zeta$, where $\zeta = 1,2,3,4$ is the index of the Dirac function. The prime in formulae (41) and (42) means the substitution of x by x'. If we denote as $H(x)$ the Dirac energy operator with the Coulomb potential of the core

$$H(x) = c\boldsymbol{\alpha}\cdot\mathbf{p} + mc\alpha_4 - \frac{(n+1)\,e^2}{r}, \tag{43}$$

then the expression for the energy of the atom can be written in the form

$$\begin{aligned} W = &\sum_{i=1}^{n+1}\int \overline{\varphi_i}(x)\,H(x)\,\varphi_i(x)\,dx + \\ &+\frac{e^2}{2}\iint \frac{\varrho(xx)\,\varrho(x'x') - \varrho(xx')\,\varrho(x'x)}{|\mathbf{r}-\mathbf{r}'|}dxdx' - \end{aligned}$$

$$-\frac{e^2}{4}\iint\frac{\mathbf{s}(xx)\cdot\mathbf{s}(x'x')-\mathbf{s}(xx')\cdot\mathbf{s}(x'x)}{\mid\mathbf{r}-\mathbf{r}'\mid}dxdx'-$$
$$-\frac{e^2}{4}\iint\frac{[(\mathbf{r}-\mathbf{r}')\cdot\mathbf{s}(xx)][(\mathbf{r}-\mathbf{r}')\cdot\mathbf{s}(x'x')]}{\mid\mathbf{r}-\mathbf{r}'\mid^3}dxdx'+$$
$$+\frac{e^2}{4}\iint\frac{[(\mathbf{r}-\mathbf{r}')\cdot\mathbf{s}(x'x)][(\mathbf{r}-\mathbf{r}')\cdot\mathbf{s}(xx')]}{\mid\mathbf{r}-\mathbf{r}'\mid^3}dxdx'. \tag{44}$$

As we already did in §1, we can single out the wave function of the valence electron $\varphi_{n+1}(x)=\psi(x)=\psi(\mathbf{r}\zeta)$ and, according to the equation (6), we can decompose the energy W in two parts. In order to write the energy W' of the valence electron in a more simple form, we introduce the following notations. We set

$$\langle\zeta'\mid\varrho^0\mid\zeta\rangle=\sum_{i=1}^{n}\overline{\varphi_i}(\mathbf{r}'\zeta')\,\varphi(\mathbf{r}\zeta) \tag{45}$$

and denote

$$\varrho^0=\varrho^0(\mathbf{r}',\mathbf{r}) \tag{46}$$

the four-row matrix with the entries (45). The part of the current vector s connected to the internal electrons is

$$\mathbf{s}_x^0=\alpha_x\varrho^0,\quad \mathbf{s}_y^0=\alpha_y\varrho^0,\quad \mathbf{s}_z^0=\alpha_z\varrho^0. \tag{47}$$

We also introduce

$$\overline{\overline{\varrho^0}}(\mathbf{r})=\text{Spur}\,\varrho^0(\mathbf{r},\mathbf{r}),\quad \overline{\overline{s^0}}(\mathbf{r})=\text{Spur}\,\mathbf{s}^0(\mathbf{r},\mathbf{r}), \tag{48}$$

where $\text{Spur}\varrho^0(\mathbf{r},\mathbf{r})$ denotes the sum of diagonal elements of the matrix $\varrho^0(\mathbf{r},\mathbf{r})$. Then, using the notation $\overline{\overline{\varrho^0}}(\mathbf{r})$, we set

$$V(\mathbf{r})=e^2\int\frac{\overline{\overline{\varrho^0}}(\mathbf{r}')}{\mid\mathbf{r}-\mathbf{r}'\mid}d\tau'. \tag{49}$$

This quantity has the same value as in (8); it is proportional to the scalar potential of internal electrons. Because for the closed shell the value of $\overline{\overline{s^0}}(\mathbf{r})$ is equal to zero,[12] we cannot write down the corresponding expression for the vector potential. We denote by γ the matrix

$$\gamma=\frac{\alpha\cdot(\mathbf{r}-\mathbf{r}')}{\mid\mathbf{r}-\mathbf{r}'\mid}. \tag{50}$$

[12]It follows from the expression for the matrix ϱ^0 being approximately expressed by the Schrödinger wave functions (see below). Cancelling $\overline{\overline{s^0}}(\mathbf{r})$ is already clear from the symmetry reasons. (*V. Fock*)

Dropping the index ζ corresponding to internal degrees of freedom of an electron, we will write simply $\psi(\mathbf{r})$ for the wave function $\psi_{n+1} = \psi$ of the valence electron. With these notations, the expression for the energy W' of the valence electron is

$$W' = \int \overline{\psi}(\mathbf{r}) H\psi(\mathbf{r}) d\tau + \int \overline{\psi}(\mathbf{r}) V(\mathbf{r}) \psi(\mathbf{r}) d\tau -$$
$$-e^2 \iint \frac{\int \overline{\psi}(\mathbf{r}) \varrho^0 \psi(\mathbf{r}')}{|\mathbf{r}-\mathbf{r}'|} d\tau d\tau' +$$
$$+\frac{e^2}{2} \iint \frac{\int \overline{\psi}(\mathbf{r}) \left[(\alpha\varrho^0\alpha) + \gamma\varrho^0\gamma\right] \psi(\mathbf{r}')}{|\mathbf{r}-\mathbf{r}'|} d\tau d\tau'. \quad (51)$$

Variation of this expression over $\psi(\mathbf{r})$ leads to the following equation for $\psi(\mathbf{r})$:

$$[H + V(\mathbf{r})]\psi - e^2 \int \frac{\varrho^0 \psi(\mathbf{r}')}{|\mathbf{r}-\mathbf{r}'|} d\tau' +$$
$$+\frac{e^2}{2} \int \frac{\left[(\alpha\varrho^0\alpha) + \gamma\varrho^0\gamma\right] \psi(\mathbf{r}')}{|\mathbf{r}-\mathbf{r}'|} d\tau' = W\psi. \quad (52)$$

Here in the first term the operator $H + V$ is the ordinary Dirac operator in a central field (the Coulomb field of the core, which is screened by the internal electrons). The other terms originate from the quantum exchange, namely, the second term is from the "Coulomb" exchange and the third is due to the Breit correction on retardation.

For further derivation it is necessary to define matrix ϱ^0 in equation (45), which describes the internal electrons. Here we come across a difficulty because for the internal electrons we can assume only the Schrödinger wave functions or the Pauli wave functions with the spin to be known. Though herefrom one can determine approximately Dirac wave functions, it is not enough for a definite derivation of a quantum exchange correction for the doublet splitting.

The structure of the matrix ϱ^0 can be understood in the following way. We take for the Dirac matrices his expressions:

$$\alpha_x = \varrho_1\sigma_1, \quad \alpha_y = \varrho_1\sigma_2, \quad \alpha_z = \varrho_1\sigma_3, \quad \alpha_4 = \varrho_3 \quad (53)$$

and denote by $\sigma_x^0, \sigma_y^0, \sigma_z^0$ the two-row Pauli matrices

$$\sigma_x^0 = \begin{pmatrix} 0 & 1 \\ 1 & 0 \end{pmatrix}, \quad \sigma_y^0 = \begin{pmatrix} 0 & -i \\ i & 0 \end{pmatrix}, \quad \sigma_z^0 = \begin{pmatrix} 1 & 0 \\ 0 & -1 \end{pmatrix}. \quad (54)$$

As is known, for the states with the positive energy two low components of the Dirac wave function will be small compared with two upper components; moreover, the following approximate equations will take place:

$$\left.\begin{aligned}\psi_3 &= \frac{1}{2mc}\left[\left(p_x - ip_y\right)\psi_2 + p_z\psi_1\right],\\ \psi_4 &= \frac{1}{2mc}\left[\left(p_x + ip_y\right)\psi_1 + p_z\psi_2\right]\end{aligned}\right\} \tag{55}$$

or, in a more symbolic form,

$$\begin{pmatrix}\psi_3\\ \psi_4\end{pmatrix} = \frac{P}{2mc}\begin{pmatrix}\psi_1\\ \psi_2\end{pmatrix}, \tag{56}$$

where P denotes the operator

$$P = \sigma_x^0 p_x + \sigma_y^0 p_y + \sigma_z^0 p_z. \tag{57}$$

Let us write ϱ^0 as a "supermatrix"

$$\varrho^0 = \begin{pmatrix}\varrho_{11} & \varrho_{12}\\ \varrho_{21} & \varrho_{22}\end{pmatrix}, \tag{58}$$

where the quantities ϱ_{ij} are two-row matrices. It is easy to determine the order of magnitude of these matrices: ϱ_{12} and ϱ_{21} will be of the order $\mid \frac{P}{2mc} \mid$ and ϱ_{22} is of the order $\frac{P^2}{4m^2c^2}$ with respect to ϱ_{11}. If we suggest that the first two Dirac functions approximately coincide with the Pauli functions, we will get

$$\langle\sigma \mid \varrho_{11} \mid \sigma'\rangle = \varrho\left(\mathbf{r}\,'\mathbf{r}\right)\delta_{\sigma\sigma'}, \tag{59}$$

where $\varrho\left(\mathbf{r}\,'\mathbf{r}\right)$ is of the Schrödinger expression (11), or, more correctly, it follows from (11) by permutation of $\mathbf{r}$ with $\mathbf{r}'$. Applying the operator P, we can get herefrom approximate expressions for $\varrho_{12}, \varrho_{21}$ and ϱ_{22}. Unfortunately, one cannot be sure that the relation (59) is correct up to the terms of the order $\frac{P^2}{4m^2c^2}$ inclusively. We can only state that in the matrix ϱ^0 the elements of the order of unity are the same as in the matrix

$$\varrho^0 = \varrho\left(\mathbf{r}'\mathbf{r}\right)\frac{1+\varrho_3}{2} = \varrho\left(\mathbf{r}'\mathbf{r}\right)\frac{1+\alpha_4}{2}, \tag{60}$$

which follows from (58) if one substitutes ϱ_{11} by (59) and puts zero instead of $\varrho_{12}, \varrho_{21}$ and ϱ_{22}. As a consequence of this uncertainty, we will not continue calculations, which are based on the Breit formulation of the many-body problem, but we give here some results following from the different (more or less arbitrary) assumptions about the form of the Dirac equation with the quantum exchange.

If one inserts (60) into (52) and neglects interaction of the retardation with the quantum exchange, then the following equation for the energy is obtained:[13]

$$\left[c\left(\boldsymbol{\alpha}\cdot\mathbf{p}\right)+mc^2\alpha_4+U\left(r\right)-\frac{1+\alpha_4}{2}A\right]\psi=W\psi, \tag{61}$$

where it is written for brevity

$$U\left(r\right)=V\left(r\right)-\frac{\left(n+1\right)e^2}{r}, \tag{62}$$

and the operator A has its previous value (16).

It is possible to show that if the wave equation has the form (61), then the term with the quantum exchange does not give any contribution to the doublet splitting of the terms. In order to prove this property, we note first that in the case of equation (61) the law of areas holds in the former Dirac form. Therefore, we can transform equation (61) in an ordinary way to the polar coordinates. We denote as k the quantum number, which is connected with the spectroscopic quantum numbers i and j by relations

$$l=\left|k-\frac{1}{2}\right|-\frac{1}{2},\quad j=\left|k\right|-\frac{1}{2},\quad\left(k=\pm1,\pm2,\ldots\right). \tag{63}$$

Then for the radial functions $f_1\left(r\right)$ and $f_2\left(r\right)$, the following system of the integro-differential equations is obtained:

$$\left.\begin{aligned}&\hbar c\left(\frac{df_2}{dr}+\frac{k}{r}f_2\right)+\left[mc^2+U\left(r\right)\right]f_1-\int_0^\infty G_{k-1}f_1\left(r'\right)dr'=Wf_1,\\&\hbar c\left(-\frac{df_1}{dr}+\frac{k}{r}f_1\right)+\left[-mc^2+U\left(r\right)\right]f_2=Wf_2\,.\end{aligned}\right\} \tag{64}$$

Here for positive k the quantity G_k has its previous value (26) and for negative k it is defined by

$$G_{-k}=G_{k-1}. \tag{65}$$

[13]It is interesting that in expression (61) the matrix α_4 is multiplied beside the rest-energy mc^2 also by the operator of the exchange energy A. But it would be premature to make herefrom a conclusion about a connection between the rest-energy and the energy of the quantum exchange. (*V. Fock*)

From (64), after simple algebra we obtained the following well-known[14] formula for the doublet splitting:

$$\Delta W = W(k) - W(-k+1) = \frac{\hbar^2}{4m^2c^2}(2k-1)\int\limits_0^\infty \frac{1}{r}\frac{dU}{dr}f_1^2 dr. \quad (66)$$

Here one can take any solution of the corresponding Schrödinger equation (28) as $f_1(r)$.

Equation (64) differs from other possible forms of the Dirac equation with the quantum exchange in that it gives the splitting formula, which does not depend on the quantum exchange. Therefore, if it is desirable to complete the wave equation with additional terms (for example, according to formula (52)), it is convenient to consider (61) as a non-perturbed equation. If we denote the perturbation as $H_2\psi$, then the following quantity is added to expression (66) for the doublet splitting:

$$\Delta_1 W = \langle k \mid H_1 \mid k\rangle - \langle -k+1 \mid H_1 \mid -k+1\rangle . \quad (67)$$

If we set, e.g.,

$$H_1\psi = \lambda\frac{1-\sigma_4}{2}A\psi, \quad (68)$$

where λ is a constant,[15] we arrive at

$$(k \mid H_1 \mid k) = \lambda\int\limits_0^\infty\int\limits_0^\infty f_2^0(r)\, G_k(r,r')\, f_2^0(r')\, dr dr', \quad (69)$$

where

$$f_2^0(r) = -\frac{\hbar}{2mc}\left(\frac{df_1}{dr} - \frac{k}{r}f_1\right) \quad (70)$$

is an approximate value of the second radial function (cf. equation (64)).

The results of the investigation of the Dirac equation with the quantum exchange, developed in this section, are not final since we do not come to a definite formula for the doublet splitting.[16] Nevertheless, we

[14] See, e.g., V.A. Fock, Basic quantum mechanics (Nachala kvantovoj mekhaniki) (in Russian), Leningrad, 1932, p. 225. (*V. Fock*)

[15] Such a term with $\lambda = 2$ is obtained if one introduces expression (52) in formula (60) and neglects the correction on retardation. (*V. Fock*)

[16] See also M. Johnson and G. Breit, Phys. Rev. **44**, 77, 1933, July 15. (*V. Fock*)

would like to emphasize that the quantum exchange can give a rather significant correction to expression (66). Hence, in one way or another it is necessary to take this correction into account. If, on the one hand, it is difficult to do this rigorously, then, on the other hand, a theory which does not take this term into account at all is even less satisfactory. A clarification of the form of the correction due to the quantum exchange could be useful to some extent in the relativistic many-body theory.

Leningrad
State Optical Institute
Sector of Spectroscopy
the USSR Academy of Sciences

Translated by Yu. Yu. Dmitriev

34-1
On the Numerical Solution of Generalized Equations of the Self-consistent Field

V.A. Fock and Mary J. Petrashen

Received 3 February 1934

JETP **4**, N 4, 295–325, 1934
Phys. Zs. Sowjetunion **6**, N 4, 368–414, 1934 (English version)

The determination of the stationary states of many-electron atoms is bound up with the solution of the many-body problem. An approximation to the solution of this problem was suggested by Hartree. His method, known as that of the self-consistent field, was generalized by Fock. The method of the self-consistent field, as well as its generalization, is based on the assumption that the wave function of a whole atom can be expressed approximately in terms of one-electron wave functions. Hartree assumed the wave function of the atom to be a mere product of electron wave functions and, consequently, did not take into consideration the symmetry properties of the wave function claimed by Pauli's exclusion principle. The generalization of the method of the self-consistent field aims at taking the Pauli principle into account. For this purpose the wave function of the atom was assumed to have the form of a finite sum of products of one-electron wave functions; for an atom with one series electron this sum reduces to a product of two determinants formed of the Schrödinger one-electron wave functions.

It was shown by Fock[1] that the equations of the self-consistent field as well as the generalized equations can be deduced from a variational principle. By introducing the wave function with proper symmetry the greatest possible precision was attained compatible with the representation of the wave function of the whole atom in terms of the one-electron wave functions. Therefore the generalized method of the self-consistent field is doubtless preferable to the original. But, until now, no use of its advantage was made, because of the mathematical difficulties involved in its numerical application. These difficulties, which are connected with the character of generalized equations of the self-consistent field, made

[1]V. Fock, *Näherungsmethode zur Lösung des quantunmechanishen Mehrkörperproblems*, Zs. Phys. **61**, 126, 1930. (See [30-2] in this book. (*Editors*))

the usual methods of numerical integration inapplicable. The advantages of the generalized method as well as its difficulties were referred to by various authors.[2]

In the present paper a method of solution of the generalized equations of the self-consistent field will be suggested that enables us to avoid the difficulties mentioned above and to find the one-electron wave functions with desirable precision. This method was developed when carrying out the calculations for the Na atom. Some results of these calculations will be given at the end of the present paper.

1 Form of Equations

For atoms consisting of closed electronic shells and a series electron we may write the wave function of an electron in the form

$$\Psi = \frac{1}{r} f_i(r) Y_{l_i}(\vartheta, \varphi), \tag{1.1}$$

r being the distance from the nucleus given in atomic units and Y a spherical harmonic dependent on angles ϑ and φ only. The radial function is supposed to be normal, that is

$$\int_0^\infty [f_i(r)]^2 \, dr = 1. \tag{1.2}$$

The variables, thus, will be assumed to be separated and radial functions will be considered only.

For a given atom there will be as many radial functions as there are electronic shells constituting the atom in question plus one series-electron wave function. Radial functions $f_i(r)$ and $f_k(r)$ for which the quantum numbers l_i and l_k are not the same must be orthogonal:

$$\int_0^\infty f_i(r) f_k(r) dr = 0, \qquad i \neq k. \tag{1.2*}$$

As was mentioned above, the generalized equations of the self-consistent field (referred to later simply as generalized equations) satisfied by the

[2] J. McDougall, *The Calculation of the Terms of the Optical Spectrum of an Atom with One Series Electron*, Proc. Roy. Soc. **A 138**, 550, 1933; D.R. Hartree and M.M. Black, *A Theoretical Investigation of the Oxygen Atom in Various States of Ionization*, Proc. Roy. Soc. **A 139**, 311, 1933; D.R. Hartree, *Results of Calculation of Atomic Wave Functions*, Proc. Roy. Soc. **A 141**, 282, 1933.

radial wave functions are deducible from a variation principle by varying the energy of the atom.

The general form of the energy of an $(n+1)$-electron system can be easily written by introducing besides the electronic coordinates the spin variable, which acquires two values.

Expressing according to equation (1.1) the wave functions of the electrons in terms of radial functions $f_i(r)$ and spherical harmonics $Y(\vartheta, \varphi)$ and performing the integration over the angular coordinates ϑ and φ, we obtain an expression for the energy in terms of the radial functions. The variation of this expression leads to the system of generalized equations.

The form of this system depends on the way in which we proceed. We may vary all the functions $f_i(r)$, the series electron wave function included, simultaneously or, as was pointed out in a paper by Fock,[3] we may proceed as follows. First, having constructed the expression for the energy of the ionized atom, we may obtain the equations for the wave functions of the core shells by varying this expression and, then, by varying the expression for the energy of a whole (not ionized) atom we may obtain the equation for the series electron wave function. The latter way is preferable for its simplicity. The wave functions of the core electrons will be determined thus once for all and for the determination of the series electron wave function we shall have a linear integro-differential equation.

In general, the expression for the energy of an atom in terms of one-electron wave functions is rather complicated. As an example, the expressions for the energies of the Li atom and Na$^+$ atom will be given in Section 3.

The system of equations obtained for the core electron wave functions will have the form

$$-C_i \frac{d^2 f_i}{dr^2} + Q_{ii}(r) f_i = \lambda_i f_i - \sum_k{}' Q_{ik} f_k \qquad (i = 1, 2 \ldots q). \tag{1.3}$$

In these equations q denotes the number of core radial functions; the summation is extended over the values $k = 1, 2, ..., i-1, i+1, ..., q$ (the omission of the terms with $k = i$ being noted by the sign $'$); the term with $k = i$ is transferred to the left-hand side. C_i are positive constants and Q_{ik}, which will be referred to as the coefficients of the equation, are

[3] V. Fock, *Über Austauschenergie*, Zs. Phys. **81**, 195, 1933. (See also [33-2] in this book. (*Editors*))

expressed linearly in terms of the quantities

$$F_l^{ik}(r) = \frac{1}{2l+1}\left\{\frac{1}{R^{l+1}}\int_0^r f_i(r')f_k(r')r'^l dr' + \right.$$

$$\left. + r^l \int_r^\infty f_l(r')f_k(r')r'^{-(l+1)}dr'\right\}, \tag{1.4}$$

where $Q_{ii}(r)$ includes terms corresponding to the Coulomb field and terms of the form $C_i l_i(l_i+1)/r^2$. Expression (1.4) can be written in the form

$$F_l^{ik}(r) = \int_0^\infty f_i(r')f_k(r')K_l(r,r')dr', \tag{1.4*}$$

where

$$\left.\begin{aligned} K_l(r,r') &= \frac{1}{2l+1}\frac{r'^l}{r^{l-1}} \qquad \text{for} \quad r' \leq r, \\ K_l(r,r') &= \frac{1}{2l+1}\frac{r^l}{r'^{l-1}} \qquad \text{for} \quad r' \geq r\,. \end{aligned}\right\} \tag{1.5}$$

The coefficients $Q_{ik}(r)$ are symmetric in the suffixes i and k, that is,

$$Q_{ik}(r) = Q_{ki}(r). \tag{1.6}$$

Hartree's equations, which do not take into consideration the quantum exchange, may be deduced from equation (1.3) by omitting the "non-diagonal" coefficients, i.e., the sum $\sum'$. The series electron wave equation is of the form

$$-\frac{1}{2}\frac{d^2 f}{dr^2} + \left[\frac{l(l+1)}{2r^2} - \frac{Z}{r} + V(r)\right] f = \lambda f - \sum_{k=1}^{q} Q_k(r) f_k, \tag{1.7}$$

l being the quantum number of the series electron, and the series electron wave function f_{q+1} being denoted by $f(r)$ gives the value of the screening potential for core electrons and is independent of f. The coefficients $Q_k = Q_{k,q+1}$ result from applying a linear integral operator to the function f, and therefore (1.7) is a linear integro-differential equation for the series electron wave function. The energy level (term) of the series electron is given, in atomic units, by the characteristic value λ of this equation.

The quantities

$$E_i = \frac{\lambda_i}{2C_i} \tag{1.8}$$

may be compared with core electron energy levels (X-ray terms) though a close agreement cannot be expected, equation (1.3) referring to the ionized atom.

When considering the series electron wave equation (1.7) as an equation for the characteristic functions of the linear integro-differential operator, this equation will be satisfied besides by the series electron wave function f, by those of the core electron wave functions f which correspond to the same quantum number l, the quantities E_i being their characteristic values. Therefore the series electron function f must be orthogonal to the core electron function f_i with the same quantum number l.

It must be pointed out that the equation for the series electron has the form (1.7) only when the core wave functions involved in $V(r)$, $Q_k(r)$ and in the right-hand side of the equation rigorously satisfy the system (1.3). But when the system (1.3) is satisfied by these functions only approximately, the solution of equation (1.7) may prove to be not orthogonal to the core wave functions with the same quantum number l. In order to obtain an orthogonal solution it is necessary in this case to add to the right-hand side of (1.7) a linear combination of the core wave functions with constant coefficients, i.e., to write instead of (1.7)

$$-\frac{1}{2}\frac{d^2f}{dr^2}+\left[\frac{l(l+1)}{2r^2}-\frac{Z}{r}+V(r)\right]f=\lambda f-\sum_{k=1}^{q}Q_k(r)f_k+\sum_i \mu_i f_i, \quad (1.7^*)$$

the summation being extended over all values of i for which the quantum number l_i is equal to l.

The constants μ_i are determined from the conditions of orthogonality of the solution sought for to the functions $f_i(r)$ and are expressed, as may be easily verified, in the form

$$\mu_i = \frac{1}{2}C_i\int_0^\infty \varphi_i(r)f(r)dr, \quad (1.9)$$

$\varphi_i(r)$ being, but for an additive term of the form $\lambda_i f_i$, the difference between the left- and the right-hand sides of the corresponding equation (1.3),

$$\varphi_i(r) \equiv -C_i\frac{d^2f_i}{dr^2}+\sum_{k=1}^{q}Q_{ik}(r)f_k \quad (1.10)$$

(summation being taken over k without omissions). It is easily seen that, provided equations (1.3) are rigorously satisfied, we have

$$\varphi_i(r)=\lambda_i f_i \quad (1.11)$$

and owing to the orthogonality conditions, the quantities μ_i vanish.

As an example, let us consider the wave equation for the Na$^+$-atom (atomic number Z=11). There will be three functions $f_i(r)$ ($q=3$). The equations (1.3) take the form

$$-\frac{d^2 f_1}{dr^2}+2\left(-\frac{11}{r}+V(r)-F_0^{11}\right)f_1=\lambda_1 f_1+2F_0^{21}f_2+6F_1^{31}f_3, \quad (1.12a)$$

$$-\frac{d^2 f_2}{dr^2}+2\left(-\frac{11}{r}+V(r)-F_0^{22}\right)f_2=\lambda_2 f_2+2F_0^{21}f_1+6F_1^{32}f_3, \quad (1.12b)$$

$$-3\frac{d^2 f_1}{dr^2}+6\left(-\frac{11}{r}+\frac{1}{r^2}+V(r)-F_0^{33}-2F_2^{33}\right)f_2=\lambda_3 f_3+$$

$$+2F_1^{31}f_1+6F_1^{32}f_2 \quad (1.12c)$$

where the screening potential $V(r)$ is

$$V(r)=2F_0^{11}+2F_0^{22}+6F_0^{33}. \quad (1.13)$$

The equation for the series electron wave function takes the form

$$-\frac{1}{2}\frac{d^2 f_4}{dr^2}+\left[\frac{l(l+1)}{2r^2}-\frac{11}{r}+V(r)\right]f_4=\lambda_4 f_4+F_l^{14}f_1+$$

$$+F_l^{24}f_2+3\left[\frac{l+1}{2l+1}F_{l+1}^{34}+\frac{l}{2l+1}F_{l-1}^{34}\right]f_3. \quad (1.14)$$

In order to express the right-hand side of this equation in the form of a linear operator acting on f_4, let us write

$$G_l(r,r')=\left[f_1(r)f_1(r')+f_2(r)f_2(r')\right]K_l(r,r')+$$

$$+3f_3(r)f_3(r')\left[\frac{l+1}{2l+1}K_{l+1}(r,r')+\frac{l}{2l+1}K_{l-1}(r,r')\right], \quad (1.15)$$

where K has the former meaning (1.5). We shall have

$$-\frac{1}{2}\frac{d^2 f_4}{dr^2}+\left[\frac{l(l+1)}{2r^2}-\frac{11}{r}+V(r)\right]f_4=\lambda_4 f_4+\int_0^\infty G_l(r,r')f_4(r')dr'. \quad (1.16)$$

Thus, the series electron wave function of the sodium atom satisfies the linear integro-differential equation (1.16).

2 Method of Successive Approximations

Equations (1.3), which in the general case form a nonlinear system [and also equation (1.7) which is linear], are evidently soluble only by the method of successive approximations. For this purpose a fair initial approximation is needed. The most convenient method of constructing initial approximation is based on the variation principle and is a generalization of the Ritz method. This method will be considered in Section 2.

In cases when Hartree's wave functions are known it is sometimes possible to take the orthogonal combinations as initial approximation. It is to be noticed, however, that when Hartree's wave functions do not satisfy the orthogonality conditions even approximately, the orthogonalization process may alter their character considerably, introducing sometimes superfluous nodes; in such cases it is besides the purpose to take Hartree's wave functions as initial approximation.

In order to obtain more precise wave functions after the initial approximation are found, we have to apply the method of successive approximations. The application of this method is based on two properties of the system in question: first, that a change in the function $f_i(r)$ does not produce a considerable alteration in the general character of the coefficient Q_{ik} [the diagonal coefficients Q_{ii} being the most stable] and, second, that the right-hand side of (1.3) is small as compared with the terms of the left-hand side.

Owing to the first of the properties mentioned, it is possible to consider approximately (1.3) as a self-adjoint system of linear differential equations. Indeed, let us denote by $Q^*_{ik}(r)$ the coefficients calculated by means of the wave functions $f^*_i(r)$ of the initial approximation. Substituting these approximate values $Q^*_{ik}(r)$ for the coefficients $Q_{ik}(r)$ in (1.3) we obtain a linear system

$$-C_i \frac{d^2 f_i}{dr^2} + Q^*_{ii}(r) f_i = \lambda_i f_i - \sum{}' Q^*_{ik}(r) f_k(r). \tag{2.1}$$

It is to be expected that the solution of this system will be a better approximation than the initial one. Considering the solution obtained as a new initial approximation and solving again an equation of the type of (2.1), we would obtain the next more precise approximation. Repeating this process adequately, we would obtain functions $f_i(r)$ with the precision desired.

As a matter of fact, however, the direct numerical integration of the system (2.1) is, in practice, inapplicable on account of the extreme instability of this process. Therefore the system (2.1) has to be solved by another method based, besides the first, on the second of the above-mentioned properties of the system (comparative smallness of terms in the right-hand side of the equations). This property affords the possibility to consider these equations not as a system but as separate equations loosely bound. It is therefore admissible to substitute the approximate values $f^*(r)$ of the required function not only in the coefficients $Q_{ik}(r)$, but also in the terms of the right-hand side of the equations, and thus to obtain a set of separate inhomogeneous differential equations of the second-order. An equation of the same type is obtained from the integro-differential equation (1.7) or (1.6) for the series electron. Consequently, the construction of each approximation reduces to the solution of these inhomogeneous equations. Usual methods of numerical integration are inapplicable to such equations owing to the instability of the process in this case, but the solutions may be found by means of the generalized Green's function.

Let us consider the equation for one of the functions for the core electrons. Let f_i^0 be a solution of the linear homogeneous equation

$$-C_i\frac{d^2 f_i^0}{dr^2} + Q_{ii}^*(r) f_i^0 = \lambda_i^0 f_i^0, \tag{2.2}$$

where λ_i^0 is the eigenvalue of the left-hand side operator. We replace equations (1.3) or (2.1) approximately by

$$-C_i\frac{d^2 f_i}{dr^2} + Q_{ii}^*(r) f_i - \lambda_i^0 f_i = \omega_i(r), \tag{2.3}$$

where $\omega_i(r)$ is known from the preceding approximation and is equal to

$$\omega_i(r) = \lambda_i' f_i^* - {\sum_k}' Q_{ik}^* f_k^*, \tag{2.4}$$

the constant λ_i' being defined by the condition of orthogonality of expression (2.4) to the solution of the homogeneous equation f_i^0

$$\int \omega_i(r) f_i^0(r) dr = 0. \tag{2.5}$$

The solution of the equation of the form (2.3) by means of the generalized Green's function will be considered in Section 4.

It is essential that the process of constructing the Green's function as well as the solution, by means of this function, of the equation of the form

(2.3) are stable and practically performable processes. On constructing thus a particular solution of (2.3), we obtain a general solution by adding a term of the form $Cf_i^0(r)$, the constant C being determined by the normalization condition.

The integro-differential equations (1.7) and (1.8) are solved analogously, the integral operator being regarded as a perturbation.

An appropriate number of approximations being constructed, the parameter λ of the equation becomes

$$\lambda = \lambda^0 + \lambda', \tag{2.6}$$

where λ^0 is the characteristic value of the corresponding homogeneous differential equation and λ' is determined by conditions analogous to (2.4) and (2.5). The value of λ^0 may be compared with the value of the energy level, quantum exchange not being considered, and the value of λ' gives the corresponding correction. A more precise value of the energy level may be obtained by substituting a final approximation of f in the definite integral expression for the energy of a series electron.

When constructing every successive approximation f_i it is necessary to keep in mind the orthogonality conditions (1.2*). The rigorous solution of the initial system (1.3) satisfies these conditions automatically, but the approximate one may not satisfy them, and the functions f_i^* must be consequently orthogonalized before being substituted in $Q_{ik}(r)$ and $Q_{ii}(r)$. The orthogonalization, in general, spoils the solution, but this effect is the smaller the nearer the approximation in question is to the rigorous solution. Therefore it was considered best, instead of applying the usual Schwartz method of orthogonalization, to construct a linear combination of the form

$$f_i = Af_i^* + Bf_i^0 + \sum_k c_k f_k, \tag{*}$$

where f_i^* is the solution to be orthogonalized, f_i^0 is the solution of the corresponding homogeneous equation and f_k are the appropriate wave functions involved in the orthogonality conditions. To determine the constants A, B and c_k we use, first of all, the orthogonality and normalization conditions. But the number of these conditions being less than that of the constants involved in (*), they are insufficient for this determination. Therefore some additional condition must be imposed. This condition must be chosen in general so as to reduce the inevitable spoiling of the differential equation to a minimum, and in particular so as not to alter considerably the general character of the solution in the vicinity of $r = 0$.

For Schwartz's method $B = 0$; but the constants c_k are then comparatively large, so that this method involves the addition of considerable terms proportional to the functions which do not satisfy the equation. It is advisable to choose the constants so as to minimize the quantity $\sum_k c_k^2$. Usually this reduces approximately to the omission in (*) of the term which involves that function f_k for which the scalar product $\int f_i^* f_k dr$ has the largest value.

A linear combination of the form (*) will satisfy approximately an inhomogeneous equation with the right-hand side increased A times. The latter circumstance does not spoil the equation in the proper sense of the word. It was established practically (and it follows from theoretical considerations to some extent) that the coefficients Q_{ik} of the series electron equation, computed for different approximations, are nearly proportional and the corresponding right-hand sides of the equation differ by a nearly constant factor only. Moreover, usually the revised right-hand side of the equation differs from the preceding one by a factor, which is very near to A. Therefore the construction of such a linear combination for f is quite admissible and may be considered as the best way of orthogonalizing the solution obtained.

It must be noticed that although the coefficient $Q_{ii}^*(r)$ in the homogeneous equation (2.2) for a core electron is changed with each approximation, it is not necessary to solve this equation and to construct the Green's function every time anew. It is sufficient to solve this equation once for the initial approximation and thereupon to include in the right-hand side of the inhomogeneous equation the difference between the initial and subsequent values of $Q_{ii}^*(r)$ obtained as the construction of successive approximations proceeds. For the hydrogen-like functions the solution of the homogeneous equation can be avoided by taking the corresponding equation for a hydrogen-like ion. The difference between the coefficient of f_i in the equation of the hydrogen-like ion and in the equation in question must also be included in the right-hand side. Such a way of proceeding does not upset the convergence of the process and shortens the calculations considerably.

In order to estimate the approximation $f_i^*(r)$ we may proceed as follows. By means of f_i and of the corresponding coefficients Q_{ik}^* let us construct an expression analogous to (1.10):

$$\varphi_i(r) = -C_i \frac{d^2 f_i^*}{dr^2} + \sum_{k=1}^{q} Q_{ik}^* f_k^*. \tag{2.7}$$

If the functions f_i^* were rigorous solutions of the initial system of equations, the expressions $\varphi_i(r)$ would be strictly proportional to $f_i^*(r)$ and if we were to construct the function

$$\varphi_i^*(r) = \varphi_i(r) - \lambda_i f_i^*(r), \tag{2.8}$$

where λ_i is determined from the orthogonality condition

$$\int_0^\infty \varphi_i^*(r) f_i^*(r) dr = 0, \tag{2.9}$$

this function would vanish identically. But $f_i^*(r)$ being an approximate solution only, $\varphi_i^*(r)$ will differ from zero and its value affords a convenient estimation of the precision of the approximation. This precision may be characterized by the smallness of the number

$$\sigma_i = \int_0^\infty [\varphi_i^*(r)]^2 dr. \tag{2.10}$$

The approximation may be considered as quite satisfactory when σ_i does not exceed 0.001 or 0.0001. Until this limit for σ_i is attained, the process must be repeated, calculation beginning each time for that function for which the value of σ_i happens to be the largest.

We may now summarize the method recommended in the following way.

In order to get the initial approximation the Ritz method must be applied or the orthogonalized Hartree function may be taken. In order to obtain every successive approximation the system of equations (1.3) and the integro-differential equation (1.7) must be replaced by separate inhomogeneous equations of the form (2.3). These equations are solved by means of the generalized Green's function. The solution obtained, being orthogonalized, serves for calculations of the revised right-hand sides of the equations. Next the equations are solved again and the process is repeated until the desired precision is attained. Separate parts of the method adopted will be discussed in the following sections.

3 Construction of the Initial Approximation

As was remarked in Section 1, the generalized equations are deduced from the variation principle. A direct method of solving variational problems (i.e., a method which does not involve the construction of an equation for varied functions) was developed by Ritz. Its general idea is simple enough. A complete set of functions is chosen, each function satisfying the same boundary conditions as f. The latter is supposed to have the form

$$\varphi = c_1 u_1 + c_2 u_2 + \ldots + c_n u_n, \tag{3.*}$$

c_i being constant adjustable parameters. Substituting such an expression for f in the integral to be minimized one finds an expression for the integral as a function of parameters c_i. These parameters are then determined so as to make the integral a minimum, the additional relations between c_i resulting from the normalization and orthogonality conditions for f being taken into consideration.

When constructing the initial approximation of the wave functions, it is expedient to generalize Ritz's method replacing the expression (3.*) for the function f, which is linear in the adjustable parameters, by such an expression which contains them in a nonlinear way and permits us to take into account the characteristic properties of the required function at $r = 0$ and $r = \infty$. According to the form of the variation equations, for small r the function $f_i(r)$ behaves like r^{l_i+1}, l_i being the quantum number of this function. At $r = \infty$, $f_i(r)$ behaves like a product of an exponential $\exp(-\beta_i r)$ with a power of r. The coefficient β_i is related to the equation parameter λ but is not known to us beforehand. Furthermore, $f_i(r)$ with the same quantum number l_i must be orthogonal to each other. The simplest form of a function with such properties is

$$f_i(r) = r^{l_i+1} e^{-\beta_i r} P(r), \tag{3.1}$$

$P_i(r)$ being a polynomial of the lowest degree compatible with orthogonality conditions. Making use of these conditions as well as of the normalization, all the coefficients of the polynomials $P_i(r)$ can be expressed in terms of the exponential coefficients β_i. The values of β_i thus remain the only parameters adjustable, their number being equal to the number of functions f_i. The integration over r being performed, we find for the energy, which is to be minimized, an expression in the form of a rational function of the parameters β_i. Although this expression for W will be complicated enough, the determination of β_i from its minimum condition is quite feasible. The functions $f_i(r)$ thus constructed are fit to be taken as initial approximation for numerical integration. Practice has shown that the introduction of other constants, besides β_i (for example, the increase of the degree of the polynomial $P_i(r)$), renders the calculations very complicated and laborious, and does not give any considerable advantage in precision. Therefore it may be recommended to vary the parameters β_i only.

In order to illustrate the method of constructing the initial approximation given above, the calculations for the Li and Na^+ atoms will be given.

The Li atom has only one core electronic shell, which consists of two electrons and is described by a radial function $f_1(r)$ with quantum numbers $n_1 = 1$, $l_1 = 0$. The series electron is described by the function $f_2(r)$ with arbitrary quantum numbers n and l; when $l = 0$, the function $f_2(r)$ must be orthogonal to $f_1(r)$. The energy of the Li atom, expressed in terms of $f_1(r)$ and $f_2(r)$, has the form

$$W = T + U_1 + U_2, \tag{3.2}$$

where T is the kinetic energy of the electrons

$$T = \int_0^\infty \left[\left(\frac{df_1}{dr}\right)^2 + \frac{1}{2}\left(\frac{df_2}{dr}\right)^2 \frac{l(l+1)}{2r^2} f_2^2 \right] dr; \tag{3.3}$$

U_1 the energy of the interaction of the electrons and the nucleus

$$U_1 = -\int_0^\infty (2f_1^2 + f_2^2)\frac{3}{r}dr; \tag{3.4}$$

and U_2 the energy of the mutual interaction of the electrons

$$U_2 = \int_0^\infty \left[\frac{1}{2}(2f_1^2 + f_2^2)(2F_0^{11} + F_0^{22}) - f_1^2 F_0^{11} - \frac{1}{2} f_2^{22} - f_1 f_2 F_l^{12}\right] dr. \tag{3.5}$$

In the latter formula the coefficients F_l^{ik} have the meaning (1.4). In the following we shall consider the case $n = 2$, $l = 0$ only. According to what has been said above, we shall write approximate expressions for $f_1(r)$ and $f_2(r)$ in the form

$$f_1(r) = are^{-\alpha r}; \quad f_2(r) = b\left[r - \frac{1}{3}(\alpha+\beta)r\right] e^{-\beta r}. \tag{3.6}$$

Then the orthogonality condition will be satisfied identically and the normalization condition gives

$$a^2 = 4\alpha^3; \quad b^2 = \frac{12\beta^5}{\alpha^2 - \alpha\beta + \beta^2}. \tag{3.7}$$

Substituting f_1 and f_2 in the expression for the energy and performing the integration, we get

$$T = \alpha^2 + \frac{1}{6}\beta^2 + \frac{\beta^2}{\alpha^2 - \alpha\beta + \beta^2}; \tag{3.8}$$

$$U_1 = -6\alpha - \frac{3}{2}\beta + \frac{3}{2}\beta^2 \frac{\alpha - 2\beta}{\alpha^2 - \alpha\beta + \beta^2}; \tag{3.9}$$

$$U_2 = \frac{21}{8}\alpha - \frac{2\alpha^3}{(\alpha+\beta)^2} - -\frac{\alpha^4\beta(3\alpha+\beta)}{(\alpha+\beta)^3(\alpha^2 - \alpha\beta + \beta^2)} - \\ - \frac{4\alpha^3\beta^5}{(\alpha+\beta)^5(\alpha^2 - \alpha\beta + \beta^2)}. \tag{3.10}$$

The kinetic energy T will be a homogeneous function of the second degree and the potential energy U_1 and U_2 will be a homogeneous function of the first degree in α and β. Writing $\beta = \lambda\alpha$, we have

$$T = \alpha^2\varphi_1(\lambda); \quad U_1 + U_2 = -\alpha\varphi_2(\lambda), \tag{3.11}$$

φ_1 and φ_2 being rational functions of λ.

Equating to zero the derivatives of the complete energy $W = T + U_1 + U_2$ with respect to α and λ, for the determination of these parameters we obtain a system of two equations

$$2\alpha\varphi_1(\lambda) - \varphi_2(\lambda) = 0; \qquad \alpha\varphi_1' - \varphi_2'(\lambda) = 0. \tag{3.12}$$

Eliminating α, we have for λ

$$\frac{\varphi_1'(\lambda)}{\varphi_1(\lambda)} - \frac{2\varphi_2'(\lambda)}{\varphi_2(\lambda)} = 0. \tag{3.13}$$

The root of this equation is $\lambda = 0.2846$; the corresponding values of α and β are

$$\alpha = 2.70; \qquad \beta = 0.87. \tag{3.14}$$

With these values of α and β, the functions $f_1(r)$ and $f_2(r)$ will have the form

$$f_1(r) = 8.87re^{-2.7r}; \qquad f_2(r) = 1.025(r - 1.19r^2)e^{-0.87r}, \tag{3.15}$$

and for the energy of the normal state of lithium we have, in atomic units,

$$W = 7.425\,\text{a.u.} \tag{3.16}$$

This value can be compared with the experimental data. In order to express it in volts, it is necessary to multiply it by twice the ionization potential of the hydrogen atom, that is by a factor of 27.09. The multiplication being performed, we obtain

$$W = -201.1\,\text{V}. \tag{3.17}$$

On the other hand, the experimental value of the first ionization potential of the lithium atom is equal to 5.36 and of the second 75.28 volts, whereas the energy of the twice ionized Li atom is equal to nine times the ionization potential of the hydrogen, i.e., 121.90 volts. In summary, we have

$$-(5.36 + 75.28 + 121.90) = -202.54\,\text{volts},$$

so that the theoretical value of the energy obtained by this rather crude method agrees within 0.9 % with that observed.

Proceeding to the Na^+ atom, let us write down the expression for its energy. We have as before

$$W = T + U_1 + U_2, \tag{3.18}$$

where

$$T = \int_0^\infty \left[\left(\frac{df_1}{dr}\right)^2 + \left(\frac{df_2}{dr}\right)^2 + 3\left(\frac{df_3}{dr}\right)^2 \frac{6}{2r^2} f_3^2 \right] dr, \tag{3.19}$$

$$U_1 = -\int_0^\infty (2f_1^2 + f_2^2 + 6f_3^2)\frac{11}{r}dr, \tag{3.20}$$

$$U_2 = \int_0^\infty [\frac{1}{2}(2f_1^2 + f_2^2 + 6f_3^2)(2F_0^{11} + F_0^{22} + 6F_0^{33}) - f_1^2F_0^{11} -$$

$$-f_2^2F_0^{22} - 3f_3^2(F_0^{33} + 2F_2^{33}) - 2f_1f_2F_0^{21} - 6f_1f_3F_1^{31} - 6f_2f_3F_1^{32}]dr. \tag{3.21}$$

We shall seek the wave functions f_1, f_2 and f_3 in the form

$$f_1(r) = are^{-\alpha r}; \quad f_2(r) = b\left[r - \frac{1}{3}(\alpha+\beta)r^2\right]e^{-\beta r}; \quad f_3(r) = cr^2e^{-\gamma r}, \tag{3.22}$$

where the constants a, b and c, determined from the normalization condition, have the values

$$a = 2\sqrt{\alpha^3}; \quad b = \sqrt{\frac{12\beta^5}{\alpha^2 - \alpha\beta + \beta^2}}; \quad c = \sqrt{\frac{4}{3}\gamma^5}. \tag{3.23}$$

The substitution of f into the expression (3.18) for W leads to a rather complicated algebraic function of α, β, γ. Equating to zero the partial derivatives of $W(\alpha, \beta, \gamma)$ with respect to α, β and γ we obtain for the determination of these parameters a system of three algebraic equations. These equations, in spite of their complicated form, are solved easily enough. For this purpose a subdivision of the terms of the equations into primary and secondary terms, based upon the fact that β and γ are nearly equal and small in comparison with α, must be performed. At first, the secondary terms are to be neglected. Taking then any approximate values for two of the parameters, we can determine the third from the corresponding equation. After repeating such calculations several times, we obtain rather precise values α_0, β_0 and γ_0.

In order to verify the calculations, we write $\alpha = k\alpha_0$, $\beta = k\beta_0$, $\gamma = k\gamma_0$ in the expression for W. The energy W will become a polynomial of the second degree in k. This polynomial must be rendered minimum by a value of k, which only slightly differs from unity. The nearness of k to unity provides a check on the calculations. In such a way, we obtain

$$\alpha = 10.68; \quad \beta = 4.22; \quad \gamma = 3.49. \tag{3.24}$$

For the approximate wave functions we shall have

$$\begin{aligned} f_1^* &= 69.804\, r\, e^{-10.68\, r}; \\ f_2^* &= 13.602(r - 4.967\, r^2)e^{-4.22\, r}; \\ f_3^* &= 26.276\, r^2 e^{-3.49\, r}. \end{aligned} \tag{3.25}$$

The energy of the ionized atom of sodium, thus calculated, is

$$W = -160.9. \tag{3.26}$$

This value can be compared with the value

$$W = -161.8 \tag{3.27}$$

calculated by means of more precise wave functions f_i that were obtained by numerical integration (Section 7).

The agreement is close enough. The comparison of the functions (3.25) with the more precise solutions of the system of equations (see Fig. 1) shows that in general these functions reproduce the character of the solutions sufficiently well. At any rate they are good enough as initial approximation. The field of the Na^+ atom, which corresponds to these functions, deviates only slightly from that obtained by Hartree.

Thus, the above-given variational method for constructing the initial approximation of the wave functions proved correct. Besides that, it has the advantage of leading to analytic expressions for the wave functions. These expressions certainly afford a more precise approximation to the wave functions than those suggested before[4] and calculations of wave functions by other methods when attaining the same precision were far more laborious and time consuming.[5] It must be pointed out, however, that such analytical expressions may be of use when only the general character of the wave functions is of importance, as for instance in the calculation of the energy. In other cases, especially in calculating the doublet splitting, when the behavior of the functions in the vicinity of $r = 0$ is important, the approximation given by such analytical expressions is quite insufficient.

4 On the Stability of the Process of Numerical Integration

As was mentioned above, the principal condition for the applicability of any process of numerical integration of differential equations consists in its stability.

We shall consider a process of integration stable if a small change in the solution at some point has no considerable influence on its successive values and if this influence decreases as the integration proceeds. Otherwise, when a change in an initial value produces a considerable and ever increasing error, we shall consider the process unstable.

In order to illustrate the concept of stability let us consider the following example. Suppose we are integrating the equation

$$\frac{d^2y}{dx^2} = y$$

outward from $x = 0$ in the direction of increasing x. The process will be stable if the integral sought for is of the form $y = A\exp(x) + B\exp(-x)$, $(A \neq 0)$, and increases as x increases. But if we are looking for an integral of the type $\exp(-x)$ which decreases as x increases, the process will be unstable because the equation possesses a second integral, which increases with x. Let us consider the latter case more closely.

Let $y(0) = 1$; $y'(0) = -1$ be the initial conditions; then the exact value of the required function is $y = \exp(-x)$. Suppose that integrating we have reached $x = x_1$ with errors Δy_1 and $\Delta y_1'$ in the values of the

[4] L. Pauling, *The Theoretical Prediction of the Physical Properties of Many Electron Atoms and Ions*, Proc. Roy. Soc. A **114**, 18, 1927; C. Zener, *Analytic Atomic Wave Functions*, Phys. Rev. **36**, 51, 1930; J. Slater, *Atomic Shielding Constants*, Phys. Rev. **36**, 57, 1930; *Analytic Atomic Wave Functions*, Phys. Rev. **42**, 1932.

[5] J. Sugiura, Phil. Mag. **4**, 495, 1926.

function and its derivative, respectively, that is to say, we have

$$y_1 = e^{-x_1} + \Delta y_1,$$

$$y_1' = -e^{-x_1} + \Delta y_1'.$$

Even if the following integration could be carried out with ideal precision, we should obtain instead of $y = \exp(-x)$ the function

$$y = e^{-x}\left[1 + \frac{\Delta y_1 - \Delta y_2'}{2} e^{x_1}\right] + \frac{\Delta y_1 + \Delta y_1'}{2} e^{x - x_1}.$$

The last term in this expression increases indefinitely, no matter how small Δy_1 and $\Delta y_1'$ might be (if only by chance we shall not have $\Delta y_1 + \Delta y_1' = 0$), whereas the exact value of the required function decreases. Therefore the error will quickly surpass the exact value of the function and we shall have a function which has nothing in common with the exact one.

This example proves that stability is quite an indispensable condition in order that the process of integration be applicable.

It is evident from the preceding discussion that stability will take place when *of all the integrals of the given equation or system of equations the required integral is that most rapidly increasing or most slowly decreasing* when integration proceeds in the given direction. On the contrary, if there exists a single integral that increases more rapidly or decreases slower than the one sought for, the process will be unstable and, consequently, inapplicable.

In the discussed example of the equation $y'' = y$ the stability may be attained by integrating inward. The same way will be appropriate in the general case of a linear homogeneous differential equation of the second order. When an integral which decreases as x increases is required, one may proceed as follows. Calculating from the asymptotic expression the values of the function and its derivative for a sufficiently large value of x the integration can be carried out inward up to a certain point x_0 near the largest maximum of the modulus of the function.

In the remaining interval the integration is carried out outward, i.e., from $x = 0$ in the direction of increasing values of x. No difficulties bound up with instability will arise, except in the case of a function with many zeros within the interval of integration. In the latter case the error may increase considerably and in order to avoid it, it is necessary to adopt other methods.[6] The function obtained by integrating outward from

[6]See, e.g., V. Fock, DAN **1**, 241, 1934 ([34-2] in this book).

$x = 0$ up to $x = x_0$ will differ from that obtained by integrating inward by a constant factor. The value of this constant can be determined by comparing solutions at $x = x_0$. We shall return to this question in Section 6.

The reversal of the direction of integration is of use not only in the case of one linear homogeneous equation of the second order, but also in one important particular case of an inhomogeneous equation. This case arises when the integral sought for is of the same character at one of the ends of the interval as the right-hand side of the equation (the term independent of the function and its derivative), and at the other end may not satisfy the boundary condition. Then the integration begun from that end of the interval on which the boundary conditions are satisfied will be stable. But the inhomogeneous equation of the general type, as well as a system of equations, cannot be solved in this way. Indeed, a system of the type (2.1) possesses solutions that behave at infinity as $\exp(-\beta_i r)$, β_i being different for different solutions. The inhomogeneous equation of the form (2.3) possesses not only the solutions of the same character as the solutions of the homogeneous equation, but also such as the term $\omega(r)$. Therefore, there exists generally an integral which increases in the given direction more rapidly than that sought for and makes the process unstable.

It is necessary therefore to reduce all the calculations to the solution of homogeneous equations and of the particular type of inhomogeneous equation for which the integration is stable. This will be attained, as was shown in Section 2, by reducing, first, the solution of the system to the solution of separate equations of the types (2.3) and (2.2) and by applying then a method based on the properties of the generalized Green's function to the solution of the inhomogeneous equation (2.3). This method will be discussed in the following section.

5 Properties of the Generalized Green's Function and Its Construction

Let us rewrite equations (2.2) and (2.3) omitting the suffixes of the functions and coefficients for simplicity:

$$-C\frac{d^2 f^0}{dr^2} + Q(r)f^0 - \lambda^0 f^0 = 0; \tag{5.1}$$

$$-C\frac{d^2 f}{dr^2} + Q(r)f - \lambda^0 f = \omega(r). \tag{5.2}$$

The function $\omega(r)$ in this equation is defined by (2.4); this function is known from the preceding approximation and is orthogonal to f'. Our aim is to solve equation (5.2).

Let us denote by $L(f)$ the operator

$$L(f) = -\frac{d}{dr}\left(P(r)\frac{df}{dr}\right) + Q(r)f. \tag{5.3}$$

If we take $P(r) = C$, equations (5.1) and (5.2) may be written as

$$L(f^0) - \lambda^0 f^0 = 0; \tag{5.4}$$

$$L(f) - \lambda^0 f = \omega(r). \tag{5.5}$$

Let $y_1(r)$ and $y_2(r)$ be those solutions of the equation

$$L(y) - \lambda^0 y = F^0(r) \tag{5.6}$$

that satisfy the following conditions. The solution $y_1(r)$ must satisfy the boundary conditions at $r = 0$; at $r = \infty$ it can be infinite. The solution $y_2(r)$ must satisfy the boundary conditions at $r = \infty$ and may not satisfy them at $r = 0$.

It must be emphasized that the process of constructing the functions $y_1(r)$ and $y_2(r)$ by numerical integration is stable.

The functions $y_1(r)$ and $y_2(r)$ are defined by these conditions only up to an additive term of the form $af^0(r)$. In order to determine them uniquely it is convenient (in the case of the equation of the type considered) to impose additional conditions upon them by requiring that

$$\lim_{r\to 0}\frac{y_1(r)}{f^0(r)} = 0; \tag{5.7}$$

$$\int_0^\infty y_2(r)f^0(r)dr = 0. \tag{5.8}$$

These conditions are of course not indispensable. The difference

$$y_2(r) - y_1(r) = Y(r) \tag{5.9}$$

is, obviously, such a solution of the homogeneous equation (5.4) that satisfies the boundary conditions neither at $r = 0$ nor at $r = \infty$. According to the properties of a solution of a homogeneous equation, the quantity

$$D = P(r)\left[f^0(r)\frac{dY}{dr} - Y(r)\frac{df^0}{dr}\right] \tag{5.10}$$

is constant. We shall show that this constant is

$$D = \int_o^\infty [f^0(r)]^2 \, dr, \tag{5.10*}$$

i.e., is unity if f^0 is normalized. To establish the interference, it must be noticed that for any function y we have identically

$$L(y) - \lambda^0 y = -\frac{1}{f^0(r)} \frac{d}{dr}\left[P f^0 \frac{dy}{dr} - P y \frac{df^0}{dr}\right]. \tag{5.11}$$

Substituting $y = y_1$ and $y = y_2$ and making use of (5.6) we obtain

$$\frac{d}{dr} P \left(f^0 \frac{dy_i}{dr} - y_i \frac{df^0}{dr}\right) = -[f^0(r)]^2 \qquad (i = 1, 2). \tag{5.12}$$

Hence, owing to the boundary conditions, we infer that

$$P\left(f^0 \frac{dy_2}{dr} - y_2 \frac{df^0}{dr}\right) = \int_r^\infty [f^0]^2 dr;$$

$$P\left(f^0 \frac{dy_1}{dr} - y_1 \frac{df^0}{dr}\right) = -\int_0^r [f^0]^2 dr. \tag{5.13}$$

Subtracting the two equations, we obtain (5.10). In terms of the functions $y_1(r), y_2(r)$ and $f_o(r)$ the generalized Green's function can be expressed as follows:

$$\begin{aligned} G(r, r') &= y_1(r) f^0(r') + y_2(r') f^0(r), \quad \text{if} \quad r < r'; \\ G(r, r') &= y_1(r') f^0(r) + y_2(r) f^0(r'), \quad \text{if} \quad r > r'. \end{aligned} \tag{5.14}$$

This function is symmetric in r and r', is continuous and at every point, except for $r = r'$, satisfies the equation

$$L(G) - \lambda^0 G = f^0(r) f^0(r') \quad (r \neq r'), \tag{5.15}$$

and at $r = r'$ the quantity $P(r)\frac{dG}{dr}$ has a discontinuity

$$P(r')\left[\left(\frac{\partial G}{\partial r}\right)_{r=r'+0} - \left(\frac{\partial G}{\partial r}\right)_{r=r'-0}\right] = D = 1. \tag{5.16}$$

Properties (5.15) and (5.16) of the function G may be written in the form

$$L(G) - \lambda^0 G = f^0(r') f^0(r') - \delta(r - r'), \tag{5.17}$$

$\delta(r - r')$ being the Dirac improper delta-function. Indeed, integrating the left-hand side of (5.17), we have

$$\lim_{\varepsilon\to 0}\int_{r'-\varepsilon}^{r'+\varepsilon}[L(G)-\lambda^0 G]dr = -\lim_{\varepsilon\to 0}\int_{r'-\varepsilon}^{r'+\varepsilon}\frac{d}{dr}\left(P\frac{dG}{dr}\right)dr =$$

$$= -P(r')\left[\left(\frac{\partial G}{\partial r}\right)_{r=r'+0} - \left(\frac{\partial G}{\partial r}\right)_{r=r'-0}\right] = -D = -1.$$

The inference of (5.17) is at once established by noticing that according to the fundamental property of the δ-function the right-hand side of (5.17) provides the same value. Functions $y_1(r)$ and $y_2(r)$ being determined apart from an additive term of the form $af^0(r)$, the generalized Green's function will be determined apart from an additive term of the form $af^0(r)f^0(r')$. The additive term may be chosen so that $G(r,r')$ will be orthogonal to $f^0(r')$ identically in r. Let

$$\Omega(r) = \int_0^\infty G(r,r')f^0(r')dr'. \tag{5.18}$$

Substituting its expression for G, we have

$$\Omega(r) = y_2(r)\int_0^r [f^0(r')]^2 dr' + y_1(r)\int_r^\infty [f^0(r')]^2 dr' +$$

$$+f^0(r)\left[\int_0^r f^0(r')y_1(r')dr' + \int_r^\infty f^0(r')y_2(r')dr'\right]. \tag{5.19}$$

Multiplying both sides of (5.17) by $f_0(r')dr'$ and integrating over r' we obtain in virtue of (5.18)

$$L(\Omega) - \lambda^0\Omega = 0, \tag{5.20}$$

i.e., the function $\Omega(r)$ satisfies the homogeneous equation. Since $\Omega(r)$ satisfies the boundary conditions also, we have

$$\Omega(r) = Af^0(r). \tag{5.21}$$

In order to determine the constant A we can make use of the identity

$$y_2(r)\int_0^r [f^0(r')]^2 dr' + y_1(r)\int_r^\infty [f^0(r')]^2 dr' =$$

$$= f^0(r)P(r)\left(y_1\frac{dy_2}{dr} - y_2\frac{dy_1}{dr}\right), \tag{5.22}$$

which follows from (5.13). Substituting (5.22) into (5.19) and equating to (5.21) we obtain

$$A = \int_0^r f^0(r')dr' + \int_r^\infty f^0(r')dr' + P(r)\left(y_1\frac{dy_2}{dr} - y_2\frac{dy_1}{dr}\right). \quad (5.23)$$

Differentiating, it is easy to verify that the right-hand side of (5.23) is constant. It may be easily proved that, provided the conditions (5.7) and (5.8) are satisfied, the expression for A vanishes. We shall have, consequently,

$$\Omega(r) = \int_0^\infty G(r,r')f^0(r')dr' = 0. \quad (5.24)$$

The validity of this identity, provided the conditions (5.7) and (5.8) are satisfied, results directly from (5.21) and (5.9).

By means of the generalized Green's function constructed above, we can write the particular solution $z(r)$ of the inhomogeneous equation (5.5) in the form

$$-z(r) = \int_0^\infty G(r,r')\omega(r')dr', \quad (5.25)$$

or more explicitly

$$-z(r) = y_2(r)\int_0^r f^0(r')\omega(r')dr' + y_1(r)\int_r^\infty f^0(r')\omega(r')dr' +$$

$$+f^0\left[(r)\int_0^r y_1(r')\omega(r')dr' + \int_r^\infty y_2(r')\omega(r')dr'\right]. \quad (5.26)$$

In virtue of

$$\int_0^\infty f^0(r')\omega(r')dr' = 0 \quad (5.27)$$

the particular solution obtained is orthogonal to $f_0(r)$ independently of the conditions (5.7) and (5.8). Indeed, according to (5.25), we have

$$-\int_0^\infty z(r')f^0(r')dr' = \int_0^\infty\int_0^\infty G(r,r')f^0(r)\omega(r')drdr' =$$

$$= \int_0^\infty \Omega(r')\omega(r')dr' = A\int_0^\infty f^0(r')\omega(r')dr' = 0. \quad (5.28)$$

This property of the particular solution provides a close check on the calculations. Owing to the same identity (5.27) we have

$$\int_0^r f^0(r')\omega(r')dr' = -\int_r^\infty f^0(r')\omega(r')dr' \quad (5.27^*)$$

and, consequently, it is sufficient to evaluate only three of the four integrals involved in (5.26). The general solution of (5) has the form

$$f(r) = z(r) + Af^0(r), \tag{5.29}$$

the arbitrary constant A being determined from the normalization condition or from the conditions of orthogonality of $f(r)$ to a given function.

In order to illustrate the formulae given in this section the following analytical example may be of use. Let

$$L(y) = -\frac{d}{dr}\left(r\frac{dy}{dr}\right) + \left(\frac{r}{4} - \frac{1}{2}\right)y; \quad \lambda^0 = 0. \tag{5.30}$$

As limiting conditions the requirement is imposed that the solution must be finite at $r = 0$ and must decrease as $r \to \infty$. We may write

$$f^0(r) = e^{-\frac{r}{2}}; \quad y_1(r) = e^{-\frac{r}{2}} \int_0^r \frac{e^{-t} - 1}{t} dt; \quad y_2(r) = e^{-\frac{r}{2}}(\log r + \gamma),$$

γ being the Euler constant ($\gamma = 0.5772156649$). Then the conditions (5.7) and (5.8) will be satisfied and for the Green's function constructed according to (5.14), the expression (5.18) will vanish identically.

6 Comments on the Numerical Integration of Equations

In previous sections it was shown how the solution of the inhomogeneous equation (5.5) can be constructed by means of the solutions of the corresponding homogeneous equation (5.4) and the equations (5.6). Methods of numerical integration of such differential equations are known. The best are those devised by Adams and Störmer[7] and Cowell's method, developed by Hartree.[8] The first (Adams–Störmer's method) is very convenient when applied to the second-order equations with the first derivative absent, but in the case of the reduction to a system of first-order equations, thc Cowell method is preferable. An account of the methods referred to may be found in the papers just quoted; here we

[7] E.T. Whittaker and G. Robinson, The Calculus of Observations, 1928.

[8] D.R. Hartree, *A Practical Method for the Numerical Solution of Differential Equations*, Mem. and Proc. Manchester Lit. and Phil. Soc. **77**, 91, 1933. We are much indebted to Professor Hartree for having kindly sent us in 1931 a typewritten manuscript that contained the main part of the results of his paper just quoted. (*Authors*)

should like only to emphasize the convenience of small intervals and of such a scheme of integration in which the differences tabulated as the work proceeds could provide a close check on the numerical work.

In Section 4 it was pointed out that in order to make the process of numerical integration stable, the integration must be carried out outward from $r = 0$ and inward from $r = \infty$ to some intermediate point. But for the equation of the type considered the ends of the interval of integration are singular points. Therefore in the vicinity of $r = 0$ and for sufficiently large values of r it is necessary to look out for solutions in series.

Let us consider the determination of such power series. Let us take the homogeneous equation which corresponds to the series electron wave function. It may be written in the form

$$\frac{d^2 f}{dr^2} = \left[\frac{l(l+1)}{r^2} - \frac{2Z}{r} + 2V(r) - 2\lambda\right] f. \tag{6.1}$$

For any value of λ, equation (6.1) possesses a system of two linearly independent integrals f_0 and f_∞, the first of which, f_0, satisfies the boundary condition at $r = 0$ and the second f_∞ at $r = \infty$. The characteristic values of (6.1) are determined as such values of λ for which the two solutions f_0 and f_∞ (apart from an arbitrary multiplying constant in each) are the same. It is necessary therefore to determine f_0 and f_∞, for different (trial) values of λ. The considerations by which one may be guided in the choice of these trial values of λ will be stated below; until then let us consider the values of λ as assigned.

The function $V(r)$ is usually given only as a tabulated function (from the initial approximation). Using this table, it is possible to construct such a polynomial of a degree not higher than the fourth, that it reproduces the behavior of $V(r)$ for small r (for instance, up to $r = 0.02$). Then, by means of the method of undetermined coefficients we can obtain for $f_0(r)$ an expansion of the form

$$f_0(r) = r^{l+1} + a_1 r^{l+2} + a_2 r^{l+3}. \tag{6.2}$$

It is convenient to take the coefficient of the lowest power of r equal to unity. This expansion enables us to obtain a set of initial values (for instance, four values) of $f_0(r)$ to begin the numerical integration. It must be noticed that practically the initial values of $f_0(r)$ are almost independent of λ (within a certain range of λ) and, consequently, it is possible to use the same initial values of $f_0(r)$ for different λ.

In order to start a table of $f_\infty(r)$ we have to compute some initial values by means of an asymptotic expression of this function, which can be obtained as follows. The asymptotic expressions of the coefficients $F_s^{ik}(r)$ are of the form

$$F_s^{ik}(r) = \frac{C_s^{ik}}{r^{s+1}}) \left(r^{\alpha_i+\alpha_k-2} e^{-(\beta_i+\beta_k)r} \right), \tag{6.3}$$

where

$$C_s^{ik} = \frac{1}{2s+1} \int_0^\infty f_i(r) f_k(r) r^s dr, \tag{6.4}$$

and α_i and β_i are constants involved in the asymptotic expressions of the functions $f(r)$:

$$f_i(r) \cong M_i r^{\alpha_i} e^{-\beta_i r}.$$

Therefore $V(r)$ will have as asymptotic expression a polynomial in $\frac{1}{r}$:

$$V(r) \cong \frac{N}{r} + ... + \frac{K}{r^k}, \tag{6.5}$$

the difference between $V(r)$ and the polynomial on the right-hand side of (6.5) decreasing at $r \to \infty$ faster than any finite power of $\frac{1}{r}$. The coefficients $N, ..., K$ in (6.5) are known,[9] being obtained from the integrals (6.4) which are evaluated for the initial approximation of $f_i(r)$. Thus, the asymptotic expression of the coefficient of f in (6.1) may be considered to be known. The inferior limit of the range of validity of this asymptotic expression can be settled by comparison with the tabulated values of the coefficient.

The substitution of expression (6.5) for $V(r)$ in (6.1) being performed, we may write this equation in the form

$$\frac{d^2 f}{dr^2} = \left(\beta^2 - \frac{2\alpha\beta}{r} + ... + \frac{2K}{r^k} \right) f, \tag{6.6}$$

where

$$\beta = \sqrt{-2\lambda}; \quad \alpha = \frac{Z-N}{\sqrt{-2\lambda}}. \tag{6.7}$$

The asymptotic expression for the solution of this equation will have the form

$$f_\infty = e^{-\beta r} r^\alpha \left(1 + \frac{c_1}{r} + \frac{c_2}{r^2} + ... \right), \tag{6.8}$$

the coefficients c_1, c_2 being determined from a recurrence formula that we obtain by substituting (6.8) in (6.6). It is convenient to take the

[9] The number N is an integer equal to the charge of the screening electronic shells (in the equation for the series electron $N = Z - 1$).

coefficient of the highest power of r in f_∞ equal to unity. Usually the first terms of (6.8) decrease sufficiently fast for such values of r only, which exceed considerably the inferior limit of the range validity of formula (6.5). The initial values of $f_\infty(r)$ are obtained from formula (6.8); for smaller r the values of $f_\infty(r)$ are found by numerical integration.

The initial values of the solutions $y_1(r)$ and $y_2(r)$ of an equation of the type (5.6) are obtained analogously. For the function $y_1(r)$ the integration is carried out outward from $r = 0$ and for $y_2(r)$ inward from $r = \infty$. If we assume the parameter λ in (6.1) to be equal to its characteristic value, the normalized solution $f^0(r)$ of the homogeneous equation will be connected with the functions $f_0(r)$ and $f_\infty(r)$ by the relations

$$f^0(r) = B_0 f_0(r) = B_0 r^{l+1} + \dots ; \tag{6.9}$$

$$f^0(r) = B_\infty f_\infty(r) = B_\infty e^{-\beta r} r^\alpha (1 + \dots). \tag{6.10}$$

Equation (5.6) becomes

$$-\frac{1}{2}\frac{d^2 y}{dr^2} + \left[\frac{l(l+1)}{2r^2} - \frac{Z}{r} + V(r) - \lambda^0\right] y = f^0(r). \qquad 6.11)$$

Substituting the expression (6.9) in the right-hand side of this equation and making use of the condition (5.7) we obtain for $y_1(r)$ an expansion of the form

$$y_1(r) = -\frac{B_0}{2l+3} r^{l+3} + \dots , \tag{6.12}$$

which is to be used for the calculation of the initial values of this function.

For computation of the initial values of $y_2(r)$ for sufficiently large r we can use its asymptotic expansion, which will differ from that of f_∞ only by involving logarithmic terms. For large r the equation for $y_2(r)$ has the form

$$\frac{d^2 y_2}{dr^2} = \left(\beta^2 - \frac{2\alpha\beta}{r} + \dots + \frac{2K}{r^k}\right) y_2 = B_\infty e^{-\beta r} r^\alpha (1 + \frac{c_1}{r} + \dots). \tag{6.13}$$

The right-hand side of this equation being a solution of the homogeneous equation, we infer easily that the asymptotic expansion for $y_2(r)$ has the form

$$y_2(r) = B_\infty \frac{\alpha}{\beta^2} f_\infty(r) \log r + B_\infty e^{-\beta r} r^{\alpha+1} \frac{1}{\beta}\left(1 + \frac{d_1}{r} + \frac{d_2}{r^2} + \dots\right). \tag{6.14}$$

The constant d_1 remains indeterminate; it may be taken equal to zero. The constants d_2, d_3 are expressible in terms of d_1 and of the coefficients of the equation. For their determination, recurrence formulae may be easily constructed.

As a check on the calculation of $y_1(r)$ and $y_2(r)$ formula (5.9) and (5.10) can be used.

In the preceding consideration the parameter λ in (6.1) was supposed to be given. Let us now consider how to determine this constant from the condition of the existence of a solution of equation (6.1) satisfying the boundary conditions, that is, how to determine the characteristic values of the equation.

Proceeding from the initial values obtained from (6.2) and (6.8), we can find, by means of numerical integration, the functions $f_0(r)$ and $f_\infty(r)$ for every value of λ and construct the Wronskian

$$f_\infty(r)\frac{df_0}{dr} - f_0(r)\frac{df_\infty}{dr} = D(\lambda). \tag{6.15}$$

It will be a function of λ only and independent of r (the latter may be used as a check on calculations). Values of λ for which $D(\lambda) = 0$ are characteristic values of the equation. Having calculated $D(\lambda)$ for some values of λ, the required one may be found by interpolation, the most convenient independent variable being not λ itself, but the quantity

$$\nu = \frac{Z}{\sqrt{-2\lambda}}, \tag{6.16}$$

or any expression proportional to it.

In order to be able to discuss the general character of the function

$$D(\lambda) = D^*(\nu), \tag{6.17}$$

let us construct its analytic expression for the case of hydrogen-like ions, the screening potential $V(r)$ in (6.1) being then equal to zero. The calculation affords

$$\begin{aligned} D^*(\nu) &= \frac{\Gamma(2l+1)}{\Gamma(l+1-\nu)}\left(\frac{\nu}{2Z}\right)^{l+\nu} = \\ &= \frac{1}{\pi}\sin\pi(\nu-l)\Gamma(\nu-l)\Gamma(2l+2)\left(\frac{\nu}{2Z}\right)^{l+\nu}. \end{aligned} \tag{6.18}$$

In this particular case the roots of $D^*(\nu)$ will be

$$\nu_1 = l+1, \quad \nu_2 = l+2, \dots . \tag{6.19}$$

In the general case $V(r) \neq 0$, the function $D^*(\nu)$ will have the same character qualitatively, that is to say, $D^*(\nu) > 0$ when $\nu < \nu_1$; $D^*(\nu) < 0$

when $\nu_1 < \nu < \nu_2$, and so on. This provides the possibility of judging in what direction ν (or λ) must be changed. Assume that we look for the second root ($\nu = \nu_2$), and for the value of $\nu = \nu'$ taken (which we suppose to be near ν_2) we have obtained $D^*(\nu') = D' > 0$); then we must diminish ν. Suppose further that for the diminished value $\nu + \nu''$ we have obtained $D^*(\nu'') = D'' < 0$. We may use then the linear interpolation and take

$$\nu''' \cong \frac{\nu' D'' - \nu'' D'}{D'' - D'} \tag{6.20}$$

as the next approximation. The linear interpolation is valid here, for D^* is nearly linear in ν within the limits considered. As the first approximate value of λ we may take the result of the substitution of the initial approximation to f in the expression

$$\lambda = \frac{1}{2}\int_0^\infty \left(\frac{df}{dr}\right)^2 dr + \int_0^\infty \left[\frac{l(l+1)}{2r^2} - \frac{Z}{r} + V(r)\right] f^2 dr, \tag{6.21}$$

which is obtained from the equation.

The method of determination of the characteristic values developed above is theoretically simpler and in practice more convenient than the method of comparing the logarithmic derivatives of f_0 and f_∞ in an arbitrary chosen point, used by Hartree. It provides the possibility of determining precisely the characteristic value λ after taking only a few (usually three or four) trial values.

As was mentioned in Section 2, the integration of the homogeneous equation and the construction of the functions $y_1(r)$ and $y_2(r)$ must be carried out only once for each case, on including the differences of the coefficients of the equation for every successive approximation in the right-hand side of the equation.

7 Some Numerical Results for Sodium

Some results of the calculation for the Na atom that were carried out by M.I. Petrashen, together with A.R. Krichagina, will be given in this section.

Equations for core-electron wave functions $f_1(r)$, $f_2(r)$ and $f_3(r)$ of the Na atom are given in Section 1 (formulae (12a), (12b), (12c)). The initial approximation to these functions was found by the Ritz method developed in Section 3. Analytical expressions of approximate functions $f_1^*(r)$, $f_2^*(r)$ and $f_3^*(r)$ are given in the same place (formula (3.25)). In Fig. 1 the curves of f_i^* versus r are drawn by a broken line. The curves for the final approximations for these functions, obtained by the method of numerical integration given above, are drawn on the same Fig. 1 by a solid line, and the values of these functions are given in Table 1 (the first three columns). The quantities σ_i defined in Section 2 (formula (2.10)), the values of which provide an estimation of the quality of the approximations for the final approximation, turned out to be

$$\sigma_1 \cong 0.0005; \quad \sigma_2 \cong 0.0025; \quad \sigma_3 \cong 0.0021.$$

For the first approximations they were several thousand times larger, so that their values provide a very sensitive estimation for the precision of the solution of (3.12). The values of the X-ray terms E_i connected with the parameters λ_i by the relations

$$E_1 = \frac{1}{2}\lambda_1; \quad E_2 = \frac{1}{2}\lambda_2; \quad \frac{1}{6}\lambda_3$$

are given in Table 7, where the experimental values of X-ray terms are also given. A good agreement could not be expected, however, as mentioned in Section 1.

Seven coefficients that correspond to the core electrons are given in the first seven columns of Table 3 and are plotted in Fig. 5. The screening potential is given in the last column of Table 3. In Fig. 10 this potential $V(r)$ (upper solid line) is compared with those of Hartree (dotted line) and Fermi (lower solid line). The latter was computed not for a ion, but for a neutral atom. In Table 3 every coefficient is tabulated up to such a point at which its asymptotic expression, given below the respective column, becomes valid.

Values of separate parts of the expression of the energy of the ionized Na atom are given in Table 8.

It is interesting to find out what part of the whole energy of the atom is formed by the energy of the quantum exchange. Strictly speaking, for this purpose it would be necessary to compare the expression of the energy without quantum exchange terms computed by means of Hartree functions with the expression of the energy, quantum exchange terms included, computed by means of our functions. But having no Hartree's functions, we were able only to compare these expressions calculated by means of our functions. In the expression

$$U_2 = U_2' + U_2''$$

for the mutual electron potential energy, the additive term U_2'' results from quantum exchange. This term turns out to be 2.05, while the total energy is −161.8. Thus the quantum exchange energy gives only 2% of the total energy value and is, consequently, of no great importance. On the contrary, in the differences of the energies (optical terms) the correction on the quantum exchange is very important.

In Table 7 the term values for all states n_l considered are given. The first column contains the experimental values and the second the characteristic values of the integro-differential operator. The characteristic values λ^0 of the corresponding homogeneous differential equation are given in the third column. By comparing these quantities we see that for term values the quantum exchange correction is of great importance. While λ^0 differs from the observed term value by as much as 10 to 20%, the corresponding deviation for the integro-differential equation does not exceed 2%, which is a rather satisfactory result.

The wave functions for different states of the series electron are given in the last two columns of Table 1 (states 3_0 and 3_1) and in Table 2 (states 4_0 and 4_1) and are plotted in Fig. 2 and Fig. 3, respectively.

It may be of interest to compare the solution of the integro-differential equation (1.16), which takes the quantum exchange into consideration, with the solution of the homogeneous differential equation obtained by the omission of the integral operator which describes the quantum exchange. In Table 6 and Fig. 4 such a comparison for the function f_4 when $n = 3$, $l = 1$ is given. The difference between f_4^0 and corresponding solution f_4 of the integro-differential equation is very considerable.

The coefficients F_l^{ik} for the considered states of the series electron are given in Table 4 and Table 5 and are plotted versus r in Figs. 5, 6, 7 and 8.

Table 1.

r	f_1	f_2	f_3	f_4 $(n=3, l=0)$	f_4 $(n=3, l=1)$
0.00	0.0000	0.000	0.000	0.0000	0.0000
0.02	1.1288	0.270	0.014	0.0463	−0.0013
0.04	1.8154	0.427	0.051	0.0731	−0.0046
0.06	2.1925	0.499	0.104	0.0854	−0.0093
0.08	2.3568	0.509	0.166	0.0868	−0.0149
0.10	2.3778	0.474	0.235	0.0804	−0.0210
0.12	2.3055	0.407	0.307	0.0685	−0.0274
0.14	2.1755	0.317	0.380	0.0529	−0.0338
0.16	2.0127	0.214	0.4511	0.0347	−0.0402
0.18	1.8344	0.103	0.5203	0.0149	−0.0462
0.20	1.6530	−0.012	0.5863	−0.0053	−0.0520
0.22	1.4754	−0.126	0.6485	−0.0252	−0.0574
0.24	1.3071	−0.238	0.7066	−0.0444	−0.0623
0.26	1.1508	−0.346	0.7602	−0.0630	−0.0669
0.28	1.0080	−0.447	0.8092	−0.0803	−0.0709
0.30	0.8791	−0.542	0.8536	−0.0962	−0.0745
0.32	0.7638	−0.630	0.8934	−0.1108	−0.0776
0.34	0.6615	−0.710	0.9286	−0.1244	−0.0803
0.36	0.5714	−0.783	0.9596	−0.1356	−0.0825
0.38	0.4924	−0.848	0.9864	−0.1456	−0.0843
0.40	0.4236	−0.905	1.0094	−0.1542	−0.0858
0.45	0.2886	−1.020	1.0506	−0.1694	−0.0876
0.50	0.1961	−1.095	1.0723	−0.1765	−0.0874
0.55	0.1328	−1.138	1.0780	−0.1769	−0.0854
0.60	0.0903	−1.155	1.0705	−0.1719	−0.0819
0.65	0.0617	−1.150	1.0529	−0.1621	−0.0771
0.70	0.0424	−1.130	1.0274	−0.1484	−0.0714
0.75	0.0294	−1.098	0.9959	−0.1321	−0.0648
0.80	0.0205	−1.056	0.9600	−0.1134	−0.0575
0.85	0.0144	−1.010	0.9210	−0.0932	−0.0497
0.90	0.0104	−0.959	0.8799	−0.0716	−0.0414
0.95	0.0075	−0.906	0.8379	−0.0489	−0.0327
1.0	0.0051	−0.852	0.7962	−0.0260	−0.0237
1.1	0.0014	−0.746	0.7131	0.0204	−0.0050
1.2	0.0000	−0.645	0.6337	0.0667	0.0142
1.3	–	−0.554	0.5587	0.1117	0.0338
1.4	–	−0.472	0.4900	0.1548	0.0536
1.5	–	−0.401	0.4278	0.1955	0.0734

Table 1 (continued).

r	f_1	f_2	f_3	f_4 $(n=3, l=0)$	f_4 $(n=3, l=1)$
1.6	0.0000	−0.339	0.3719	0.2336	0.0931
1.7	–	−0.285	0.3220	0.2691	0.1126
1.8	–	−0.240	0.2777	0.3019	0.1318
1.9	–	−0.201	0.2380	0.3320	0.1506
2.0	–	−0.168	0.2050	0.3592	0.1691
2.2	–	−0.118	0.1493	0.4060	0.2047
2.4	–	−0.082	0.1076	0.4429	0.2378
2.6	–	−0.058	0.0773	0.4709	0.2684
2.8	–	−0.042	0.0549	0.4908	0.2964
3.0	–	−0.030	0.0388	0.5036	0.3215
3.2	–	−0.021	0.0273	0.5103	0.3438
3.4	–	−0.015	0.0191	0.5116	0.3631
3.6	–	−0.010	0.0134	0.5082	0.3796
3.8	–	−0.006	0.0095	0.5010	0.3933
4.0	–	−0.003	0.0065	0.4908	0.4043
4.4	–	−0.000	0.0026	0.4639	0.4192
4.8	–	–	0.0000	0.4294	0.4251
5.2	–	–	–	0.3921	0.4234
5.6	–	–	–	0.3535	0.4155
6.0	–	–	–	0.3149	0.4026
6.4	–	–	–	0.2780	0.3858
6.8	–	–	–	0.2432	0.3662
7.2	–	–	–	0.2110	0.3447
7.6	–	–	–	0.1820	0.3220
8.0	–	–	–	0.1561	0.2990
8.8	–	–	–	0.1133	0.2533
9.6	–	–	–	0.0810	0.2104
10.4	–	–	–	0.0571	0.1719
11.2	–	–	–	0.0399	0.1384
12.0	–	–	–	0.0275	0.1103
12.8	–	–	–	0.0189	0.0866
13.6	–	–	–	0.0129	0.0674
14.4	–	–	–	0.0000	0.0519
15.2	–	–	–	–	0.0396
16.0	–	–	–	–	0.0300
16.8	–	–	–	–	0.0226
17.6	–	–	–	–	0.0168
18.4	–	–	–	–	0.0125

Table 2.

r	f_4 ($n = 4$, $l = 0$)	f_4 ($n = 4$, $l = 1$)	r	f_4 ($n = 4$, $l = 0$)	f_4 ($n = 4$, $l = 1$)	r	f_4 ($n = 4$, $l = 1$)	f_4 ($n = 4$, $l = 1$)
0.00	0.0000	0.00000	0.85	−0.0423	0.0283	6.4	−0.1616	−0.0534
0.02	0.0220	0.00077	0.90	−0.0322	0.0231	6.8	−0.2022	−0.0189
0.04	0.0335	0.00282	0.95	−0.0216	0.0177	7.2	−0.2379	0.0161
0.06	0.0392	0.00569	1.0	−0.0107	0.0121	7.6	−0.2683	0.0506
0.08	0.0399	0.00913	1.1	0.0112	0.0007	8.0	−0.2936	0.0843
0.10	0.0369	0.0129	1.2	0.0331	−0.0111	8.8	−0.3290	0.1463
0.12	0.0314	0.0168	1.3	0.0542	−0.0228	9.6	−0.3462	0.1993
0.14	0.0242	0.0207	1.4	0.0744	−0.0346	10.4	−0.3484	0.2415
0.16	0.0159	0.0246	1.5	0.0933	−0.0463	11.2	−0.3389	0.2727
0.18	0.0070	0.0283	1.6	0.1109	−0.0578	12.0	−0.3211	0.2934
0.20	−0.0023	0.0318	1.7	0.1270	−0.0691	12.8	−0.2977	0.3044
0.22	−0.0114	0.0350	1.8	0.1416	−0.0801	13.6	−0.2711	0.3072
0.24	−0.0203	0.0380	1.9	0.1547	−0.0907	14.4	−0.2433	0.3031
0.26	−0.0288	0.0408	2.0	0.1661	−0.1010	15.2	−0.2155	0.2936
0.28	−00368	0.0432	2.2	0.1845	−0.1201	16.0	−0.1888	0.2800
0.30	−0.0442	0.0454	2.4	0.1969	−0.1373	16.8	−0.1639	0.2663
0.32	−0.0509	0.0472	2.6	0.2035	−0.1523	17.6	−0.1411	0.2450
0.34	−0.0570	0.0488	2.8	0.2048	−0.1650	18.4	−0.1205	0.2255
0.36	−0.0623	0.0501	3.0	0.2013	−0.1754	19.2	−0.1023	0.2058
0.38	−0.0670	0.0511	3.2	0.1936	−0.1833	20.0	−0.0863	0.1863
0.40	−0.0710	0.0520	3.4	0.1819	−0.1888	20.8	−0.0725	0.1674
0.45	−0.0781	0.0530	3.6	0.1670	−0.1920	22.4	−0.0504	0.1327
0.50	−0.0815	0.0526	3.8	0.1493	−0.1929	24.0	−0.0347	0.1028
0.55	−0.0818	0.0512	4.0	0.1292	−0.1915	25.6	−0.0235	0.0782
0.60	−0.0794	0.0489	4.4	0.0838	−0.1826	27.2	−0.0156	0.0585
0.65	−0.0748	0.0458	4.8	0.0340	−0.1664	28.8	−0.0103	0.0431
0.70	−0.0685	0.0421	5.2	−0.0174	−0.1442	30.4	−0.0067	0.0312
0.75	−0.0607	0.0379	5.6	−0.0681	−0.1171	32.0	–	0.0223
0.80	−0.0519	0.0333	6.0	−0.1166	−0.0865	33.6	–	0.0158

Table 3.
F_l^{ik}

r	F_0^{11}	F_0^{22}	F_0^{33}	F_0^{21}	F_1^{31}	F_1^{23}	F_2^{33}	$V(r)$
0.00	10.613	1.835	1.722	1.591	0.000	0.000	0.000	35.228
0.02	10.348	1.820	1.722	1.528	0.062	−0.010	0.003	34.668
0.04	9.748	1.786	1.721	1.384	0.132	−0.024	0.009	33.394
0.06	9.010	1.744	1.720	1.208	0.189	−0.040	0.016	31.823
0.08	8.242	1.703	1.718	1.029	0.228	−0.056	0.025	30.198
0.10	7.505	1.665	1.716	0.8622	0.2554	−0.072	0.035	28.636
0.12	6.826	1.633	1.713	0.7132	0.2715	−0.088	0.045	27.196
0.14	6.213	1.606	1.709	0.5842	0.2792	−0.106	0.055	25.892
0.16	5.668	1.583	1.704	0.4751	0.2803	−0.126	0.065	24.726
0.18	5.189	1.566	1.697	0.3841	0.2763	−0.147	0.074	23.691
0.20	4.767	1.551	1.688	0.3090	0.2687	−0.167	0.083	22.764
0.22	4.396	1.539	1.678	0.2478	0.2587	−0.186	0.092	21.938
0.24	4.072	1.529	1.666	0.1982	0.2469	−0.205	0.101	21.198
0.26	3.786	1.520	1.653	0.1581	0.2341	−0.222	0.108	20.530
0.28	3.533	1.512	1.638	0.1260	0.2210	−0.237	0.115	19.918
0.30	3.310	1.504	1.622	0.1002	0.2078	−0.251	0.121	19.359
0.32	3.110	1.496	1.605	0.0797	0.1949	−0.264	0.127	18.841
0.34	2.932	1.487	1.587	0.0633	0.1824	−0.274	0.132	18.359
0.36	2.773	1.477	1.569	0.0500	0.1705	−0.284	0.136	17.914
0.38	2.629	1.467	1.549	0.0400	0.1593	−0.291	0.139	17.486
0.40	2.499	1.456	1.530	0.0317	0.1487	−0.298	0.1414	17.089
0.45	2.221	1.424	1.478	0.0179	0.1253	−0.3082	0.1451	16.158
0.50	2.000	1.387	1.424	0.0102	0.1059	−0.3114	0.1455	15.317
0.55	1.818	1.346	1.370	0.0058	0.0901	−0.3094	0.1433	14.545
0.60	1.666	1.302	1.315	0.0033	0.0771	−0.3032	0.1392	13.828
0.65	1.538	1.257	1.262	0.0019	0.0666	−0.2938	0.1338	13.162
0.70	1.428	1.210	1.210	0.0011	0.0578	−0.2824	0.1274	12.537
0.75	0.000	1.164	1.161	0.0007	0.0507	−0.2697	0.1205	11.957
0.80	–	1.118	1.113	0.0004	0.0447	−0.2564	0.1134	11.414
0.85	–	1.074	1.067	0.0002	0.0397	−0.2427	0.1061	10.903
0.90	–	1.031	1.024	0.0001	0.0356	−0.2291	0.0990	10.428
0.95	–	0.990	0.983	0.0001	0.0320	−0.2158	0.0922	9.982
1.0	–	0.951	0.944	0.0000	0.0289	−0.2030	0.0856	9.567
1.1	–	0.879	0.873	–	0.0240	−0.1792	0.0734	8.815
1.2	–	0.815	0.810	–	0.0202	−0.1580	0.0627	8.157
1.3	–	0.758	0.754	–	0.0172	−0.1395	0.0536	7.579
1.4	–	0.707	0.704	–	0.0149	−0.1234	0.0457	7.069

Table 3 (continued).
F_l^{ik}

r	F_0^{11}	F_0^{22}	F_0^{33}	F_0^{21}	F_1^{31}	F_1^{23}	F_2^{33}	$V(r)$
1.5		0.662	0.660		0.0130	−0.1096	0.0391	6.619
1.6		0.622	0.621		0.0114	−0.0977	0.0335	6.219
1.7		0.586	0.585		–	−0.0874	0.0288	5.862
1.8		0.554	0.554		–	−0.0786	0.0248	5.542
1.9		0.526	0.525		–	−0.0709	0.0215	5.254
2.0		0.499	0.499		–	−0.0642	0.0187	4.994
2.2		0.454	0.454		–	−0.0533	0.0143	4.542
2.4		0.416	0.416		–	−0.0449	0.0112	4.165
2.6		–	0.384		–	−0.0383	0.0088	3.845
2.8		–	0.357		–	−0.0330	0.0071	3.570
3.0		–	0.333		–	−0.0288	0.0058	3.333
3.2		–	0.312		–	–	0.0048	–
3.4		–			–	–	0.0040	–
	$\frac{1}{r}$	$\frac{1}{r}$	$\frac{1}{r}$	0	$\frac{0.0293}{r^2}$	$\frac{-0.2590}{r^2}$	$\frac{0.1560}{r^3}$	$\frac{10}{r}$

Table 4.

$n=3 \quad l=0$ $\qquad F_l^{ik} \qquad$ $n=3 \quad l=1$

r	F_0^{14}	F_0^{24}	F_1^{34}	F_0^{34}	F_2^{14}	F_1^{14}	F_1^{24}
0.00	0.267	0.189	0.0000	−0.0928	0.0000	0.0000	0.0000
0.02	0.257	0.186	−0.0012	−0.0928	0.0000	−0.0064	0.0001
0.04	0.232	0.180	−0.0027	−0.0928	−0.0001	−0.0122	0.0005
0.06	0.202	0.173	−0.0043	−0.0928	−0.0004	−0.0168	0.0009
0.08	0.172	0.166	−0.0060	−0.0927	−0.0009	−0.0203	0.0020
0.10	0.143	0.1596	−0.0078	−0.0926	−0.0019	−0.0226	0.0034
0.12	0.1178	0.1541	−0.0101	−0.0923	−0.0030	−0.0240	0.0048
0.14	0.0959	0.1495	−0.0129	−0.0919	−0.0040	−0.0247	0.0066
0.16	0.0774	0.1458	−0.0158	−0.0913	−0.0050	−0.0247	0.0081
0.18	0.0621	0.1427	−0.0187	−0.0906	−0.0059	−0.0243	0.0097
0.20	0.0495	0.1402	−0.0215	−0.08*g*8	−0.0068	−0.02362	0.0112
0.22	0.0393	0.1382	−0.0242	−0.0889	−0.0074	−0.02270	0.0126
0.24	0.0311	0.1365	−00267	−0.0878	−0.0081	−0.02162	0.0140
0.26	0.0246	0.1350	−0.0289	−0.0866	−0.0087	−0.02047	0.0152
0.28	0.0194	0.1335	−0.0309	−0.0853	−0.0092	−0.01929	0.0162
0.30	0.0152	0.1321	−0.0326	−0.0839	−0.0097	−0.01811	0.0172
0.32	0.0119	0.1307	−0.0340	−0.0825	−0.0100	−0.01695	0.0180
0.34	0.0093	0.1292	−0.0351	−0.0809	−0.0108	−0.01584	0.0186
0.36	0.0072	0.1275	−0.0360	−0.0793	−0.0106	−0.01478	0.0191
0.38	0.0057	0.1257	−0.0365	−0.0776	−0.0107	−0.01377	0.0195
0.4	0.0044	0.1238	−0.03690	−0.07582	−0.01077	−0.01284	0.01977
0.45	0.0024	0.1183	−0.03680	−0.07131	−0.01071	−0.01077	0.01994
0.5	0.0013	0.1120	−0.03560	−0.06668	−0.01036	−0.00907	0.01952
0.55	0.0007	0.1051	−0.03355	−0.06204	−0.00979	−0.00769	0.01868
0.6	0.0004	0.0978	−0.03090	−0.05747	−0.00907	−0.00656	0.01752
0.65	0.0002	0.0904	−0.02788	−0.05303	−0.00824	−0.00565	0.01616
0.7	0.0001	0.0830	−0.02466	−0.04879	−0.00730	−0.00491	0.01468
0.75	0.0000	0.0757	−0.02137	−0.04478	−0.00645	−0.00429	0.01316
0.8	–	0.0688	−0.01813	−0.04096	−0.00555	−0.00378	0.01164
0.85	–	0.0623	−0.01498	−0.03740	−0.00467	−0.00335	0.01015
0.9	–	0.0562	−0.01199	−0.03409	−0.00383	−0.00299	0.00873
0.95	–	0.0505	−0.00920	−0.03102	−0.00304	−0.00269	0.00740
1.0	–	0.0453	−0.00663	−0.02817	−0.00231	−0.00243	0.00617
1.1	–	0.0362	−0.00218	−0.02317	−0.00102	−0.00201	0.00399
1.2	–	0.0287	0.00138	−0.01898	0.00004	−0.00169	0.00221
1.3	–	0.0226	0.00415	−0.01548	0.00088	–	0.00078
1.4	–	0.0177	0.00622	−0.01259	0.00153	–	−0.00033

Table 4 (continued).

$n=3 \quad l=0$ $\qquad F_l^{ik} \qquad$ $n=3 \quad l=1$

r	F_0^{14}	F_0^{24}	F_1^{34}	F_0^{34}	F_2^{14}	F_1^{14}	F_1^{24}
1.5	–	0.0139	0.00771	−0.01021	0.00201	–	−0.00118
1.6	–	0.0108	0.00873	−0.00826	0.00235	–	−0.00181
1.7	–	0.0084	0.00938	−0.0666	0.00258	–	−0.00226
1.8	–	0.0065	0.00974	−0.0537	0.00271	–	−0.00271
1.9	–	0.0051	0.00987	−0.0431	0.00277	–	−0.00277
2.0	–	0.0039	0.00984	−0.0346	0.00275	–	−0.00288
2.2	–	0.0025	0.00942	−0.0220	0.00266	–	−0.00294
2.4	–	0.0015	0.00876	−0.0139	0.00245	–	−0.00284
2.6	–	0.0008	0.00800	−0.0087	0.00221	–	−0.00267
2.8	–	0.0004	0.00725	−0.0053	0.00195	–	−0.00247
3.0	–	0.0002	0.00653	−0.0032	0.00171	–	−0.00226
3.2	–	0.0001	0.00588	−0.00019	0.00150	–	−0.00205
3.4	–	0.0000	0.00529	−0.00011	0.00130	–	−0.00186
3.6	–	–	0.00477	−0.00006	0.00113	–	−0.00168
3.8	–	–	0.00431	−0.00003	0.000982	–	−0.00145
4.0	–	–	0.00391	−0.00001	0.000855	–	−0.00131
4.2	–	–	0.00356	–	0.000746	–	−0.00125
4.4	–	–	0.00325	–	0.000652	–	−0.00114
4.6	–	–	–	–	0.000572	–	−0.00104
4.8	–	–	–	–	0.000505	–	−0.00096
	0	0	$\frac{0.0629}{r^2}$	0	$\frac{0.0558}{r^3}$	$\frac{-0.00243}{r^2}$	$\frac{-0.00220}{r^2}$

Table 5.

$n=4 \quad l=0 \qquad F_l^{ik} \qquad n=4 \quad l=1$

r	F_0^{14}	F_0^{24}	F_1^{34}	F_0^{34}	F_1^{14}	F_1^{24}	F_2^{34}
0.00	0.124	0.0865	0.00000	0.0555	0.0000	0.00000	0.00000
0.02	0.119	0.0850	−0.00025	0.0555	0.0030	−0.00020	0.00020
0.04	0.107	0.0820	−0.00068	0.0554	0.0070	−0.00055	0.00035
0.06	0.0933	0.07884	−0.00128	0.0554	0.0102	−0.00090	0.00063
0.08	0.07925	0.07563	−0.00214	0.0554	0.0124	−0.00140	0.00104
0.10	0.06610	0.07270	−0.00322	0.0553	0.0138	−0.00215	0.00151
0.12	0.05412	0.07017	−0.00442	0.0552	0.0147	−0.00299	0.00203
0.14	0.04436	0.06807	−0.00577	0.0548	0.0151	−0.00390	0.00256
0.16	0.03588	0.06638	−0.00712	0.0545	0.0151	−0.00484	0.00308
0.18	0.02884	0.06500	−0.00848	0.0510	0.0149	−0.005773	0.00356
0.20	0.02305	0.06390	−0.00979	0.0535	0.01445	−0.006680	0.00406
0.22	0.01836	0.06300	−0.01103	0.0530	0.01385	−0.007541	0.00450
0.24	0.01458	0.06225	−0.01217	0.05234	0.01325	−0.008342	0.004408
0.26	0.01154	0.06158	−0.01321	0.05158	0.01255	−0.009073	0.005265
0.28	0.00911	0.06093	−0.01414	005075	0.01183	−0.009718	0.005579
0.30	0.00717	0.06030	−0.01490	0.04987	0.01111	−0.01027	0.005843
0.32	0.00562	0.05966	−0.01555	0.04894	0.01021	−0.01075	0.006056
034	0.00441	0.05897	−0.01607	0.04797	0.009727	−0.01113	0.006224
0.36	0.00344	0.05823	−0.01646	0.04698	0.009078	−0.01143	0.006347
0.38	0.00269	0.05740	−0.01674	0.04595	0.008463	−0.01164	0.006429
0.40	0.00210	0.05650	−0.01690	004490	0.007880	−0.01179	0.006462
0.45	0.00114	005400	−0.01685	0.04216	0.006600	−0.01186	0.006409
0.50	000062	0.05112	−0.01629	0.03934	0.005550	−0.01158	0.006178
0.55	0.00035	0.04795	−0.01533	0.03653	0.004698	−0.01104	0.005816
0.60	0.00020	0.04460	−0.01410	0.03377	0.004008	−0.01032	0.005362
0.65	0.00012	0.04119	−0.01270	0.03111	0.003448	−0.009477	0.004849
0.70	0.00009	0.03779	−0.01120	0.02856	0.002990	−0.008569	0.004303
0.75	0.00007	0.03447	−0.009673	0.02615	0.002615	−0.007636	0.003747
0.80	0.00005	0.03129	−0.008167	0.02389	0.002304	−0.006709	0.003196
0.85	0.00003	0.02828	−0.006715	0.02178	0.002044	−0.005811	0.002663
0.90	0.00001	0.02547	−0.005341	0.01981	0.001824	−0.004957	0.002157
0.95	0.00000	0.02285	−0.004060	0.01799	0.001638	−0.004157	0.001681
1.0	–	0.02046	−0.002878	0.01631	0.001478	−0.003420	0.001242
1.1	–	0.01628	−0.000834	0.01335	0.001220	−0.002129	0.000473

Table 5 (continued).

$n=4 \quad l=0$ $\qquad F_l^{ik}$ $\qquad n=4 \quad l=1$

r	F_0^{14}	F_0^{24}	F_1^{34}	F_0^{34}	F_1^{14}	F_1^{24}	F_2^{34}
1.2	–	0.012845	0.000800	0.01088	–	−0.001079	−0.000152
1.3	–	0.010076	0.002061	0.00883	–	−0.000250	−0.000642
1.4	–	0.007866	0.002999	0.00714	–	0.000389	−0.001014
1.5	–	0.006117	0.003668	0.00576	–	0.000871	−0.001285
1.6	–	0.004741	0.004118	0.00463	–	0.001222	−0.001472
1.7	–	0.003666	0.004395	0.00371	–	0.001470	−0.001519
1.8	–	0.002828	0.004536	0.00297	–	0.001636	−0.001655
1.9	–	0.002178	0.004573	0.00237	–	0.001738	−0.001677
2.0	–	0.001675	0.004535	0.00187	–	0.001790	−0.001688
2.2	–	0.000987	0.004306	0.00117	–	0.001794	−0.001581
2.4	–	0.000580	0.003971	0.000729	–	0.001716	−0.001444
2.6	–	0.000340	0.003602	0.000450	–	0.001600	−0.001289
2.8	–	0.000200	0.003237	0.000275	–	0.001469	−0.001133
3.0	–	0.000120	0.002897	0.000173	–	0.001336	−0.0009873
3.2	–	0.000076	0.002592	0.000100	–	0.001209	−0.0008556
3.4	–	0.000052	0.002321	0000059	–	0.001091	−0.0007399
3.6	–	0.000040	0.002085	0.000033	–	0.0009842	−0.0006395
3.8	–	0.000032	0.001879	0.000021	–	0.0008887	−0.0005534
4.0	–	–	0.001699	0.000012	–	0.0008042	−0.0004800
4.2	–	–	0.001543	0.000007	–	0.0007302	−0.0004174
4.4	–	–	0.001407	0.000005	–	0.0006657	−0.0003641
4.6	–	–	0.001287	0.000004	–	0.0006091	−0.0003189
4.8	–	–	0.001182	0.000004	–	0.0005594	−0.0002806
5.0	–	–	–	–	–	0.0005156	−0.0002482
	0	0	$\frac{0.02723}{r^2}$	0	$\frac{0.001475}{r^2}$	$\frac{0.01289}{r^2}$	$\frac{-0.03103}{r^3}$

Table 6.
$n = 3 \quad l = 1$

r	f_4^0	f_4	r	f_4^0	f_4
0.00	0.0000	−0.0000	1.8	−0.0821	0.1318
0.04	−00093	−0.0046	2.0	−0.0494	0.1691
0.08	−0.0305	−0.0149	2.4	0.0175	0.2378
0.12	−0.0566	−0.0374	2.8	0.0835	0.2964
0.16	−0.0838	−0.0402	3.2	0.1458	0.3438
0.20	−0.1097	−0.0520	3.6	0.2024	0.3791
0.24	−0.1332	−0.0623	4.0	0.2517	0.4043
0.28	−0.1537	−0.0709	4.4	0.2932	0.4192
0.32	−0.1710	−0.0776	4.8	0.3266	0.4251
0.36	−0.1851	−0.0825	5.2	0.3520	0.4234
0.4	−0.1964	−0.0858	5.6	0.3700	0.4155
0.5	−0.2142	−0.0874	6.0	0.3812	0.4026
0.6	−0.2208	−0.0819	6.4	0.3863	0.3858
0.7	−0.2201	−0.0714	7.2	0.3814	0.3447
0.8	−0.2147	−0.0575	8.0	0.3614	0.2990
0.9	−0.2062	−0.0414	8.8	0.3318	0.2533
1.0	−0, 1958	−0.0237	9.6	0.2972	0.2104
1.2	−0.1713	−0.0142	10.4	0.2608	0.1719
1.4	−0.1435	0.0536	11.2	0.2250	0.1384
1.6	−0.1136	0.0931	12.0	0.1914	0.1103

Table 7.

n_l	Eexp	E_i	λ^0
1_0	−39.4	−40.6	–
2_0	−2.16	− 3.00	–
2_1	−1.04	− 1.83	–
3_0	− 0.1888	− 0.186	−0.15795
3_1	− 0.1115	− 0.1094	−0.08895
4_0	− 0.0716	− 0.0703	−0.06422
4_1	− 0.05095	− 0.0501	−0.04386

Table 8.
Energy of Na^+

T	U_1	U_2'	U_2''	W
162.1	−387.5	65.66	−2.05	−161.8

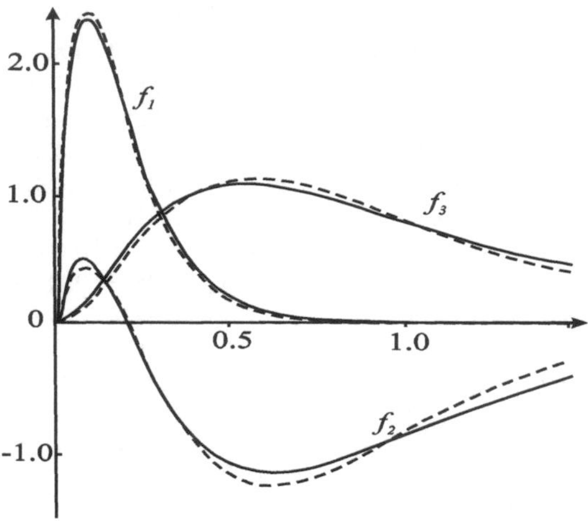

Fig. 1. Core electron wave functions.

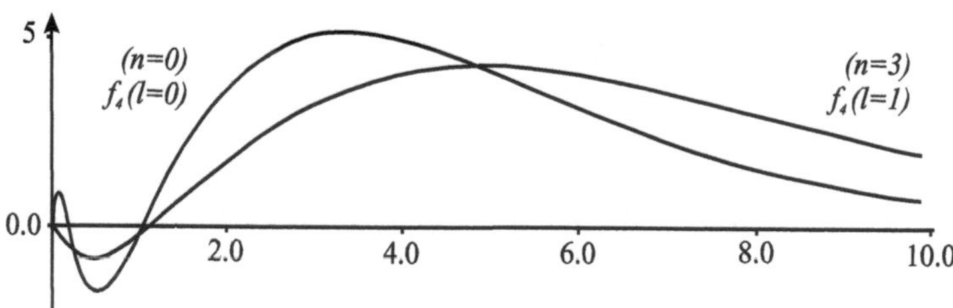

Fig. 2. Series electron wave function.

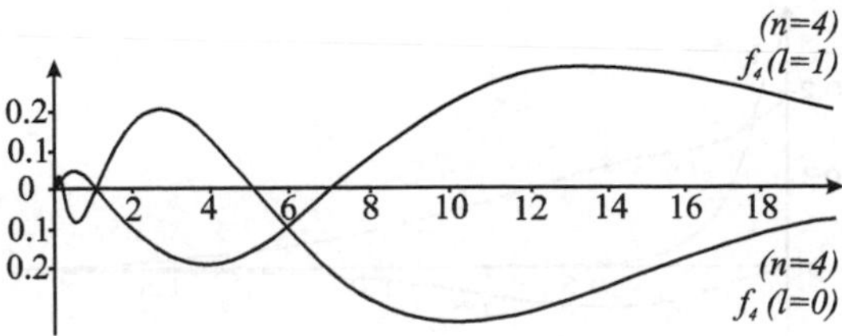

Fig. 3. Series electron wave function.

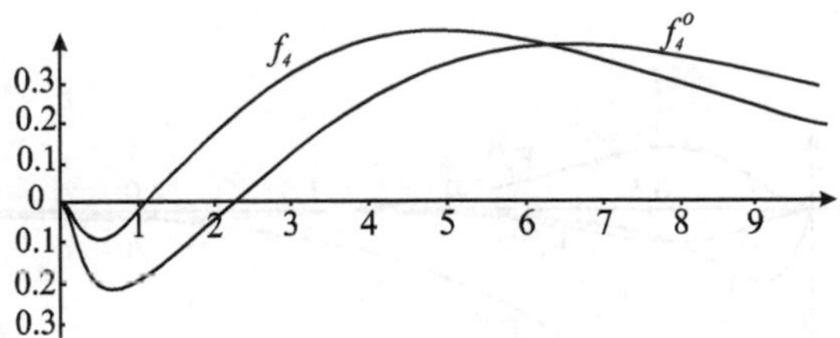

Fig. 4. Wave function f_4 for $n = 3$, $l = 1$ and the solution of the corresponding homogeneous differential equation.

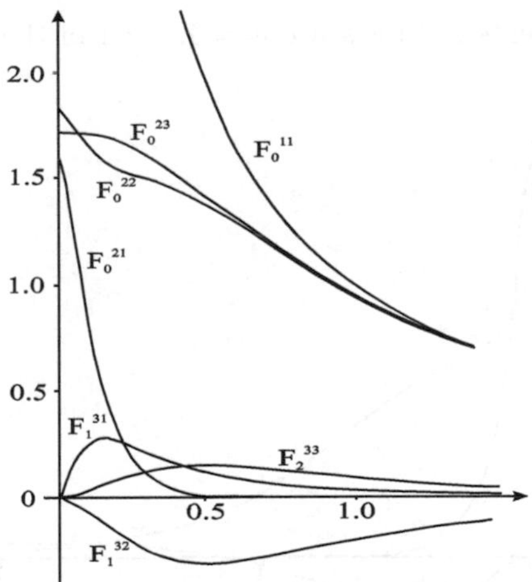

Fig. 5. Coefficients F_l^{ik} for the core electrons.

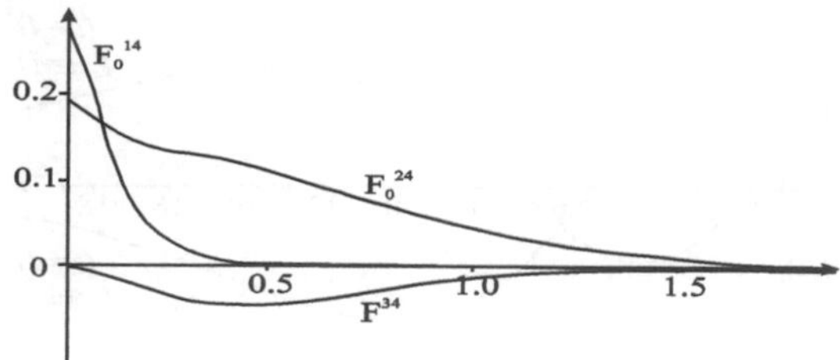

Fig. 6. Coefficients F_l^{i4} for state $n = 3$, $l = 0$ of the series electron.

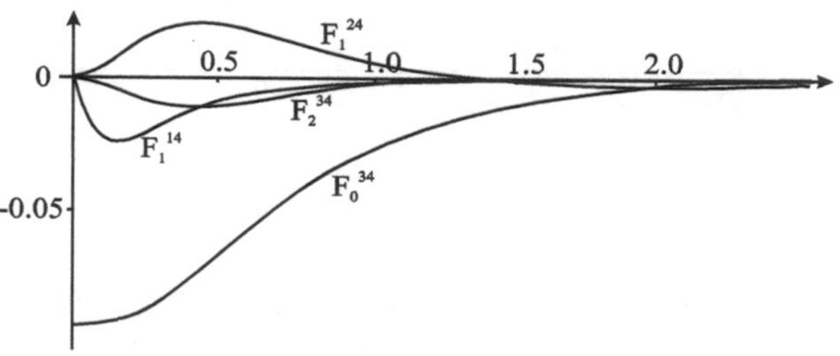

Fig. 7. Coefficients F_l^{i4} for state $n = 3$, $l = 1$ of the series electron.

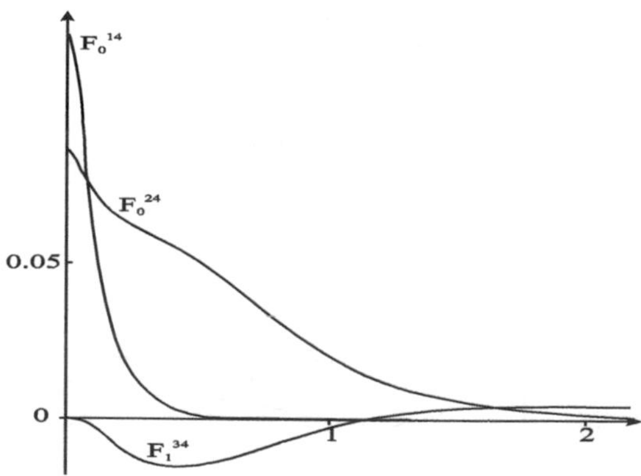

Fig. 8. Coefficients F_l^{i4} for state $n = 4$, $l = 0$ of the series electron.

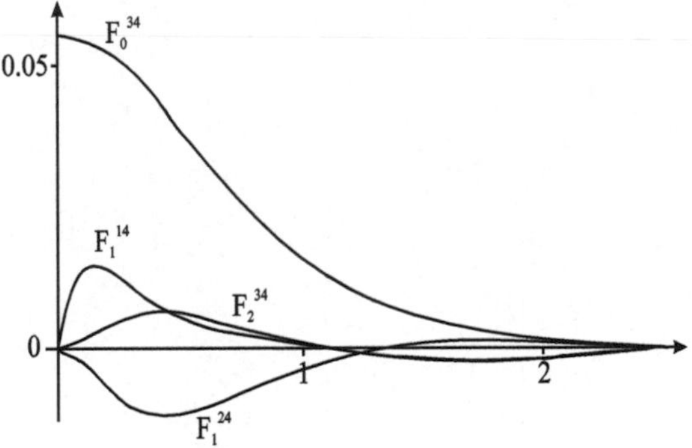

Fig. 9. Coefficients F_l^{i4} for state $n = 4$, $l = 1$ of the series electron.

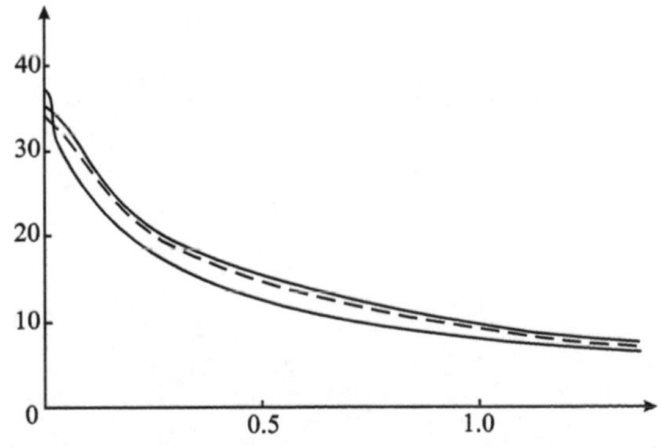

Fig. 10. The screening potential.

State Optical Institute,
Leningrad

Typeset by E.D. Trifonov

34-2
An Approximate Representation of the Wave Functions of Penetrating Orbits

V.A. Fock

Presented by Academician S.I. Vavilow 21 January 1934

DAN **1**, N 5, 241, 1934
TOI **10**, N 96, 1, 1934

Let us denote $R_{nl}(r)$ the radial wave function of the valence electron in an atom and let $f(r)$ denote the function $f(r) = rR_{nl}(r)$. If we neglect the quantum exchange forces, the function $f(r)$ satisfies the differential equation

$$\frac{d^2 f}{dr^2} + \left\{ \frac{2m}{\hbar^2} \left[E + \frac{Ze^2}{r} - V(r) \right] - \frac{l(l+1)}{r^2} \right\} f = 0 \tag{1}$$

and the normalization condition

$$\int_0^{\infty} |\, f(r)\, |^2\, dr = 1. \tag{2}$$

The aim of the present article is to give an approximate representation of the function $f(r)$ for the case when the principal quantum number is very big and the azimuth quantum number l is finite or equals zero.[1] We set

$$p_1(r) = \frac{2m}{\hbar^2} \left[E + \frac{Ze^2}{r} - V(r) \right], \quad p_2(r) = -\frac{l(l+1)}{r^2}. \tag{3}$$

The coefficient at f in eq. (1) will be equal to $p(r) = p_1(r) + p_2(r)$. We denote as r_1 the root of the function $p_1(r)$, which will be approximately equal to $r_1 = -\frac{e^2}{E}$. For $r < r_1$, the function is $p_1(r) > 0$ and for $r > r_1$ it is $p_1(r) < 0$. For $r \leq r_1$, we set

$$x = \int_0^r \sqrt{p_1(r)}dr, \quad x_1 = \int_0^{r_1} \sqrt{p_1(r)}dr, \tag{4}$$

[1] If the numbers l and $n - l - 1$ are both big, one can apply a somewhat modified Lanczos method [1].

and for $r \geq r_1$, we introduce a new variable

$$\xi = \int_{r_1}^{r} \sqrt{-p_1(r)}dr. \tag{5}$$

Further, we set

$$f^*(r) = A\sqrt{\frac{dr}{dx}} \cdot \sqrt{x}J_{2l+1}(x) \tag{6}$$

and show that when r is less than r_1, but is not too close to r_1, the function f^* satisfies the equation which differs a little from (1). The equation for f^* has the form

$$\frac{d^2 f^*}{dr^2} + [p_1^*(r) + p_2^*(r)] f^* = 0, \tag{7}$$

where $p_1^*(r)$ has the previous value and $p_2^*(r)$ is equal to

$$p_2^*(r) = -\left[4l(l+1) + \frac{3}{4}\right]\left(\frac{1}{x}\frac{dx}{dr}\right)^2 - \frac{ds}{dr} - s^2, \tag{8}$$

with

$$s = -\frac{1}{4}\frac{d}{dr}\ln p_1(r). \tag{9}$$

When r is not too small, the main term in the coefficient at f^* in (7) is $p_1(r)$ and it is the same term as in the exact equation (1). For very small values of r, we have

$$\left.\begin{aligned} x &= 2\sqrt{\frac{2mZe^2}{\hbar^2}} \cdot \sqrt{r}\left(1 - \tfrac{1}{6}r\frac{V(0) - E}{Ze^2} + \ldots\right), \\ s &= \frac{1}{4r} + \frac{V(0) - E}{4Ze^2} + \ldots, \end{aligned}\right\} \tag{10}$$

from where, due to (8), we get

$$p_2^*(r) = -\frac{l(l+1)}{r^2} + \frac{2}{3}\frac{V(0) - E}{Ze^2} \cdot \frac{l(l+1)}{r} + \ldots. \tag{11}$$

Thus, when $l \neq 0$, the function $p_2^*(r)$ has the same singularity at $r = 0$ as $p_2(r)$ and for $l = 0$ the quantity $p_2^*(0)$ remains finite. Hence, eq. (7) differs very little from (1) from $r = 0$ up to certain r, which is less than r_1.

In the vicinity of $r = r_1$, (7) ceases to be correct since here $p_2^*(r)$ becomes infinite. Therefore, in this region we have to look for an approximate expression for $f(r)$ in a somewhat different form. Applying the method, which is analogous to that used by Lanczos [1] when he investigated the Stark effect in strong fields, we can approximately express $f(r)$ by means of the Bessel functions of the order one third. At $r \leq r_1$, we set

$$f^{**}(r) = B\sqrt{\frac{dr}{dx}} \cdot \sqrt{x_1 - x}\left\{\cos\frac{\pi}{6}J_{\frac{1}{3}}(x_1 - x) - \sin\frac{\pi}{6}Y_{\frac{1}{3}}(x_1 - x)\right\} \tag{12a}$$

and for $r \geq r_1$

$$f^{**}(r) = \frac{B}{\pi}\sqrt{\frac{dr}{d\zeta}}\sqrt{\zeta}K_{\frac{1}{3}}(\zeta), \tag{12b}$$

where ζ has the value (5). The function $f^{**}(r)$ satisfies the equation

$$\frac{d^2 f^{**}}{dr^2} + [p_1(r) + p_2^{**}(r)] f^{**} = 0, \tag{13}$$

where

$$p_2^{**}(r) = \frac{5}{36}\left(\frac{1}{x_1 - x}\frac{dx}{dr}\right)^2 - \frac{ds}{dr} - s^2, \tag{14}$$

and s has the previous value (9). It is not difficult to check that at $r = r_1$ when $x = x_1$ the quantity $p_2^{**}(r)$ remains finitc. Sincc, generally speaking, this quantity as well as the quantity $p_2(r)$ in the exact equation is small compared with $p_1(r)$, then (13) differs very little from the exact equation (1). Therefore, its solution f^{**} has to differ a little from the solution of the exact equation.

Thus, for the required wave function $f(r)$ we have two approximate representations, namely, (6) and (12a) or (12b). First representation is justified from $r = 0$ up to some mean value r, which is less than r_1 and the second one is valid from this mean value r up to $r = \infty$. In the middle region where both x and x_1 are big, both representations are justified and here they have to coincide approximately. Indeed, substituting the Bessel functions by their asymptotic values, we have

$$f^{*}(r) \cong A\sqrt{\frac{2}{\pi}}\sqrt{\frac{dr}{dx}}\cos\left(x - l\pi - \frac{3}{4}\pi\right), \tag{15}$$

$$f^{**}(r) \cong B\sqrt{\frac{2}{\pi}}\sqrt{\frac{dr}{dx}}\cos\left(x_1 - x - \frac{1}{4}\pi\right). \tag{16}$$

These two expressions should be identically equal and it can be only if the sum of the arguments in the cosines is equal to an integer multiple of π, i.e., to be equal to $n_r\pi$ where n_r is the radial quantum number. From here, it follows:

$$x_1 = \frac{\sqrt{2m}}{\hbar}\int_0^{r_1}\sqrt{E + \frac{Ze^2}{r} - V(r)}dr = (n_r + l + 1) = n\pi, \qquad (17)$$

where n is equal to the principal quantum number.

Thus, we have obtained the Bohr quantization condition, which was applied in old quantum mechanics but with integer quantum numbers. Therefore, the Kramers quantization rule with half-integer quantum numbers is not universal and is applicable only if the singular points of the function $p(r)$ (the coefficient at f in (1)) are placed far from its roots.

Substituting in (15) and (16) instead of x_1 its value $n\pi$, we get identical expressions if the constants A and B are connected by the relation

$$A = B\cdot(-1)^{n-l-1}. \qquad (18)$$

The constant A (up to its sign) is determined from the normalization condition (2). For an approximate determination of A, we use the fact that the main contribution to integral (2) is given by that interval where the asymptotic expressions (15) and (16) are valid. As the upper limit of the integration, we can take the value $r = r_1$ instead of $r = \infty$ and after the substitution of (15) in (2) we can change the square of the cosine to the mean value $\frac{1}{2}$. Then, we get

$$\frac{1}{\pi}A^2\cdot\int_0^{r_1}\frac{dr}{\sqrt{p_1(r)}} = 1. \qquad (19)$$

If we take expression (3) for $p_1(r)$ and consider the number n in (17) as a continuous parameter, we can write (17) as follows:

$$A^2 = \frac{m}{\hbar^2}\frac{dE}{dn}. \qquad (20)$$

From here, we easily get the expression

$$\lim_{r\to 0}\left(\frac{R_{nl}(r)}{r^l}\right)^2 \cong \frac{2m}{\hbar^2}\frac{dE}{dn}\frac{\left(\frac{2Z}{a}\right)^{2l+1}}{(2l+1)!}, \qquad (21)$$

where $a = \frac{\hbar^2}{me^2}$ is the radius of the Bohr hydrogen orbit. A particular case of this expression (for $l = 0$) is used by Fermi and Segré [2] in their work on the magnetic moments of nuclei but it was given there without derivation. For a hydrogen ion and for $l = 0$, formula (21) gives the exact expression

$$| R_{n0} |^2 = \frac{4Z^3}{a^3 n^3}. \tag{22}$$

In conclusion, we note that expression (6) is applicable also in the case of the continuous spectrum if the energy parameter E is sufficiently small. In this case, formula (6) is justified uniformly for all values of r from $r = 0$ up to $r = \infty$. If $E = 0$, then for the hydrogen ion the expression (6) is exact.

References

1. C. Lanczos, Zs. Phys. **65**, 204, 1931.
2. E. Fermi and E. Segré, Zs. Phys. **62**, 729, 1933.

State Optical Institute
Leningrad

Translated by Yu. Yu. Dmitriev

34-3*
On Quantum Electrodynamics

V.A. Fock

Phys. Zs. Sowjetunion **6**, 425, 1934,
Fock57, pp. 88–123

Introduction

The aim of the present paper[1] is to elaborate and simplify the mathematical foundations for quantum electrodynamics. This theory, created by Dirac[2] and Heisenberg and Pauli,[3] presents a consistent generalization of classical electrodynamics, answering the correspondence principle [...].[4] The difficulties of the theory arising from the infinite self-energy of an electron are well known [...]. Nevertheless, it should not be forgotten that all the results of the quantum theory of radiation, which presents an immediate application of the correspondence principle, can be obtained from quantum electrodynamics. But we often should waive the mathematical strictness. The physical sense of "illegal" mathematical operations used there (truncating an expansion in a power series by a certain power of the fine structure constant α, throwing off divergent integrals, etc.) is that employing them one tries to eliminate the discrepancies arising from an imperfection of the theory.

The results that were originally deduced in a quite different way and only subsequently were justified by quantum electrodynamics (as for example the Breit and Möller formulae for electron interaction) can also be obtained from quantum electrodynamics.

The difficulties of quantum electrodynamics raised doubts on the correctness of this theory; even a question was discussed if it is suitable

[1] The work was reported in November 1932 at the theoretical seminar in Leningrad Physico-Technical Institute.

[2] P.A.M. Dirac, *Relativistic Quantum Mechanics*, Proc. Roy. Soc. **A 136**, 453, 1932; P. Dirac, V. Fock and B. Podolsky, *On Quantum Electrodynamics*, Sow. Phys. **2**, 468, 1932.

[3] W. Heisenberg and W. Pauli, *Zur Quantendynamik den Wellenfelder*, Zs. Phys. **59**, 1, 1929; *Zur Quantentheorie der Wellenfelder*, 2, Zs. Phys. **59**, 168, 1929.

[4] Throughout this paper the symbol [...] shows the parts of the original text of 1934 that were abridged by the author in Fock57. (*Editors*)

at all in the quantum region.[5] But this latter doubt turned out to be groundless. In a fundamental work[6] Bohr and Rosenfield elucidated the region where the theory was applicable. This paper has shown that always when we may omit the atomic structure of a measuring device and when examined distances and wavelengths considerably exceed the classical electron radius, the application of quantum electrodynamics to the problem of measurability of the electromagnetic field leads to reasonable results.

Since the limits of applicability for the theory may be considered as established, first of all it is necessary to pay attention to the creation of its mathematical foundations for its further development. The paper is devoted to this problem [...].

I Introducing the Functionals for the Case of Bose Statistics

When stating the theory of the second quantization and also in the cases of its applications, one usually introduces the point spectrum even when a physical problem demands the consideration of states belonging to the continuous spectrum. In these cases the problem is altered to the point spectrum (introducing "boxes," periodical conditions etc.) and only toward the end of the calculations it is returned by the limiting procedure to the initial problem. This roundabout way is on no account necessary since it is often very awkward and complicates the calculations. Subsequently, for the case of the Bose statistics we propose a method that allows one to operate in the same way both with states of the point and with states of the continuous spectrum; in addition, it has an advantage that the interrelation of the second quantization and the configuration space is transparent there. This method is based on applying the generating functions (functionals) and is a generalization of the method developed by the author in one of his previous works.[7] It is known that the quantized wave function $\Psi(x)$ in the case of the Bose

[5] L. Landau and R. Peierls, *Erweiterung des Unbestimmtheistprintzips für die relativistische Quantentheorie*, Zs. Phys. **69**, 56, 1931.

[6] N. Bohr and L. Rosenfeld, *Zur Frage der Messbarkeit der elektromagnetischen Feldgrößen*, Kgl. Danske Vid. Selskab, Math.-Fys. Meddleser **12**, 8, 1933.

[7] V. Fock, *Verallgemeinerung und Lösung der Diracschen statistischen Gleichung*, Zs. Phys. **49**, 339, 1928. (See [28-3*] in this book. (*Editors*))

statistics satisfies the commutation relations

$$\begin{aligned} \Psi(x')\Psi^\dagger(x) - \Psi^\dagger(x)\Psi(x') &= \delta(x-x'), \\ \Psi(x')\Psi(x) - \Psi(x)\Psi(x') &= 0, \end{aligned} \tag{1}$$

where x denotes all independent variables. The quantity $\Psi(x)$ can be considered as an operator, which acts on a sequence of the wave functions

$$\left\{ \begin{array}{l} \text{const} \\ \psi(x_1) \\ \psi(x_1 x_2) \\ \psi(x_1 x_2 x_3) \\ \dots\dots\dots\dots \end{array} \right\} \tag{2}$$

in the configuration spaces of $0, 1, 2, \ldots n, \ldots$ dimensions and transforms this sequence into the following one:[8]

$$\Psi(x) \left\{ \begin{array}{l} \text{const} \\ \psi(x_1) \\ \psi(x_1 x_2) \\ \dots\dots \end{array} \right\} = \left\{ \begin{array}{c} \psi(x) \\ \sqrt{2}\psi(x x_1) \\ \sqrt{3}\psi(x x_1 x_2) \\ \dots\dots\dots \end{array} \right\}. \tag{3}$$

All the functions ψ are supposed to be symmetric. The conjugated operator $\Psi^\dagger(x)$ transforms sequence (2) into an analogous sequence $\psi'(x_1), \psi'(x_1 x_2), \ldots$, where the transformed function of the n-th subspace can be expressed in terms of the previous function of the $(n-1)$-th subspace in the following way:

$$\begin{aligned} &\psi'(x_1\, x_2\, \ldots x_n) = \frac{1}{\sqrt{n}}[\delta(x_1 - x)\psi(x_2\, x_3\, \ldots\, x_n) + \ldots \\ &+\delta(x_k - x)\psi(x_1 \ldots x_{k-1}\, x_{k+1} \ldots x_n) + \ldots \\ &+\delta(x_n - x)\psi(x_1 \ldots x_{n-1})]. \end{aligned} \tag{4}$$

On the other hand, for the operator $\Psi(x)$ one can choose a representation where the conjugated operator $\Psi^\dagger(x)$ corresponds to multiplication by some function $\overline{b}(x)$.[9] A functional Ω of the function $\overline{b}(x)$ will be then an object on which the operators Ψ and $\Psi^\dagger$ act. The form of this functional

---

[8] V. Fock. *Konfigurationsraum und zweite Quantelung*, Zs. Phys. **75**, 622–647 (1932). (*V. Fock*) (See also [32-2*] in this book. (*Editors*))

[9] It would be more consistent to denote this function by $\overline{\psi}(x)$ but we prefer $\overline{b}(x)$, not to use the symbol ψ in too many meanings. *(V. Fock)*

will be completely determined by the sequence of functions (2). We show now that the functional Ω has the form

$$\Omega = \psi_0 + \frac{1}{\sqrt{1!}}\int \psi(x_1)\overline{b}(x_1)dx_1 + \\ + \frac{1}{\sqrt{2!}}\int\int \psi(x_1\ x_2)\overline{b}(x_1)\overline{b}(x_2)dx_1\ dx_2 + ... \tag{5}$$

or

$$\Omega = \sum_{n=0}^{\infty}\Omega_n, \tag{6}$$

where Ω_n has the form of the n-multiple integral

$$\Omega_n = \frac{1}{\sqrt{n!}}\int ... \int \psi(x_1 ...\ x_n)\overline{b}(x_1)...\ \overline{b}(x_n)dx_1 ...\ dx_n. \tag{7}$$

We need first of all to give the definition of the functional derivative $\frac{\delta'\Omega}{\delta\overline{b}(x)}$ of Ω with respect to $\overline{b}(x)$. The general definition is

$$\frac{\delta'\Omega}{\delta\overline{b}(x)} = \lim_{\eta\to 0}\frac{1}{\eta}\{\Omega[\overline{b}(x') + \eta\delta(x'-x)] - \Omega[\overline{b}(x')]\}, \tag{8}$$

where we denoted the current coordinate in $\Omega = \Omega[\overline{b}(x')]$ by x'. For practical purposes a different definition equivalent to (8) is, however, more useful. Let us write the variation of $\delta\Omega$ with respect to $\overline{b}$ in the form of a single integral

$$\delta\Omega = \int A\delta\overline{b}(x)dx. \tag{9}$$

Then the functional derivative is equal to the coefficient by $\delta\overline{b}(x)$ in the integrand

$$\frac{\delta'\Omega}{\delta\overline{b}(x)} = A\,. \tag{9*}$$

Let us suppose, for example,

$$\Omega = \overline{b}(x') = \int \delta(x_1 - x')\overline{b}(x_1)dx_1;$$

then

$$\frac{\delta'\overline{b}(x')}{\delta\overline{b}(x)} = \delta(x - x'). \tag{10}$$

It follows from the definition of the functional derivative that for an arbitrary functional

$$\frac{\delta'}{\delta\bar{b}(x')}\bar{b}(x)\Omega - \bar{b}(x)\frac{\delta'\Omega}{\delta\bar{b}(x')} = \delta(x - x')\Omega. \qquad (11)$$

Hence in the present representation multiplication by $\bar{b}(x)$ corresponds to the operator $\Psi^\dagger(x)$, while the operator $\Psi(x)$ means a functional derivative with respect to $\bar{b}(x)$.

We shall prove, moreover, that the functional Ω indeed may be expressed in terms of the wave functions $\psi(x_1x_2...x_n)$ of the separate subspaces by formula (5). It is sufficient to show that for the sequence of coefficients (2) in series (5), transformation (3) corresponds to a functional derivative of Ω with respect to $\bar{b}(x)$ while transformation (4) corresponds to multiplication Ω by $\bar{b}(x)$.

Let us consider expression (7) for Ω_n and form the variation of Ω_n with respect to $\bar{b}(x)$. We have

$$\delta\Omega_n = \frac{1}{\sqrt{n!}}\cdot n\int \psi(x\ x_1\ ...\ x_{n-1})\cdot$$
$$\cdot\ \delta\bar{b}(x)\bar{b}(x_1)\ ...\ \bar{b}(x_{n-1})dx\ dx_1\ ...\ dx_{n-1}.$$

According to the definition (9), the functional derivative is equal to

$$\frac{\delta'\Omega_n}{\delta\bar{b}(x)} = \frac{1}{\sqrt{(n-1)!}}\sqrt{n}\int \psi(x\ x_1\ ...\ x_{n-1})\cdot$$
$$\cdot\ \bar{b}(x_1)\ ...\ \bar{b}\ (x_{n-1})\ dx_1\ ...\ dx_{n-1} \qquad (12)$$

or

$$\frac{\delta'\Omega_n}{\delta\bar{b}(x)} = \Omega'_{n-1} = \frac{1}{\sqrt{(n-1)!}}\int \psi'(x_1\ ...\ x_{n-1})\cdot$$
$$\cdot\ \bar{b}(x_1)\ ...\ \bar{b}(x_{n-1})dx_1\ ...\ dx_{n-1},$$

where the new $(n-1)$-dimensional wave function $\psi'(x_1\ ...\ x_{n-1})$ may be expressed in terms of the previous n-dimensional one as follows:

$$\psi'(x_1\ ...\ x_{n-1}) = \sqrt{n}\psi(x\ x_1\ ...\ x_{n-1}).$$

But this is just the transformation that the operator $\Psi(x)$ in (3) generates. Since this formula is valid for any n, we really have

$$\Psi(x)\Omega = \frac{\delta'\Omega}{\delta\bar{b}(x)}. \qquad (13)$$

On the other hand, multiplying Ω_{n-1} by

$$\bar{b}(x) = \int \delta(x_n - x)\bar{b}(x_n)dx_n$$

we have

$$\bar{b}(x)\Omega_{n-1} = \\ = \frac{1}{\sqrt{(n-1)!}} \int \delta(x_n - x)\psi(x_1 ... x_{n-1})\bar{b}(x_1)...\bar{b}(x_n)dx_1...dx_n$$

or, symmetrizing the function $\delta(x_n - x)\psi(x_1 \,...\, x_{n-1})$ with respect to $x_1 \,...\, x_n$

$$\bar{b}(x)\Omega_{n-1} = \frac{1}{\sqrt{n!}} \int \psi'(x_1...x_n)\bar{b}(x_1)...\bar{b}(x_n)dx_1...dx_n,$$

where ψ' means (4). Since the formula obtained is valid for any n, we have

$$\Psi^\dagger(x)\Omega = \bar{b}(x)\Omega. \tag{14}$$

Thus, our statement is proved. The coefficients $\psi_0, \psi(x_1), \psi(x_1x_2)$ in series (5) may be really identified with the wave functions in appropriate subspaces of the configuration space.

In the second quantization theory the operator of number of particles plays an important role:

$$\mathbf{n} = \int \Psi^\dagger(x)\Psi(x)dx\,. \tag{15}$$

Let us show that the quantity Ω_n (7) is an eigenfunctional of this operator. Indeed, taking into account (12), we have

$$\mathbf{n}\Omega_n = \int \bar{b}(x)\frac{\delta'\Omega_n}{\delta\bar{b}(x)}dx = n\Omega_n. \tag{16}$$

Thus, we can consider series (6) as an expansion of an arbitrary functional Ω in the series of the eigenfunctionals of the operator $\mathbf{n}$.

By means of (13) and (14), any operator of the second quantization theory can be represented as an operator in the configuration space. For instance, for the operator

$$L = \int \Psi^\dagger(x)L(x)\Psi(x)dx, \tag{17}$$

we obtain

$$L\Omega = \int \overline{b}(x)L(x)\frac{\delta'\Omega}{\delta\overline{b}(x)}dx. \tag{17*}$$

Substituting here series (6) for Ω and using (2), one can get for $L\Omega$ a similar series

$$L\Omega = \sum_{n=0}^{\infty} \Omega'_n,$$

where by Ω'_n we denote

$$\Omega'_n = \frac{1}{\sqrt{n!}}n \int [L(x)\psi(x\,x_1 \ldots x_{n-1})]\overline{b}(x)\overline{b}(x_1)\ldots\overline{b}(x_{n-1})dx\,dx_1\ldots dx_{n-1}.$$

Writing here x_n instead of x and symmetrizing the integrand with respect to $x_1 \ldots x_n$, one can write the functional Ω'_n in the form (7) where the function $\psi(x_1 \ldots x_n)$ is replaced by

$$\psi'(x_1 \ldots x_n) = [L(x_1) + L(x_2) + \ldots + L(x_n)]\psi(x_1 \ldots x_n). \tag{17**}$$

The latter equation gives the representation of the operator L in the configuration space.

It is possible also to find the representation in the configuration space for an operator not commuting with **n**. It goes without saying that for this case the function $\psi'(x_1 \ldots x_n)$ is expressed not only by the function $\psi(x_1 \ldots x_n)$, but also by the wave functions in the subspaces of the different number of dimensions.

The scalar product of two functionals Ω and Ω' can be determined by means of the wave functions in the proper subspaces, namely:

$$(\Omega, \Omega') = \overline{\psi}_0\psi'_0 + \sum_{n=1}^{\infty}\int \overline{\psi}(x_1 \ldots x_n)\psi'(x_1 \ldots x_n)dx_1 \ldots dx_n. \tag{18}$$

Then the orthogonality condition is

$$(\Omega, \Omega') = 0 \tag{19}$$

and the normalization condition reads as

$$(\Omega, \Omega) = 1 \qquad \text{or} \qquad (\Omega, \Omega) = \text{const.} \tag{19*}$$

To determine the relation with formulae of our previous work, let us consider briefly how to write the expression for Ω_n when the variable x runs only the discrete values $x^{(1)}, x^{(2)} \ldots$. Let us write $\overline{b}_r$ instead of

$\bar{b}(x^{(r)})$ and $c(r_1 \ldots r_n)$ instead of $\psi(x^{(r_1)}, x^{(r_2)} \ldots x^{(r_n)})$ and replace the integral by the sum in (7). Then for Ω_n we obtain:

$$\Omega_n = \frac{1}{\sqrt{n!}} \sum_{r_1 r_2 \ldots r_n} c(r_1\, r_2 \ldots r_n) \bar{b}_{r_1}\, \bar{b}_{r_2} \ldots \bar{b}_{r_n}. \tag{20}$$

In this sum the term containing n_1 times the factor $\bar{b}$, then n_2 times the factor $\bar{b}_2$, etc. appears $\frac{n!}{n_1! n_2! \ldots}$ times. Thus, if we suppose

$$c(r_1\, r_2 \ldots r_n) = c^*(n_1\, n_2 \ldots),$$

then it will be

$$\Omega_n = \frac{1}{\sqrt{n!}} \sum_{n_1\, n_2 \ldots} \frac{n!}{n_1! n_2! \ldots} c^*(n_1\, n_2 \ldots) \bar{b}_1^{n_1}\, \bar{b}_2^{n_2} \ldots .$$

If we suppose

$$\sqrt{\frac{n!}{n_1!\, n_2! \ldots}} \cdot c^*(n_1\, n_2 \ldots) = f(n_1\, n_2),$$

then it will be

$$\Omega_n = \sum_{n_1\, n_2 \ldots} \frac{1}{\sqrt{n_1!\, n_2! \ldots}} f(n_1\, n_2 \ldots)\, \bar{b}_1^{n_1}\, \bar{b}_2^{n_2} \ldots . \tag{21}$$

Here the integers $n_1, n_2, \ldots$ obey the condition

$$n_1 + n_2 + \ldots = n. \tag{22}$$

If we reject this condition, we obtain from (21) the expression for the general functional Ω. Formula (21) up to notations coincides with the corresponding formula of our cited work.[10]

II Electrodynamics of Vacuum

In the case of vacuum, application of the quantization rules for the electromagnetic field does not meet special difficulties. To establish the equations of motion for the field,[11] let us express the electromagnetic field $\mathbf{E}$ and $\mathbf{H}$ as usual by the scalar and vector potentials

$$\mathbf{E} = -\mathrm{grad}\Phi - \frac{1}{c}\frac{\partial \mathbf{A}}{\partial t}; \quad \mathbf{H} = \mathrm{curl}\mathbf{A}. \tag{1}$$

[10]See [28-3*] in this book, eqn. (54). (*V. Fock, 1957*)

[11]V. Fock and B. Podolsky, *On the Quantization of Electromagnetic Waves and an Interaction of Charges on Dirac's Theory.* See [32-4*] in this book. (*Editors*)

We consider the components of the potential (Φ, A_1, A_2, A_3) as coordinates of the field (Q, Q_1, Q_2, Q_3) and take as a Lagrangian function for the field the expression

$$L = \frac{1}{2}(\mathbf{E}^2 - \mathbf{H}^2) - \frac{1}{2}\left(\mathrm{div}\mathbf{A} + \frac{1}{c}\frac{\partial \Phi}{\partial t}\right)^2, \tag{2}$$

which differs from the Fermi expression only by the four-dimensional divergence.[12] The momenta canonically conjugated to the field coordinates (Q_0, Q_1, Q_2, Q_3) have the form

$$\left.\begin{aligned} P_0 &= -\tfrac{1}{c}\left(\mathrm{div}\mathbf{A} + \tfrac{1}{c}\tfrac{\partial\Phi}{\partial t}\right), \\ (P_1, P_2, P_3) &= \mathbf{P} = -\tfrac{1}{c}\mathbf{E}. \end{aligned}\right\} \tag{3}$$

For the Hamilton function H we arrive at

$$H = \frac{c^2}{2}(\mathbf{P}^2 - \mathbf{P}_0^2) + \frac{1}{2}(\mathrm{curl}\mathbf{A})^2 - cP_0\mathrm{div}\mathbf{A} - c\mathbf{P}\mathrm{grad}\Phi. \tag{4}$$

The Hamilton equations of motion have the form

$$\begin{aligned} \dot{\mathbf{A}} &= c^2\mathbf{P} - c\ \mathrm{grad}\Phi, \\ \dot{\Phi} &= -c^2 P_0 - c\ \mathrm{div}\mathbf{A}, \\ \dot{\mathbf{P}} &= \Delta\mathbf{A} - \mathrm{grad}\ \mathrm{div}\mathbf{A} - c\ \mathrm{grad}P_0, \\ \dot{P}_0 &= -c\ \mathrm{div}\mathbf{P}, \end{aligned} \tag{5}$$

where the point means the time derivative. Eliminating the momenta, we obtain the d'Alembert equation for the potential components. To get the Maxwell equations for vacuum in the case of the classical field theory, it is necessary to assume $P_0 = 0$.

Coming to the field quantization, we should take into account that the wave functions of light quanta are not real field variables but complex quantities that can be obtained from the former by the Fourier expansion. For some real field variable, satisfying the d'Alembert equation, we write the expansion in the Fourier integral as follows:

$$\begin{aligned} F(\mathbf{r}, t) = {} & \frac{1}{(2\pi)^{3/2}} \int F(\mathbf{k}) e^{-ickt + i\mathbf{kr}} (d\mathbf{k}) + \\ & + \frac{1}{(2\pi)^{3/2}} \int F^\dagger(\mathbf{k}) e^{ickt - i\mathbf{kr}} (d\mathbf{k}). \end{aligned} \tag{6}$$

[12] E. Fermi. Rend. Lincei **(6) 9**, 881 (1929).

Here we denote by k the magnitude of the wave vector $\mathbf{k}$; $(d\mathbf{k})$ stands for $dk_1\ dk_2\ dk_3$. Since the quantities (k, k_1, k_2, k_3) are components of a four-dimensional vector (namely, the zero vector), the expressions $\frac{d\mathbf{k}}{k}$ and $k\delta(\mathbf{k}-\mathbf{k}')$ are invariant under Lorentz transformations. Hence the quantities $kF(\mathbf{k})$, i.e., the amplitudes of the components of the field $F(\mathbf{r}, t)$ multiplied by k, are transformed as these components themselves.

The first integral in (6) taken separately is a component of the wave function for a light quantum. The amplitude $F(\mathbf{k})$ can be considered as a wave function in the momentum space of light quanta in the Heisenberg representation. The corresponding wave function in the Schrödinger representation is then

$$F'(\mathbf{k}, t) = F(\mathbf{k})e^{-ickt}.$$

The amplitudes of the field components satisfy the next equations of motion:

$$\left.\begin{aligned} \mathbf{P}(\mathbf{k}) &= \tfrac{i}{c}[\mathbf{k}\Phi(\mathbf{k}) - k\mathbf{A}(\mathbf{k})] = -\tfrac{1}{c}\mathbf{E}(\mathbf{k}) \\ P_0(\mathbf{k}) &= \tfrac{i}{c}[k\Phi(\mathbf{k}) - \mathbf{k}\mathbf{A}(\mathbf{k})]. \end{aligned}\right\} \tag{7}$$

To pass on from the usual amplitudes (the "small" field) to quantized amplitudes (the "big" field),[13] one has to establish commutation relations for the latter. Applying the usual quantization rules, we obtain

$$\left.\begin{aligned} \Phi^\dagger(\mathbf{k})\Phi(\mathbf{k}') - \Phi(\mathbf{k}')\Phi^\dagger(\mathbf{k}) &= \tfrac{\hbar c}{2k}\delta(\mathbf{k}-\mathbf{k}') \\ A_l^\dagger(\mathbf{k})A_m(\mathbf{k}') - A_m(\mathbf{k}')A_l^\dagger(\mathbf{k}) &= -\tfrac{\hbar c}{2k}\delta_{lm}\delta(\mathbf{k}-\mathbf{k}') \end{aligned}\right\}, \tag{8}$$

whereas all the components $A_1(\mathbf{k}), A_2(\mathbf{k}), A_3(\mathbf{k}), \Phi(\mathbf{k})$ of the quantized wave function commute with each other. The relativistic invariance of the commutation relations takes place because the quantities $k\Phi(\mathbf{k})$, $kA(\mathbf{k})$ are the components of the four-vector, and $k\delta(\mathbf{k}-\mathbf{k}')$ is an invariant.

The volume integral of the Hamilton function (4), presented as an integral in the momentum space, is

$$\begin{aligned} H = \int [&\mathbf{A}^\dagger(\mathbf{k})\mathbf{A}(\mathbf{k}) + \mathbf{A}(\mathbf{k})\mathbf{A}^\dagger(\mathbf{k}) - \Phi(\mathbf{k})\Phi^\dagger(\mathbf{k}) - \\ &- \Phi^\dagger(\mathbf{k})\Phi(\mathbf{k})]k^2(d\mathbf{k}). \end{aligned}$$

[13]The terms "small" and "big" were introduced by Pauli: W. Pauli. Zs. Phys. **80**, 373 (1933). (*V. Fock*)

To eliminate the so-called zero energy and also the negative eigenvalues, we have to change the order of the factors there and write

$$H = 2\int [\mathbf{A}^\dagger(\mathbf{k})\mathbf{A}(\mathbf{k}) - \Phi(\mathbf{k})\Phi^\dagger(\mathbf{k})k^2(d\mathbf{k}). \tag{9}$$

This expression is just the energy operator for the radiation field. Under Lorentz transformation the operator H behaves as a time component of a four-vector, the spatial components of which are

$$\mathbf{G} = \frac{2}{c}\int \mathbf{k}[\mathbf{A}^\dagger(\mathbf{k})\mathbf{A}(\mathbf{k}) - \Phi(\mathbf{k})\Phi^\dagger(\mathbf{k})]k(d\mathbf{k}) \tag{10}$$

and represent the quantity of motion (momentum) of the field.

Simultaneously with the vector of energy and momentum, one can consider the invariant

$$N = \frac{2}{\hbar c}\int [\mathbf{A}^\dagger(\mathbf{k})\mathbf{A}(\mathbf{k}) - \Phi(\mathbf{k})\Phi^\dagger(\mathbf{k})]k(d\mathbf{k}), \tag{11}$$

which has to be interpreted as the operator of the whole number of quanta. If we introduce the density in the momentum space, corresponding to expression (11),

$$n(\mathbf{k}) = \frac{2}{\hbar c}[\mathbf{A}^\dagger(\mathbf{k})\mathbf{A}(\mathbf{k}) - \Phi(\mathbf{k})\Phi^\dagger(\mathbf{k})]k, \tag{12}$$

the vector of energy and momentum can be rewritten in the form

$$H = \hbar c\int kn(\mathbf{k})(d\mathbf{k}), \tag{9*}$$

$$\mathbf{G} = \hbar\int \mathbf{k}n(\mathbf{k})(d\mathbf{k}). \tag{10*}$$

The operator $n(\mathbf{k})$ is not positively definite itself, but owning to the supplementary conditions that will be established below, the mathematical expectation of the quantities N and H in any physically possible state of the field is positive (or equal to zero).

As was mentioned above, the Heisenberg representation of the wave functions $\Phi(\mathbf{k})$, $\mathbf{A}(\mathbf{k})$ is connected with the Schrödinger representation $\Phi'(\mathbf{k}, t)$ by the relations

$$\Phi'(\mathbf{k}, t) = \Phi(\mathbf{k})e^{-ickt}; \qquad \mathbf{A}'(\mathbf{k}, t) = \mathbf{A}(\mathbf{k})e^{-ickt}. \tag{13}$$

In the Heisenberg representation the functional Ω, describing a state of the field, contains as argumental functions the following ones:

$$\Phi(\mathbf{k}), \quad \overline{A}_1(\mathbf{k}), \quad \overline{A}_2(\mathbf{k}), \quad \overline{A}_3(\mathbf{k}). \tag{14}$$

In this case, it does not depend explicitly on time; therefore, the wave equation is reduced to the form $\frac{\partial\Omega}{\partial t} = 0$.

In the Schrödinger representation the quantities $\Phi'(\mathbf{k}, t)$ and $\overline{\mathbf{A}}'(\mathbf{k}, t)$ serve as argumental functions in the functional Ω and the wave equation becomes:

$$\left(H - i\hbar\frac{\partial}{\partial t}\right)\Omega = 0. \tag{15}$$

The choice of quantities (14) as argumental functions in the functional Ω is determined solely by commutation relations (8). The operators $\Phi^\dagger(\mathbf{k})$ and $A_l(\mathbf{k})$ then have the form

$$\Phi^\dagger(\mathbf{k})\Omega = \frac{\hbar c}{2k}\frac{\delta'\Omega}{\delta\Phi(\mathbf{k})}; \qquad A_l(\mathbf{k})\Omega = \frac{\hbar c}{2k}\frac{\delta'\Omega}{\delta\overline{A}_l(\mathbf{k})}. \tag{16}$$

Unlike the case of space rotation, canonical transformation of non-commutative operators correspond to Lorentz transformation but not just a linear substitution of the argumental functions (14).

Since the operators $\Phi'(\mathbf{k}, t)$ and $\Phi'^\dagger(\mathbf{k}, t)$ etc. (in the Schrödinger representation) satisfy the same commutation relations (8) as the operators $\Phi(\mathbf{k})$ and $\Phi^\dagger(\mathbf{k})$ (in the Heisenberg representation), for the first ones formulae (16) also are valid. If, instead of $\Phi(\mathbf{k}, t)$, we write again $\Phi(\mathbf{k})$, etc., the wave equation (16) written explicitly has the form:

$$H\Omega = \hbar c \int \left[\sum_{l=1}^{3} \overline{A}_l(\mathbf{k})\frac{\delta'\Omega}{\delta\overline{A}_l(\mathbf{k})} - \Phi(\mathbf{k})\frac{\delta'\Omega}{\delta\Phi(\mathbf{k})}\right] k(d\mathbf{k}) = i\hbar\frac{\partial\Omega}{\partial t}. \tag{17}$$

Further, we use mainly the Heisenberg representation.

Now we should consider the question of those supplementary conditions for the functional Ω that correspond to the classical relation $P_0 = 0$. Following Dirac, we interpret this equation in the sense of two conditions that consist of the equality of the amplitude $P_0(\mathbf{k})$ and its conjugated $P_0^\dagger(\mathbf{k})$ to zero:

$$P_0(\mathbf{k})\Omega = \frac{i}{c}[k\Phi(\mathbf{k}) - \mathbf{k}\mathbf{A}(\mathbf{k})]\Omega = 0; \tag{18}$$

$$P_0^\dagger(\mathbf{k})\Omega = -\frac{i}{c}[k\Phi^\dagger(\mathbf{k}) - \mathbf{k}\mathbf{A}^\dagger(\mathbf{k})]\Omega = 0. \tag{18*}$$

Both these conditions are relativistic invariants. They are compatible since for arbitrary values $\mathbf{k}$ and $\mathbf{k}'$, including coinciding ones, the following operator equation takes place:

$$P_0^\dagger(\mathbf{k})P_0(\mathbf{k}') - P_0(\mathbf{k}')P_0^\dagger\mathbf{k}) = 0. \tag{19}$$

Further, conditions (18) and (18*) are compatible also with the wave equation that is especially easy to see in the Heisenberg representation where the wave equation has the simple form $\frac{\partial\Omega}{\partial t} = 0$.

Now it is easy to obtain the general form of the functional Ω, satisfying this system of equations. For this, let us pick out a component $\theta(\mathbf{k})$ in the direction of the wave vector $\mathbf{k}$ from $\mathbf{A}(\mathbf{k})$ and suppose

$$\theta(\mathbf{k}) = \frac{\mathbf{k}\mathbf{A}(\mathbf{k})}{k}; \tag{20}$$

$$\mathbf{B}(\mathbf{k}) = \mathbf{A}(\mathbf{k}) - \mathbf{k}\frac{\mathbf{k}\cdot\mathbf{A}(\mathbf{k})}{k^2}. \tag{21}$$

As the argumental functions in Ω, we take five quantities:

$$\Phi(\mathbf{k}), \quad \overline{\theta}(\mathbf{k}), \quad \overline{b}_1(\mathbf{k}), \quad \overline{b}_2(\mathbf{k}), \quad \overline{b}_3(\mathbf{k}).$$

The last three of them are connected with one another by the identity

$$\mathbf{k}\overline{\mathbf{B}}(\mathbf{k}) = 0.$$

In new notations, equations (18) and (18*) can be written as

$$\left[\Phi(\mathbf{k}) - \theta(\mathbf{k})\right]\Omega = 0; \tag{22}$$

$$\left[\Phi^\dagger(\mathbf{k}) - \theta^\dagger(\mathbf{k})\right]\Omega = 0. \tag{22*}$$

Since the quantity θ is one of the components of the vector potential, the commutation relations for θ and $\theta^\dagger$ are the same as for their certain component and its conjugate one. Therefore, the operator $\theta(\mathbf{k})$ has the value

$$\theta(\mathbf{k})\Omega = \frac{\hbar c}{2k}\frac{\delta'\Omega}{\delta\overline{\theta}(\mathbf{k})}. \tag{23}$$

Hence, the explicitly written equations (22) and (22*) form

$$\left.\begin{aligned} \frac{\hbar c}{2k}\frac{\delta'\Omega}{\delta\Phi(\mathbf{k})} - \overline{\theta}(\mathbf{k})\Omega = 0, \\ \frac{\hbar c}{2k}\frac{\delta'\Omega}{\delta\overline{\theta}(\mathbf{k})} - \Phi(\mathbf{k})\Omega = 0. \end{aligned}\right\} \tag{24}$$

These equations give

$$\Omega = e^{\chi}\Omega^0\left[\overline{\mathbf{B}}(\mathbf{k})\right], \tag{25}$$

where the functional χ has the following value:

$$\chi = \frac{2}{\hbar c}\int \overline{\theta}(\mathbf{k})\Phi(\mathbf{k})k(d\mathbf{k}) \tag{26}$$

and Ω_0 is an arbitrary functional, which now does not contain $\overline{\theta}(\mathbf{k})$ and $\Phi(\mathbf{k})$. Thus, the supplementary conditions allowed us to reduce the general functional Ω, depending on four argumental functions (14), to a functional Ω_0 depending only on two argumental functions that are the two orthogonal to the wave vector components of the vector potential.

This reduction corresponds to the fact that an orthogonal light quantum at a fixed momentum is characterized by two numbers, let us say two possible states of polarization. The functional (25) also satisfies the wave equation; thus, it describes the very general state of the radiation field in vacuum.

By the supplementary conditions, one can exclude the quantities θ and Φ from operator (11) of a light quanta number and from the operator of energy and momentum. Let us consider the quantity $n(\mathbf{k})$ — equation (12) — and write it in the form

$$n(\mathbf{k}) = n'(\mathbf{k}) + n''(\mathbf{k}), \tag{27}$$

where we supposed

$$n'(\mathbf{k}) = \frac{2}{\hbar ck}(\mathbf{k}\cdot\mathbf{A}^{\dagger}(\mathbf{k}))\cdot(\mathbf{k}\cdot\mathbf{A}(\mathbf{k})), \tag{28}$$

or differently

$$n'(\mathbf{k}) = \frac{2}{\hbar ck}(\mathbf{k}\cdot\mathbf{B}^{\dagger}(\mathbf{k}))\cdot(\mathbf{k}\cdot\mathbf{B}(\mathbf{k})) \tag{28*}$$

and also

$$n''(\mathbf{k}) = \frac{2k}{\hbar c}[\theta^{\dagger}(\mathbf{k})\theta(\mathbf{k}) - \Phi(\mathbf{k})\Phi^{\dagger}(\mathbf{k})]. \tag{29}$$

Writing the latter quantity as

$$n''(\mathbf{k}) = \frac{2k}{\hbar c}\{\theta^\dagger(\mathbf{k})[\theta(\mathbf{k}) - \Phi(\mathbf{k})] + \Phi(\mathbf{k})[\theta^\dagger(\mathbf{k}) - \Phi^\dagger(\mathbf{k})]\} \tag{29*}$$

and taking into account equations (22) and (22*), we make sure that the operator $n''(\mathbf{k})$ being applied to any possible (i.e., satisfying the supplementary conditions) functional gives zero. But the operator $n'(\mathbf{k})$ depends only on the quantities $\mathbf{B}$ and, moreover, it is positively definite. That proves our statement about the positive sign of the mathematical expectation and the number of light quanta.

To establish the connection with reasoning in the first part, let us introduce quantities b and $b^\dagger$ as applied to our task. Let $\mathbf{e}(\mathbf{k},1)$ and $\mathbf{e}(\mathbf{k},2)$ be two unit vectors orthogonal both to each other and to the wave vector $\mathbf{k}$. These unit vectors correspond to two states of polarization. Since, according to (21), the vector $\mathbf{B}(\mathbf{k})$ is also orthogonal to the wave vector, we can put

$$\mathbf{B}(\mathbf{k}) = \sqrt{\frac{\hbar c}{2k}} \sum_{j=1}^{2} \mathbf{e}(\mathbf{k},j) b(\mathbf{k},j). \tag{30}$$

Based on formulae (8) and (21), it is easy to show that the operators $b(\mathbf{k},j)$ thus satisfy the commutation relations

$$\left.\begin{aligned} b(\mathbf{k}',j')b^\dagger(\mathbf{k},j) - b^\dagger(\mathbf{k},j)b(\mathbf{k}',j') &= \delta_{j,j'}\delta(\mathbf{k}-\mathbf{k}') \\ b(\mathbf{k}',j')b(\mathbf{k},j) - b(\mathbf{k},j)b(\mathbf{k}',j') &= 0 \end{aligned}\right\}. \tag{31}$$

These relations coincide with (**I**, 1) if under the variable x one means an aggregate of three continuous variables k_1, k_2, k_3 and one discrete variable j, taking only two values j = 1, 2. Therefore, the whole mathematical theory of the functionals developed in Section I can be applied to our task. In particular, the form of the functional Ω_0 is given by formula (**I**, 5), where ψ is the wave function in the momentum space of light quanta. In the special case when there are no light quanta, functional (25) is reduced to

$$\Omega = \exp\left(\frac{2}{\hbar c}\int \overline{\theta}(\mathbf{k})\Phi(\mathbf{k})k(d\mathbf{k})\right). \tag{25*}$$

It goes without saying that the quantum-free state is a relativistic invariant conception though expression (25) is not an invariant.

Let us express now the operators $n(\mathbf{k}), N, H, \mathbf{G}$ by the introduced quantities b and $b^\dagger$. If we omit the part of these operators "vanishing" due to the supplementary conditions (22) and denote the rest as $n', N', H', \mathbf{G}'$ correspondingly, we obtain

$$n'(\mathbf{k}) = b^\dagger(\mathbf{k},1)b(\mathbf{k},1) + b^\dagger(\mathbf{k},2)b(\mathbf{k},2); \tag{32}$$

$$N' = \sum_j \int b^\dagger(\mathbf{k},j)b(\mathbf{k},j)(d\mathbf{k}); \tag{33}$$

$$H' = \sum_j \int \hbar ck b^\dagger(\mathbf{k},j)b(\mathbf{k},j)(d\mathbf{k}); \tag{34}$$

$$\mathbf{G}' = \sum_j \int \hbar \mathbf{k} b^\dagger(\mathbf{k},j)b(\mathbf{k},j)(d\mathbf{k}). \tag{35}$$

It follows from the general theory of the second quantization that the operator N' of the light quanta number has the integer eigenvalues 0, 1, 2, ... The operator for the number of light quanta with the wave number $\mathbf{k}$ from the interval $(\mathbf{k}, \mathbf{k}+\Delta\mathbf{k})$

$$N'(\mathbf{k}, \Delta\mathbf{k}) = \sum_j \int_{\mathbf{k}}^{\mathbf{k}+\Delta\mathbf{k}} b^\dagger(\mathbf{k},j)b(\mathbf{k},j)(d\mathbf{k}) \tag{36}$$

has the same eigenvalues. To show it, let us introduce the "normalized eigen differentials"

$$b_j = \frac{1}{\sqrt{(\Delta\mathbf{k})}} \int_{\mathbf{k}}^{\mathbf{k}+\Delta\mathbf{k}} b(\mathbf{k},j)(d\mathbf{k}), \tag{37}$$

which, as it follows from (31), satisfy the commutation relations

$$\left.\begin{aligned} b_{j'}b_j^\dagger - b_j^\dagger b_{j'} &= \delta_{jj'}, \\ b_{j'}b_j - b_j b_{j'} &= 0. \end{aligned}\right\} \tag{38}$$

Expression (36) then becomes

$$N'(\mathbf{k}, \Delta\mathbf{k}) = \sum_j b_j^\dagger b_j\,, \tag{36*}$$

and it follows from the theory of harmonic oscillator that the operator (36) has the eigenvalues 0, 1, 2, ...

In the same way one can examine the operators $H'(\mathbf{k}, \Delta\mathbf{k})$ and $\mathbf{G}'(\mathbf{k}, \Delta\mathbf{k})$ of the energy and momentum of the light quanta with the wave vector from an interval $(\mathbf{k}, \mathbf{k}+\Delta\mathbf{k})$ and show that these operators determined analogous to (36) have the eigenvalues $n\hbar ck$ and correspondingly $n\hbar\mathbf{k}$ where $n = 0, 1, 2, \ldots$ as it should be.

On the Measurability of the Fourier Components for the Field

We shall derive now a relation expressing the "complementarity" (in the spirit of Bohr) of the idea of light quanta with the conception of the classically measurable amplitude of the electromagnetic field.

The complex amplitude $F(\mathbf{k})$ of the field component is not, properly speaking, a measurable physical magnitude, for the real part of the amplitude does not commute with its imaginary part. But in a region of large quantum numbers this non-commutability becomes inessential and the amplitudes behave like classical quantities, which means that they turn out to be roughly measurable. Let us consider in detail the matter here and prove the next suggestion: if we take the amplitudes, averaged over an interval $(\mathbf{k}, \mathbf{k} + \Delta\mathbf{k})$, and calculate by their means the energy of radiation, related to this interval, the uncertainty of E will always exceed one energy quantum $\hbar ck$ of the corresponding frequency.

First of all, let us note that it does not matter whether we use the amplitudes for the strength of the field or some auxiliary quantities, e.g., the quantities $b(\mathbf{k}, j)$ from (30). We take advantage of the latter and write down the operator of the radiation energy in an interval $(\mathbf{k}, \mathbf{k}+\Delta\mathbf{k})$ in the form

$$H'(\mathbf{k}, \Delta\mathbf{k}) = \int_{\mathbf{k}}^{\mathbf{k}+\Delta\mathbf{k}} \hbar ck \left[b^\dagger(\mathbf{k}, 1)b(\mathbf{k}, 1) + b^\dagger(\mathbf{k}, 2)b(\mathbf{k}, 2)\right] (d\mathbf{k}). \quad (39)$$

Supposing now that the real and imaginary parts of the averaged amplitude

$$\overline{\overline{b}}_j = \frac{1}{(\Delta\mathbf{k})} \int_{\mathbf{k}}^{\mathbf{k}+\Delta\mathbf{k}} b(\mathbf{k}, j)(d\mathbf{k}) = \frac{1}{\sqrt{\Delta\mathbf{k}}} b_j \quad (40)$$

are measured with an accuracy allowed by the uncertainty relation, let us calculate the energy E by the formula

$$E = H'(\mathbf{k}, \Delta\mathbf{k}) = \hbar ck(\overline{\overline{b^\dagger}}_1 \overline{\overline{b}}_1 + \overline{\overline{b^\dagger}}_2 \overline{\overline{b}}_2)(\Delta\mathbf{k}) = \hbar ck(b_1^\dagger b_1 + b_2^\dagger b_2). \quad (39^*)$$

One can ask what the uncertainty of the energy is.

This problem can be easy resolved, since it can be reduced to the analogous problem for an harmonic oscillator. Substituting in the formula

$$H_0 = \frac{1}{2m}p^2 + \frac{1}{2}m\omega^2 q^2 - \frac{1}{2}\hbar\omega \tag{41}$$

for the energy of the oscillator (without zero-energy) the expression

$$b = \frac{p - im\omega q}{\sqrt{2m\omega\hbar}}; \qquad b^\dagger = \frac{p + im\omega q}{\sqrt{2m\omega\hbar}}, \tag{42}$$

we get from the commutation relations for p and q that

$$bb^\dagger - b^\dagger b = 1 \tag{43}$$

and the energy of the oscillator is equal to

$$H_0 = \hbar\omega b^\dagger b. \tag{41*}$$

The above-given expression (39*) for the energy E is the sum of two terms of the form (41*) where $\omega = ck$.

If we measure at the same time the momentum and coordinate of the harmonic oscillator with an uncertainty Δp and Δq, the uncertainty of the energy, calculated by these p and q, is at least

$$\Delta E_0 = \frac{1}{2m}(\Delta p)^2 + \frac{1}{2}m\omega^2(\Delta q)^2. \tag{44}$$

In the most favorable case, Δp and Δq are connected by the relation $\Delta p \Delta q = \frac{\hbar}{2}$; then the uncertainty of the energy ΔE satisfies the inequality

$$\Delta E_0 \geq \frac{1}{2}\hbar\omega. \tag{45}$$

The suggestion, formulated above, for the radiation energy E (39) follows from here.

In the course of our reasoning we told about *measuring* the field; but such measuring is not possible without an interaction with matter. Nevertheless, this circumstance on no account depreciates our reasoning for we are concerned with the purely kinematical properties of the field. Really, they were based only on the commutation relations that stay the same in the presence of matter.

The question about the measurability of the field itself averaged over some region of the space-time (but not its Fourier components) was exhaustively investigated by Bohr and Rosenfeld in the work quoted in the introduction.

III Interaction with Matter. Initial Equations

We shall consider a quantum mechanical system, consisting of an indefinite number of light quanta and of a fixed number n of material particles of masses m_s and charges ε_s ($s = 1, 2, ...n$). Thus, we do not consider the case of a variable number of particles (Dirac's hole theory of the positron). To be definite, we shall talk about electrons, though the form of the wave equation for material particles still remains arbitrary. We shall describe this quantum mechanical system by means of the following independent variables: the field — by its amplitudes (as in the previous section); the particles — by their coordinates $\mathbf{r}_s = (x_s, y_s, z_s)$; times t_s and variables for their inner degrees of freedom. It means that, following Dirac, we introduce for every particle its special time; time for light quanta we shall note by t. Only toward the end of calculations we put $t_s = t$. The operators for the momentum and kinetic energy have the form

$$\left.\begin{aligned} P_x^{(s)} &= -i\hbar\frac{\partial}{\partial x_s} - \frac{\varepsilon_s}{c}A_x(\mathbf{r}_s, t_s), \\ P_y^{(s)} &= -i\hbar\frac{\partial}{\partial y_s} - \frac{\varepsilon_s}{c}A_y(\mathbf{r}_s, t_s), \\ P_z^{(s)} &= -i\hbar\frac{\partial}{\partial z_s} - \frac{\varepsilon_s}{c}A_z(\mathbf{r}_s, t_s), \end{aligned}\right\} \tag{1}$$

$$T^{(s)} = i\hbar\frac{\partial}{\partial t_s} - \varepsilon_s\Phi(\mathbf{r}_s, t_s). \tag{2}$$

The operators related to different particles mutually commute. The operators related to the same particle satisfy the commutation relations

$$\left.\begin{aligned} TP_x - P_xT &= i\hbar\varepsilon E_x, & P_yP_z - P_zP_y &= i\hbar\varepsilon H_x, \\ TP_y - P_yT &= i\hbar\varepsilon E_y, & P_zP_x - P_xP_z &= i\hbar\varepsilon H_y, \\ TP_z - P_zT &= i\hbar\varepsilon E_z, & P_xP_y - P_yP_z &= i\hbar\varepsilon H_z \end{aligned}\right\} \tag{3}$$

(the index s is omitted here). Let us denote as $\sigma_x^{(s)}, \sigma_y^{(s)}, \sigma_z^{(s)}$ the components of the spin for the s-th electron and construct the operator

$$P^{(s)} = \sigma_x^{(s)}P_x^{(s)} + \sigma_y^{(s)}P_y^{(s)} + \sigma_z^{(s)}P_z^{(s)}. \tag{4}$$

According to Dirac, for each electron we can write its own wave equation. In the non-relativistic case, it has the form[14]

$$\frac{1}{2m_s}(P^{(s)})^2\Psi = T^{(s)}\Psi \qquad (s = 1, 2, ...n) \tag{5}$$

and for the relativistic case

$$\left[c\varrho_a^{(s)}P^{(s)} + m_s c^2 \varrho_c^{(s)}\right]\Psi = T^{(s)}\Psi \quad (s = 1, 2, ...n), \tag{6}$$

where we denoted as $\varrho_a^{(s)}, \varrho_b^{(s)}, \varrho_c^{(s)}$ the operators related to the second inner degree of freedom for the s-th electron (Dirac's $\varrho_1, \varrho_2, \varrho_3$).

The wave equation (5) or (6) may be considered as a certain supplementary condition for the wave functional Ψ; this supplementary condition gives the dependence between the momentum and the energy that is needed in the corresponding (relativistic or non-relativistic) theory.

Besides the wave equations for the wave functional, supplementary conditions corresponding to the equality to zero of the divergence of the four-dimensional potential, i.e., equations (**II**, 18) or (**II**, 22), have to exist as well. We write down these conditions in the form[15]

$$C(\mathbf{k})\Psi = 0, \qquad C^\dagger(\mathbf{k})\Psi = 0, \tag{7}$$

where C($\mathbf{k}$) denotes the following operator:

$$C(\mathbf{k}) = ik[\theta(\mathbf{k}) - \Phi(\mathbf{k})] + \frac{i}{2(2\pi)^{3/2}k}\sum_{s=1}^{n}\varepsilon_s e^{ickt_s - i\mathbf{k}\mathbf{r}_s}. \tag{8}$$

Considering C($\mathbf{k}$) as the amplitude of a certain field C($\mathbf{r}$,t), we have this field in the form

$$C(\mathbf{r}, t) = \mathrm{div}\mathbf{A} + \frac{1}{c}\frac{\partial\Phi}{\partial t} - \sum_{s=1}^{n}\frac{\varepsilon_s}{4\pi}\Delta(X - X_s), \tag{9}$$

where $X = (\mathbf{r}, t)$ and $X_s = (\mathbf{r}_s, t_s)$ are four-dimensional vectors and $\Delta(X)$ is the so-called invariant delta function

$$\Delta(X) = \frac{1}{r}[\delta(r + ct) - \delta(r - ct)]. \tag{10}$$

[14]The letter Ψ here means the wave functional but not the quantized function as it was in Section I. (*V. Fock*)

[15]P.A.M. Dirac, V.A. Fock and B. Podolsky, *On Quantum Electrodynamics*, Sow. Phys. **2**, 468–479 (1932). (*V. Fock*). (See also [32-5*] in this book. (*Editors*))

The form of the terms proportional to the charges ε_s in the expression for the quantity C(**k**) follows from the demand of commutating this quantity with operators (1) and (2) for the components of the four-dimensional energy–momentum vector. One can easily verify that $C(\mathbf{k})$ commutes with $C^\dagger(\mathbf{k})$.

Thus, the wave functional Ψ of the n-body problem must satisfy n wave equations of the form (5) or (6) and both of the supplementary conditions of the form (7). Since supplementary conditions (7) do not contain derivatives of Ψ with respect to time, the initial state of the quantum mechanical system must also satisfy them. It is possible to say that only those states have the physical sense that satisfy condition (7). Further, it follows herefrom that only those quantities are physically measurable that (i.e., operators of theirs) commute with $C(\mathbf{k})$ and $C^\dagger(\mathbf{k})$. These are the so-called gradient invariant quantities and first of all quantities (1), (2) and (3).

Eliminating the Supplementary Conditions

By virtue of the supplementary conditions (7), one can get in the general form the dependence of the wave functional Ψ on its argumental functions $\overline{\theta}(\mathbf{k})$ and $\Phi(\mathbf{k})$ in absolutely the same way as for the case without matter.[16]

Putting for brevity

$$f = \frac{1}{(2\pi)^{3/2}} \sum_{s=1}^{n} \varepsilon_s e^{i\varphi_s}, \qquad \varphi_s = ckt_s - \mathbf{k}\mathbf{r}_s, \tag{11}$$

we can write down both equations (7) as

$$\left.\begin{aligned} \left[\theta(\mathbf{k}) - \Phi(\mathbf{k}) + \frac{1}{2k^2} f\right]\Psi = 0, \\ \left[\theta^\dagger(\mathbf{k}) - \Phi^\dagger(\mathbf{k}) + \frac{1}{2k^2} f^\dagger\right]\Psi = 0. \end{aligned}\right\} \tag{12}$$

If we take here the operators (II, 16) and (II, 23) for $\Phi^\dagger(\mathbf{k})$ and $\theta(\mathbf{k})$,

[16] L. Rosenfeld, *La théorie quantique des champs*, Annales de l'institut H. Poincaré. Paris, 1931, and Enrico Fermi, *Quantum Theory of Radiation*, Rev. Mod. Phys., January, 1932.

formula (12) will get the form

$$\left.\begin{aligned}\frac{\delta'\Psi}{\delta\Phi(\mathbf{k})} &= \Big(\frac{2k}{\hbar c}\overline{\theta}(\mathbf{k}) + \frac{1}{\hbar ck}\overline{f}\Big)\Psi,\\ \frac{\delta'\Psi}{\delta\overline{\theta}(\mathbf{k})} &= \Big(\frac{2k}{\hbar c}\Phi(\mathbf{k}) - \frac{1}{\hbar ck}f\Big)\Psi.\end{aligned}\right\} \tag{13}$$

The solution of these equations is

$$\Psi = e^{\chi}\Omega(\overline{\mathbf{B}}(\mathbf{k})), \tag{14}$$

where χ abbreviates the functional

$$\begin{aligned}\chi = {} & \frac{2}{\hbar c}\int \overline{\theta}(\mathbf{k})\Phi(\mathbf{k})k(d\mathbf{k}) + \\ & + \frac{1}{\hbar c}\int \Phi(\mathbf{k})\frac{\overline{f}}{k}(d\mathbf{k}) - \frac{1}{\hbar c}\int \overline{\theta}(\mathbf{k})\frac{f}{k}(d\mathbf{k}) + \chi'.\end{aligned} \tag{15}$$

The additional term χ' in formula (15) is a function of coordinates and time of n particles and will be defined hereafter. The functional Ω in (14) does not already depend on the argumental functions of $\overline{\theta}(\mathbf{k})$ and $\Phi(\mathbf{k})$.

The transformation carried out here can be also written in the operator form. It may be considered as a non-unitary canonical transformation that sends the operator L acting on the functional Ψ to an operator L' acting on the functional Ω, where the operator L' is connected with L by the relation

$$L' = e^{-\chi}Le^{\chi}. \tag{16}$$

In this formula the symbol χ has to be considered as an operator obtained from (15) by the replacement of $\overline{\theta}(\mathbf{k})$ by $\theta^{\dagger}(\mathbf{k})$.

Let us investigate now the transformations of our operators, in particular, operators (1) and (2) for the energy and momentum. The quantities $\mathbf{B}(\mathbf{k})$ and $\mathbf{B}^{\dagger}(\mathbf{k})$ keep invariant under transformations (16) because χ commutes with them. The quantities $\theta^{\dagger}(\mathbf{k})$ and $\Phi(\mathbf{k})$ of course also keep invariant. For $\theta(\mathbf{k})$ and $\Phi^{\dagger}(\mathbf{k})$ we have the formulae of transformation

$$e^{-\chi}\theta(\mathbf{k})e^{\chi} = \theta(\mathbf{k}) + \frac{\hbar c}{2k}\frac{\delta'\chi}{\delta\overline{\theta}(\mathbf{k})} = \theta(\mathbf{k}) + \Phi(\mathbf{k}) - \frac{1}{2k^2}f, \tag{17}$$

$$e^{-\chi}\Phi^{\dagger}(\mathbf{k})e^{\chi} = \Phi^{\dagger}(\mathbf{k}) + \frac{\hbar c}{2k}\frac{\delta'\chi}{\delta\Phi(\mathbf{k})} = \Phi^{\dagger}(\mathbf{k}) + \theta^{\dagger}(\mathbf{k}) + \frac{1}{2k^2}f^{\dagger}. \tag{18}$$

From formula (17) we obtain

$$e^{-\chi}\mathbf{A}(\mathbf{k})e^{\chi} = \mathbf{A}(\mathbf{k}) + \frac{\mathbf{k}}{k}\Phi(\mathbf{k}) - \frac{\mathbf{k}}{2k^2}f. \tag{19}$$

It follows from here and (18) that for $\mathbf{A}(\mathbf{r}_s, t_s)$ and $\Phi(\mathbf{r}_s, t_s)$

$$\begin{aligned} e^{-\chi}\mathbf{A}(\mathbf{r}_s, t_s)e^{\chi} &= \mathbf{A}(\mathbf{r}_s, t_s) + \frac{1}{(2\pi)^{3/2}}\int \frac{\mathbf{k}}{k}\Phi(\mathbf{k})e^{-i\varphi_s}(d\mathbf{k}) - \\ &- \frac{1}{2(2\pi)^{3/2}}\int \frac{\mathbf{k}}{k^3} f e^{-i\varphi_s}(d\mathbf{k}); \qquad (20) \\ e^{-\chi}\Phi(\mathbf{r}_s, t_s)e^{\chi} &= \Phi(\mathbf{r}_s, t_s) + \frac{1}{(2\pi)^{3/2}}\int \theta^{\dagger}(\mathbf{K})e^{i\varphi_s}(d\mathbf{k}) + \\ &+ \frac{1}{2}\frac{1}{(2\pi)^{3/2}}\int \frac{1}{k^2} f^{\dagger} e^{i\varphi_s}(d\mathbf{k}). \qquad (21) \end{aligned}$$

The last integral in (21) contains an infinite constant that appears from the term with $e^{-i\varphi_s}$ in the sum $f^{\dagger}$. The operators

$$p_x^{(s)} = -i\hbar\frac{\partial}{\partial x_s}; \ldots; \; i\hbar\frac{\partial}{\partial t_s}$$

are transformed as

$$e^{-\chi}p_x^{(s)}e^{\chi} = p_x^{(s)} - i\hbar\frac{\partial\chi}{\partial x_s}, \tag{22}$$

$$e^{-\chi}i\hbar\frac{\partial}{\partial t_s}e^{\chi} = i\hbar\frac{\partial}{\partial t_s} + i\hbar\frac{\partial\chi}{\partial t_s}. \tag{23}$$

With the help of the formulae obtained, operators (1) and (2) for the energy and momentum can be transformed in the following way. Let us construct, according to (**II**, 6), a vector that corresponds to the amplitude $\mathbf{B}(\mathbf{k})$ (equation (**II**, 21)),

$$\mathbf{B}(\mathbf{r}, t) = \frac{1}{(2\pi)^{3/2}}\int \mathbf{B}(\mathbf{k})e^{-i\varphi}(d\mathbf{k}) + \frac{1}{(2\pi)^{3/2}}\int \mathbf{B}^{\dagger}(\mathbf{k})e^{i\varphi}(d\mathbf{k}). \tag{24}$$

This vector may be interpreted as a vector potential normalized by the condition

$$\mathrm{div}\mathbf{B} = 0. \tag{25}$$

We get then

$$e^{-\chi}\Big(p_x^{(s)} - \frac{\varepsilon_s}{c}A_x(\mathbf{r}_s, t_s)\Big)e^{\chi} = p_x^{(s)} - \frac{\varepsilon_s}{c}B_x(\mathbf{r}_s, t_s) -$$

$$-i\hbar\frac{\partial\chi'}{\partial x_s}-\frac{\varepsilon_s}{c}\frac{1}{(2\pi)^{3/2}}\int\theta(\mathbf{k})\frac{k_x}{k}e^{-i\varphi_s}(d\mathbf{k})+$$
$$+\frac{1}{2}\frac{\varepsilon_s}{c}\frac{1}{(2\pi)^{3/2}}\int\frac{k_x}{k^3}fe^{-i\varphi_s}(d\mathbf{k}); \tag{26}$$

$$e^{-\chi}T^{(s)}e^{\chi}=i\hbar\frac{\partial}{\partial t_s}+i\hbar\frac{\partial\chi'}{\partial t_s}-\frac{\varepsilon_s}{(2\pi)^{3/2}}\int Phi^{\dagger}(\mathbf{k})e^{i\varphi_s}(d\mathbf{k})-$$
$$-\frac{1}{2}\frac{\varepsilon_s}{(2\pi)^{3/2}}\int\frac{f^{\dagger}}{k^2}e^{i\varphi_s}(d\mathbf{k}). \tag{27}$$

Since the transformed operators (26) and (27) act only on those functionals Ω that do not contain the argumental functions $\theta(\mathbf{k})$ and $\Phi(\mathbf{k})$ and consequently satisfy the conditions

$$\left.\begin{aligned}\theta(\mathbf{k})\Omega&=\frac{\hbar c}{2k}\cdot\frac{\delta'\Omega}{\delta\overline{\theta}(\mathbf{k})}=0,\\ \Phi^{\dagger}(\mathbf{k})\Omega&=\frac{\hbar c}{2k}\cdot\frac{\delta'\Omega}{\delta\Phi(\mathbf{k})}=0,\end{aligned}\right\} \tag{28}$$

we can omit the integrals containing the amplitudes $\theta(\mathbf{k})$ and $\Phi^{\dagger}(\mathbf{k})$ in formulae (26) and (27) (these integrals are non-Hermitian operators). Really, two other integrals are also complex quantities. But one can fit the still undetermined function χ so the imaginary part of these quantities would be canceled by the terms $-i\hbar\frac{\partial\chi'}{\partial x_s}$ and $i\hbar\frac{\partial\chi'}{\partial t_s}$. For this we put

$$\chi'=-\frac{1}{4\hbar c}\int\frac{f^{\dagger}f}{k^3}(d\mathbf{k})+\text{const}. \tag{29}$$

The integral diverges when $k=0$. Thus, it is determined only up to an additive constant. Calculating it we have

$$\chi'=-\frac{1}{4\hbar c}\cdot\frac{1}{(2\pi)^3}\sum_{uv}\varepsilon_u\varepsilon_v F(X_u-X_v)+\text{const}, \tag{30}$$

where X means a four-dimensional vector ($\mathbf{r}$,t) and $F(x)$ has the following meaning:

$$F(X)=F(\mathbf{r},t)=\int\frac{\cos(crt-\mathbf{kr})}{k^3}(d\mathbf{k})+\text{const}=$$
$$=-\frac{2\pi}{r}\{(r+ct)\log|r+ct|+(r-ct)\log|r-ct|\}. \tag{31}$$

If we consider the latter expression (with logarithms) in (30) as F and omit the terms with $u=v$ as well as the constant, the quantity χ' is then an entirely determined function of coordinates $\mathbf{r}_s$ and times t_s. Its derivatives formally coincide with the derivatives of expression (29). For these derivatives, one can get

$$i\hbar\frac{\partial\chi'}{\partial x_s} = i\frac{\varepsilon_s}{2c}\cdot\frac{1}{(2\pi)^3}\int\frac{k_x}{k^3}\sum_u \varepsilon_u \sin(\varphi_u-\varphi_s)(d\mathbf{k}), \tag{32}$$

$$-i\hbar\frac{\partial\chi'}{\partial t_s} = i\frac{\varepsilon_s}{2}\cdot\frac{1}{(2\pi)^3}\int\frac{1}{k^2}\sum_u \varepsilon_u \sin(\varphi_u-\varphi_s)(d\mathbf{k}), \tag{33}$$

which coincide with the imaginary part of the integrals forming (26) and (27), the last terms of the right-hand side. Taking into account relation (28), we can consider the operator

$$\begin{aligned} P_x'^{(s)} &= p_x^{(s)} - \frac{\varepsilon}{c}B_x(\mathbf{r}_s,t_s) + \\ &+ \frac{\varepsilon_s}{2c}\frac{1}{(2\pi)^3}\sum_u \varepsilon_u\int\frac{k_x}{k^3}\cos(\varphi_u-\varphi_s)(d\mathbf{k}), \end{aligned} \tag{34}$$

$$T'^{(s)} = i\hbar\frac{\partial}{\partial t_s} - \frac{\varepsilon_s}{2}\frac{1}{(2\pi)^3}\sum_u \varepsilon_u\int\frac{1}{k^2}\cos(\varphi_u-\varphi_s)(d\mathbf{k}) \tag{35}$$

as transformed operators of the momentum and kinetic energy for an s-th particle. In the sum standing in (35), the term with $u = s$ is an infinite constant. These expressions can be written in a more plain form by introducing the quantities

$$V_s = \sum_u \frac{\varepsilon_u}{(2\pi)^3}\int\frac{\sin(\varphi_s-\varphi_u)}{k^3}(d\mathbf{k}). \tag{36}$$

Then

$$P_x'^{(s)} = p_x^{(s)} - \frac{\varepsilon_s}{c}B_x(\mathbf{r}_s,t_s) - \frac{\varepsilon_s}{2c}\frac{\partial V_s}{\partial x_s}, \tag{37}$$

$$T'^{(s)} = i\hbar\frac{\partial}{\partial t_s} - \frac{\varepsilon_s}{2c}\frac{\partial V_s}{\partial t_s}. \tag{38}$$

The integrals in formula (36) are easy to calculate. Let us write expression (36) in the form

$$V_s = \sum_u \frac{\varepsilon_u}{4\pi}V(X_s - X_u), \tag{39}$$

where $V(X) = V(\mathbf{r},t)$ abbreviates the integral

$$V(\mathbf{r},t) = \frac{1}{2\pi^2}\int\frac{\sin(ckt-\mathbf{kr})}{k^3}(d\mathbf{k}). \tag{40}$$

This integral is

$$V(\mathbf{r},t) = \frac{1}{2r}(|r+ct| - |r-ct|) \tag{41}$$

or, introducing the auxiliary function $\xi(x)$, defined by

$$\left.\begin{array}{lll} \xi(x) = -1 & \text{for} & x \leq -1, \\ \xi(x) = x & \text{for} & -1 \leq x \leq 1, \\ \xi(x) = 1 & \text{for} & 1 \leq x \ , \end{array}\right\} \tag{42}$$

we have

$$V(\mathbf{r},t) = \xi\Big(\frac{ct}{r}\Big). \tag{43}$$

Substituting (41) or (43) for V($\mathbf{r}$,t) in the sum (39) and calculating the derivatives of V_s entering formulae (37) and (38), for $P_x'^{(s)}$ and $T'^{(s)}$ we get

$$P_x'^{(s)} = p_x^{(s)} - \frac{\varepsilon_s}{c}B_x(\mathbf{r}_s,t_s) + \frac{\varepsilon_s}{2}\sum_u{}' \frac{\varepsilon_u}{4\pi}\frac{(x_s-x_u)(t_s-t_u)}{|\mathbf{r}_s-\mathbf{r}_u|^3}, \tag{44}$$

$$T'^{(s)} = i\hbar\frac{\partial}{\partial t_s} - \frac{\varepsilon}{2}\sum_u{}' \frac{\varepsilon_u}{4\pi|\mathbf{r}_s-\mathbf{r}_u|}. \tag{45}$$

While doing this, one has to bear in mind that the function $V(\mathbf{r},t)$ differs from a constant only for a spatially similar interval $(\mathbf{r},t)$. Therefore, summing in (44) and (45) runs only over those particles the intervals of which with respect to the given particle are spatially similar. Consequently, only pairs of the particles with a spatially similar interval contribute to the interaction, which corresponds to excluded "longitudinal light quanta" as a result of our transformation.

The wave equations for the functional Ω have the same form as the above-listed equations for the functional Ψ, namely:

$$\frac{1}{2m_s}(P'^{(s)})^2\Omega = T'^{(s)}\Omega \qquad (s=1,\dots n) \tag{46}$$

for the nonrelativistic case and

$$[c\varrho_a^{(s)}P'^{(s)} + m_sc^2\varrho_c^{(s)}]\Omega = T'^{(s)}\Omega \qquad (s=1,\dots n) \tag{47}$$

for the relativistic case. Here $P'^{(s)}$ denotes operator (4) composed of operators (37), while $T'^{(s)}$ means (38). Thus, the supplementary conditions (7) are satisfied automatically if the functional Ω besides the variables

q_s of material particles depends only on the argumental functions $\overline{\mathbf{B}}(\mathbf{k})$ (or on the functions $\overline{b}(\mathbf{k}, j)$ connected with them according to (**II**, 30)).

Introducing the functional derivatives of Ω to $\overline{b}(\mathbf{k}, j)$ it is easy to write down the wave equations (47) or (46) in the explicit form; but we will not do it here.

Let us note that if we consider only one of any equations (46) or (47) (let us say s-th) the form of the operators $P_x'^{(s)}$ and $T'^{(s)}$ allows a further simplification. With the help of a suitable canonical transformation (depending on the number s of the equation considered) one can get rid of the term $-\frac{\varepsilon_s}{2c}\frac{\partial V_s}{\partial t_s}$ in expression (37) for $P_x'^{(s)}$; as a result, the term $-\frac{\varepsilon_s}{2c}\frac{\partial V_s}{\partial t_s}$ in (38) is doubled. Then, omitting the index s, we have

$$P_x'' = p_x - \frac{\varepsilon}{c} B_x(\mathbf{r}, t),$$

$$T'' = i\hbar\frac{\partial}{\partial t} - {\sum_u}' \frac{\varepsilon\varepsilon_u}{4\pi|\mathbf{r} - \mathbf{r}_u|},$$

where summing runs over all the electrons for which the intervals with respect to the chosen s-th electron are spatially similar. Putting all the times equal to one another we get the sum reduced to the potential energy of the chosen electron in the electrostatic field of all the other particles.[17]

The Wave Equation in the Case of Coinciding Times

Let us write down our equations for the case when all particle times t_s and the time t of light quanta coincide with one another.[18] The common time is denoted by T for a while, but later we shall use the letter t again. We have

$$\frac{\partial}{\partial t} + \sum_{n=1}^{n} \frac{\partial}{\partial t}.$$

[17] As it was noted by Bloch (F. Bloch, Sow. Phys. **5**, 301 (1934)), in the case of non-coinciding times the wave equations for the different particles are compatible only if all the intervals between the particles are of the spatial character. (*V. Fock*)

[18] Let us note that for coinciding times quantity (30) can be reduced to

$$\chi' = +\frac{\alpha}{\pi} \sum_{u<v} \log|\mathbf{r}_u - \mathbf{r}_v|,$$

where α is the fine structure constant (here all quantities ε_s are supposed to be equal to the electron charge). Due to (14) it follows herefrom that at $\mathbf{r}_u = \mathbf{r}_v$ the functional Ψ is equal to zero, even if Ω is finite. (*V. Fock*)

Thus, to get an expression for the derivatives $\frac{\partial\Omega}{\partial T}$ of the wave functional Ω with respect to the common time T, we must sum up the right-hand sides of the separate wave equations (46) or (47). According to (45), the operator for the sum of the kinetic energies for the separate particles is

$$\sum_{n=1}^{n} T'^{(s)} = i\hbar\frac{\partial}{\partial t} - \frac{1}{2}\sum_{uv}\frac{\varepsilon_u\varepsilon_v}{4\pi|\mathbf{r}_u - \mathbf{r}_v|} \tag{48}$$

(here instead of T we write t again). To get the final expression, we have to omit here the terms with $u = v$; all these terms correspond to the self-energy of the electrons. The remaining terms give just the Coulomb potential energy of the electrons.

To compose the left-hand side of the wave equation we need to use (37) or (44) for $P_x'^{(s)}$. In our case these expressions become simpler, since for coinciding times the quantities V_s and their spatial derivatives disappear. Then we have simply

$$P_x'^{(s)} = p_x^{(s)} - \frac{\varepsilon}{c}B_x(\mathbf{r}_s, t). \tag{49}$$

We restrict ourselves by transforming relativistic equations (47). Denoting by D_s the Dirac operator in the absence of a field applied to the coordinates of an s-th electron

$$D_s = c(\alpha_x^{(s)}p_x^{(s)} + \alpha_y^{(s)}p_y^{(s)} + \alpha_z^{(s)}p_z^{(s)}) + m_s c^2\alpha_4^{(s)}, \tag{50}$$

we have the wave equation for the functional Ω in the form

$$\sum_{s=1}^{n}(D_s - \varepsilon_s\boldsymbol{\alpha}_s\cdot\mathbf{B}(\mathbf{r}_s, t))\Omega + \frac{1}{2}\sum_{uv}\frac{\varepsilon_u\varepsilon_v}{4\pi|\mathbf{r}_u - \mathbf{r}_v|}\Omega = i\hbar\frac{\partial\Omega}{\partial t}. \tag{51}$$

We should transform the operator containing the vector potential $\mathbf{B}$ and write it down in a more explicit form. Substituting equation (**II**, 30) for the amplitude $\mathbf{B}(\mathbf{k})$ in formula (24) for $\mathbf{B}(\mathbf{r}$,t) we get

$$\sum_{s=1}^{n}\varepsilon_s\boldsymbol{\alpha}_s\cdot\mathbf{B}(\mathbf{r}_s, t) = \sum_{j=1}^{2}\int(d\mathbf{k})\{G^\dagger(\mathbf{k}, j)b(\mathbf{k}, j) + $$
$$+ G(\mathbf{k}, j)b^\dagger(\mathbf{k}, j)\}, \tag{52}$$

where for brevity we put

$$G(\mathbf{k}, j)\frac{1}{(2\pi)^{3/2}}\sqrt{\frac{\hbar c}{2k}}\sum_{s=1}^{n}\varepsilon_s(\boldsymbol{\alpha}_s\cdot\mathbf{e}(\mathbf{k}, j))e^{ickt - i\mathbf{k}\mathbf{r}_s}. \tag{53}$$

Further, denoting by H the usual operator of energy for the n-body problem acting in the configuration space of n particles

$$H = \sum_{s=1}^{n} D_s + \frac{1}{2} \sum_{uv} \frac{\varepsilon_u \varepsilon_v}{4\pi |\mathbf{r}_u - \mathbf{r}_v|}, \tag{54}$$

we get the wave equation in the form

$$H\Omega - i\hbar \frac{\partial \Omega}{\partial t} = \sum_j \int (\mathbf{k}) \{ G^\dagger(\mathbf{k}, j) b(\mathbf{k}, j) + + G(\mathbf{k}, j) b^\dagger(\mathbf{k}, j) \} \Omega. \tag{55}$$

If we consider Ω as a functional of $\bar{b}(\mathbf{k}, j)$, the previous equation may be written as

$$H\Omega - i\hbar \frac{\partial \Omega}{\partial t} = \sum_j \int (d\mathbf{k}) \Big\{ G^\dagger(\mathbf{k}, j) \frac{\delta' \Omega}{\delta \bar{b}(\mathbf{k}, j)} + G(\mathbf{k}, j) \bar{b}(\mathbf{k}, j) \Omega \Big\}. \tag{56}$$

This is an explicit form of the wave equation for the functional Ω which, according to quantum electrodynamics, describes a system of n electrons together with an indefinite number of light quanta. Mathematically, wave equation (55) or (56) is an equivalent of other forms of the wave equation used in the literature but it is much simpler and easier to handle.

The coefficients G and $G^\dagger$ of the wave equation depend explicitly on time; but[19] this time dependence can be excluded by a canonical transformation. Thus, we get the representation of the operators, which corresponds to the one used by Heisenberg and Pauli.

If we denote by L^0 some operator in this representation and by L the same operator in the initial representation, such a canonical transformation reads as

$$L^0 = e^{-iwt} L e^{iwt}, \tag{57}$$

where w is the energy operator of light quanta divided by $\hbar$ (**II**, 34):

$$w = c \sum_j \int k b^\dagger(\mathbf{k}, j) b(\mathbf{k}, j) (d\mathbf{k}). \tag{58}$$

It follows from the commutation relations for the quantities $b(\mathbf{k}, j)$ and $b^\dagger(\mathbf{k}, j)$ that

$$e^{-iwt} b e^{iwt} = e^{ickt} b, \qquad e^{-iwt} b^\dagger e^{iwt} = e^{-ickt} b^\dagger. \tag{59}$$

[19]L. Rosenfeld, Zs. Phys. **76**, 729 (1932).

Thus, after the transformation the operator $b(\mathbf{k}, j)$ is multiplied by the factor e^{ickt}. But the same factor gives a time dependence of the quantity G defined by (53). We have

$$G(\mathbf{k}, j) = e^{ickt} G_0(\mathbf{k}, j). \tag{60}$$

Therefore, in the products $Gb^\dagger$ and $G^\dagger b$ entering the wave equation the factors $e^{\pm ickt}$ are canceled. Further, we have the relation

$$e^{-iwt}\Big(-i\hbar\frac{\partial}{\partial t}\Big)e^{iwt} = -i\hbar\frac{\partial}{\partial t} + \hbar w. \tag{61}$$

As a result the wave equation for the transformed functional, which we denote again as Ω (instead of Ω_0), takes the form

$$H\Omega + \hbar c\sum_j \int kb^\dagger b(d\mathbf{k})\Omega - i\hbar\frac{\partial\Omega}{\partial t} = \\ = \sum_j \int (d\mathbf{k})\{G_0^\dagger b + G_0 b^\dagger\}\Omega. \tag{62}$$

This form of the wave equation differs from (55) because, first, the energy of light quanta enters here explicitly and, second, the coefficients of the equation do not depend on time.

In the case of one electron, with the help of the analogous transformation

$$L^0 = e^{-iwt + i\mathbf{qr}} L e^{iwt - i\mathbf{qr}}, \tag{63}$$

where

$$\mathbf{q} = \sum_j \int d(\mathbf{k})\mathbf{k} b^\dagger(\mathbf{k}, j) b(\mathbf{k}, j), \tag{64}$$

we can also exclude the coordinates of the electron.[20] For the transformed wave functional Ω, we get

$$\{c\boldsymbol{\alpha}\cdot(\mathbf{p} - \hbar\mathbf{q}) + mc^2\alpha_4 + hw\}\Omega - i\hbar\frac{\partial\Omega}{\partial t} = \\ = \frac{\varepsilon}{(2\pi)^{3/2}}\sum_j \int (d\mathbf{k})\sqrt{\frac{\hbar c}{2k}}\boldsymbol{\alpha}\cdot\mathbf{e}(\mathbf{k}, j)\{b(\mathbf{k}, j) + b^\dagger(\mathbf{k}, j)\}\Omega. \tag{65}$$

Because of the term with $b^\dagger$ a paradoxical result follows from (65) that it is not possible to have a stationary state without light quanta, i.e.,

[20] W. Heisenberg, Zs. Phys. **65**, 4, 1930.

even an electron moving freely and uniformly must radiate. Taking this into account, it might have been physically more correct to omit the whole right-hand side of (65). Then the momentum conservation law in its usual form (the Compton effect) follows from the equation thus obtained:

$$\{c\boldsymbol{\alpha}(\mathbf{p}-\hbar\mathbf{q})+mc^2\alpha_4+\hbar w\}\Omega - i\hbar\frac{\partial\Omega}{\partial t}=0. \tag{66}$$

Equations in the Momentum Space of Light Quanta

For most applications, it is sufficient to consider a finite number of light quanta. In this case, it would be useful to pass to wave functions in the momentum space of light quanta. Since we consider a system consisting of light quanta and n particles (electrons), the wave functions in question

$$\psi_q=\psi_q(\mathbf{r}_1\zeta\ \dots\ \mathbf{r}_n\zeta;\ \mathbf{k}_1 j_1,\ \dots\ \mathbf{k}_q j_q) \tag{67}$$

contain together with variables $(\mathbf{k}_u j_u)$ of light quanta the variables $(\mathbf{r}_s\zeta_s)$ of all n electrons. These wave functions are symmetric with respect to the variables $(\mathbf{k}_u j_u)$ and asymmetric with respect to the variables $(\mathbf{r}_s\zeta_s)$. Further, for brevity we omit electron variables $(\mathbf{r}_s\zeta_s)$ and write

$$\psi_q=\psi_q(\mathbf{k}_1 j_1,\ \dots\ \mathbf{k}_q j_q) \tag{67*}$$

instead of (67). The expansion of the wave functional Ω in terms of the eigenfunctionals Ω_q for the number of the light quanta operator

$$N=\sum_j\int(d\mathbf{k})b^\dagger(\mathbf{k},j)b(\mathbf{k},j) \tag{68}$$

has the form

$$\Omega=\sum_{q=0}^{\infty}\Omega_q, \tag{69}$$

where we denote

$$\Omega_q=\frac{1}{\sqrt{q!}}\sum_{j_1\cdots j_q}\int(d\mathbf{k}_1)\dots(d\mathbf{k}_q)\psi_q(\mathbf{k}_1 j_1\dots\mathbf{k}_q j_q)\bar{b}(\mathbf{k}_1 j_1)\dots\bar{b}(\mathbf{k}_q j_q). \tag{70}$$

If we truncate expansion (69) by a finite number of terms, it means that we restrict ourselves by considering a finite number of light quanta. Substituting series (69) in wave equation (62) we get the following equations for wave functions (67):

$$
\begin{aligned}
&H\psi_q + \hbar c(k_1 + \ldots + k_q)\psi_q - i\hbar\frac{\partial\psi_q}{\partial t} = \\
= &\sqrt{q+1}\sum_j \int (d\mathbf{k})G_0^\dagger(\mathbf{k}j)\psi_{q+1}(\mathbf{k}j, \mathbf{k}_1 j_1, \ldots, \mathbf{k}_q j_q) + \\
&+\frac{1}{\sqrt{q}}\{G_0(\mathbf{k}_1 j_1)\psi_{q-1}(\mathbf{k}_2 j_2 \ldots \mathbf{k}_q j_q) + \ldots \\
&+G_0(\mathbf{k}_q j_q)\psi_{q-1}(\mathbf{k}_1 j_1 \ldots \mathbf{k}_{q-1} j_{q-1})\}. \qquad (71)
\end{aligned}
$$

Certainly, the same equations could have been obtained in a more formal way if in (62) we make a transition from the second quantization of operators to the configuration space considered in Section 1 (here, to the momentum space).

We will not write down here the system of equations that arises from wave equation (55) or (56), for it obviously follows from (21) by canceling out the terms with the energy of light quanta and replacing G_0 by G.

Some Applications[21]

If we truncate expansion (69) already at the first term, i.e., suppose simply $\Omega = \Omega_0 + \Omega_1$, then for ψ_0 and ψ_1 we have the equations

$$H\psi_0 - i\hbar\frac{\partial\psi_0}{\partial t} = \sum_j \int (d\mathbf{k})G_0^\dagger(\mathbf{k}, j)\psi_1(\mathbf{k}, j), \qquad (72)$$

$$H\psi_1 + \hbar ck\psi_1 - i\hbar\frac{\partial\psi_1}{\partial t} = G_0(\mathbf{k}, j)\psi_0. \qquad (72^*)$$

The Breit and Möller formulae. Let us show that this system of equations contains both Breit and Möller formulae.

We will derive first the Breit formula. To this end, let us suppose that in (72*) we can omit the terms $H\psi_1 - i\hbar\frac{\partial\psi_1}{\partial t}$ in comparison with $\hbar k\psi_1$. Then we get

$$\psi_1(\mathbf{k}, j) = \frac{1}{\hbar ck}G_0(\mathbf{k}, j)\psi_0. \qquad (73)$$

[21]After the general form of the functional as infinite series (6) is established, the below method based on considering the finite number of the series terms seems very natural. In the works of I.E. Tamm (1945 and further) this method was applied to the meson theory and named as "the method of truncating equations by particle number." In the literature it is also known as "the Tamm–Dankov method." (*V. Fock*)

Substituting this value of ψ_1 in (72) and calculating the integral

$$\sum_j \int (d\mathbf{k}) \frac{1}{\hbar ck} G^\dagger(\mathbf{k}, j) G(\mathbf{k}, j) =$$

$$= E_0 + \sum_{u>v} \frac{\varepsilon_u \varepsilon_v}{8\pi} \left\{ \frac{\boldsymbol{\alpha}_u \cdot \boldsymbol{\alpha}_v}{|\mathbf{r}_u - \mathbf{r}_v|} + \frac{[\boldsymbol{\alpha}_u \cdot (\mathbf{r}_u - \mathbf{r}_v)][\boldsymbol{\alpha}_v \cdot (\mathbf{r}_u - \mathbf{r}_v)]}{|\mathbf{r}_u - \mathbf{r}_v|^3} \right\}, \quad (74)$$

we immediately arrive at the famous Breit formula. (The infinite self-energy E_0, which corresponds to the terms proportional to the square of charges, should be omitted now.)

Let us recall here that, as Breit showed, expression (74) should be considered as a perturbation energy and be taken into account only in the first approximation; transition to the second approximation leads to wrong results.

Based on another principle, the approximation method where higher powers of charges are neglected leads to the Möller formula.[22]

As is generally known, for deriving this formula we need to take as a zero approximation for ψ_0 the expression corresponding to a free motion of two electrons and then to calculate the matrix element of the interaction energy.

We shall use the representation of the wave functions in the momentum space. We take the function ψ_0 to be equal to

$$\psi_0 = e^{-\frac{i}{\hbar}Wt} \psi_0{}^0 = e^{-\frac{i}{\hbar}Wt} \delta(\mathbf{p}_1 - \mathbf{p}_1{}^0) \delta(\mathbf{p}_2 - \mathbf{p}_2{}^0) \psi(\mathbf{p}_1{}^0, \mathbf{p}_2{}^0), \quad (75)$$

where $\psi(\mathbf{p}_1, \mathbf{p}_2)$ is the general solution of the equations

$$D_1\psi = W_1\psi; \qquad D_2\psi = W_2\psi. \quad (76)$$

At a fixed momentum, each of these equations has four solutions. Therefore, we should, properly speaking, write in detail $\psi(\mathbf{p}_1, \mathbf{p}_2, s_1, s_2)$ instead of $\psi(\mathbf{p}_1, \mathbf{p}_2)$, where the variable $s = 1, 2, 3, 4$ marks the number of a solution. Nevertheless, we shall omit the variable s for brevity. The quantity in the exponent in (75) is a sum of the kinetic energies of both particles

$$W = W_1 + W_2. \quad (77)$$

We denote quantities W_1, W_2, D_1, D_2, W, corresponding to momenta $\mathbf{p}_1{}^0$ and $\mathbf{p}_2{}^0$, as $W_1{}^0, W_2{}^0$, etc.

[22] C. Möller, Zs. Phys. **70**, 786 (1931).

In the Möller formula only that matrix element of interaction energy enters that corresponds to the conservation law $W = W^0$. Thus, we can put in our formulae

$$W = W_1 + W_2 = W_1{}^0 + W_2{}^0 = W^0 . \tag{78}$$

The dependence of $\psi_1(\mathbf{k}, j)$ on time is obviously the same as of ψ_0. We have

$$\psi_1(\mathbf{k}, j) = e^{-\frac{i}{\hbar}Wt}\psi_1{}^0(\mathbf{k}, j). \tag{79}$$

Substituting this expression in (74) and neglecting there the Coulomb terms, we get

$$(D_1 + D_2 + \hbar ck - W)\psi_1{}^0(\mathbf{k}, j) = G_0(\mathbf{k}, j)\psi_0{}^0 =$$

$$= \frac{1}{(2\pi)^{3/2}}\sqrt{\frac{\hbar c}{2k}}\{\varepsilon_1(\boldsymbol{\alpha}_1 \cdot \mathbf{e}(\mathbf{k}, j))\delta(\mathbf{p}_1 + h\mathbf{k} - \mathbf{p}_1{}^0)\delta(\mathbf{p}_2 - \mathbf{p}_2{}^0) +$$

$$+\varepsilon_2(\boldsymbol{\alpha}_2 \cdot \mathbf{e}(\mathbf{k}, j))\delta(\mathbf{p}_1 - \mathbf{p}_1{}^0)\delta(\mathbf{p}_2 + h\mathbf{k} - \mathbf{p}_2{}^0)\}\psi(\mathbf{p}_1{}^0, \mathbf{p}_2{}^0). \tag{80}$$

We solve this equation with respect to $\psi_1{}^0(\mathbf{k}, j)$. Solving becomes easier because the function ψ satisfies equations (76). We obtain:

$$\psi_1{}^0(\mathbf{k}, j) = \frac{1}{2\pi^{3/2}}\sqrt{\frac{\hbar c}{2k}} \cdot$$

$$\cdot\{(D_1 + \hbar ck - W_1{}^0)^{-1}\varepsilon_1(\boldsymbol{\alpha}_1\mathbf{e})\delta(\mathbf{p}_1 + h\mathbf{k} - \mathbf{p}_1{}^0)\delta(\mathbf{p}_2 - \mathbf{p}_2{}^0) +$$

$$+(D_2 + \hbar ck - W_2{}^0)^{-1}\varepsilon_2(\boldsymbol{\alpha}_2\mathbf{e})\delta(\mathbf{p}_1 - \mathbf{p}_1{}^0)\delta(\mathbf{p}_2 + h\mathbf{k} - \mathbf{p}_2{}^0)\} \cdot$$

$$\cdot\psi(\mathbf{p}_1{}^0, \mathbf{p}_2{}^0). \tag{81}$$

The desirable matrix element of the interaction energy is a sum of the matrix element

$$\langle \mathbf{p}_1, \mathbf{p}_2 | U' | \mathbf{p}_1{}^0, \mathbf{p}_2{}^0 \rangle =$$

$$= \frac{\varepsilon_1\varepsilon_2}{\hbar(2\pi)^3} \cdot \delta(\mathbf{p}_1 + \mathbf{p}_2 - \mathbf{p}_1{}^0 - \mathbf{p}_2{}^0)\frac{\overline{\psi}(\mathbf{p}_1, \mathbf{p}_2)\psi(\mathbf{p}_1{}^0, \mathbf{p}_2{}^0)}{|\mathbf{p}_1 - \mathbf{p}_1{}^0|^2} \tag{82}$$

for the Coulomb energy, which already entered the operator H (see (72*)) and the term in the expression

$$-\overline{\psi}(\mathbf{p}_1, \mathbf{p}_2)\sum_j \int (d\mathbf{k})G_0^\dagger(\mathbf{k}j)\psi_1{}^0(\mathbf{k}, j) \tag{83}$$

proportional to the product $\varepsilon_1\varepsilon_2$. The terms proportional to ${\varepsilon_1}^2$ and ${\varepsilon_2}^2$ are infinite and must be omitted. After some calculations, where (76), (78) and the relation

$$\sum_j [\boldsymbol{\alpha}_1 \cdot \mathbf{e}(\mathbf{k}, j)]\ [\boldsymbol{\alpha}_2 \cdot \mathbf{e}(\mathbf{k}, j)] = \boldsymbol{\alpha}_1 \cdot \boldsymbol{\alpha}_2 - \frac{1}{k^2}(\boldsymbol{\alpha}_1 \cdot \mathbf{k})(\boldsymbol{\alpha}_2 \cdot \mathbf{k}) \tag{84}$$

are used, we get the following expression for the term proportional to $\varepsilon_1\varepsilon_2$:

$$\langle \mathbf{p}_1, \mathbf{p}_2 | U'' | {\mathbf{p}_1}^0, {\mathbf{p}_2}^0 \rangle =$$

$$= -\frac{\varepsilon_1\varepsilon_2}{\hbar(2\pi)^3} \cdot \frac{\delta(\mathbf{p}_1 + \mathbf{p}_2 - {\mathbf{p}_1}^0 - {\mathbf{p}_2}^0)}{(\mathbf{p}_1 - {\mathbf{p}_1}^0)^2 - \frac{1}{c^2}(W_1 - {W_1}^0)^2} \cdot \tag{85}$$

$$\cdot \overline{\psi}(\mathbf{p}_1, \mathbf{p}_2) \Big\{ \boldsymbol{\alpha}_1 \cdot \boldsymbol{\alpha}_2 - \frac{[\boldsymbol{\alpha}_1 \cdot (\mathbf{p}_1 - {\mathbf{p}_1}^0)]\ [\boldsymbol{\alpha}_2 \cdot (\mathbf{p}_1 - {\mathbf{p}_1}^0)]}{|\mathbf{p}_1 - {\mathbf{p}_1}^0|^2} \Big\} \psi({\mathbf{p}_1}^0, {\mathbf{p}_2}^0).$$

By virtue of the momentum (the function δ as a factor) and energy (78) conservation law, we can replace here $\mathbf{p}_1 - {\mathbf{p}_1}^0$ by $\mathbf{p}_2 - {\mathbf{p}_2}^0$ and $W_1 - {W_1}^0$ by $W_2 - {W_2}^0$. It also follows from relation (76) that

$$\overline{\psi}(\mathbf{p}_1, \mathbf{p}_2)\ [\boldsymbol{\alpha}_1 \cdot (\mathbf{p}_1 - {\mathbf{p}_1}^0)]\ [\boldsymbol{\alpha}_2 \cdot (\mathbf{p}_2 - {\mathbf{p}_2}^0)]\ \psi({\mathbf{p}_1}^0, {\mathbf{p}_2}^0) =$$

$$= \frac{1}{c^2}(W_1 - {W_1}^0)(W_2 - {W_2}^0)\overline{\psi}(\mathbf{p}_1, \mathbf{p}_2)\psi({\mathbf{p}_1}^0, {\mathbf{p}_2}^0). \tag{86}$$

Taking into account (78) and (76) for the sum of the matrix elements (82) and (85), we get

$$\langle \mathbf{p}_1, \mathbf{p}_2 | U | {\mathbf{p}_1}^0, {\mathbf{p}_2}^0 \rangle = \frac{\varepsilon_1\varepsilon_2}{\hbar(2\pi)^3} \delta(\mathbf{p}_1 + \mathbf{p}_2 - {\mathbf{p}_1}^0 - {\mathbf{p}_2}^0) \cdot$$

$$\cdot \frac{\overline{\psi}(\mathbf{p}_1, \mathbf{p}_2)(1 - \boldsymbol{\alpha}_1 \cdot \boldsymbol{\alpha}_2)\psi({\mathbf{p}_1}^0, {\mathbf{p}_2}^0)}{(\mathbf{p}_1 - {\mathbf{p}_1}^0)^2 - \frac{1}{c^2}(W_1 - {W_1}^0)^2}. \tag{87}$$

This is the formula proposed by Möller for the matrix element of the two-electron interaction energy. Thus, we showed that this formula like the Breit formula follows from our system of equations (72). However, while deriving both these formulae, we had to throw out some infinite terms.

The natural width of spectral lines. As the final example, let us consider a derivation of the famous formulae for the natural width of spectral lines. In this case, we again can base our reasoning on (72). For the present, H is the operator of the energy of an atom; the field of the nucleus is included in H. Let us denote the eigenfunctions by u_n and eigenvalues by E_n:

$$Hu_n = E_n u_n. \tag{88}$$

We expand entering (72) the functions ψ_0 and $\psi_1(\mathbf{k}, j)$ in terms of u_n

$$\psi_0 = \sum_n a_n u_n, \tag{89}$$

$$\psi_1(\mathbf{k}, j) = \sum_n c_n(\mathbf{k}, j) u_n. \tag{90}$$

Further, we restrict ourselves by considering only one term in expansions (89) and (90); but at first let us write down the complete expansions. We have to calculate a matrix element of the operator G_0 entering (72). Doing this, we suppose that the wavelength of considered light is large as compared to atomic sizes. We may replace then in the expression for G_0 all the factors $e^{-i\mathbf{k}\mathbf{r}_s}$ by a factor $e^{-i\mathbf{k}\mathbf{r}}$, where $\mathbf{r}$ is, say, the radius vector of atom's centre of mass. We have then

$$\langle n|G_0|n'\rangle = e^{-i\mathbf{k}\mathbf{r}} \frac{1}{(2\pi)^{3/2}} \sqrt{\frac{\hbar}{2ck}} \dot{\mathbf{D}}_{nn'} \cdot \mathbf{e}(\mathbf{k}, j), \tag{91}$$

where $\dot{\mathbf{D}}_{nn'}$ means the matrix element for the time derivative of an atomic electric moment. Introducing expansions (89) and (90) and also quantities (91) in formulae (72), we get the system of equations

$$E_n a_n - i\hbar \dot{a}_n =$$
$$= \frac{1}{(2\pi)^{3/2}} \sum \int (d\mathbf{k}) \sqrt{\frac{\hbar}{2ck}} e^{i\mathbf{r}\mathbf{k}} \sum_{n'} [\mathbf{e}(\mathbf{k}, j) \cdot \dot{\mathbf{D}}_{nn'}] c_{n'}(\mathbf{k}, j), \tag{92}$$

$$(E_n + \hbar ck) c_n(\mathbf{k}, j) - i\hbar c_n(\mathbf{k}, j) =$$
$$= \frac{1}{(2\pi)^{3/2}} \sqrt{\frac{\hbar}{2ck}} e^{-i\mathbf{k}\mathbf{r}} \sum_{n'} [\mathbf{e}(\mathbf{k}, j) \cdot \dot{\mathbf{D}}_{nn'}] a_{n'}. \tag{92*}$$

We restrict ourselves now to the consideration of two following states:
1. The excited state of an atom ($n = 2$) without light quanta;
2. The ground state of an atom ($n = 1$) with one light quantum.
Therefore, we assume all the quantities a_n and c_n except a_2 and $c_1(\mathbf{k}, j)$

to be zeros. Since the diagonal matrix elements $\mathbf{D}_{nn'}$ are zeros, in our case equations (92) take the form

$$E_2 a_2 - i\hbar \dot{a}_2 = \frac{1}{(2\pi)^{3/2}} \sum_j \int (d\mathbf{k}) \sqrt{\frac{\hbar}{2ck}} e^{i\mathbf{kr}} \mathbf{e}(\mathbf{k}, j) \cdot \dot{\mathbf{D}}_{21} c_1(\mathbf{k}, j), \quad (93)$$

$$(E_1 + \hbar ck) c_1(\mathbf{k}, j) - i\hbar \dot{c}_1(\mathbf{k}, j) = \frac{1}{(2\pi)^{3/2}} \sqrt{\frac{\hbar}{2ck}} e^{-i\mathbf{kr}} \cdot \dot{\mathbf{D}}_{12} a_2. \quad (93^*)$$

As the initial conditions, we take

$$a_2 = 1; \quad c_1(\mathbf{k}, j) = 0 \qquad \text{at} \quad t = 0 \quad (94)$$

and put the quantity a_2 equal to

$$a_2 = e^{-\frac{i}{\hbar} E_2 t - \gamma t}, \quad (95)$$

where the constant γ has to be determined from the system of equations (93). If we substitute expression (95) in (93*), then for $c_1(\mathbf{k}, j)$ we get a differential equation, the solution of which, satisfying the initial conditions (94), has the form

$$c_1(\mathbf{k}, j) = \frac{1}{(2\pi)^3} \cdot \frac{1}{\sqrt{2\hbar ck}} e^{-i\mathbf{kr}} \mathbf{e}(\mathbf{k}, j) \cdot \dot{\mathbf{D}}_{12} \cdot \cdot e^{-\frac{i}{\hbar}(E_0 + \hbar ck)t} \cdot \frac{e^{i(ck-\omega)t - \gamma t} - 1}{ck - \omega + i\gamma}. \quad (96)$$

Now we should substitute expressions (95) and (96) for a_2 and for $c_1(\mathbf{k}, j)$ in equation (93). We obtain

$$i\hbar\gamma = \frac{1}{(2\pi)^3} \frac{1}{2c} \sum_j \int \frac{(d\mathbf{k})}{k} (\mathbf{e}(\mathbf{k}, j) \cdot \dot{\mathbf{D}}_{12})(\mathbf{e}(\mathbf{k}, j) \dot{\mathbf{D}}_{12}) \cdot \cdot \frac{1 - e^{-i(ck-\omega)t - \gamma t}}{ck - \omega + i\gamma}. \quad (97)$$

Obviously, this equation cannot be satisfied rigorously; indeed, its right-hand side depends on t and at $t = 0$ vanishes, whereas its left-hand side is constant. Moreover, it is not possible to extend the integral with respect to the wave vector of the light quantum over the whole momentum space (because then it diverges and the suggestion made while deriving formula (91) is not implemented). We should extend this integral only over the region of the resonance $k = \frac{\omega}{c}$. Taking this into account, one can show that for large values of ωt the right-hand side of equation (97) is, indeed, nearly constant. Integrating (to be more exact, averaging) over all directions of the wave vector gives

$$\text{Average value}\left(\sum_j (\mathbf{e}(\mathbf{k},j)\dot{\mathbf{D}}_{12})(\mathbf{e}(\mathbf{k},j)\mathbf{D}_{12})\right) = \frac{2}{3}|\dot{\mathbf{D}}_{12}|^2. \tag{98}$$

Therefore, equation (97) takes the form

$$i\hbar\gamma = \frac{1}{(2\pi)^3}\frac{4\pi}{3c}|\dot{\mathbf{D}}_{12}|^2 \cdot \int kdk \frac{1-e^{-i(ck-\omega)t+\gamma t}}{ck-\omega+i\gamma}. \tag{99}$$

Replacing here the factor k in the integrand by $\frac{\omega}{c}$ and applying the formula

$$\lim_{X\to\infty}\int_{-X}^{+X}\frac{dx}{x+iy}(1-e^{-i(x+iy)}) = \pi i, \tag{100}$$

we obtain for γ

$$\gamma = \frac{1}{2\pi}\frac{\omega}{3\hbar c^3}|\dot{\mathbf{D}}_{12}|^2 = \frac{1}{2\pi}\frac{\omega^3}{3\hbar c^3}|\mathbf{D}_{12}|^2. \tag{101}$$

Coming from the Heaviside units used here for the electric moment to the usual electrostatic units, we obtain an expression for γ coinciding with the usual one.

In the simplest case considered the calculations were carried out with great detail. We did it to show by an example how to apply the mathematical approach proposed in this paper. In the same way, one can carry out the calculations for more complicated cases when several light quanta are considered. Then in expansion (69) of the wave functional we should retain several terms and in equation (71) take several wave functions ψ_q. On the other hand, if it is desirable to take into account an uncertain number of light quanta with fixed frequencies, it is possible to introduce quantities b_j by formulae (**II**, 37) and replace the functional derivatives by the usual ones.

The mathematical approach developed here is sufficiently flexible to cover all the problems to which quantum electrodynamics is applicable at all.

Translated by Yu. V. Novozhilov

35-1
Hydrogen Atom and Non-Euclidean Geometry[1]

V.A. Fock

Received on 3 April 1935

Izv AN, 169, 1935,
Zs. Phys., **98**, N 3–4, 145, 1935

As is known, the hydrogen atom energy levels depend on the principal quantum number n only and are independent of the azimuthal quantum number l. One can say (by using the common but not very suitable term) that there is degeneracy (i.e., the multiplicity of the level) with respect to the azimuthal quantum number. On the other hand, there is a general rule according to which the multiplicity of the Schrödinger equation eigenvalues relates to the invariance of the equation under a definite group of transformations. So, for example, the invariance under a common rotation group (spherical symmetry) leads to the energy level independence on the magnetic quantum number m. Therefore, it should be expected that the energy level independence of the azimuthal quantum number is explained by the existence of a certain transformation group that is more general than the three-dimensional rotation group. Up to now, this group of transformations for the Schrödinger equation has not been found. In the present work, we shall show that this group is equivalent to the rotation group in the four-dimensional Euclidean space.

1. Let us write the Schrödinger equation for a hydrogen atom in the momentum representation. Since the Coulomb potential energy operator $-\frac{Ze^2}{r}$ in the momentum representation is the integral operator, the Schrödinger equation will be an integral equation of the form

$$\frac{1}{2m}\mathbf{p}^2\psi(\mathbf{p}) - \frac{Ze^2}{2\pi^2 h}\int \frac{\psi(\mathbf{p}')(d\mathbf{p}')}{\mid \mathbf{p}-\mathbf{p}' \mid^2} = E\psi(\mathbf{p}), \qquad (1)$$

[1]Reported on February 8, 1935 at a theoretical seminar of the Physical Institute of Leningrad State University, and on March 23, 1935 at the Session of the USSR Academy of Sciences in Moscow.

where $(d\mathbf{p}') = dp'_x dp'_y dp'_z$ is an elementary volume in the momentum space. First we consider the discrete spectrum, for which the energy E is negative, and denote mean-square momentum as p_0:

$$p_0 = \sqrt{-2mE}. \tag{2}$$

We shall treat components of momentum $\mathbf{p}$ divided by p_0 as rectangular coordinates on the hyperplane which is stereographic a projection of a sphere in the four-dimensional Euclidean space. The rectangular coordinates of a certain point on the sphere will be

$$\begin{aligned}
\xi &= \frac{2p_0 p_x}{p_0^2 + p^2} = \sin\alpha \sin\vartheta \cos\varphi, \\
\eta &= \frac{2p_0 p_y}{p_0^2 + p^2} = \sin\alpha \sin\vartheta \sin\varphi, \\
\zeta &= \frac{2p_0 p_z}{p_0^2 + p^2} = \sin\alpha \cos\vartheta, \\
\chi &= \frac{p_0^2 - p^2}{p_0^2 + p^2} = \cos\alpha,
\end{aligned} \tag{3}$$

being

$$\xi^2 + \eta^2 + \zeta^2 + \chi^2 = 1. \tag{3*}$$

The angles α, θ, and φ are spherical coordinates of a point on a hypersphere. At the same time, angles θ and φ are ordinary spherical coordinates characterizing the momentum direction. The surface element of a hypersphere is equal to

$$d\Omega = \sin^2 \alpha d\alpha \sin\vartheta d\vartheta d\varphi. \tag{4}$$

It is related to the volume element in the momentum space by

$$(d\mathbf{p}') = dp'_x dp'_y dp'_z = p^2 dp \sin\vartheta d\vartheta d\varphi = \frac{1}{8p_0^3}(p_0^2 + p^2)^3 d\Omega. \tag{5}$$

Let us denote for brevity

$$\lambda = \frac{Zme^2}{hp_0} = \frac{Zme^2}{h\sqrt{-2mE}} \tag{6}$$

and introduce instead of $\psi(\mathbf{p})$ the function

$$\Psi(\alpha, \vartheta, \varphi) = \frac{\pi}{\sqrt{8}} p_0^{-\frac{5}{2}} (p_0^2 + p^2)^2 \psi(\mathbf{p}). \tag{7}$$

Here the factor is chosen to fulfill the normalization condition[2]

$$\frac{1}{2\pi^2}\int |\,\Psi(\alpha,\vartheta,\varphi)\,|^2\,d\Omega = \int \frac{p^2+p_0^2}{2p_0^2}\,|\,\psi(\mathbf{p})\,|^2\,(d\mathbf{p}) =$$

$$= \int |\,\psi(\mathbf{p})\,|^2\,(d\mathbf{p}) = 1. \tag{*}$$

In new notations, the Schrödinger equation (1) takes the form

$$\Psi(\alpha,\vartheta,\varphi) = \frac{\lambda}{2\pi^2}\int \frac{\Psi(\alpha',\vartheta',\varphi')}{4\sin^2\frac{\omega}{2}}d\Omega', \tag{8}$$

where $2\sin\frac{\omega}{2}$ is the length of a chord and ω is the length of a great circle arc joining the points α, θ, φ and α', θ', φ' on the four-dimensional sphere, so that

$$4\sin^2\frac{\omega}{2} = (\xi-\xi')^2+(\eta-\eta')^2+(\zeta-\zeta')^2+(\chi-\chi')^2 \tag{9}$$

or

$$\cos\omega = \cos\alpha\cos\alpha' + \sin\alpha\sin\alpha'\cos\gamma, \tag{10}$$

$\cos\gamma$ being of common value

$$\cos\gamma = \cos\vartheta\cos\vartheta' + \sin t\sin\vartheta'\cos(\varphi-\varphi'). \tag{10*}$$

Equation (8) is nothing else but an integral equation for spherical functions of the four-dimensional sphere. To prove this, we need to recall some basic notions of the four-dimensional potential theory. Let us put

$$x_1 = r\xi, \qquad x_2 = r\eta, \qquad x_3 = r\zeta, \qquad x_4 = r\chi \tag{11}$$

and consider the Laplace equation

$$\frac{\partial^2 u}{\partial x_1^2}+\frac{\partial^2 u}{\partial x_2^2}+\frac{\partial^2 u}{\partial x_3^2}+\frac{\partial^2 u}{\partial x_4^2} = 0. \tag{12}$$

Let us introduce the function

$$G = \frac{1}{2R^2}+\frac{1}{2R_1^2}, \tag{13}$$

where R^2 and R_1^2 are

$$R^2 = r^2 + r'^2 - 2rr'\cos\omega; \qquad R_1^2 = 1 - 2rr'\cos\omega + r^2r'^2. \tag{14}$$

[2]Recall that the four-dimensional sphere surface area is equal to $2\pi^2$ so that, e.g., the function $\Psi = 1$ fits this condition. (*V. Fock*)

This function can be called the Green's function of the third kind because it satisfies the boundary condition on the sphere surface

$$\frac{\partial G}{\partial r'} + G = 0 \quad \text{at} \quad r' = 1. \tag{15}$$

By the Green theorem the function, which is harmonic inside the sphere, can be expressed through the boundary value of $u + \frac{\partial u}{\partial r}$ as

$$u(x_1, x_2, x_3, x_4) = \frac{1}{2\pi^2} \int \left(u + \frac{\partial u}{\partial r'} \right)_{r'=1} G d\Omega'. \tag{16}$$

Let u be a homogeneous harmonic polynomial of degree $n-1$

$$u(x_1, x_2, x_3, x_4) = r^{n-1} \Psi_n(\alpha, \vartheta, \varphi) \quad (n = 1, 2, 3, \ldots). \tag{17}$$

Then we shall have

$$\left(u + \frac{\partial u}{\partial r} \right)_{r=1} = nu = n\Psi_n(\alpha, \vartheta, \varphi). \tag{18}$$

Putting these expressions into (16) and using (13) and (14) for $r' = 1$, one gets

$$r^{n-1} \Psi_n(\alpha, \vartheta, \varphi) = \frac{n}{2\pi^2} \int \frac{\Psi_n(\alpha', \vartheta', \varphi')}{1 - 2r\cos\omega + r^2} d\Omega'. \tag{19}$$

This equation is valid also for $r = 1$ and in this case can be reduced to (8); the parameter λ is equal to integer n

$$\lambda = \frac{Zme^2}{h\sqrt{-2mE}} = n, \quad (n = 1, 2, 3, \ldots), \tag{20}$$

and obviously is the principal quantum number.

Hence, we showed the solution of the Schrödinger equation to be the four-dimensional spherical function.[3] So we found the Schrödinger equation transformation group as well. Obviously, this group is equivalent to the four-dimensional rotation group.

[3] Applications of four-dimensional spherical functions to the theory of a spherical top has been given by Hund [F. Hund, Göttinger Nachr., M.-Phys. Kl., 1927, 465]. (*V. Fock*)

2. We will give the following explicit expression for the four-dimensional spherical function

$$\Psi_{nlm}(\alpha,\vartheta,\varphi) = \Pi_l(n,\alpha)Y_{lm}(\vartheta,\varphi), \tag{21}$$

where l and m have the usual meaning of azimuthal and magnetic quantum numbers and $Y_{lm}(\theta,\varphi)$ is a standard spherical function normalized by the condition

$$\frac{1}{4\pi}\int_0^\pi \sin\vartheta d\vartheta \int_0^{2\pi} \mid \Psi_{lm}(\vartheta,\varphi) \mid^2 d\varphi = 1. \tag{22}$$

If we set

$$M_l^2 = n^2(n^2-1)\dots(n^2-l^2), \tag{23}$$

then the function normalized according to

$$\frac{2}{\pi}\int_0^\pi \Pi_l^2(n,\alpha)\sin^2\alpha d\alpha = 1 \tag{24}$$

can be represented either in the integral form

$$\Pi_l(n,\alpha) = \frac{M_l}{\sin^{l+1}\alpha}\int_o^\alpha \cos n\beta \cdot \frac{(\cos\beta - \cos\alpha)^l}{l!} d\beta, \tag{25}$$

or as a derivative

$$\Pi_l(n,\alpha) = \frac{\sin^l\alpha}{M_l}\frac{d^{l+1}\cos n\alpha}{d\cos\alpha^{l+1}}. \tag{25*}$$

At $l=0$, both formulae give

$$\Pi_0(n,\alpha) = \frac{\sin n\alpha}{\sin\alpha}. \tag{26}$$

Note that these formulae give the determination of the function $\Pi_l(n,\alpha)$ also valid for complex n (continuous spectrum). The function Π_l satisfies the following relations:

$$-\frac{d\Pi_l}{d\alpha} + l\cot\alpha\,\Pi_l = \sqrt{n^2-(l+1)^2}\,\Pi_{l+1}; \tag{27a}$$

$$\frac{d\Pi_l}{d\alpha} + (l+1)\cot\alpha\,\Pi_l = \sqrt{n^2-1)}\,\Pi_{l-1}, \tag{27b}$$

where from a differential equation it follows[4]

$$\frac{d^2\Pi_l}{d\alpha^2} + 2\cot\alpha\frac{d\Pi_l}{d\alpha} - \frac{l(l+1)}{\sin^2\alpha}\Pi_l + (n^2-1)\Pi_l = 0. \tag{28}$$

3. Let us turn now to the derivation of the addition theorem for four-dimensional spherical functions. Equation (19) is an identity relative to r. If we expand the integrand as a power series in r,

$$\frac{1}{1-2r\cos\omega+r^2} = \sum_{k=1}^{\infty} r^{k-1}\frac{\sin k\omega}{\sin\omega}, \tag{29}$$

and compare the coefficients in the right- and left-hand sides of (19), we get

$$\frac{n}{2\pi^2}\int \Psi_n(\alpha',\vartheta',\varphi')\frac{\sin k\omega}{\sin\omega}d\Omega' = \delta_{kn}\Psi_n(\alpha,\vartheta,\varphi). \tag{30}$$

But the quantity $n\frac{\sin n\omega}{\sin\omega}$ as a function of α, θ, φ is a four-dimensional spherical function that can be expanded in series of functions $\Psi_{nlm}(\alpha', \theta',\varphi')$. The expansion coefficients are determined by relation (30) (at $k=n$). Thus, we arrive at the addition theorem

$$n\frac{\sin n\omega}{\sin\omega} = \sum_{l=0}^{n-1}\sum_{m=-l}^{l}\overline{\Psi}_{nlm}(\alpha,\vartheta,\varphi)\Psi_{nlm}(\alpha',\vartheta',\varphi'). \tag{31}$$

If we substitute here relation (21) and use the well-known addition theorem for three-dimensional spherical functions, then we can rewrite (31) in the form

$$n\frac{\sin n\omega}{\sin\omega} = \sum_{l=o}^{\infty}\Pi_l(n,\alpha)\Pi_l(n,\alpha')(2l+1)P_l(\cos\gamma), \tag{32}$$

where P_l is the Legendre polynomial, and $\cos\gamma$ is determined by (10*). We wrote the summation limits as 0 and ∞. For integer n, series (32) is truncated at $l=n-1$, but it is possible to prove that addition theorem (32) is also valid for the complex values of n.

[4]In his work on the wave equation for the Kepler problem, Hylleraas [E. Hylleraas, Zs. Phys. **74**, 216, 1932] derived a differential equation [see (9g) and (10b) of his paper], which can be easily reduced to that for the four-dimensional spherical functions in stereographic projection. [With the kind permission of E. Hylleraas, we correct here the following misprints in his works: in the last term of equation (9f) as well as of (9g) the magnitude E must have a factor 4]. (*V. Fock*)

4. We investigated integral equation (1) for the case of a discrete spectrum. For a continuous spectrum one needs to consider (instead of a four-dimensional sphere) a two-sheet hyperboloid in the four-dimensional pseudo-Euclidean space. One sheet of the hyperboloid corresponds to the momentum inside the interval $0 < p < \sqrt{2mE}$ and the other one corresponds to the interval $\sqrt{2mE} < p < \infty$. In this case, the Schrödinger equation can be written as a system of two integral equations connecting the values of the desirable functions on both sheets of the hyperboloid.

We can give the following geometrical interpretation of our results. In the momentum space in the case of a discrete spectrum, there is *the Riemann geometry* with the constant positive curvature and in the case of a continuous spectrum there is *the Lobachevskii geometry* with the constant negative curvature.

The geometrical interpretation of the Schrödinger equation for a continuous spectrum is less descriptive than that for the case of a discrete spectrum. Therefore, it is reasonable first to obtain the formulae for the point spectrum and only in the final result take the principal quantum number to be imaginary. This can be done because $\Pi_l(n, \alpha)$ are analytical functions of n and α, which for the complex values of n and α differ from those for a discrete spectrum by constant factors only.[5]

5. Now let us list briefly the problems to which the presented theory of hydrogen-like atoms can be applied.[6] In many applications, e.g., in the theory of the Compton effect by bounded electrons [2, 3] as well as in the theory of inelastic electron collisions by atoms [4], in the mathematical sense the problem is reduced to the following. The projection norm [5] of the given function φ on the subspace of the Hilbert space, characterized by the principal quantum number n, needs to be found. In other words, one needs to find the sum

$$N = \int |P_n \varphi|^2 \, d\tau = \sum_{l,m} | \int \overline{\psi}_{nlm} \, \varphi d\tau \, |^2 . \tag{33}$$

In calculation of this expression, the main difficulty is to sum over l, especially for the cases when the level belongs to the continuous spectrum (imaginary n) and the sum is of an infinite number of terms. Sometimes it is possible to reduce the sum to the integral by using the parabolic

[5] See [1] V.A. Fock, Basic quantum mechanics (Nachala kvantovoj mekhaniki), Leningrad, Kubuch, 1932, 162 (in Russian), eqs. (16) and (17); eq. (17) should be corrected, namely, factor $\sqrt{2}$ should be omitted. (*V. Fock*)

[6] We propose to treat these problems in a separate paper in more detail; it will be published in Phys. Zs. Sowjetunion. (*V. Fock*)

quantum numbers and make the summation in an explicit form, but corresponding manipulations turn out to be extremely complicated.

If, however, we use the group symmetry of the Schrödinger equation and the addition theorem (31) for its eigenfunctions, then the calculation is drastically simplified. The expression for the whole sum often appears to be simpler than that for its individual term.

Similar simplifications appear in the application of our theory to the calculation of the projection norm of a certain operator $\mathbf{L}$ on the n-subspace, i.e., to the calculation of the double sum

$$N(L) = \sum_{lm} \sum_{l'm'} \mid \int \overline{\psi}_{nlm} L \psi_{nl'm'} d\tau \mid^2 . \tag{34}$$

The expressions of the type (34) arise, e.g., in the calculation of the so-called atomic form-factors. Here the operator $\mathbf{L}$ in the momentum representation is of the form

$$L = e^{-\mathbf{k}\frac{\partial}{\partial \mathbf{p}}}; \qquad L\psi(\mathbf{p}) = \psi(\mathbf{p} - \mathbf{k}). \tag{35}$$

The calculation of (33) and (34) is based on its invariance to the choice of an orthogonal system of functions Ψ_{nlm} in the n-space. The orthogonal transformation of coordinates ξ, η, ζ, χ (four-dimensional rotation) corresponds to the introduction of a new orthogonal system of functions only and, therefore, does not change the values of sums (33) and (34). But this four-dimensional rotation can be chosen in such a manner so as to simplify integrals in these formulae or to put them to zero (except a definite number of them). Note that in (34) we can apply two different rotations of arguments of functions Ψ_{nlm} and $\Psi_{nl'm'}$ that correspond to two independent orthogonal substitutions of these functions.

6. The expression for projection $\mathbf{P}_n\varphi$ of function φ onto the n-subspace, entering (33), is of the form

$$\mathbf{P}_n\varphi = \sum_{lm} \psi_{nlm} \int \overline{\psi}_{nlm} \varphi d\tau. \tag{36}$$

In the momentum representation, a kernel of the projector $\mathbf{P}_n$ is

$$\varrho_n(\mathbf{p}', \mathbf{p}) = \sum_{lm} \overline{\psi}_{nlm}(\mathbf{p}') \psi_{nlm}(\mathbf{p}). \tag{37}$$

By relation (7) we can express the function Ψ_{nlm} through the four-dimensional spherical functions. Since the mean-square momentum p_0

depends on the principal quantum number n, we shall denote it by p_n and rewrite (7) as

$$\Psi_{nlm}(\alpha,\vartheta,\varphi) = \frac{\pi}{\sqrt{8}} p_n^{-5/2}(p_n^2+p^2)^2\psi_{nlm}(\mathbf{p}). \tag{38}$$

By substituting this expression into (37) and using the addition theorem (31), we get

$$\varrho_n(\mathbf{p}',\mathbf{p}) = \frac{8p_n^5}{\pi^2(p^2+p_n^2)^2(p'^2+p_n^2)^2} \cdot n\frac{\sin n\omega}{\sin\omega}. \tag{39}$$

In the particular case of $\mathbf{p}'=\mathbf{p}$, a simple result follows:

$$\varrho_n(\mathbf{p},\mathbf{p}) = \frac{8p_n^5}{\pi^2(p^2+p_n^2)^4}. \tag{40}$$

The integral

$$4\pi\int \varrho_n(\mathbf{p},\mathbf{p})p^2dp = n^2 \tag{41}$$

is equal to the dimension of a given subspace.

7. The success of Bohr's scheme for the Mendeleev periodic system and good applicability of the Rietz formula for energy levels show that the assumption that the atomic electrons are in the Coulomb field can give a satisfactory approximation.

Therefore, it is interesting to consider the following simplified variant of the Bohr model of an atom. The atomic electrons are subdivided on "large layers"; all electrons with the same principal quantum number n belong to the large n-layer. Electrons of the large n-layer are described by hydrogen-like functions with the effective nuclear charge Z_n.

Instead of Z_n, one can also introduce the mean-square momentum p_n connected with Z_n by

$$Z_n = np_n\frac{a}{h} \qquad (a \text{ is "the hydrogen radius"}). \tag{42}$$

Under these assumptions, one can express the atomic energy as a function of the nuclear charge and parameters p_n and then determine these parameters from a variational principle. Here one needs to take into account the following. Although our wave functions of electrons belonging to the same large n-layer are orthogonal to each other, they are not orthogonal to the wave functions of other large layers. Therefore, in this approximation we have to neglect the quantum exchange energy between electrons of different large layers and take into account the exchange energy between electrons of the same large layer only.

Application of this calculation method to atoms with two large layers gave very satisfactory results. For a Na^+ atom ($Z = 11$) we obtained the following values of the parameters p_1 and p_2 (in atomic units):

$$p_1 = 10.63; \qquad p_2 = 3.45 \qquad (Z = 11), \tag{43}$$

while for $Al^{+++}(Z = 14)$ they are

$$p_1 = 12.62; \qquad p_2 = 4.45. \tag{44}$$

A simple analytical expression follows for a screening potential in this method. If we substitute in it the above obtained numerical values of p_1 and p_2, then its value practically does not differ from the self-consistent field obtained in Hartree's method by an incomparably complete way of numerical integration of the system of differential equations. For the sodium atom, our analytical result gives the value of the potential lying between the self-consistent potential obtained with and without quantum exchange [6].

Therefore, it is very probable that the accuracy of the proposed method for many-electron atoms will be pretty high at least for not very heavy atoms.

So far as this description corresponds to reality, the "mixed charge density" in the momentum space can be represented by a sum of expressions (39) for different large layers in an atom. The knowledge of the mixed charge density — as it has been stressed by Dirac [7] — allows one to give answers to all atomic problems, in particular on the light scattering by atoms and the inelastic electron scattering (atomic form-factors). For example, we present here an explicit expression of the atomic factor F_n for a large n-layer. In atomic units, we have

$$F_n = \int e^{i\mathbf{kr}} \varrho_n(\mathbf{r}, \mathbf{r})(d\tau) = \int \varrho_n(\mathbf{p}, \mathbf{p} - \mathbf{k})(d\mathbf{k}). \tag{45}$$

If we substitute here instead of $\varrho_n(\mathbf{p}, \mathbf{p} - \mathbf{k})$ its cxpression that follows from (39), then the integral can be expressed in an explicit form. Putting for brevity

$$x = \frac{4p_n^2 - k^2}{4p_n^2 + k^2}, \tag{46}$$

we obtain

$$F_n(x) = \frac{1}{4n^2} T_n'(x)(1 + x)^2 \{P_n'(x) + P_{n-1}'(x)\}, \tag{47}$$

where $T_n'(x)$ is a derivative of the Chebyshev polynomial with the minimum deviation from zero

$$T_n(x) = \cos(n \arccos x) \tag{48}$$

and P_n' is a derivative of the Legendre polynomial $P_n(x)$. Obviously, for $k = 0$ it will be $x = 1$ and $F_n(1) = n^2$.

The sum of expressions (40) over all large layers in an atom is proportional to the charge density in the momentum space. It can be compared with that obtained by the Fermi statistical model of an atom; the latter appears to be less accurate than ours. For Ne ($Z = 10$) and Na^+ ($Z = 11$) atoms; we obtained a good coincidence for large values of p, while for small p ($p < 2$ in atomic units) Fermi's result gives a too high value of the charge density.

Note in conclusion that though our method can be applied rigorously to atoms with complete large layers only, it can serve as a basis for calculation of an atom with uncompleted layers.

References

1. V.A. Fock, Basic quantum mechanics (Nachala kvantovoj mekhaniki), Leningrad, Kubuch, 1932 (in Russian).
2. G. Wentzel, Zs. Phys. **68**, 348, 1929.
3. F. Bloch, *Contribution to the Theory of the Compton Lines*, Phys. Rev. **46**, 674, 1934.
4. H. Bethe, *Zur Theory des Durchgangs shneller Korpuskularstrahlen durch Materie*, Ann. Phys. **5**, 325, 1930.
5. J. v. Neumann, Mathematische Grundlagen der Quantenmechanik, Berlin, Jul. Springer, 1932.
6. V. Fock and M. Petrashen, *On the Numerical Solution of the Generalized Equation of the Self-Consistent Field*, Phys. Zs. Sowjetunion **6**, 368, 1934. (See also [34-1] in this book. (*Editors*))
7. P.A.M. Dirac, *Note on Interpretation of the Density Matrix in the Many-Electron Problem*, Proc. Cambr. Phil. Soc. **28**, 240, 1931.

Moscow
P.N. Lebedev Physical Institute
the USSR Academy of Sciences,
Physical Institute
Leningrad State University

Translated by E.D. Trifonov

35-2
Extremal Problems in Quantum Theory

V.A. Fock

Leningrad

UFN **15**, 341, 1935

1 Introduction

The subject of my talk is the relations between the mathematical theory of extremal problems and the physical theory of quanta.[1]

The theory of quanta, or quantum mechanics, studies phenomena in the world of bodies of atomic scale. It turned out that the usual notions of motion and the usual methods of description of a mechanical system were inapplicable for such bodies. One needed to find another formulation of the basic laws of nature instead of the old one based on observations of bodies containing a very large number of atoms. The drastic revision of all usual notions of mathematics and physics was required. As a result of this revision and due to the efforts of several scientists such formulation has been found and the theory, which is as elegant and complete as the Newton mechanics is, has been created. This new theory — the new quantum mechanics — contains the Newton mechanics as a limiting case. Although some particular questions and problems remain open, the basic principles and the mathematical technique of the new theory can be considered to be completely established.

The role of this mathematical technique is played by the field of mathematics dealing with linear operators, their characteristic functions and numbers, i.e., the class of problems that is usually called by the German word *Eigenwertprobleme*.

[1] Note added in proof. Recall that this talk was given in 1930. At present (1935), the notions described herein are commonly known. The description, which is the closest to the present talk, is given in the book V.A. Fock, Basic Quantum Mechanics (Nachala Kvantovoj Mekhaniki), Leningrad, Kubuch, 1932 (in Russian). (*V.A. Fock*)

2 Description of Physical Quantities

The first question to be answered in the new theory is the question how a given physical quantity should be described, i.e., to which mathematical notion should it be assigned. In classical mechanics, numbers were assigned to physical quantities such as coordinates and velocities of particles. The numbers represented the values that the corresponding quantities were able to take.

In quantum mechanics, it is a certain *linear operator* rather than a number that is assigned to any physical quantity. In order to clarify the meaning of such an assignment, consider an example taken from the classical (not quantum) physics. The equation for the amplitude of oscillations of a membrane as it is well known reads as

$$\frac{\partial^2 u}{\partial x^2} + \frac{\partial^2 u}{\partial y^2} + \lambda u = 0, \text{ or } -\Delta u = \lambda u,$$

where Δ is the Laplace operator and the parameter λ is related to the oscillation frequency by the equality

$$\lambda = \frac{\omega^2}{a^2},$$

where a is a constant.

In quantum mechanics, such relations would be interpreted as if the frequency squared is described by the operator $-a^2\Delta$. Or, in other words, the operator ω^2 equals $-a^2\Delta$.

As in this example, the characteristic numbers λ of the operator of a given quantity λ are interpreted in quantum mechanics as the values that this quantity is able to take. One can also say that the fundamental function $u(x, y, \omega)$ describes the state of the membrane with oscillation frequency ω.

As in the above, in quantum mechanics the fundamental function for a given characteristic number λ' is said to describe such a state of a physical system for which the quantity λ is equal to λ'.

3 Construction of Quantum Operators

The question arises how one can find the operator corresponding to a given physical quantity. This question can be solved by analogy with classical mechanics. In quantum mechanics, one managed to find an operation analogous to the Poisson bracket $[f, g]$ of two quantities f, g. If we denote the corresponding operators by the same letters, we get

$$[f, g] \to \frac{2\pi i}{h}(fg - gf),$$

where h is a certain constant having the dimension of the action and numerically equal to

$$h = 6.55 \cdot 10^{-27} \text{erg} \cdot \text{s}.$$

Consider the coordinate x and the corresponding "momentum," i.e., the component of the momentum p_x. According to classical mechanics, the Poisson bracket of these two quantities equals unity; this equality is transferred to quantum mechanics. We have

$$\frac{2\pi i}{h}(p_x x - x p_x) = 1,$$

where by x and p_x we already understand corresponding operators. This relation can be satisfied assuming that x is the operator of multiplication by the coordinate x, and p_x is the operator

$$p_x = \frac{h}{2\pi i}\frac{\partial}{\partial x}$$

and analogously for the other two coordinates

$$p_y = \frac{h}{2\pi i}\frac{\partial}{\partial y}, \quad p_z = \frac{h}{2\pi i}\frac{\partial}{\partial z}.$$

Therefore, we found the operators for the components of the particle's momentum. More complicated operators are constructed out of them by analogy with classical mechanics; for instance, the operators for the rotational momentum are

$$m_x = y p_z - z p_y = \frac{h}{2\pi i}\left(y\frac{\partial}{\partial z} - z\frac{\partial}{\partial y}\right),$$

$$m_y = z p_x - x p_z = \frac{h}{2\pi i}\left(z\frac{\partial}{\partial x} - x\frac{\partial}{\partial z}\right),$$

$$m_z = x p_y - y p_x = \frac{h}{2\pi i}\left(x\frac{\partial}{\partial y} - y\frac{\partial}{\partial x}\right),$$

and the operator of the full energy of a particle with the potential energy $U(x, y, z)$ is given as

$$H = \frac{1}{2m}(p_x^2 + p_y^2 + p_z^2) = U(x, y, z) = -\frac{h^2}{8\pi^2 m}\Delta + U(x, y, z).$$

4 Examples of Physical Interpretations of Operators

Let us find characteristic numbers and fundamental functions of some of these operators.

The operators p_x, p_y, p_z commute with one another and, hence, have common fundamental functions. If we denote their characteristic numbers, as it used to be in quantum mechanics, by the same letters with primes, then their common fundamental function writes down as

$$\psi = e^{\frac{2\pi i}{h}(xp'_x + yp'_y + zp'_z)}.$$

In order that the function ψ remains finite in the whole space, it is necessary and sufficient that p'_x, p'_y, p'_z are real. Physically, it means that the momentum components may take only real values.

The function ψ describes the state of a particle with the momentum having definite values of components p'_x, p'_y, p'_z.

The operators m_x, m_y, m_z do not commute with one another and have no common fundamental functions (except for the constant one). It means that there exists no state of a particle such that three or even only two of the angular momentum components have definite nonzero values.

If we construct the operator for the total angular momentum squared $M^2 = m_x^2 + m_y^2 + m_z^2$, we can assure ourselves that it commutes with each of the operators m_x, m_y, m_z. Therefore, the operators M^2 and, for instance, m_z have common fundamental functions. It means physically that the full angular momentum and its component along the z-axis may have definite values. Let us find the fundamental function describing such a state of a particle. In polar coordinates r, θ, φ, the operators m_z and M^2 read as

$$m_z\psi = \frac{h}{2\pi i}\frac{\partial\psi}{\partial\varphi} = m'_z\psi,$$

$$M^2\psi = -\frac{h^2}{4\pi^2}\left[\frac{1}{\sin^2\theta}\frac{\partial}{\partial\theta}\left(\sin\theta\frac{\partial\psi}{\partial\theta}\right) + \frac{1}{\sin^2\theta}\frac{\partial^2\psi}{\partial\varphi^2}\right] = M'^2\psi.$$

These are the known equations for the spherical functions. Their solution satisfying the single-valuedness condition is

$$\psi = f(r)e^{im\varphi}P_l^{|m|}(\cos\theta).$$

Here l and m are integers:

$$l = 0, 1, 2, \ldots; \qquad m = -l, -l+1, \ldots, l.$$

The characteristic numbers of our operators can be expressed via l and m in the following way:

$$m'_z = \frac{h}{2\pi} m, \qquad M'^2 = \frac{h^2}{4\pi^2} l(l+1).$$

Therefore, we obtained the values, which the angular momentum can take, and the expression for the function describing the corresponding state of a particle.

5 Statistical Interpretation of the Fundamental Functions

It is clear from the examples considered which meaning could be given to the description of a state of a particle by the function of coordinates $\psi(x, y, z)$. This function admits the following statistical interpretation. The quantity

$$\frac{|\psi(x, y, z)|^2 dx dy dz}{\iiint |\psi(x, y, z)|^2 dx dy dz}$$

is the probability that if one measures the coordinates of a particle in the state ψ, then one gets them within the limits $(x, x + dx), (y, y + dy), (z, z + dz)$. For a normalized function, such probability reads in a simpler way:

$$|\psi(x, y, z)|^2 dx dy dz = |\psi|^2 dV.$$

The total sum of the probabilities is obviously equal to unity.

Let $\psi(x, y, z; \lambda')$ describe the state of a particle for which λ equals λ'; in other words, let $\psi(x, y, z; \lambda')$ be the fundamental function of the operator λ corresponding to the characteristic number λ'. Then the probability of obtaining λ' as a result of the measurement for a particle described by some other function $\varphi(x, y, z)$ is given by

$$w(\lambda') = \left| \int \overline{\varphi} \psi dV \right|^2,$$

provided that both functions φ and ψ are normalized. The completeness theorem manifests itself in the fact that the values λ', λ'' etc. are the only possible values.

Assume that φ is the fundamental function of the same operator as that for the function ψ but corresponding to another characteristic number λ''; then

$$\left|\int \overline{\psi}(x,y,z;\lambda'')\psi(x,y,z;\lambda')dV\right|^2 = 0.$$

Therefore, the orthogonality expresses the fact that the values $\lambda = \lambda'$ and $\lambda = \lambda''$ are incompatible.

One can pass from the probabilities to the expectation values. The expectation value of λ in the state φ is

$$\text{e.v. } \lambda = \sum_{\lambda'} \lambda' w(\lambda'),$$

and by the completeness theorem this equation is equivalent to

$$\text{e.v. } \lambda = \int \overline{\varphi} L \varphi dV,$$

where L is the operator of the quantity λ.

In the continuous spectrum case the integral of the absolute value squared of $\psi(x,y,z;\lambda')$ diverges. It is an expression of the fact that the probability for the quantity λ to be exactly equal to λ' is exactly zero. The standard normalization condition for the continuous spectrum consists of the following. One introduces the square integrable functions

$$\psi(x,y,z;\lambda,\Delta\lambda) = \int_{\lambda'}^{\lambda'+\Delta\lambda} \psi(x,y,z;\lambda)d\lambda,$$

satisfying the condition

$$\lim_{\Delta\lambda\to 0} \frac{1}{\Delta\lambda} \int |\psi(x,y,z;\lambda,\Delta\lambda)|^2 dV = 1.$$

It corresponds to the fact that the probability for the quantity λ to belong to the interval $(\lambda', \lambda' + \Delta\lambda)$ is finite and equal to

$$|\psi(x,y,z;\lambda,\Delta\lambda)|^2 \Delta\lambda$$

provided that $\Delta\lambda \neq 0$ and the coordinates have the given values x, y, z.

We see, therefore, that a certain physical notion corresponds to each notion from the mathematical theory of linear operators. One can make a dictionary to translate the notions from the mathematical language to the physical one. Such a dictionary would look as follows:

Mathematics	Physics
Linear operator L	Physical quantity λ
Characteristic numbers λ'	Values taken by the physical quantity
Fundamental function for the characteristic number λ	State of a mechanical system for which λ equals λ'
Commutativity of operators	Simultaneous observability of physical quantities
Absolute value squared $\|\psi\|^2$ of the wave function	Probability density
Normalization $\int \|\psi\|^2 dV = 1$	Total sum of probabilities equals 1
Orthogonality $\int \overline{\varphi}\psi dV = 1$	Incompatibility of states ψ and φ
Completeness of the system of fundamental functions $\psi(\mathbf{r}; \lambda')$	Values $\lambda = \lambda', \lambda'' \ldots$ are the only possible ones
Integral $\int \overline{\psi} L \psi dV$	Expectation value of quantity λ in state ψ
Square of the decomposition coefficient of $\varphi(\mathbf{r})$ by $\psi(\mathbf{r}, \lambda')$, $\left\|\int \overline{\psi}(\mathbf{r}, \lambda')\varphi(\mathbf{r}) dV\right\|^2$	Probability of the equality $\lambda = \lambda'$ in the state φ
Normalization for the continuous spectrum	Finite probability of the inequality $\lambda' < \lambda < \lambda' + \Delta\lambda$

Apart from the operators acting on the functions of the coordinates of one particle considered here, quantum mechanics considers operators of other types, e.g., finite and infinite matrices. But in any case, the interpretation of their characteristic numbers and fundamental function remains unchanged.

6 Conclusion

My main aim in this talk was not only to describe the basics of quantum mechanics, but rather to acquaint the reader with the language of this theory. The mathematical technique of the quantum theory is the theory of linear operators. And if I managed to give an impression of the close relation between both theories, the aim of my talk has been achieved.

Translated by V. V. Fock

36-1
The Fundamental Significance of Approximate Methods in Theoretical Physics

V.A. Fock

Leningrad

UFN **16**, N 8, 1070, 1936

1. The aim of theoretical physics is the mathematical formulation of the laws of nature. This problem is closely related to but by no means equal to that of mathematical physics — to solve equations put forward by theoretical physics. Equations of theoretical physics can never be absolutely accurate: one always has to ignore this or that secondary factor while deriving them.

The correct account of all really essential factors for a physical problem is called physical rigor. Physical rigor is necessary for solving physical problems as common mathematical rigor does for solving the problems of the analysis. One of the most important and difficult questions in each particular field of theoretical physics is to what extent the requirements of physical and mathematical rigor are satisfied when the problems in a given field are posed and solved. An investigation of this kind is especially important in the field of laws of elementary physics — the field which is close to the theory of the structure of matter. However, the attempt to give such kind of analysis of physical rigor to the existing formulation of the elementary physical laws is out of the scope of this paper. I should like to dwell upon another question closely connected with it, i.e., upon the formation of a physical notion itself.

Assume that we have a more general physical theory and a more particular one, the particular theory being included into the general one as a particular case. The transition from the particular theory to the general one is certainly related to the introduction of new physical notions. Usually we pay attention to this aspect of notion formation, namely, to the acquirement of new concepts during a generalization of the theory. However, the development of physics during the recent decades has shown that the opposite process, such as the rejection of old physical notions, is closely related to the generalization of the theory.

Consider, e.g., Newton classical mechanics and the special relativity mechanics (the special relativity theory appears in this case to be more general, compared with Newton mechanics). In Newton mechanics we deal with absolute time; the notion of the simultaneity of two events does not require any special assumptions (a fixed coordinate system and so on) and in this sense it is an absolute notion. The notion of simultaneity is lost in the relativity theory. It might be impossible to say without special remarks which of two events has taken place earlier, and which later. The simultaneity becomes an approximate notion, applicable only in the cases when it is possible to neglect the time interval needed for the light to pass the distance between the points, where the discussed events take place (for events on the Earth and on the Sun this time interval is about 500 sec.). Thus generalization of physical theories is related not only to new notions, but also to the denial of old ones. It is important now to draw attention to the following psychological factor: the rejection of old, usual physical notions is much more difficult than the assimilation of new notions, which are not related to such kind of rejection.

In order to clarify such a necessity, we shall try to follow the process of the formation of notions in the direction opposite in a sense to the historic development of a theory. If one follows this direction, one should proceed from the currently most general physical theory; we convince ourselves that at each simplification, at each transition to a more particular theory newer and newer physical notions arise. But, therefore, it becomes clear that the opposite process — the transition from a more particular theory to a more general one — must be related to the rejection of some physical notions.

We choose such a method of consideration — sequential simplification of a theory — since it is easier to follow from the mathematical point of view. The transition to a simplified theory means, as a matter of fact, the usage of one or another approximate method based on the possibility to ignore one or another secondary factor, one or another small quantity in the problem. The validity of given approximate methods — then physical rigor — can be estimated by means of usual mathematical inequalities, which characterize the negligibility of the quantities omitted in the situation. However, this "error evaluation" in common sense gives at the same time the applicability criteria for physical notions connected with this approximate method. Thus, a more difficult logical, or perhaps philosophical question concerning the applicability of some physical notions acquires here a concrete mathematical expression.

Thereby, the fundamental significance of the approximate methods in theoretical physics is clarified.

2. As we have already mentioned, the equations of theoretical physics are never absolutely correct. Even the most general theory at the present period of the development of physics cannot aspire to be universal because it contains a number of physical neglects in itself. That is why to start to formulate such a theory, first of all, it is necessary to establish what are these neglects and what are the applicability limitations of basic physical notions operated by this theory. One of the most general existing physical theories appears to be quantum electrodynamics. This theory does not examine the nature of the atomic structure of matter, but takes it as an experimental fact. The structure of material particles is not examined, either, and these particles are characterized by certain constants, by their charge and mass, in particular. The mass of a particle is an approximate notion. According to the relativity theory the mass m is related to the energy mc^2, but if two particles interact so powerfully that their energy becomes comparable to mc^2, then the notion of mass loses its definite meaning. Such a considerable interaction energy may be observed in nuclear processes and manifests itself in breaking rigorous mass additivity. For instance, the mass of a particle consisting of two protons and two neutrons turns out to be less than the sum of masses of the constituent particles up to a quantity equal to their interaction energy divided by c^2. Due to their small mass and large interaction energy, the notion of mass loses its meaning for electrons more often than for heavier particles (protons and neutrons); that is why it is impossible to talk about electrons and nuclei as separate particles. All these intranuclear processes are outside of the applicability domain of quantum electrodynamics.

As to the charge of a particle, it possesses a property to take values, multiple of a certain elementary charge, so the additivity is evidently observed in this case. But the number of particles itself is not constant. The experiments of recent years showed us that under the influence of sufficiently high-frequency radiation, pairs of particles with equal masses but opposite charges (electrons and positrons) can be created.

If we do not examine intranuclear processes, are not interested in the structure of particles of matter and ignore the possibility of the creation of pairs, then we can use quantum electrodynamics in such a form as it was developed, e.g., in the works of Heisenberg and Pauli or Dirac, Podolsky and myself.

The possibility of the creation of pairs can be taken into account in the framework of the existing theory.

The idea of the quantum electrodynamics is that a group of a given number of n particles of matter and indefinite number of light quanta is considered as one system. This system consisting of matter and electromagnetic field possesses an infinite number of degrees of freedom. Denote by x the variables related to a single particle of matter (the coordinates and the so-called spin), and by k the variables related to a light quantum (a wave vector and a polarization of the corresponding plane wave), then the states of such a system can be described by means of a sequence of functions

$$\psi_0, \psi_1, \ldots, \psi_N, \ldots \ , \tag{1}$$

where

$$\psi_N = \psi(x_1, x_2, , \ldots, x_n; k_1, k_2, \ldots k_n) \tag{2}$$

and the function ψ_N should be symmetric with respect to the variables $k_1, k_2, \ldots, k_N$. Instead of the sequence of functions, one may consider, as I have already shown in [1], *one* quantity, i.e., a functional Ω of a certain auxiliary function $\overline{b}(k)$. This functional looks like

$$\Omega = \psi_0 + \sum_{N=1}^{\infty} \frac{1}{\sqrt{N!}} \int \psi_N \overline{b}(k_1) \ldots \overline{b}(k_N) dk_1 \ldots dk_N. \tag{3}$$

For the physical interpretation of this functional, it is sufficient to give an expression for the scalar product of two functionals Ω and Ω'. This scalar product (Ω, Ω') has the same physical meaning as the scalar product

$$(\psi, \psi') = \int \overline{\psi} \psi' d\tau$$

of two usual wave functions in quantum mechanics.

If the functional Ω looks like (3) and the functional Ω' is made in a similar way out of the functions ψ'_N, then their scalar product is

$$(\Omega, \Omega') = \int dx_1 \ldots dx_n \left\{ \overline{\psi}_0 \psi'_0 + \sum_{N=1}^{\infty} \int \overline{\psi}_N \psi'_N dk_1 \ldots dk_n \right\}. \tag{4}$$

Assume $\Omega' = \Omega$, so that $\psi'_N = \psi_N$; then individual terms of the expression may be interpreted as probabilities.

For instance, the integral

$$\int dx_1 \ldots dx_n \int \overline{\psi}_N \psi_N dk_1 \ldots dk_N \tag{5}$$

is the probability of the existence of exactly N light quanta.

The time dependence of the functional Ω is determined by the equation

$$H\omega - i\hbar\frac{\partial\Omega}{\partial t} = \sqrt{\alpha}\int\left\{G^{\dagger}(k)\frac{\delta'\Omega}{\delta\bar{b}(k)} + G(k)\bar{b}(k)\Omega\right\}dk, \qquad (6)$$

where $\frac{\delta'\Omega}{\delta\bar{b}(k)}$ is the functional derivative of Ω by $\bar{b}(k)$, defined as a coefficient in the expression for the variation

$$\delta\Omega = \int\frac{\delta'\Omega}{\delta\bar{b}(k)}\delta\bar{b}(k)dk\,. \qquad (7)$$

The quantities $G(k)$ and $G(k)$ are certain given operators. The operator H from the left-hand side of equation (6) is the usual full energy operator for the system of n particles of matter

$$H = \sum_{s=1}^{n} T_s + \sum_{u>v=1}^{n}\frac{e_u e_v}{|\mathbf{r}_u - \mathbf{r}_v|}. \qquad (8)$$

The right-hand side of equation (6) represents the interaction energy operator between the particles and the radiation (the latter energy is taken with the opposite sign). Here α is the so-called fine structure constant

$$\alpha = \frac{e^2}{\hbar c} = \frac{1}{137.3}\,. \qquad (9)$$

The derivation procedure of equation (6) is mathematically not rigorous; computations of the operator H (8) lead, e.g., to a double sum, where terms with $u = v$ exist. One has to cross these parts out because they represent an infinitely large constant. Such kinds of assumptions that result from the concepts, lying in the basis of the theory, are undoubtedly its serious defects. Nevertheless, in a certain approximate theory these tricks are physically necessary. As a matter of fact, they represent a rough method, using which the defects of the first formulation, accounting for the role of the light quanta with very large energy in a wrong way, are corrected.[1]

Not only the derivation of wave equation (6) but also its solution cannot be called mathematically rigorous. Formally viewed, these approximated methods are based on the fact that the parameter $\sqrt{\alpha}$ in the

[1] The premises of the theory fail when the energy of a quantum $\hbar\nu$ becomes of the order of $\frac{mc^2}{\alpha}$, where m is the electron mass. Nevertheless, deriving the quantum electrodynamics formulae, it is necessary to integrate over ν up to infinity. (*V. Fock*)

right-hand side is small. From the physical point of view, it corresponds to the secondary role of the interaction of particles with radiation with respect to the interaction between particles themselves; this fact has been checked experimentally. However, the attempt to look for a solution of the wave equation as a power series in $\sqrt{\alpha}$ leads to reasonable results only if one takes several terms of this series, neglecting terms of the order greater than α or α^2. It is impossible to give any physical meaning to the other terms because they contain expressions like products of a small coefficient and a divergent integral. Thus, in order to obtain minimal physical rigor here, we need to give up a mathematical one. Nevertheless, despite all its drawbacks, the theory based on wave equation (6) describes quite well a broad class of physical notions. It reproduces the processes of radiation of the atoms and molecules with great precision and lets one determine not only the frequencies but also the natural width of the spectral lines. It leads to the right formula for the dispersion of light by free electrons (the Klein–Nishina formula). Finally, it gives corrections to the Coulomb law for the interaction between electrons; these corrections come from radiation and absorption of light quanta by electrons (in the classical theory, the corresponding corrections are interpreted as accounting for retarding potentials).

3. From quantum electrodynamics, the main ideas of which we have tried to convey, we will now proceed to the usual relativity theory.

The main subject of quantum electrodynamics is to formulate the laws of interaction of particles of matter with radiation. When this interaction is taken into account, an excited atom (i.e., an atom possessing not the least possible energy, but a larger one) can emit a light quantum and transit to a lower energy level. The duration of the atom dwelling on the exited state is limited, so strictly speaking this state is not stationary (it is the natural width of spectrum lines that is related to this fact). If we neglect the interaction of atoms and radiation, *then we come to the new physical notion of an atom's stationary state.* We have now an opportunity to speak about an atom as being a mechanical system in its proper meaning. Thus, *the notion of a mechanical system* arises that is related to the neglect of the interaction between matter and radiation. But we can do one step more: we can neglect relativistic corrections. This neglect is valid if the velocities of the constituent parts of the system are small in comparison with the speed of light:

$$\frac{v}{c} \ll 1. \tag{10}$$

For electrons in an atom the ratio v/c is of the order of

$$\frac{v}{c} \sim \alpha = \frac{1}{137.03}. \tag{11}$$

Therefore, for electrons the validity of this neglect is related to the neglect of radiation. (Recall that in formula (6) the interaction energy of particles of matter with radiation has a factor of $\sqrt{\alpha}$.)

Assumption of new physical notions concerning separation of the space and time and absolute simultaneity is related to the neglect of the corrections related to the relativity theory. These neglects simplify our problems: from quantum electrodynamics we proceed to quantum mechanics (in its proper meaning) and to the Schrödinger theory. Formally the neglect of radiation corresponds to crossing out the right-hand side of equation (6) and to the substitution of the functional Ω by its zeroth expansion term ψ_0. Then, by neglecting the relativistic corrections we will come to the usual Schrödinger equation,

$$H\psi = i\hbar\frac{\partial\psi}{\partial t}, \tag{12}$$

where the energy operator looks as (8). If one adds the potential energy of the particles in an external field (which is omitted in equation (8)) to this operator, it takes the form

$$H = -\sum_{s=1}^{n}\frac{\hbar^2}{2m_s}\Delta_s + \sum_{s=1}^{n}U_s(x_s) + \sum_{u>v=1}^{n}\frac{e_u e_v}{|\mathbf{r}_u - \mathbf{r}_v|}, \tag{13}$$

where Δ_s is the Laplace operator acting on the coordinates of the particle number s. The theory based on the Schrödinger theory is much more satisfactory from the mathematical point of view than quantum electrodynamics. In this case, the problems can be formulated quite rigorously (in the mathematical meaning). The nonrigor assumed during its solving comes only from the difficulty of the problem but not from the shortage of the initial equations as it takes place in quantum electrodynamics. From the physical point of view this theory — quantum mechanics in its proper meaning — presents a closed logical scheme operating with notions defined sufficiently rigorously. In all the cases when the basic assumptions of the theory are fulfilled (where the most important of them have been already mentioned) the calculations made on the basis of quantum mechanics are in complete agreement with the experiment, if only the latter has been performed with sufficient precision. The assumptions of the theory are fulfilled in most of the problems related to atoms and molecules. Therefore, quantum mechanics spreads all over the theory of structure and interaction between atoms and molecules, including chemistry.

According to quantum mechanics, the state of a mechanical system is described by the wave function ψ, satisfying the Schrödinger equation (12). The knowledge of the wave function gives us all the information about the system, which can be obtained as a result of a certain maximally complete experiment on it. Under the maximally complete experiment we understand such a one, which gives the values of all mechanical quantities that can be measured simultaneously. (Measurements of different quantities may interfere with one another.) The maximally complete experiments may be different according to the selection of quantities measured. One can also say that the physical meaning of the wave function consists in representing the collection of information about the system obtained as a result of a certain maximally complete experiment. The meaning of wave equation (12) is that it allows one to proceed from experimental data or information related to the initial moment (the initial value of the wave function) to the information related to a later moment (the value of ψ at the moment t). The information written down in the form of a wave function also allows one to calculate *the probabilities* of different results of later measurements of different quantities, and also the *mean value* of such measurement results.

Here we shall not dwell upon the question about the recording of experimental data for the case when this experiment is not maximally complete[2] but we shall rather pay attention to the following circumstance important to our analysis of the formation process of a physical notion.

In quantum mechanics, the notion of a system state is merged with the notion of the maximally complete data or information obtainable about the state. It is related to the fact that the laws of quantum mechanics lead to the conclusion of *impossibility of an objective description of the detailed behavior of physical processes.* In fact, the description by means of the wave function is not objective in the common meaning.[3] It can be shown especially clearly by example of the so-called dispersing wavelet. According to quantum mechanics, a free particle with the initial position and momentum known with a precision allowed by the

[2]The mathematical approach to this notion was investigated in detail in the book by J. v. Neumann, [2]. (*V. Fock*)

[3]See the articles by A. Einstein and N. Bohr [3] that are published together with an introduction by the author under the common title "Is it possible to consider the quantum mechanical description to be a maximally complete one?" (*V. Fock*)

Heisenberg inequality

$$\Delta p \Delta x \geq \hbar \tag{14}$$

is described by the function ψ being a wavelet, i.e., it is essentially nonzero only within some nonsharply bounded space domain. This wavelet is a superposition of plane waves, whose directions and frequencies are not quite equal. The Schrödinger equation shows that this wavelet disperses with time, i.e., the wave function becomes distinct from zero in the more and more wide volume. As the particle is free, it is obvious that there are no objective changes to it. It is only the loss of our information about the localization of the particle in space that takes place. (Speaking the classical mechanics language, the free motion of a particle is unstable in the meaning of Ljapunov.) It is clear, therefore, that the description of a state by the wave function is not objective. Nevertheless, this description reproduces everything that we can obtain as a result of a maximally complete measurement, and that is why it is an expression of the objectively existing laws of nature. The necessity of just this biased description comes from the impossibility to coordinate in another way the wave and corpuscular nature of matter equally definitely established in experiment.

The other characteristic feature of quantum mechanics is that the knowledge of a state of a whole system (the wave function), for instance of a whole atom, does not imply the knowledge of states of particular parts of the system (wave functions), for instance, of particular electrons in an atom. The experiment, which is maximally complete with respect to the whole system, is not such a one with respect to the system parts.

Therefore, in quantum mechanics there exist such notions as mass and charge of particles and the notions of the mechanical system in its proper meaning (i.e., independent from radiation), of the system stationary state (for example, of an atom or molecule), but there are no such notions as the states of particular particles forming the system (electrons in the atom) and of the objective description of the system behavior.

4. We have clarified what are the physical notions operated by quantum mechanics. We proceed now to the consideration of approximated methods of solving quantum mechanical problems and to clarifying the arising physical notions. The problem of the determination of the stationary states of a system reduces to solving the equations like

$$H\psi = E\psi, \tag{15}$$

i.e., to finding eigenfunctions of the energy operator H. If a system consists of identical particles, for example, of electrons in the field of an atomic nucleus, the wave function ψ has to satisfy, besides equation (15), the demand of antisymmetry. It has to change the sign when two arguments, corresponding to two respective electrons, are exchanged. For example,

$$\psi(x_1, x_2, \ldots, x_n) = -\psi(x_2, x_1, \ldots, x_n). \tag{16}$$

This condition, which is a form of the Pauli principle, is compatible with equation (15) because in the case of identical particles the energy operator H is symmetric. In the case of electrons, each argument x of the function ψ is a collection of three space coordinates x, y and z and one more variable σ takes only two values. This variable in a certain meaning corresponds to the electron orientation (the so-called spin). So, in the case of n electrons it is necessary to find one function of the variables

$$x_s, y_s, z_s, \sigma_s \quad (s = 1, 2, \ldots n) \tag{17}$$

or, which is the same, 2^n functions of 3^n variables

$$x_1, y_1, z_1; x_2, y_2, z_2; \ldots x_n, y_n, z_n\,. \tag{18}$$

In order to imagine the level of difficulty of this problem better, it is enough to recall that, e.g., for the sodium atom $n = 11$, it is necessary to find $2^{11} = 2048$ functions of 33 variables. For a copper atom, $n = 29$ and one needs to deal with about half a billion functions of 87 variables.

It is obvious that the exact solution of this problem is impossible and one needs to call upon approximate methods.

One of the main methods used in quantum mechanics to solve a problem of this type is the famous Rietz method, or rather its modifications. It is based on a reformulation of the problem as a certain variational problem: we have to find the minimum of the integral

$$W = \int \overline{\psi} H \psi d\tau, \tag{19}$$

which gives the atomic energy on condition that

$$\int \overline{\psi}\psi d\tau = 1, \tag{20}$$

i.e., when the wave function is restricted to be normalized.

The function ψ to be found can be approximately presented as a linear combination of products of functions $\varphi_s(x_u)$, depending on the variables of one electron each. In order to satisfy the Pauli principle (16), the antisymmetric combination has to be taken,

$$\psi = \frac{1}{\sqrt{n!}} \sum \varepsilon_{\alpha_1 \cdots \alpha_n} \varphi_{\alpha_1}(x_1) \dots \varphi_{\alpha_n}(x_n), \tag{21}$$

where ε is the quantity antisymmetric with respect to the indices and such that $\varepsilon_{123\cdots n} = 1$. The functions $\varphi_s(x_u)$ can be assumed to be orthogonal and normalized:

$$\int \overline{\varphi}_s(x) \varphi_r(x) dx = \delta_{sr} \, . \tag{22}$$

If we substitute (19) into expression (21) for ψ (which can also be written down as a determinant), we get the energy W of the atom expressed in terms of the functions $\varphi_s(x)$. Varying the energy W with respect to $\varphi_s(x)$ under additional conditions (22), we get the equation of the form

$$-\frac{\hbar^2}{2m} \Delta \varphi_s(x) + U_s(x) \varphi_s(x) - A_s \varphi_s(x) = E_s \varphi_s(x), \tag{23}$$

where $U_s(x)$ are some functions of coordinates, depending on all functions to be determined except the function $\varphi_s(x)$. The $U_s(x)$ plays the role of the potential energy. The symbol A denotes some integral operator, also depending on all functions to be found except the function $\varphi_s(x)$.

Thus, if all the functions except $\varphi_s(x)$ are assumed to be known, then for the $\varphi_s(x)$ one gets the linear, integro-differential equation.

Equations (23) were first derived by me in 1929–1930.

Note that in the calculated cases (lithium atom, sodium atom etc.) the energy levels obtained in this way are very close to the experiment: the error is less than 1–2%. Thus the approximated method in question gives a precise enough formulation of our very complicated physical problem.

What is the fundamental significance of the method? The wave function of the whole atom is approximately expressed via the wave functions of single electrons. But it means that we have obtained a *new physical notion*, namely, that the electrons in an atom are also in certain states. (As we have already mentioned, in the precise Schrödinger theory this notion does not exist.)

This notion is especially important because it is the basis of the Bohr scheme of the atom's electron shells, explaining the structure of the Mendeleev periodic system.

But that is not everything. Besides the term $U_s(x)$, which can be interpreted as the potential energy, originating from the atom's nuclei and from the other electrons, our equations (23) have one more term, namely, A_s. This term represents the new kind of energy, unknown both for classical mechanics and for the exact Schrödinger theory, namely, the so-called quantum exchange energy. (This name came from the fact that the term A_s comes from the Pauli principle, which is a consequence of the identity of electrons, i.e., of the fact that nothing changes if two electrons get interchanged.)

Therefore, we have obtained one more physical notion — *the notion of the quantum exchange energy.* This notion plays a rather important role both in the theory of atoms (the rejection of the term A_s increases the error from 1–2% to 20–30%), and especially in the theory of molecules (as far as molecules are concerned, this notion was first introduced by Heisenberg and Heitler). It appears that the existence of homeopolar molecules (consisting of identical atoms) is impossible to explain without taking into account the quantum exchange energy. Therefore, this new notion plays the decisive role in quantum chemistry.

5. The further and perhaps most important approximated method of solving quantum mechanical problems appears to be the perturbation theory. As is widely known, its main idea is that the Hilbert space, characterizing the variety of all possible system states and having an infinite dimension, is substituted by a certain finite dimensional subspace. This subspace is chosen in such a way that it corresponds exactly to the states of the system, playing the most important role in a given problem. If the dimension of this subspace is still large, then to deal with it one needs to use some special methods, usually the group theory methods.

Simplifications in such studies come first from the property of antisymmetry of the wave function, and second from the fact that while the wave function itself depends on spin variables, the energy operator in the Schrödinger approximation does not. It allowed Dirac to formulate the problem of the perturbation theory by introducing operators (finite matrices) acting on spin variables. These operators could be interpreted as spin moments of the momentum of single electrons, and the corresponding terms of the expression for the energy — *as the interaction energy of these spin moments.* This *new physical notion* turns out to be very useful in interpreting the spectra of complicated atoms.

One of the most amazing applications of the perturbation theory based on the same ideas is the spin invariant theory proposed by Weyl. The mathematical scheme of this theory is equivalent to the scheme of chemical substances, coming from the valency theory.[4] Thus, such an application of the approximated methods of the perturbation theory leads to the *new physical* (or chemical, if you wish) *notion of chemical valency.*

Proceed now to the boundary domain between quantum and classical mechanics and consider for simplicity the Schrödinger equation for one particle

$$-\frac{\hbar^2}{2m}\Delta\psi + U(x,y,z)\psi = i\hbar\frac{\partial\psi}{\partial t}. \tag{24}$$

If the corresponding classical motion of a particle of matter is such that

$$\frac{mv^3}{w} \gg \hbar, \tag{25}$$

where v is the particle velocity and w is its acceleration, then the solution of the Schrödinger equation (24) can be expressed with a great accuracy by the solution of the classical Hamilton–Jacobi equation

$$\frac{1}{2m}(\mathrm{grad}S)^2 + U(x,y,z) - \frac{\partial S}{\partial t} = 0. \tag{26}$$

If S is the complete integral of equation (26), depending on three arbitrary constants c_1, c_2 and c_3, then the wave function ψ can be approximately expressed via S in the following way:

$$\psi = \sqrt{\left\|\frac{\partial^2 S}{\partial x_i \cdot \partial x_k}\right\|}e^{\frac{i}{\hbar}S}. \tag{27}$$

For the stationary states, one can assume

$$S = -Et + V(x,y,z) \tag{28}$$

and the exponential function can be substituted by the expression

$$e^{-\frac{i}{\hbar}Et}\cos\left(\frac{V}{\hbar}+\alpha\right), \tag{29}$$

where α is a constant phase. This approximated method, belonging to Wentzel and Brillouin, leads us to the so-called *old quantum mechanics* and all physical notions related to it. From the boundary conditions for the wave functions characteristic of this theory, one can obtain the *quantization rules.*

[4]See, e.g., the paper [4] by M. Born. (*V. Fock*)

Finally, in some phenomena the restrictions imposed by the Heisenberg inequalities (14) are inessential. As a result, the wave character of matter recedes into the background. Then classical mechanics with its notions, usual for us, should be applied. The new (however, historically old) *physical notions of the behavior of objective processes, of a particle trajectory, of the fact that any physical quantity has always a definite value* come into play.

Our review, in which the course of the historical development of mechanics and physics is inverted to some extent, would probably also help to understand the historical development of the physical notions as well. We wanted to show that any physical theory, any physical notion, is, as a matter of fact, an approximation. Each great progress of physical science is related not only to the creation of new notions, but also to the critical revision of old ones. And if it is proven that some of the old notions are inapplicable to a newly discovered one, then it is necessary to give them up without regret.

References

1. V. Fock, *Zur Quantenelektrodynamik*, Sow. Phys. **6**, 425, 1934. (See [34-3*] in this book. (*Editors*))
2. J. v. Neumann, Mathematische Grundlagen der Quantenmechanik, Berlin, Springer, 1932.
3. V. Fock, A. Einstein, B. Podolsky and N. Rosen, N. Bohr, UFN **16**, 436, 1936.
4. M. Born, Chemical Bond and Quantum Mechanics (Khimicheskaya svjaz' i kvantovaja meckhanika), ONTIVU Kharkov, 1932 (in Russian). (Translated from: M. Born, Ergebnisse der Exakten Naturwissenschaften **10**, 387–452, Springer, Berlin, 1931. (*Editors*))

Translated by V.V. Fock

37-1*
The Method of Functionals in Quantum Electrodynamics

V. Fock

UZ LGU **17**, 267, 1937
Fock57, pp.124–140

Introduction

The Schrödinger equation for a material system (atom or molecule) gives for the energy of the system a set of levels corresponding to stationary states. To be specific, only the ground level remains strictly stationary; for a system in an excited state, a probability to pass to the ground state appears, and such a transition is accompanied by radiation. So the concept of a purely mechanical system independent of radiation is approximate. We get a better approximation to reality if we consider an atom or a molecule together with radiation, i.e., in combination with the light quanta, of which the radiation is composed. While the energy of a mechanical system itself can be carried out by radiation or increased due to the absorption of light quanta, the energy of the total system composed of matter and light will remain unchanged, so this system is conservative.

Therefore, we need a coherent theory that treats matter and radiation as a uniform conservative system. This theory known as quantum electrodynamics has been developing since 1929 by the joint efforts of many physicists, primarily Heisenberg, Pauli and Dirac. Soviet physicists, including the author of this paper, have also participated in the elaboration of the theory.

The main difficulty to be overcome in the construction of this theory is that the number of light quanta can vary during the process; they can be absorbed or emitted. So, a system with an indefinite number of particles should be considered. The mathematical theory allowing one to treat such systems is known as the theory of the second quantization. Its basic ideas will be given below.

The further step in the development of quantum electrodynamics consists of the explanation of electrostatic forces. Actually, the electromagnetic field cannot be treated only as radiation; besides the radiation there exist electrostatic forces, also. True, the latter were included in the Schrödinger equation explicitly and, hence, already taken into account. However, according to the relativity theory, the electromagnetic field represents something entire, so its separation into electrostatic forces and forces of radiation is not invariant. The problem arises in deriving both forces from the same quantum concepts. This problem was also solved by quantum electrodynamics.

Quantum electrodynamics represents the complete physical theory with correct answers to many problems referring to the interaction between matter and radiation. This theory can be extended to the variable number of material particles (creation and annihilation of electrons and positrons). Meanwhile there are some essential difficulties in modern quantum electrodynamics that formally manifest themselves in the failure to obtain a rigorous solution of its equations due to the divergence of some integrals and other difficulties of mathematical origin.

1 Description of a System with Indefinite Number of Particles by Functionals

a) Functionals and quantized operators

Consider a system of particles obeying the so-called Bose statistics. Generally speaking, such particles will be of integer and zero spin, for instance, light quanta and α-particles.

Denote by k all the variables referring to the degrees of freedom of one particle. For a light quantum, the letter k will denote three components of the wave vector k_x, k_y, k_z and a variable j defining the state of polarization. As is known in quantum mechanics the state of a system with the definite number n of such particles is described by a wave function

$$\psi_n = \psi_n(k_1, k_2, \ldots, k_n). \tag{1.1}$$

The mentioned condition that the particles obey the Bose statistics implies that the wave function ψ_n is symmetric with respect to its arguments $k_1, k_2, \ldots, k_n$. Let us consider an arbitrary, auxiliary function $\overline{b}(k)$ and construct the expression

$$\Omega_n = c_n \int \psi_n(k_1, \ldots, k_n)\, \overline{b}(k_1) \ldots \overline{b}(k_n)\, dk_1 \ldots dk_n, \tag{1.2}$$

where c_n is a numerical coefficient depending only on index n. Obviously, if the wave function ψ_n is fixed, the value of Ω_n will also be fixed for any arbitrary auxiliary function $\overline{b}(k)$. And vice versa, due to its symmetry the wave function ψ_n will be defined if the value of Ω_n is fixed for some $\overline{b}(k)$.

We will treat Ω_n as the *functional* depending on the function $\overline{b}(k)$. Hence, to describe the state of an n-particle system, one can introduce the functional Ω_n instead of the wave function ψ_n. This mathematical trick is analogous to treating a quadratic form instead of the tables of its coefficients or to the method of "generating functions" popular in the analysis. In quantum mechanics, this approach was originally applied by the author of this article.[1]

For our problem, the introduction of functionals of the type Ω_n has some advantages because one can easily generalize the approach to a system with an indefinite number of particles. To do this, it is sufficient to consider the sum

$$\Omega = \sum_{n=0}^{\infty} \Omega_n, \tag{1.3}$$

where Ω_n is given by (1.2).

Let us, besides the functional Ω_n expressed in terms of a function ψ_n, have another functional Ω'_n formed with the help of a function ψ'_n. We define a scalar product of two functionals as

$$(\Omega_n, \Omega'_n) = \int \overline{\psi}_n(k_1, \ldots, k_n)\psi'_n(k_1, \ldots, k_n)\, dk_1 \ldots dk_n. \tag{1.4}$$

The physical meaning of the latter expression is known from quantum mechanics. For instance, if ψ'_n is a result of action on ψ_n of some operator $L^{(n)}$,

$$\psi'_n = L^{(n)}\psi_n, \tag{1.5}$$

having a physical meaning for a n-particle system, then expression (1.4) defines the expectation value of the physical quantity represented by $L^{(n)}$. It is supposed that the wave function ψ_n is normalized to one, so

$$(\Omega_n, \Omega_n) = \int \overline{\psi}_n \psi_n\, dk_1 \ldots dk_n = 1. \tag{1.6}$$

For the functional of a more general form (1.3), the scalar product is defined as a sum of scalar products of the corresponding terms:

$$(\Omega, \Omega') = \sum_{n=0}^{\infty} (\Omega_n, \Omega'_n). \tag{1.7}$$

[1] V. Fock, Zs. Phys. **49**, 339, 1928.

Particularly, when $\Omega_n = \Omega'_n$ and, therefore, all the functions ψ_n and ψ'_n coincide, we obtain the normalization condition by equating to unity the expression

$$(\Omega, \Omega) = \sum_{n=0}^{\infty} (\Omega_n, \Omega_n) = 1. \tag{1.8}$$

An individual term (Ω_n, Ω_n) of the latter expression represents a probability for finding exactly n particles. The normalization condition (1.8) claims that the sum of this probability is equal to one.

Consider an operator L for some quantity possessing the additivity, e.g., kinetic energy, angular momentum or electric dipole moment. Then we will have

$$L^{(n)} = L(k_1) + L(k_2) + \ldots + L(k_n). \tag{1.9}$$

Let the functions ψ'_n in the functional Ω'_n be equal to (1.5) and the operator $L^{(n)}$ have the value (1.9). Let us ask ourselves by which transformations the functional Ω'_n can be obtained from Ω_n.

For each term Ω'_n, we have

$$\Omega'_n = c_n \int (L(k_1) + \ldots + L(k_n)\psi_n)\, \bar{b}(k_1) \ldots \bar{b}(k_n)\, dk_1 \ldots dk_n. \tag{1.10}$$

Due to the symmetry of $\psi_n(k_1, \ldots, k_n)$, this expression splits to the sum of n equal integrals. Hence, we have

$$\Omega'_n = nc_n \int (L(k_1)\psi_n(k_1, k_2, \ldots, k_n))\, \bar{b}(k_1) \ldots \bar{b}(k_n)\, dk_1 \ldots dk_n, \tag{1.11}$$

or, briefer,

$$\Omega'_n = \int \bar{b}(k) L(k) f_n(k)\, dk, \tag{1.12}$$

where the function

$$f_n(k) = nc_n \int \psi_n(k, k_2, \ldots, k_n) \bar{b}(k_2) \ldots \bar{b}(k_n)\, dk_2 \ldots k_n \tag{1.13}$$

does not depend on the choice of operator L. To express Ω'_n in terms of Ω_n, it is sufficient to express f_n by Ω_n. To do this, we will vary an auxiliary function $\bar{b}(k)$ in Ω_n. Forming a variation, it is easy to verify that

$$\delta\Omega_n = \int f_n(k) \delta\bar{b}(k)\, dk. \tag{1.14}$$

Hence, $f_n(k)$ is nothing else but the coefficient at the variation $\delta\bar{b}(k)$ under the integral for $\delta\Omega_n$. We call this coefficient *the functional derivative* of Ω_n and denote it as $\delta'\Omega_n/\delta\bar{b}(k)$. The functional derivative of a functional Ω of a general form (1.3) is defined by the relation

$$\delta'\Omega = \int \frac{\delta\Omega}{\delta\bar{b}(k)} \delta b(k)\, dk. \tag{1.15}$$

Hence, expression (1.12) is written as

$$\Omega_n' = \int \bar{b}(k) L(k) \frac{\delta\Omega_n}{\delta\bar{b}(k)}\, dk. \tag{1.16}$$

Making a sum of expressions (1.16) over index n we obtain the required relation for the total functional Ω,

$$\Omega' = \int \bar{b}(k) L(k) \frac{\delta\Omega}{\delta\bar{b}(k)}\, dk. \tag{1.17}$$

If we denote by a symbol $b(k)$ an operator of a functional derivative

$$b(k)\Omega = \frac{\delta\Omega}{\delta\bar{b}(k)} \tag{1.18}$$

and the operation of multiplication by an auxiliary function $\bar{b}(k)$ by symbol $b^\dagger(k)$,

$$b^\dagger(k)\Omega = \bar{b}(k)\Omega, \tag{1.19}$$

then, according to (1.17), the functional Ω' can be obtained from Ω by the action of the operator

$$\mathrm{L} = \int b^\dagger(k) L(k) b(k)\, dk\,, \tag{1.20}$$

which is called a "quantized operator" L corresponding to the unquantized operator $L(k)$. With the help of a quantized operator L, relation (1.17) can be rewritten as

$$\Omega' = \mathrm{L}\Omega. \tag{1.21}$$

The expectation value of an additive physical quantity corresponding to the operator L is equal to the scalar product (Ω, Ω'), so

$$\text{E.v. L} = (\Omega, \mathrm{L}\Omega). \tag{1.22}$$

The latter expression holds irrespective of whether the number of particles is fixed or not. For instance, it allows one to calculate the energy of the light field consisting of an indefinite number of light quanta.

Similar relations can be established for the quantities, which do not obey the additivity. For example, denote by $U^{(n)}$ the interaction energy of an n-particle system

$$U^{(n)} = \sum_{r<s}^{n} U(k_r, k_s). \tag{1.23}$$

In the same way as before, when we put in correspondence a quantized operator L (1.20) to the operator $L^{(n)}$ (1.9), we can find a quantized version of $U^{(n)}$:

$$\mathrm{U} = \frac{1}{2}\int b^\dagger(k)\, b^\dagger(k')\, U(k,k')\, b(k')b(k)\, dk\, dk'. \tag{1.24}$$

The expectation value of the interaction energy can be written as

$$\text{E.v. } \mathrm{U} = (\Omega, \mathrm{U}\Omega) \tag{1.25}$$

and it is of application also for an indefinite number of particles.

b) Conjugacy of operators $b(k)$ and $b^\dagger(k)$ and their properties

Operators of functional derivative over $\overline{b}(k)$ and multiplication by $\overline{b}(k)$ were denoted as b and $b^\dagger$. Such a notation supposes that the operators are mutually conjugated. Let us show that the coefficient c_n in (1.2) undefined until now can be chosen to really satisfy this relation. We will give in detail the calculations related to the proof to illustrate how to deal with the functionals.

As is known the conjugacy condition for two operators $b(k)$ and $b^\dagger(k)$ claims the equality

$$(\Omega, b(k)\Omega') = \overline{(\Omega', b^\dagger(k)\Omega)} \tag{1.26}$$

for any two functionals Ω and Ω'. Let the functionals Ω_n defined in terms of function ψ_n according to (1.2), (1.3) and Ω'_n be defined in the same way by ψ'_n. We find the functions ψ^*_n and ψ^{**}_n corresponding to functionals $\Omega^* = b(k)\Omega'$ and $\Omega^{**} = b^\dagger(k)\Omega$. Applying in this case formulae (1.13) and (1.14), we obtain

$$\psi^*_n(k_1, \ldots, k_n) = \frac{(n+1)c_{n+1}}{c_n}\psi'_{n+1}(k, k_1, \ldots, k_n), \tag{1.27}$$

$$\psi^{**}_0 = 0,$$

$$\psi^{**}_n(k_1, \ldots k_n) = \frac{c_{n-1}}{c_n}\{\delta(k-k_1)\psi_{n-1}(k_2, k_3, \ldots k_n)\}_{\text{sym}}. \tag{1.28}$$

In the last expression $\delta(k-k_1)$ is the Dirac δ-function defined by the relation

$$f(k)=\int \delta(k-k_1)f(k_1)\,dk_1, \tag{1.29}$$

which is supposed to be valid for any function f(k). The subscript "sym" at the braces means that the expression in them is symmetrized with respect to $k_1,\ldots,k_n$ (any function ψ_n is supposed to be symmetric by the condition). Hence, the left-hand side of equality (1.26) is

$$(\Omega,b(k)\Omega')=\sum_{n=0}^{\infty}\int \overline{\psi}_n\psi_n^*\,dk_1\ldots dk_n= \tag{1.30}$$

$$=\sum_{n=0}^{\infty}\frac{(n+1)c_{n+1}}{c_n}\int \overline{\psi}_n(k_1,k_2,\ldots,k_n)\psi'_{n+1}(k,k_1,\ldots,k_n)\,k_1\ldots dk_n.$$

On the other hand, the quantity conjugated with the right-hand side of (1.26) is equal to

$$\begin{aligned}(\Omega',b^\dagger(k)\Omega)&=\sum_{n=0}^{\infty}\int \overline{\psi}'{}_{n+1}\psi_{n+1}\,dk_1\ldots k_{n+1}=\\&=\sum_{n=0}^{\infty}\frac{c_n}{c_{n+1}}\int \overline{\psi}'_{n+1}(k_{n+1},k_1,\ldots,k_n)\delta(k-k_{n+1})\times\\&\qquad\times\psi_n(k_1,\ldots,k_n)\,dk_1\ldots dk_{n+1}\end{aligned} \tag{1.31}$$

and after integration on variable k_{n+1}, we get

$$(\Omega',b^\dagger(k)\Omega)=\sum_{n=0}^{\infty}\frac{c_n}{c_{n+1}}\int \overline{\psi}'_{n+1}(k,k_1,\ldots,k_n)\psi_n(k_1,\ldots,k_n)\,dk_1\ldots dk_n. \tag{1.32}$$

The integrals in (1.30) and (1.32) are mutually complex conjugated. Expression (1.30) as a whole should be conjugated to (1.32); for this, it is necessary that all the factors in front of integrals are conjugated, hence we arrive at

$$(n+1)|c_{n+1}|^2=|c_n|^2. \tag{1.33}$$

The quantity c_n can be taken real and positive because the constant phase factor can be included into the wave function. If additionally one puts $c_0=1$, from (1.33) we get

$$c_n=\frac{1}{\sqrt{n!}}. \tag{1.34}$$

Therefore, the general form of the functional will be

$$\Omega = \sum_{n=0}^{\infty} \Omega_n, \tag{1.35}$$

where

$$\Omega_n = \frac{1}{\sqrt{n!}} \int \psi(k_1, \ldots, k_n)\overline{b}(k_1) \ldots \overline{b}(k_n)\, dk_1 \ldots dk_n. \tag{1.36}$$

Action of operator b(k) on functional Ω, i.e., its functional derivative with respect to $\overline{b}(k)$, is equivalent to the substitution of the function ψ_n by

$$\psi_n^*(k_1, \ldots, k_n) = \sqrt{n+1}\psi_{n+1}(k, k_1, \ldots, k_n) \tag{1.37}$$

and the action of operator $b^\dagger(k)$, i.e., its multiplication by $\overline{b}(k)$, will mean the substitution of ψ_n by

$$\psi_n^{**}(k_1, \ldots, k_n) = \sqrt{n}\,\{\delta(k - k_1)\psi_{n-1}(k_2, \ldots, k_n)\}_{\text{sym}}\,, \tag{1.38}$$

where $\psi_0^{**} = 0$.

Expressions (1.37) and (1.38) give the representation of operators $b(k)$ and $b^\dagger(k)$ in the space of wave functions.

Let us study the properties of these operators in more detail. Using representations (1.18) and (1.19) for $b(k)$ and $b^\dagger(k)$, consider the expression

$$b(k)b^\dagger(k)\Omega = \frac{\delta'}{\delta\overline{b}(k)}[\overline{b}(k')\Omega]. \tag{1.39}$$

The rule for the functional derivatives of a product is the same as for the usual derivatives. Hence,

$$\frac{\delta'}{\delta\overline{b}(k)}\left(\overline{b}(k')\Omega\right) = \frac{\delta'\overline{b}(k')}{\delta\overline{b}(k)}\Omega + \overline{b}(k')\frac{\delta'\Omega}{\delta\overline{b}(k)}. \tag{1.40}$$

But we have

$$\overline{b}(k') = \int \overline{b}(k)\delta(k - k')\, dk, \tag{1.41}$$

and, therefore,

$$\delta\overline{b}(k') = \int \delta\overline{b}(k)\delta(k - k')\, dk. \tag{1.42}$$

The coefficient at the variation $\delta\overline{b}(k)$ under the integral is Dirac's function $\delta(k - k')$, consequently,

$$\frac{\delta\overline{b}(k')}{\delta\overline{b}(k)} = \delta(k - k') \tag{1.43}$$

and the right-hand side of (1.40) takes the form

$$b(k)b^\dagger(k)\Omega = \delta(k-k')\Omega + b^\dagger(k)b(k)\Omega. \tag{1.44}$$

The relation for operators $b(k)$ and $b^\dagger(k)$ is as follows:

$$b(k)b^\dagger(k) - b^\dagger(k)b(k) = \delta(k-k'). \tag{1.45}$$

Obviously, besides this,

$$b(k)b(k') - b(k')b(k) = 0, \tag{1.46}$$

$$b^\dagger(k)b^\dagger(k') - b^\dagger(k')b^\dagger(k) = 0\,. \tag{1.47}$$

Equality (1.46) takes place because in the given representation operators $b^\dagger(k)$ are reduced to usual multiplication by $\overline{b}(k)$ and equality (1.47) follows from it by passage to conjugate operators.

The operators $b(k)$ are usually called quantized wave functions or quantized amplitudes, and relations (1.45), (1.46) and (1.47) are known as commutation relations.

2 Quantum Electrodynamics

a) Basic ideas

In the preceding paragraph, we considered the problem of the description of a system with an indefinite number of particles obeying Bose statistics, and no assumptions were made about the system itself or the law of its evolution in time. We dealt with a kind of kinematics of such systems. Now we should come to dynamics. To do this, we should take some specific physical system and study the law of its evolution in time.

We will treat a system consisting of matter and light. Let us assume that the number of material particles remains permanent. As is known, this is only an approximate assumption since experimental data show the creation and annihilation of electrons and positrons. But in many phenomena these effects are not essential and for them our suggestion about the permanent number of particles would be a reasonable approximation to reality.

As to light, we will treat it as an ensemble of light quanta whose number can vary. The light is only a special case of the electromagnetic field; apart from it, electrostatic forces also exist. Hence, one can give two different formulations of the basic equations of quantum electrodynamics.

In the first formulation, operators corresponding to a general form of electromagnetic field are considered, namely, the operators for vector and scalar potentials

$$A_x(\mathbf{r},t), \quad A_y(\mathbf{r},t), \quad A_z(\mathbf{r},t), \quad \Phi(\mathbf{r},t). \tag{2.1}$$

The electrostatic forces between the charged particles have not been introduced. Every particle is supposed to interact only with the surrounding field and just this field carries the interaction between the particles. This approach is of principal interest in the respect that it can be treated as a coherent consequence of the short-range action idea.

From this approach, one can deduce in a pure mathematical way the second formulation, in which instead of four operators (2.1) one considers only three field operators for three components of a vector potential

$$B_x(\mathbf{r},t), \quad B_y(\mathbf{r},t), \quad B_z(\mathbf{r},t) \tag{2.2}$$

constrained by the condition

$$\operatorname{div}\mathbf{B}(\mathbf{r},t) = 0, \tag{2.3}$$

i.e., in fact there are only two linearly independent field operators but electrostatic forces enter explicitly.

The equivalence of both approaches is itself a fact of great interest. Indeed, here for the first time the electrostatic forces are deduced from the ideas of the quantum nature of the electromagnetic field. However, we will not prove here the equivalence of both formulations, but below will follow only the second approach.

b) Energy of the light field

As is known the components of the electromagnetic field satisfy the D'Alembert equation

$$\Delta F - \frac{1}{c^2}\frac{\partial^2 F}{\partial t^2} = 0. \tag{2.4}$$

The same equation is also valid for components of vector potential $\mathbf{B}(\mathbf{r},t)$. Therefore, we can write them as the Fourier integral

$$\mathbf{B}(\mathbf{r},t) = \frac{1}{(2\pi)^{3/2}}\int \mathbf{B}(\mathbf{k})e^{-ickt+i\mathbf{kr}}\,d\mathbf{k} + \frac{1}{(2\pi)^{3/2}}\int \mathbf{B}^\dagger(\mathbf{k})e^{ickt-i\mathbf{kr}}\,d\mathbf{k}. \tag{2.5}$$

Here $\mathbf{k}$ is a wave vector with the components k_x, k_y, k_z and dk is written instead of $dk_x dk_y dk_z$. The absolute value of vector $\mathbf{k}$ is denoted as k. Due to the condition $\operatorname{div}\mathbf{B} = 0$ the amplitudes $\mathbf{B}(\mathbf{k})$ are constrained by the relation

$$\mathbf{k}\mathbf{B}(\mathbf{k}) = 0. \tag{2.6}$$

By the known formulae

$$\mathbf{E} = -\frac{1}{c}\frac{\partial \mathbf{B}}{\partial t}, \quad \mathbf{H} = \operatorname{curl}\mathbf{B} \tag{2.7}$$

all the components of the electric and magnetic fields can be given in terms of $\mathbf{B}$. Let H denote the energy of the field. In the Heaviside units, we get

$$H = \frac{1}{2}\int (\mathbf{E}^2 + \mathbf{H}^2)dV. \tag{2.8}$$

Here the energy is expressed as the volume integral. We can also represent it as the integral over the wave vector. The components of electric and magnetic fields can be written as Fourier integrals of the type (2.5) with amplitudes $\mathbf{E}(\mathbf{k})$ and $\mathbf{H}(\mathbf{k})$ related as

$$\mathbf{E}(\mathbf{k}) = ik\mathbf{B}(\mathbf{k}), \qquad \mathbf{H}(\mathbf{k}) = i[\mathbf{k}\times\mathbf{B}(\mathbf{k})]. \tag{2.9}$$

Inserting in (2.8), instead of $\mathbf{E}$ and $\mathbf{H}$, their Fourier transforms with amplitudes (2.9), applying the completeness relation and using (2.6), we get the energy of the field as

$$H = 2\int \mathbf{B}^\dagger(\mathbf{k})\mathbf{B}(\mathbf{k})k^2 d\mathbf{k}. \tag{2.10}$$

Due to (2.6), all three components of vector $\mathbf{B}(\mathbf{k})$ can be expressed by only two independent quantities. Let $\mathbf{e}(\mathbf{k},1)$ and $\mathbf{e}(\mathbf{k},2)$ be two mutually orthogonal unit vectors both perpendicular to wave vector $\mathbf{k}$. Then we can write

$$\mathbf{B}(\mathbf{k}) = B(k,1)\mathbf{e}(\mathbf{k},1) + B(k,2)\mathbf{e}(\mathbf{k},2), \tag{2.11}$$

where $B(k,1)$ and $B(k,2)$ are two scalar quantities. In their terms the energy of the field reads as

$$H = 2\int \{B^\dagger(k,1)B(k,1) + B^\dagger(k,2)B(k,2)\}k^2\, d\mathbf{k}, \tag{2.12}$$

or more briefly

$$H = 2\sum_j \int B^\dagger(\mathbf{k},j)B(\mathbf{k},j)k^2\, d\mathbf{k}, \tag{2.13}$$

where index j takes the values 1 and 2. This index can be treated as the label of a polarization state.

c) Field quantization

In the preceding paragraph, the equations of electromagnetic field were considered within the framework of classical theory. Now let us pass to quantum mechanics.

In quantum electrodynamics, we put in correspondence to the field components some operators with the eigenvalues representing the observable values of the field. These operators act on some functional, which is similar to that studied in Section 1 of this paper, and describes a set of an indefinite number of light quanta. We should derive the main properties of these operators and find a representation for them.

The field is expressed in terms of Fourier amplitudes and, therefore, the field operators are expressed by operators for amplitudes. We will study the latter because they have simpler properties.

We will consider the field as an ensemble of light quanta, or in other words as a set of oscillators. We will apply the usual rules of quantum mechanics to these oscillators and in this way obtain the quantum properties of the field.

A quantum of light is described by a wave vector and by its state of polarization. Therefore, an oscillator is introduced for each wave vector $\mathbf{k}$ (or for each infinitely small interval of its values) and for each polarization state j. We take a quantity proportional to $B(\mathbf{k}, j)$ as the complex amplitude of an oscillator.

If one includes in the amplitude its time dependence, i.e., the factor e^{-ickt}, then instead of $B(\mathbf{k}, j)$ one should take the product $B(\mathbf{k}, j)e^{-ickt}$, because just this product enters the Fourier integrals (2.5). This quantity corresponds to that complex combination $\xi = q + \dfrac{i}{\omega}\dot{q} = ae^{-i\omega t}$ of the oscillator coordinate $q = a\cos(\omega t)$ and the velocity $\dot{q} = a\omega\sin(\omega t)$, which depends on time as $e^{-i\omega t}$. According to quantum mechanics, the oscillator coordinate q and the velocity $\dot{q}$ of the same oscillator (or a momemtum $p = m\dot{q}$) do not commute and satisfy the relation

$$qp - pq = i\hbar, \tag{2.14}$$

where $\hbar$ is the Plank constant divided by 2π. For different oscillators operators p and q commute. Hence, for complex coordinates $\xi = q + \dfrac{i}{m\omega}p$ and $\xi^\dagger = q - \dfrac{i}{m\omega}p$ of the same oscillator the following relation holds:

$$\xi\xi^\dagger - \xi^\dagger\xi = \frac{2\hbar}{m\omega}, \tag{2.15}$$

while for different oscillators ξ and $\xi^\dagger$ commute. Because in our case $B(\mathbf{k},j,t) = B(\mathbf{k},j)e^{-ickt}$ play the role of ξ, they should satisfy the relations

$$B(\mathbf{k},j,t)B^\dagger(\mathbf{k}',j',t) - B^\dagger(\mathbf{k}',j',t)B(\mathbf{k},j,t) = f(\mathbf{k},j)\delta(\mathbf{k}-\mathbf{k}')\delta_{jj'}, \tag{2.16}$$

where f(**k**,j) is a function undefined until now, $\delta(\mathbf{k}-\mathbf{k}')$ is the Dirac function and $\delta_{jj'} = 1$ at $j = j'$ and $\delta_{jj'} = 0$ for $j \neq j'$. In fact, when $\mathbf{k} \neq \mathbf{k}'$ or $j \neq j'$ the right-hand side of (2.16) is zero as it should be for different oscillators. For $j = j'$ and for $\mathbf{k} = \mathbf{k}'$ the right-hand side goes to infinity, but its integral over $\mathbf{k}$ will be finite (more rigorously, one can derive (2.16) from relation (2.15) considering infinitely small intervals of $\mathbf{k}$). When instead of the function $B(\mathbf{k},j,t)$ one inserts the expression $B(\mathbf{k},j)e^{-ickt}$ into (2.16), the exponential factor disappears and for operators $B(\mathbf{k},j)$ we will have the relation

$$B(\mathbf{k},j)B^\dagger(\mathbf{k}',j') - B^\dagger(\mathbf{k}',j')B(\mathbf{k},j) = f(\mathbf{k},j)\delta(\mathbf{k}-\mathbf{k}')\delta_{jj'}. \tag{2.17}$$

Besides this,

$$B(\mathbf{k},j)B(\mathbf{k}',j') - B(\mathbf{k}',j')B(\mathbf{k},j) = 0\,, \tag{2.18}$$

since at $k \neq k'$ and $j \neq j'$ the quantities $B(\mathbf{k},j)$ and $B(\mathbf{k}',j')$ refer to the different oscillators and at $\mathbf{k} = \mathbf{k}'$ and for $j = j'$ relation (2.18) becomes an identity.

To find the form of function $f(\mathbf{k},j)$ in (2.16) and (2.18), we should consider the quantum equations of motion. The energy operator for our set of oscillators is nothing else but expression (2.13) for the energy in terms of amplitudes:

$$H = 2\sum_j \int B^\dagger(\mathbf{k},j)B(\mathbf{k},j)k^2\,d\mathbf{k}. \tag{2.19}$$

Obviously, one can also write

$$H = 2\sum_j \int B^\dagger(\mathbf{k},j,t)B(\mathbf{k},j,t)k^2\,d\mathbf{k}. \tag{2.20}$$

But we know beforehand that $B(\mathbf{k},j,t)$ depends on time via e^{-ickt}e, so

$$\frac{d}{dt}B(\mathbf{k},j,t) = -ickB(\mathbf{k},j,t). \tag{2.21}$$

On the other hand, from the quantum equations of motion we should have

$$\frac{d}{dt}B(\mathbf{k},j,t) = \frac{i}{\hbar}\{HB(\mathbf{k},j,t) - B(\mathbf{k},j,t)H\}. \tag{2.22}$$

Equating the right-hand sides of (2.21) and (2.22) we get

$$HB(\mathbf{k},j,t) - B(\mathbf{k},j,t)H = -\hbar ckB(\mathbf{k},j,t). \tag{2.23}$$

Here one can replace $B(\mathbf{k},j,t)$ by $B(\mathbf{k},j)$, so we will have

$$HB(\mathbf{k},j) - B(\mathbf{k},j)H = -\hbar ckB(\mathbf{k},j). \tag{2.24}$$

Calculating the left-hand side with the help of the definition of H and the commutation relations (2.17) and (2.19) for $B(\mathbf{k},j)$, we arrive at

$$HB(\mathbf{k},j) - B(\mathbf{k},j)H = -2k^2 f(\mathbf{k},j)B(\mathbf{k},j). \tag{2.25}$$

Comparing the right-hand sides of (2.24) and (2.25), we find $f(\mathbf{k},j)$:

$$f(\mathbf{k},j) = \frac{\hbar c}{2k}. \tag{2.26}$$

Consequently, if we put

$$b(\mathbf{k},j) = \sqrt{\frac{2k}{\hbar c}}B(\mathbf{k},j), \tag{2.27}$$

the new amplitude operators $b(\mathbf{k},j)$ will satisfy the commutation relations

$$b(\mathbf{k},j)b^\dagger(\mathbf{k}',j') - b^\dagger(\mathbf{k}',j')b(\mathbf{k},j) = \delta(\mathbf{k}-\mathbf{k}') = \delta_{jj'}, \tag{2.28}$$

$$b(\mathbf{k},j)b(\mathbf{k}',j') - b(\mathbf{k}',j')b(\mathbf{k},j) = 0, \tag{2.29}$$

$$b^\dagger(\mathbf{k},j)b^\dagger(\mathbf{k}',j') - b^\dagger(\mathbf{k}',j')b^\dagger(\mathbf{k},j) = 0. \tag{2.30}$$

These relations coincide with those given by (1.45), (1.46) and (1.47) with the only difference that in the former case the letter k denoted the set of all variables and now besides the wave vector $\mathbf{k}$ a new variable j is written explicitly. Since all field operators are expressed in terms of $b(\mathbf{k},j)$, the representation for these operators as well as the general form of a functional, to which they are applied, can be supposed to be known from the results of Section 1.

Let us write down the field operators in terms of $b(\mathbf{k}, j)$. The amplitude of a vector potential takes the form

$$\mathbf{B}(\mathbf{k}) = \sqrt{\frac{\hbar c}{2k}} \sum_j b(\mathbf{k}, j)\mathbf{e}(\mathbf{k}, j). \qquad (2.31)$$

Substituting this expression into the Fourier integral (2.5), we obtain the vector potential as a function of coordinates and time. Now the field energy reads

$$H = \sum_j \int \hbar c k b^\dagger(\mathbf{k}, j) b(\mathbf{k}, j)\, d\mathbf{k}. \qquad (2.32)$$

Finally we can introduce the operator N of the total number of light quanta

$$N = \sum_j \int b^\dagger(\mathbf{k}, j) b(\mathbf{k}, j)\, d\mathbf{k}. \qquad (2.33)$$

This operator will have the eigenvalues equal to integer $N = 0, 1, 2, \ldots$. Commutation relations for the components of the electromagnetic field, originally obtained by Heisenberg and Pauli,[2] follow from relations (2.28), (2.29), (2.30) for operators $b(\mathbf{k}, j)$. The resulting uncertainty relations for the field were studied in detail by Bohr and Rosenfeld.[3]

d) Basic equation of quantum electrodynamics

If we neglect the radiation and treat the system of material particles in an external field, the energy operator of the system becomes

$$H^0 = \sum_{q=1}^{s} (T_q + U_q) + \sum_{p>q=1}^{s} U_{pq}. \qquad (2.34)$$

Here T_q stands for the kinetic energy operator of a q-th particle, U_q for the operator of potential energy of the same particle in an external field and U_{pq} for an interaction energy operator for particles p and q. If we accept the Dirac equation for a particle, the kinetic energy operator T_q will be

$$\begin{aligned} T_q = {} & m_q c^2 \alpha_4^{(q)} + c\alpha_x^{(q)} \left[p_x^{(q)} - \frac{\varepsilon_q}{c} B_x^0\right] + \\ & + c\alpha_y^{(q)} \left[p_y^{(q)} - \frac{\varepsilon_q}{c} B_y^0\right] + c\alpha_z^{(q)} \left[p_z^{(q)} - \frac{\varepsilon_q}{c} B_z^0\right]. \end{aligned} \qquad (2.35)$$

---

[2]W. Heisenberg, W. Pauli, Zs. Phys. **59**, 1, 1929; *ibid* **59**, 168, 1929.

[3]N. Bohr, L. Rosenfeld, Kgl. Danske. Vid. Selskab. Math.–Phys. Medd. XII, 1933.

Here B_x^0, B_y^0, B_z^0 are the components of vector potential of an external field; ε_q and m_q stand for the charge and mass of a q-th particle, respectively, and $p_x^{(q)}$, $p_y^{(q)}$, $p_z^{(q)}$ are operators:

$$p_x^{(q)} = -i\hbar\frac{\partial}{\partial x_q}, \quad p_y^{(q)} = -i\hbar\frac{\partial}{\partial y_q}, \quad p_z^{(q)} = -i\hbar\frac{\partial}{\partial z_q}; \tag{2.36}$$

$\alpha_4^{(q)}$, $\alpha_x^{(q)}$, $\alpha_y^{(q)}$, $\alpha_z^{(q)}$ are the Dirac matrices.

The operator U_q is represented by

$$U_q = \varepsilon_q \Phi^0(\mathbf{r}_q, t), \tag{2.37}$$

where Φ^0 is a scalar potential of an external field. The interaction energy is

$$U_{pq} = \frac{\varepsilon_p \varepsilon_q}{4\pi|\mathbf{r}_p - \mathbf{r}_q|}. \tag{2.38}$$

The wave function ψ describing an s-particle system satisfies the equation

$$H^0\psi = i\hbar\frac{\partial\psi}{\partial t}. \tag{2.39}$$

Now let us take into account the radiation of the system by itself. To do this, we need to extend the system by including light quanta. Then its state will be described not only by a single function

$$\psi = \psi(\mathbf{r}_1, \zeta_1, \mathbf{r}_2, \zeta_2, \ldots, \mathbf{r}_s, \zeta_s) \tag{2.40}$$

but by a sequence of functions

$$\begin{aligned}
&\psi_0 = \psi_0(\mathbf{r}_1, \zeta_1, \ldots, \mathbf{r}_s, \zeta_s)\\
&\psi_1 = \psi_1(\mathbf{r}_1, \zeta_1, \ldots, \mathbf{r}_s, \zeta_s; \mathbf{k}_1, j_1)\\
&\psi_2 = \psi_2(\mathbf{r}_1, \zeta_1, \ldots, \mathbf{r}_s, \zeta_s; \mathbf{k}_1, j_1, \mathbf{k}_2, j_2)\\
&\cdots\cdots\cdots\cdots\cdots\cdots\cdots\cdots\cdots\cdots\\
&\psi_n = \psi_1(\mathbf{r}_1, \zeta_1, \ldots, \mathbf{r}_s, \zeta_s; \mathbf{k}_1, j_1, \ldots, \mathbf{k}_n, j_n)\\
&\cdots\cdots\cdots\cdots\cdots\cdots\cdots\cdots\cdots\cdots
\end{aligned} \tag{2.41}$$

These functions must be symmetric with respect to variables $\mathbf{k}_1, j_1, \ldots, \mathbf{k}_n, j_n$. If material particles obey the Pauli principle, the functions should be additionally antisymmetric with respect to variables $\mathbf{r}_1, \zeta_1, \ldots, \mathbf{r}_s, \zeta_s$. If function ψ differs from zero, there is a probability to find some number n of light quanta besides the s material particles.

Instead of a sequence of functions (2.41) *only one* quantity can be considered, namely, the functional Ω of the kind treated in Section 1.

To get the wave equation for this functional, it is sufficient to add to the vector potential $\mathbf{B}^0$ of the external field a vector potential $\mathbf{B}$ of radiation, i.e., the operator studied in the preceding paragraph, within the operator H^0. Then the operator H^0 is replaced by H:

$$H = H^0 - L, \tag{2.42}$$

where L is the operator;

$$L = \sum_{q=1}^{s} \varepsilon_q \left(\alpha_x^{(q)} B_x(\mathbf{r}_q, t) + \alpha_y^{(q)} B_y(\mathbf{r}_q, t) + \alpha_z^{(q)} B_z(\mathbf{r}_q, t) \right). \tag{2.43}$$

The wave equation for the functional Ω becomes

$$H^0\Omega - i\hbar \frac{\partial \Omega}{\partial t} = L\Omega. \tag{2.44}$$

Taken with the minus, operator L can be interpreted as the interaction energy operator of a material system with a radiation. Consider operator L in slightly more detail.

Substituting the function $B(\mathbf{r}, t)$ in (2.43) as the Fourier integral and denoting for brevity by Q_x (the component of vector $\mathbf{Q}$) the following sum

$$Q_x = \sum_{q=1}^{s} \varepsilon_q \alpha_x^{(q)} e^{-i\mathbf{k}\mathbf{r}_q}, \tag{2.45}$$

we can write

$$L = \frac{1}{(2\pi)^{3/2}} \int \mathbf{Q}^\dagger \mathbf{B}(\mathbf{k}) e^{-ickt}\, d\mathbf{k} + \frac{1}{(2\pi)^{3/2}} \int \mathbf{Q}\mathbf{B}^\dagger(\mathbf{k}) e^{ickt}\, d\mathbf{k}. \tag{2.46}$$

According to (2.31) the scalar product of vectors $\mathbf{Q}^\dagger$ and $\mathbf{B}(\mathbf{k})$ can be written in terms of $b(\mathbf{k}, j)$. We obtain

$$\mathbf{Q}^\dagger \mathbf{B}(\mathbf{k}) = \sqrt{\frac{\hbar c}{2k}} \sum_j b(\mathbf{k}, j) \sum_{q=1}^{s} \varepsilon_q \gamma^{(q)}(\mathbf{k}, j) e^{i\mathbf{k}\mathbf{r}_q}, \tag{2.47}$$

where $\boldsymbol{\gamma}^{(q)}$ stands for the matrix

$$\gamma^{(q)}(\mathbf{k}, j) = \alpha_x^{(q)} e_x(\mathbf{k}, j) + \alpha_y^{(q)} e_y(\mathbf{k}, j) + \alpha_z^{(q)} e_z(\mathbf{k}, j). \tag{2.48}$$

This matrix represents a projection of the matrix vector $\boldsymbol{\alpha}^{(q)}$ on the direction j orthogonal to the wave vector. Matrices $\boldsymbol{\gamma}^{(q)}$ have the properties similar to that of the Dirac $\boldsymbol{\alpha}^{(q)}$ matrices, for example, their squares are equal to unity. Inserting expression (2.47) into equation (2.46), we finally get the operator L as

$$L = \sum_j \int \{G^\dagger(\mathbf{k}, j) b(\mathbf{k}, j) + G(\mathbf{k}, j) b^\dagger(\mathbf{k}, j)\}\, d\mathbf{k}, \tag{2.49}$$

where the $G(\mathbf{k}, j)$ denotes the operator

$$G(\mathbf{k}, j) = \frac{e^{ickt}}{(2\pi)^{3/2}} \sqrt{\frac{\hbar c}{2k}} \sum_{q=1}^{s} \varepsilon_q \gamma^{(q)}(\mathbf{k}, j) e^{-i\mathbf{k}\mathbf{r}_q}, \tag{2.50}$$

proportional to the component of vector $\mathbf{Q}$ along the direction j.

Therefore, the wave equation can be written as

$$H^0\Omega - i\hbar\frac{\partial\Omega}{\partial t} = \sum_j \int \{G^\dagger(\mathbf{k}, j) b(\mathbf{k}, j) + G(\mathbf{k}, j) b^\dagger(\mathbf{k}, j)\}\, d\mathbf{k}\,\Omega, \tag{2.51}$$

or, if we take into account the representation of b and $b^\dagger$ derived in Section 1,

$$H^0\Omega - i\hbar\frac{\partial\Omega}{\partial t} = \sum_j \int \left\{G^\dagger(\mathbf{k}, j)\frac{\delta\Omega}{\delta\bar{b}(\mathbf{k}, j)} + G(\mathbf{k}, j)\bar{b}(\mathbf{k}, j)\Omega\right\} d\mathbf{k}. \tag{2.52}$$

It is the basic equation of quantum electrodynamics. It sums up all that is known about the interaction of light with atomic or molecular systems. Particularly, it gives a basis for the Schrödinger theory of radiation based on the correspondence principle. Besides the results derived from the Schrödinger theory, it gives a series of new results explaining, for instance, the natural line width of spectral lines that is evidence of incomplete stability of so-called stationary states. It can be applied also to the problem of light scattering by free electrons (the Klein–Nishina formula), in computing correcting terms to the Coulomb interaction between charged particles and in other areas.

As was mentioned in the Introduction, the basic equation of quantum electrodynamics can be solved only approximately. This equation does not account for the structure of material particles and their properties at rather high energies; therefore its accurate solutions hardly have a physical meaning. More exact description of matter and radiation properties should be based on essentially new ideas and this is one of the main problems of the further development in theoretical physics.

Translated by A. V. Tulub

37-2*
Proper Time in Classical and Quantum Mechanics[1]

V. Fock

Leningrad University, Physical Institute,
P.N. Lebedev Physical Institute, the USSR Academy of Sciences, Moscow

Izv. AN **4–5**, 551, 1937
Fock59, pp. 141–158

I Classical Mechanics

1. Let L^0 be the ordinary Lagrange function, from which the equation of motion of a charged particle of matter in special relativity in the presence of an external field is deduced. It is known that L^0 is given by

$$L^0 = -mc^2\sqrt{1-\beta^2} - \frac{e}{c}(x'A_x + y'A_y + z'A_z) + e\Phi, \tag{1}$$

where

$$\beta^2 = \frac{1}{c^2}(x'^2 + y'^2 + z'^2). \tag{2}$$

Here the primes denote the time derivatives. Introduce the proper time

$$\tau = \int_{t^0}^{t} \sqrt{1-\beta^2}dt, \tag{3}$$

and write down the action integral

$$S = \int_{t^0}^{t} L^0 dt. \tag{4}$$

The variation of this action gives the equations of motion in the form

$$S = \int_0^{\tau} L^0 \frac{dt}{d\tau} d\tau. \tag{5}$$

[1]Reported at the Section of Physics, the USSR Academy of Sciences on 14 March 1937.

Since the value of the upper integration limit depends on the shape of the trajectory, this upper limit should also be varied. Therefore, the proper time cannot be taken for an independent (i.e., not being a subject of variation) variable for the Lagrange function $L^0 \frac{dt}{d\tau}$.

2. However, one can introduce another Lagrange function L. Let

$$L = \frac{1}{2}m(\dot{x}^2 + \dot{y}^2 + \dot{z}^2 - c^2\dot{t}^2) - \frac{1}{2}mc^2 -$$

$$-\frac{e}{c}(\dot{x}A_x + \dot{y}A_y + \dot{z}A_z) + e\dot{t}\Phi, \tag{6}$$

where the dots denote the differentiation by the independent variable τ (which will turn out to coincide with the proper time).

Indeed, writing down the extremum condition for the integral

$$S = \int_0^\tau L dt, \tag{7}$$

where the upper limit is not varied, we obtain the ordinary Lagrange equations

$$\left.\begin{array}{l} \dfrac{d}{d\tau}\dfrac{\partial L}{\partial \dot{x}} - \dfrac{\partial L}{\partial x} = 0, \\ \dots\dots\dots\dots\dots\dots \\ \dfrac{d}{d\tau}\dfrac{\partial L}{\partial \dot{t}} - \dfrac{\partial L}{\partial t} = 0. \end{array}\right\} \tag{8}$$

Since L does not depend on τ explicitly, these equations admit the integral

$$\dot{x}^2 + \dot{y}^2 + \dot{z}^2 - c^2\dot{t}^2 = \text{const.} \tag{9}$$

The requirement that the value of the constant is $-c^2$,

$$\dot{x}^2 + \dot{y}^2 + \dot{z}^2 - c^2\dot{t}^2 = -c^2, \tag{10}$$

is equivalent to the condition that the independent variable τ is just the proper time; and the Lagrange equations turn out to be the equations of motion for a particle of matter in special relativity.

Note that using (6) we can show that

$$L = L^0 \frac{dt}{d\tau}, \tag{11}$$

where L^0 is given by (1).

3. One can easily derive the Hamilton equations of motion as well as the Hamilton–Jacobi equation for the Lagrange function (6). For the latter, one gets

$$\frac{\partial S}{\partial \tau} + \frac{1}{2m}\left[\left(\operatorname{grad} S + \frac{e}{c}\mathbf{A}\right)^2 - \frac{1}{c^2}\left(\frac{\partial S}{\partial t} - e\Phi\right)^2 + mc^2\right] = 0. \qquad (12)$$

In the theory under consideration the variables x, y, z, t play the role of coordinates, though the variable τ plays the role of the time of the classical nonrelativistic mechanics. For this reason, we can freely use the formulae known from the classical mechanics.

Let the action integral (7) expressed in terms of the variables x, y, z, t, x^0, y^0, z^0; τ be

$$S = S(x, y, z, t, x^0, y^0, z^0; \tau). \qquad (13)$$

However, τ is considered here as one of the independent variables. Therefore, the derivative $\frac{\partial S}{\partial \tau}$ is constant as it follows from the equations of motion. Assuming this constant to be zero,

$$\frac{\partial S}{\partial \tau} = 0, \qquad (14)$$

we get the condition for τ to be the proper time.

The generalized momenta canonically conjugated to the coordinates and the time can be expressed via partial derivatives of the function S:

$$p_x = \frac{\partial S}{\partial x}; \quad p_y = \frac{\partial S}{\partial y}; \quad p_z = \frac{\partial S}{\partial z}; \quad p_t = -H = \frac{\partial S}{\partial t}, \qquad (15)$$

$$p_x^0 = -\frac{\partial S}{\partial x^0}; \quad p_y^0 = -\frac{\partial S}{\partial y^0}; \quad p_z^0 = -\frac{\partial S}{\partial z^0}; \quad p_t^0 = -H^0 = -\frac{\partial S}{\partial t^0}. \qquad (16)$$

Note that if we solve equation (14) with respect to the variable τ and substitute the result into (13), we get the usual action function

$$S = S^*(x, y, z, t, x^0, y^0, z^0, t^0). \qquad (17)$$

Moreover, it obviously follows from condition (14) that

$$\frac{\partial S^*}{\partial x} = \frac{\partial S}{\partial x} + \frac{\partial S}{\partial \tau} \cdot \frac{\partial \tau}{\partial x} = \frac{\partial S}{\partial x}. \qquad (18)$$

However, in practice such exclusion of τ is inconvenient. In several problems, for example in the problem of motion of an electron in an electric and magnetic field, function (13) containing τ is rather simple, although the function S^* cannot be expressed in explicit form.

II Proper Time in the Dirac Equation

4. The Dirac equation for the electron in an electromagnetic field is

$$\left\{(\boldsymbol{\alpha}\mathbf{P}) + mc\,\alpha_4 - \frac{T}{c}\right\}\psi = 0, \tag{1}$$

where $\boldsymbol{\alpha}$ is the vector constructed out of the matrices $\alpha_1, \alpha_2, \alpha_3$; $\mathbf{P}$ is the momentum vector, i.e., the operator

$$\mathbf{P} = -i\hbar\,\mathrm{grad} + \frac{e}{c}\,\mathbf{A}, \tag{2}$$

and T is the kinetic energy operator:

$$T = i\hbar\frac{d}{dt} + e\Phi. \tag{3}$$

A solution for the Dirac equation can be represented in the form

$$\psi = \left\{(\boldsymbol{\alpha}\mathbf{P}) + mc\alpha_4 - \frac{T}{c}\right\}\Psi, \tag{4}$$

where Ψ satisfies the second-order equation

$$\left\{\mathbf{P}^2 + m^2c^2 - \frac{T^2}{c^2} + \frac{e\hbar}{c}(\boldsymbol{\sigma}\cdot\mathfrak{H}) - \frac{ie\hbar}{c}(\boldsymbol{\alpha}\cdot\mathfrak{E})\right\}\psi = 0. \tag{5}$$

This equation can be rewritten as

$$\hbar^2\Lambda\Psi = 0, \tag{6}$$

where Λ is the operator defined by

$$\Lambda\Psi = -\Box\Psi - \frac{2ie}{\hbar c}\left(\mathbf{A}\cdot\mathrm{grad}\,\Psi + \frac{\Phi}{c}\cdot\frac{\partial\Psi}{\partial t}\right) + \left\{-\frac{ie}{\hbar c}\left(\mathrm{div}\mathbf{A} + \frac{1}{c}\cdot\frac{\partial\Phi}{\partial t}\right) + \right.$$

$$\left. + \frac{e^2}{\hbar^2c^2}(\mathbf{A}^2 - \Phi^2) + \frac{m^2c^2}{\hbar^2}\right\}\Psi + \frac{e}{\hbar c}\left\{(\boldsymbol{\sigma}\cdot\mathfrak{H}) - i(\boldsymbol{\alpha}\cdot\mathfrak{E})\right\}\Psi. \tag{7}$$

One can look for a solution of equation (5) in the form of a definite integral

$$\Psi = \int_C F d\tau, \tag{8}$$

taken over the auxiliary variable τ between certain fixed limits (or along a certain contour on the complex τ-plane).

It is obvious that equation (5) would be satisfied if we impose on F the equation

$$\frac{\hbar^2}{2m}\Lambda F = i\hbar\frac{\partial F}{\partial \tau} \tag{9}$$

and choose the contour C in such a way that

$$\int_C \frac{\partial F}{\partial \tau} d\tau = F|_{\partial C} = 0. \tag{10}$$

As it will become clear below, the variable τ has the meaning of the proper time. Therefore, it is reasonable to call (9) the proper time Dirac equation.

Let

$$F = e^{\frac{i}{\hbar}S} f; \quad \Psi = \int_C e^{\frac{i}{\hbar}S} f d\tau, \tag{11}$$

where S is the classical action function satisfying (I, 12). In order to obtain the equation for f, note that

$$\Lambda F = e^{\frac{i}{\hbar}S}\Lambda' f; \quad i\hbar\frac{\partial F}{\partial \tau} = e^{\frac{i}{\hbar}S}\left(i\hbar\frac{\partial f}{\partial \tau} - \frac{\partial S}{\partial \tau} f\right), \tag{12}$$

where Λ' is obtained from Λ by substituting $\mathbf{A} + \frac{c}{e}\mathrm{grad}\, S$ instead of $\mathbf{A}$ and $\Phi - \frac{1}{e}\cdot\frac{\partial S}{\partial t}$ instead of Φ. Due to the equation for S the terms of (9), which do not contain $\hbar$, cancel each other, and finally we get the equation

$$2m\frac{df}{d\tau} = \left\{\Box S + \frac{e}{c}\left(\mathrm{div}\mathbf{A} + \frac{1}{c}\cdot\frac{\partial \Phi}{\partial t}\right)\right\} f + \\ + \frac{e}{c}\{i(\boldsymbol{\sigma}\cdot\mathfrak{H}) + (\boldsymbol{\alpha}\cdot\mathfrak{E})\} f = i\hbar\Box f. \tag{13}$$

Here $\frac{df}{d\tau}$ is the sign of the "full derivative":

$$\frac{df}{d\tau} = \frac{\partial f}{\partial \tau} + \dot{x}\frac{\partial f}{\partial x} + \dot{y}\frac{\partial f}{\partial y} + \dot{z}\frac{\partial f}{\partial z} + \dot{t}\frac{\partial f}{\partial t}, \tag{14}$$

where by $\dot{x}, \dot{y}, \dot{z}, \dot{t}$ we assume the classical expressions

$$\left.\begin{array}{l} \dot{x} = \dfrac{1}{m}\left(\dfrac{\partial S}{\partial x} + \dfrac{e}{c}A_x\right), \\ \dots\dots\dots\dots\dots\dots\dots\dots \\ \dot{t} = \dfrac{1}{mc^2}\left(\dfrac{\partial S}{\partial t} - e\Phi\right). \end{array}\right\} \tag{15}$$

5. Equation (13) is very convenient for finding approximate solutions by the Brillouin–Wentzel method. In some cases, e.g., for constant electric and magnetic fields, this method gives an exact solution. The constant $\hbar$ enters only the right-hand side of the equation. By omitting the right-hand side, we obtain the equation

$$2m\frac{df}{d\tau} = \left\{\Box S + \frac{e}{c}\left(\operatorname{div}\mathbf{A} + \frac{1}{c}\cdot\frac{\partial\Phi}{\partial t}\right)\right\} f+$$

$$+\frac{e}{c}\{i(\boldsymbol{\sigma}\cdot\mathfrak{H}) + (\boldsymbol{\alpha}\cdot\mathfrak{E})\}f = 0, \tag{16}$$

the solution of which reduces to solving a system of ordinary differential equations (rather than partial differential equations).

One can get rid of the term containing $\Box S$ in equation (16). Indeed, denote by ϱ the absolute value of the fourth order determinant of the matrix of derivatives of S taken once by x, y, z, t and once by x^0, y^0, z^0, t^0 (or, in other words, by the corresponding integration constants):

$$\varrho = \operatorname{Det}\left\|\frac{\partial^2 S}{\partial x \partial y^0}\right\|. \tag{17}$$

The quantity ϱ satisfies the continuity equation:

$$\frac{\partial\varrho}{\partial\tau}\frac{\partial S}{\partial x}(\varrho\dot{x}) + \frac{\partial}{\partial y}(\varrho\dot{y}) + \frac{\partial}{\partial z}(\varrho\dot{z}) + \frac{\partial}{\partial t}(\varrho\dot{t}) = 0. \tag{18}$$

It implies that $\sqrt{\varrho}$ satisfies the equation

$$2m\frac{d\sqrt{\varrho}}{d\tau} + \left\{\Box S + \frac{e}{c}\left(\operatorname{div}\mathbf{A} + \frac{1}{c}\cdot\frac{\partial\Phi}{\partial t}\right)\right\}\sqrt{\varrho} = 0. \tag{19}$$

Therefore, a new function f^0 defined by

$$f = \sqrt{\varrho}f^0 \tag{20}$$

satisfies (in the given approximation) the equation

$$2m\frac{df^0}{d\tau} + \frac{e}{c}i(\boldsymbol{\sigma}\cdot\mathfrak{H}) + (\boldsymbol{\alpha}\cdot\mathfrak{E})f^0 = 0\,. \tag{21}$$

For the case of a constant field we can assume that

$$\frac{\partial f^0}{\partial x} = \frac{\partial f^0}{\partial y} = \frac{\partial f^0}{\partial z} = \frac{\partial f^0}{\partial t} = 0$$

and take into account only the dependence of f^0 on τ. Then equation (21) has constant coefficients and one can easily solve it.

Since in this case ϱ is also a function of τ only, we have $\Box f = 0$ and approximated equation (16) coincides with exact equation (13). Therefore, in the case of a constant field (which will be considered below in paragraph 8) the described method gives an exact solution.

In the general case when the field is not constant one should consider $\mathfrak{E}$ and $\mathfrak{H}$ as functions of τ [expressing x, y, z, t as functions of τ using the equations of the classical trajectory (I,16)]. One should solve the system of ordinary differential equations with variable coefficients and then substitute the expressions (I,16) for $p_x^0, p_y^0, p_z^0, p_t^0$ in the solution. The result is the desired solution of the partial differential equation.

If one replaces the function S by S^* [formula (I,17)] and considers f to be independent on τ, then our formulae will give a generalization and development of the result by Pauli [4] who was the first to apply the Brillouin–Wentzel method to the Dirac equation. However, the Pauli formulae are much more complicated since he has started from the first order Dirac equation instead of the second-order one.

Note that if we apply the stationary phase method to compute the integral (11) we should put out of the integration the exponent of the value of S at the point where $\frac{\partial S}{\partial \tau} = 0$. But it is just the ordinary (containing no τ) action function S^*.

6. The obtained way of solving the Dirac equation (by integrating over the proper time) is especially suitable for solving the Cauchy problem, i.e., for computing ψ from given initial values.

Let ψ satisfy the Dirac equation (1) and the initial condition

$$\psi = \psi^0 \qquad (\text{for } t = t^0). \tag{22}$$

In order to find the function ψ it suffices to find a solution Ψ of the second-order equation $\Lambda\Psi = 0$, satisfying the conditions

$$\Psi = 0; \quad \frac{\partial \Psi}{\partial t} = -\frac{ic}{\hbar}\psi^0 = \dot{\Psi}^0 \qquad (\text{for } t = t^0). \tag{23}$$

Let us look for the function Φ in the form of the integral

$$\Psi = \int Q\dot{\Psi}^0 dV, \tag{24}$$

where

$$dV = dx^0 dy^0 dz^0, \tag{25}$$

$\dot{\Psi}^0$ is a given function of x^0, y^0, z^0 and Q is a function of both x, y, z, t and x^0, y^0, z^0, t^0.

Let

$$\xi = c^2(t - t^0)^2 - (x - x^0)^2 - (y - y^0)^2 - (z - z^0)^2 \tag{26}$$

and let $\gamma(\xi)$ be the function defined by

$$\left.\begin{aligned} \gamma(\xi) &= 1 && \text{for } \xi > 0, \\ \gamma(\xi) &= \tfrac{1}{2} && \text{for } \xi = 0, \\ \gamma(\xi) &= 0 && \text{for } \xi < 0. \end{aligned}\right\} \tag{27}$$

Its derivative $\gamma'(\xi) = \delta(\xi)$ is the Dirac delta function.

The function Q from the integral (24) is the generalized function of the form

$$Q = R\gamma(\xi) + R^*\delta(\xi), \tag{28}$$

where R and R^* are continuous functions. The function R is just the Riemann function.

If we substitute the expression for Q into the integral (24), then it splits into a sum of two integrals

$$\Psi = \int R\dot{\Psi}^0\gamma(\xi)dV + \int R^*\dot{\Psi}^0\delta(\xi)dV. \tag{29}$$

The first of them is the volume integral over the domain

$$c^2(t - t^0)^2 - (\mathbf{r} - \mathbf{r}^0)^2 \geq 0$$

(i.e., over the ball of the radius $r^* = c|t - t^0|$ and with the center at the point $\mathbf{r}^0 - \mathbf{r}$), and the second is the integral over the surface

$$c^2(t - t^0)^2 - (\mathbf{r} - \mathbf{r}^0)^2 = 0$$

(i.e., over the surface of this ball). Indeed, getting rid of the generalized functions, we have

$$\Psi = \int_V R\dot{\Psi}^0 dV + \frac{1}{2r^*}\int_S R^*\dot{\Psi}^0 dS. \tag{30}$$

Since the radius of the ball tends to zero at $t = t^0$, we obviously have $\Psi = 0$ at $t = t^0$. Moreover, the time derivative of the volume integral is zero when $t = t^0$ as well. The surface integral for small values of $t - t^0 > 0$ is

$$\frac{1}{2r^*}\int_S R^*\dot{\Psi}^0 dS = 2\pi r^*(R^*\dot{\Psi}^0)_0 = 2\pi c(t-t^0)(R^*\dot{\Psi}^0)_0. \tag{31}$$

Therefore,

$$\left(\frac{\partial\Psi}{\partial t}\right)_{t=t^0+0} = 2\pi R_0^*\dot{\Psi}^0, \tag{32}$$

where by R_0^* we denoted the value of R^* for $\mathbf{r} = \mathbf{r}^0, t = t^0$. To satisfy the initial condition (23), it is necessary that

$$R_0^* = \frac{1}{2\pi c} \tag{33}$$

independent of time and coordinates.

But the function ψ should satisfy the second-order Dirac equation, which means that the function Q should also satisfy this equation. Therefore, one has

$$\Lambda Q = \lambda[R\gamma(\xi) + R^\delta(\xi)] = 0. \tag{34}$$

Taking into account that

$$\Box\gamma(\xi) = -4\delta(\xi), \tag{35}$$

$$\Box\delta(\xi) = 0, \tag{36}$$

we see that the differentiation in (34) gives terms proportional to $\gamma(\xi)$, $\delta(\xi)$, $\delta'(\xi)$. Denote for brevity the operator M given by

$$MF = (x-x^0)\frac{\partial F}{\partial x}(y-y^0)\frac{\partial F}{\partial y}(z-z^0)\frac{\partial F}{\partial z}(t-t^0)\frac{\partial F}{\partial t}+$$

$$+\frac{ie}{\hbar c}\left\{(x-x^0)A_x + (y-y^0)A_y + (z-z^0)A_z - c(t-t^0)\Phi\right\}F. \tag{37}$$

Then expression (34) can be rewritten as

$$\Lambda(R\gamma(\xi) + R^*\delta(\xi)) =$$

$$= (\Lambda R)\gamma(\xi)\{\Lambda R^* + 4(M+1)R\}\delta(\xi) + 4(MR^*)\delta'(\xi). \tag{38}$$

This expression vanishes if we require that

$$\Lambda R = 0, \tag{39}$$

$$(M+1)R = -\frac{1}{4}\Lambda R^*, \tag{40}$$

$$MR^* = 0. \tag{41}$$

It is sufficient, however, that equation (40) is satisfied at $\xi = 0$.

It is easy to find a solution of equation (41). Indeed, let

$$\chi = \int_{(\mathbf{r^0},t^0)}^{(\mathbf{r},t)} (A_x dx + A_y dy + A_z dz - e\Phi dt), \tag{42}$$

where the integral is taken over the line connecting the points $(\mathbf{r^0}, t^0)$ and $(\mathbf{r}, t)$. Then

$$(\mathbf{r}-\mathbf{r}^0)\cdot \operatorname{grad}\chi + (t-t^0)\frac{\partial\chi}{\partial t} = (\mathbf{r}-\mathbf{r}^0)\mathbf{A} - c(t-t^0)\Phi. \tag{43}$$

Therefore, if we assume that

$$R^* = \frac{1}{2\pi i c}e^{-i\frac{e}{\hbar c}\chi}, \tag{44}$$

the equation $MR^* = 0$ will be satisfied. Moreover, condition (33) will also be satisfied.

Using this value of R^*, equations (39) and (40) allow one to determine the function R. These equations get simplified by the following substitution:

$$R = \frac{1}{2\pi i c}e^{-i\frac{e}{\hbar c}\chi}\cdot R'. \tag{45}$$

Such substitution is equivalent to the replacement of the potentials $\mathbf{A}, \Phi$ by

$$\mathbf{A}' = \mathbf{A} - \operatorname{grad}\chi; \quad \Phi' = \Phi + \frac{1}{c}\cdot\frac{\partial\chi}{\partial t}. \tag{46}$$

Note that if $\mathbf{A}$ and Φ satisfy the equations

$$\Box\mathbf{A} = 0; \quad \Box\Phi = 0; \quad \operatorname{div}\mathbf{A} + \frac{1}{c}\cdot\frac{\partial\Phi}{\partial t} = 0, \tag{47}$$

then the same equations are satisfied by $\mathbf{A}'$ and Φ'. Moreover, they satisfy the relation

$$(\mathbf{r}-\mathbf{r}^0)\mathbf{A}' - c(t-t^0)\Phi' = 0, \tag{48}$$

which follows from (43).

The new potentials are unambiguously expressible via the field. If we denote by double overbars the averaging between the two points $\mathbf{r}^0, t^0$ and $\mathbf{r}, t$ given by the formula

$$\overline{\overline{f}} = 2\int_0^1 f[\mathbf{r}^0 + (\mathbf{r} - \mathbf{r}^0)u, t^0 + (t - t^0)u]u\,du, \tag{49}$$

then we have

$$\left.\begin{aligned} \mathbf{A}' &= -\tfrac{1}{2}[(\mathbf{r} - \mathbf{r}^0)\cdot\overline{\overline{\mathfrak{H}}}] - \tfrac{1}{2}c(t - t^0)\overline{\overline{\mathfrak{E}}}, \\ \Phi' &= -\tfrac{1}{2}(\mathbf{r} - \mathbf{r}^0)\cdot\overline{\overline{\mathfrak{E}}}. \end{aligned}\right\} \tag{50}$$

Using the substitution (45), i.e., after we have introduced the new potentials, equations (39) and (40) turn into

$$\Lambda' R' = 0, \tag{51}$$

$$(L+1)R' = -\frac{1}{4}\Lambda' 1 =$$

$$= -\frac{1}{4}\left[\frac{m^2c^2}{\hbar^2} + \frac{e^2}{\hbar^2c^2}(\mathbf{A}'^2 - \Phi^2)\right] - \frac{1}{4}\cdot\frac{e}{\hbar c}[(\boldsymbol{\sigma}\cdot\mathfrak{H}) - i(\boldsymbol{\alpha}\cdot\mathfrak{E})], \tag{52}$$

where by L we denoted the operator

$$Lf = (\mathbf{r} - \mathbf{r}^0)\cdot\operatorname{grad} f + (t - t^0)\frac{\partial f}{\partial t}, \tag{53}$$

which could also be denoted by M' since it is obtained from M by introducing the new potentials.

One can get out of equation (52) not only the value of $(L+1)R'$, but also the very function R' at $\xi = 0$. Indeed, if one considers a function $f(x, y, z, t)$ as a function of the ratios $(x - x^0) : (y - y^0) : (z - z^0) : c(t - t^0)$ and of ξ and assumes that $\xi\frac{\partial f}{\partial \xi} \to 0$ then $\xi \to 0$, then the value of $(L+1)f$ at $\xi = 0$ is defined by the value of f at $\xi = 0$ and vice versa.

Consider the equation

$$(L + p)f = \varphi(\mathbf{r}, t), \tag{54}$$

where p is a positive integer. It is easy to verify that it is satisfied by the function

$$f(\mathbf{r}, t) = \int_0^1 \varphi[\mathbf{r}^0 + (\mathbf{r} - \mathbf{r}^0)u, t^0 + (t - t^0)u]u^{p-1}du. \tag{55}$$

The only solution of the homogeneous equation $(L+p)f=0$, which is regular around $\mathbf{r}=\mathbf{r}^0, t=t^0$, is just zero. Therefore, for positive values of p the function f is unambiguously defined by formula (55). Taking the right-hand side of equation (52) for φ and applying formula (55), we obtain the well-known expression for the Riemann function for the ball $\xi=0$.

7. As is well known from the classical investigations of the Cauchy problem by Hadamard [2], the Riemann function is closely related to the fundamental solution (*solution élémentaire*) of a given equation (which may be either of elliptic or hyperbolic type). The fundamental solution, e.g., the function $\frac{1}{r}$ for the Laplace equation or the function

$$1/\sqrt{c^2(t-t^0)^2-(x-x^0)^2-(y-y^0)^2}$$

for the equation

$$\frac{\partial^2 u}{\partial x^2}+\frac{\partial^2 u}{\partial y^2}-\frac{1}{c^2}\cdot\frac{\partial^2 u}{\partial t^2}=0,$$

has a singularity either at a given point or on a characteristic cone containing this point. The type of singularity depends on the particular equation. In the case when the number of independent variables is odd the fundamental solution is defined unambiguously. For the case of even independent variables there exist infinitely many fundamental solutions; these solutions have logarithmic singularities, the coefficient of the logarithm being defined unambiguously and being nothing but the Riemann function. One can also construct a fundamental solution for equations of the parabolic type. It can be obtained by a limiting procedure from the elliptic or hyperbolic cases. An elementary example is given by the function

$$u=\frac{1}{\sqrt{y}}e^{-\frac{x^2}{4y}},$$

which is the fundamental solution of the equation

$$\frac{\partial^2 u}{\partial x^2}=\frac{\partial u}{\partial y}.$$

Consider now the Dirac equation. The Riemann function for the second-order Dirac equation can be represented as an integral over the proper time variable

$$R=\int F d\tau, \tag{56}$$

where F is the fundamental solution of the "Dirac equation with proper time," i.e., the equation

$$\frac{\hbar^2}{2m}\Lambda F = i\hbar\frac{\partial F}{\partial \tau}. \tag{57}$$

The independent variables of this equation are x, y, z, t, τ. Since the number of them is odd, the fundamental solution is defined unambiguously.

Clear up the behavior of the fundamental solution around the essential singularity $\tau = 0$. For this purpose, assume as above

$$F = e^{\frac{i}{\hbar}S} f \tag{58}$$

and look for the solution as a power series over τ. The function S satisfies the Hamilton–Jacobi equation with the proper time (I, 12). Substituting the series

$$S = \frac{S_{-1}}{\tau} + S_0 + S_1\tau + S_2\tau^2 + \dots \tag{59}$$

into this equation, we get

$$(\operatorname{grad} S_{-1})^2 - \frac{1}{c^2}\left(\frac{\partial S_{-1}}{\partial t}\right)^2 = 2mS_{-1}; \tag{60}$$

therefore, we can take

$$S_{-1} = \frac{1}{2}m[(\mathbf{r} - \mathbf{r}^0)^2 - c^2(t - t^0)^2] = -\frac{1}{2}m\xi. \tag{61}$$

Using this formula for S_{-1}, one gets the equation for S_0:

$$(\mathbf{r} - \mathbf{r}^0)\cdot \operatorname{grad} S_0 + (t - t^0)\frac{\partial S_0}{\partial t} = -\frac{e}{c}\left\{\mathbf{r} - \mathbf{r}^0)\mathbf{A} - c(t - t^0)\Phi\right\}. \tag{62}$$

This equation coincides with equation (43) for χ up to the factor $-\frac{e}{c}$. Therefore, we are able to let

$$S_0 = -\frac{e}{c}\chi = -\frac{e}{c}\int_{(\mathbf{r}^0, t^0)}^{(\mathbf{r}, t)} (\mathbf{A}d\mathbf{r} - c\Phi dt). \tag{63}$$

Finally, the equation for S_1 is

$$(L + 1)S_1 = -\frac{1}{2m}\left\{m^2c^2 + \frac{e^2}{c^2}(\mathbf{A}'^2 - \Phi'^2)\right\}, \tag{64}$$

where L is given by (53). The solution for this equation can be obtained by formula (55). The further coefficients are unambiguously determined from the equations

$$(L+p)S_p = \varphi_p, \qquad (p = 2, 3, \ldots), \tag{65}$$

where the φ_p is known as far as the previous terms are determined. Finally, we have

$$S = -\frac{m\xi}{2\tau} - \frac{e}{c}\chi - \frac{1}{2}mc^2\tau - \frac{e^2\tau}{2mc^2}\int_0^1 (\mathbf{A}'^2 - \Phi'^2)du + \ldots . \tag{66}$$

For the case of a free particle, formula (66) gives the exact solution for S.

Equation (16) for f can be solved in an analogous way. We get

$$f = \frac{C}{\tau^2}\left\{1 - \frac{e\tau}{2mc}\int_0^1 (i\boldsymbol{\sigma}\cdot\mathfrak{H} + \boldsymbol{\alpha}\cdot\mathfrak{E})du + \ldots\right\}. \tag{67}$$

Let us study now what integration contour should be taken in the formula

$$R = \int e^{\frac{i}{\hbar}S} f d\tau, \tag{68}$$

in order to obtain the Riemann function. The integral (68) obviously satisfies the second-order Dirac equation. In order to be the Riemann function it should also satisfy conditions (40) or (52) on the ball (light cone) $\xi = 0$. Show that this condition is indeed satisfied if we take a closed integration contour on the complex τ-plane around the point $\tau = 0$. In order to verify this fact, note that the action function S has no pole when $\xi = 0$ and, therefore, the whole expression under integration has no essential singularity at this value of ξ. To compute the integral it is enough to take the residue at $\tau = 0$. For $\xi = 0$, we have

$$\begin{aligned} R = \exp\left(-\frac{ie}{\hbar c}\chi\right)\int_0^1 \bigg\{1 - \frac{imc^2}{2\hbar}\tau - \frac{ie^2}{2\hbar mc^2}\tau\int_0^1 (\mathbf{A}'^2 - \Phi'^2)du - \\ -\frac{e\tau}{2mc}\int_0^1 (i\boldsymbol{\sigma}\cdot\mathfrak{H} + \boldsymbol{\alpha}\cdot\mathfrak{E})du + \ldots\bigg\}\frac{C}{\tau^2}d\tau = \\ = -2\pi i C\exp\left(-\frac{ie}{\hbar c}\chi\right)\int_0^1 \bigg\{\frac{imc^2}{2\hbar} + \frac{ie^2}{2\hbar mc^2}(\mathbf{A}'^2 - \Phi^2) + \\ + \frac{e}{2mc}(i\boldsymbol{\sigma}\cdot\mathfrak{H} + \boldsymbol{\alpha}\cdot\mathfrak{E})\bigg\}du. \end{aligned} \tag{69}$$

The function R' defined by (45) is

$$R' = 2\pi^2 C\frac{\hbar c}{m}\int_0^1 \left\{\frac{m^2c^2}{\hbar^2} + \frac{e^2}{\hbar^2 c^2}(\mathbf{A}' - \Phi'^2) + \frac{e}{\hbar c}(\boldsymbol{\sigma}\cdot\mathfrak{H} - i\boldsymbol{\alpha}\cdot\mathfrak{E})\right\} du. \tag{70}$$

Therefore, if we let

$$\frac{2\pi^2\hbar c}{m}C = -\frac{1}{4}; \quad C = -\frac{m}{8\pi^2\hbar c}, \tag{71}$$

then the function R' will satisfy equation (52). It means that the integral (68) is indeed the Riemann function.

Choosing the integration path in an appropriate way, one can also get the fundamental solution of the second-order Dirac equation out of (68).

The fundamental solution U of the equation $\Lambda U = 0$ is

$$U = \frac{1}{\pi i}\left\{R\log|\xi| + \frac{R^*}{\xi}\right\} + U^*, \tag{72}$$

where R and R^* are the same as above: R is the Riemann function and R^* is defined by formula (44). The function U^* is already regular around the point $\xi = 0$. The only requirement on it is that expression (72) satisfies the equation $\Lambda U = 0$. This condition is obviously insufficient to determine it completely. It means that different fundamental solutions differ from each other by the functions U^*. They can be obtained from (68) by choosing different integration paths. The ambiguity of the fundamental solution is a reflection of the fact that the number of variables of the Dirac equation without proper time is even.

8. Consider as an example the motion of an electron in a constant electric and magnetic field and construct the corresponding Riemann function. For the simplicity of computations let us restrict ourselves to the case of collinear fields, which we assume to be parallel to the z-axis.

Assuming the potentials to be

$$A_x = -\frac{1}{2}\mathcal{H}y; \quad A_y = \frac{1}{2}\mathcal{H}x; \quad A_z = 0; \quad \Phi = -\mathcal{E}z, \tag{73}$$

we can easily solve the classical equations of motion corresponding to the Lagrange function (I, 6). For the action integral (I, 7), we get the expression

$$S = S_0 - \frac{1}{2}mc^2\tau + \frac{e\mathcal{E}}{4c}[(z-z^0)^2 - c^2(t-t^0)^2]\coth\frac{e\mathcal{E}\tau}{2mc} +$$

$$+\frac{e\mathcal{H}}{4c}[(x-x^0)^2+(y-y^0)^2]\coth\frac{e\mathcal{H}\tau}{2mc}, \tag{74}$$

where S_0 is

$$S_0=-\frac{e}{c}\chi=-\frac{e\mathcal{E}}{2}(z+z^0)(t-t^0)-\frac{e\mathcal{H}}{2c}(x^0y-y^0x), \tag{74*}$$

which does not depend on τ.

Returning to equation (13) for f, we see that it can be satisfied by a function depending on τ only. In this case it coincides with equation (16), which we have already transformed to the form (21). In our case, the determinant ϱ is equal to

$$\varrho=\frac{\text{const}}{\sin^2\frac{e\mathcal{H}\tau}{2mc}\cdot\sin\hbar^2\frac{e\mathcal{H}\tau}{2mc}}, \tag{75}$$

and the solution of equation (21) is given by

$$f^0=\exp\left(-\frac{ie}{2mc}\sigma_z\mathcal{H}\tau-\frac{e}{2mc}\alpha_z\mathcal{E}\tau\right). \tag{76}$$

Determining the constant factor of the function f from (71), we obtain

$$f=-\frac{m}{8\pi^2\hbar c}\left(\frac{e\mathcal{H}}{2mc}\right)\cdot\left(\frac{e\mathcal{E}}{2mc}\right)\frac{f^0}{\sin^2\frac{e\mathcal{H}\tau}{2mc}\cdot\sin\hbar^2\frac{e\mathcal{H}\tau}{2mc}}. \tag{77}$$

Taking this expression for f and taking the integral

$$R=\int e^{\frac{i}{\hbar}S}f\,d\tau \tag{78}$$

along a sufficiently small circle around the origin, one gets the Riemann function for the given problem.

In a particular case, when the field vanishes, one has

$$f=-\frac{m}{8\pi^2\hbar c}\cdot\frac{1}{\tau^2};\quad S=-\frac{m\xi}{2\tau}-\frac{1}{2}mc^2\tau, \tag{79}$$

and, therefore,

$$R=-\frac{m}{8\pi^2\hbar c}\int\exp\left(-\frac{im\xi}{2\hbar\tau}-\frac{imc^2}{2\hbar}\tau\right)\cdot\frac{d\tau}{\tau^2}=$$

$$=-\frac{m}{4\pi\hbar\sqrt{\xi}}J_1\left(\frac{mc}{\hbar}\sqrt{\xi}\right), \tag{80}$$

while R^* has the constant value $\frac{1}{2\pi c}$. Therefore, for the case of a vanishing field the function Q of the general theory is given by

$$Q=-\frac{m}{4\pi h\sqrt{\xi}}J_1\left(\frac{mc}{h}\sqrt{\xi}\right)\gamma(\xi)+\frac{1}{2\pi c}\delta(\xi). \tag{81}$$

III Applications to the Positron Theory

The basis of the positron theory suggested by Dirac is the consideration of the "mixed density" corresponding to the distribution of electrons among the states with positive and negative energies.

Dirac considers the mixed densities of two types, namely, R_1 and R_F, where

$$\langle \mathbf{r}, t, \zeta | R_1 | \mathbf{r}^0, t^0, \zeta^0 \rangle =$$

$$= \sum_{\text{occupied}} \psi(\mathbf{r}, t, \zeta)\overline{\psi}(\mathbf{r}^0, t^0, \zeta^0) - \sum_{\text{free}} \psi(\mathbf{r}, t, \zeta)\overline{\psi}(\mathbf{r}^0, t^0, \zeta^0), \tag{1}$$

$$\langle \mathbf{r}, t, \zeta | R_F | \mathbf{r}^0, t^0, \zeta^0 \rangle =$$

$$= \sum_{\text{occupied}} \psi(\mathbf{r}, t, \zeta)\overline{\psi}(\mathbf{r}^0, t^0, \zeta^0) + \sum_{\text{free}} \psi(\mathbf{r}, t, \zeta)\overline{\psi}(\mathbf{r}^0, t^0, \zeta^0). \tag{2}$$

Here $\psi(\mathbf{r}, t, \zeta)$ denotes the wave function of one electron depending on coordinates $\mathbf{r}$, time t, and the spin variable (component number) ζ. The sum is taken over the wave function index (the number of the state), which we omit here for brevity. The first sum is taken over all occupied states and the second over free ones.

Dirac computes (1) and (2) for electrons in the absence of a field by direct summation expressions and studies their singularity on a light cone. He constructs then analogous expressions for electrons in the presence of a field and obtains their singularities from the requirements that they satisfy the wave equations and tend to the ones already computed when the field tends to zero.

However, expressions (1) and (2) are closely related to the function that one encounters in the study of the Cauchy problem for the Dirac equation. This relation has not been mentioned by Dirac and we are going to clarify it now.

Consider first expression (2). Taking R_F as a matrix with indices ζ and ζ_0 we can rewrite it in the form

$$\langle \mathbf{r}, t | R_F | \mathbf{r}^0, t^0 \rangle = R_F(\mathbf{r}, t). \tag{3}$$

This expression satisfies the Dirac equation and turns into the kernel of the unity operator

$$\left\{ (\boldsymbol{\alpha} \cdot \mathbf{P}) + mc\alpha_4 - \frac{T}{c} \right\} R_F = 0, \tag{4}$$

$$R_F = \delta(\mathbf{r} - \mathbf{r}^0) \qquad \text{when } t = t^0. \tag{5}$$

These conditions fix the function R_F unambiguously. Hence, it is not necessary to compute it by direct summation of the series (2), but just by solving the Cauchy problem instead.

Indeed, the function Q (II, 26) which we have considered in part II satisfies the second-order Dirac equation and the condition

$$Q = 0; \quad \frac{\partial Q}{\partial t} = \delta(\mathbf{r} - \mathbf{r}_0) \qquad \text{when} \quad t = t^0. \tag{6}$$

It implies that the expression

$$R_F = -\frac{ic}{\hbar}\left\{(\boldsymbol{\alpha}\cdot\mathbf{P}) + mc\alpha_4 - \frac{T}{c}\right\}Q, \tag{7}$$

where, according to (II, 26),

$$Q = R\gamma(\xi) + R^*\delta(\xi), \tag{8}$$

satisfies equations (4) and (5). Therefore, the Dirac function R_F can be expressed via the Riemann function, which we have studied in detail in part II.

As for the function R_1, it can be expressed via the fundamental solution U like R_F can be expressed via Q:

$$R_1 = -\frac{ic}{\hbar}\left\{(\boldsymbol{\alpha}\cdot\mathbf{P}) + mc\alpha_4 - \frac{T}{c}\right\}U, \tag{9}$$

where, according to (II, 72),

$$U = \frac{1}{\pi i}\left\{R\log|\xi| + \frac{R^*}{\xi}\right\} + U^*. \tag{10}$$

One can call expression (9) the fundamental solution of the first-order Dirac equation.

An unambiguous determination of the function R_1 is possible only if the initial conditions are fixed. Fixing of these initial conditions should be made from some physical reasoning. For example, one can require that at the initial moment of time $t = t^0$ all electron states with negative kinetic energy are occupied and all states with positive energy are free. Then, which is obvious from the initial definition (1) of the function R_1, it should become the kernel of the operator of sign of kinetic energy (taken with minus). Solving the Dirac equation with these initial conditions, one obtains the value of the mixed density for an arbitrary value of t.

As a concluding remark, let us mention the physical meaning of mixed density. Which terms of it have a physical meaning and which do not? Dirac as well as Heisenberg [3] suggest to fix the singular part in the expression for R_1 and provide with physical meaning only the difference between the full expression for R_1 and the singular part. From our point of view, this procedure contains too much arbitrariness. The only safe criterion is the correspondence principle. Concerning our problem this principle means that only those expressions have a physical meaning that remains finite everywhere including the light cone when $h \to 0$ (in other words, uniformly finite with respect to ξ). The other terms should be rejected as having no physical meaning. This criterion is in agreement with the computation of the vacuum polarization made by several authors.[2] According to these computations, the additional terms of the Lagrange function for the electromagnetic field are power series with respect to $\hbar$. This fact also shows the applicability of the Wentzel–Brillouin method considered in the second part of the paper.

References

1. P.A.M. Dirac, Proc. Cambr. Phil. Soc., **30**, 150, 1934.
2. J. Hadamard, Le problème de Cauchy et les équations aux derivées partielles linéaires hyperbolique, Paris, 1932.
3. W. Heisenberg, Zs. Phys. **90**, 209, 1934.
4. W. Pauli, Helvetica Phys. Acta **5**, 179, 1932.
5. V. Weisskopf, Über die Elektrodynamik des Vakuume auf Grund der Quantenteorie des Elektrons, Køpenhagen, 1936.

Translated by V. V. Fock

[2] See, e.g., [5]. (*V. Fock*)

40-1
Incomplete Separation of Variables for Divalent Atoms

V.A. Fock, M.G. Veselov and M.I. Petrashen

Received 28 April 1940

JETP, **10**, N 7, 723, 1940

The self-consistent field method with the exchange is one of the most precise of the existing methods for a calculation of many-electron atoms. The complication which is brought into the calculation in comparison with other methods (method of the variation of parameters in analytic one-electron wave functions and the self-consistent method without exchange) is undoubtedly warranted by the higher accuracy of the calculation of univalent atoms and ions similar to them. The situation is somewhat different for atoms and ions with several valence electrons. The calculations of the beryllium atom, for example, show [1] that the discrepancy between the experimental and calculated values for the total energy exceeds several times the refinement of the correction of the energy due to the exchange in the self-consistent method. Limitations of the existing approximate methods appear to be more noticeable in the calculations of the optical terms of divalent atoms.

That comparatively big discrepancy of the experimental and theoretical results can be attributed to the disregard of electron interaction, which is expressed in the separation of variables in the atomic wave function. The greater or lesser admissibility of the separation of variables in the wave function of the many-electron atom, i.e., the introduction of one-electron states, is determined by the feature of the potential field which acts on the separate electrons and, namely, by the degree to which this field can be approximated by the spherically symmetric field. From this point of view, the previously mentioned results can be easily explained.

The inner-shell electrons of the atoms constitute a core that has the spherical symmetry (if we disregard a polarization of the core). The highest probability of the location of valence electrons is outside the core and, therefore, the interaction of the inner-shell electrons with the valence electrons can be represented to a great extent by the screening of the charge of the nucleus, which preserves the spherical symmetry of the field. Hence, it is clear that the exterior electron of a univalent atom moves in an approximately central Coulomb field and a state of the electron can be described by a hydrogen-like wave function. This is confirmed by the similarity of the optical spectra of alkaline atoms and the hydrogen atom. In the presence of two valence electrons, their interaction in the ground state of an atom violates noticeably the spherical symmetry of the potential field for each of them; hence, this restricts the applicability of the one-electron wave functions. For those excited states that can be considered as a transition of one valence electron from the ground state to an excited state, the separation of the variables gives a smaller error. This is verified, for example, by the results of the calculations of the electronic states of the molecule H_2 presented by Hylleraas [2]; in this molecule the interaction of the electrons in the ground state violates the axial symmetry stronger than in excited states.

Similar considerations can be also extended to atoms with an arbitrary number of the valence electrons.

As to inner-shell electrons, the potential field that acts on them is mainly determined by the Coulomb field of the nucleus. Mutual perturbation of the electrons from the inner shell influences the total field in the atom relatively less than similar interactions of the valent electrons on the total peripheral field where the nuclear charge is strongly screened.

A complete rejection of the one-electron wave functions in atomic calculations seems to be not possible, but on the basis of qualitative arguments stated above one can expect a significant improvement of the accuracy in the calculations of atoms with several valent electrons if one rejects the approximation of one-electron states for the valent electrons and preserves this approximation for inner-shell electrons. This separation of the variables for inner electrons influences especially little the optical frequencies and the probability of transitions, which are mainly determined by the behavior of the valence electrons.

In the present article a theory of a calculation with incomplete separation of the variables for divalent atoms is given.

1 Construction of the Wave Function

Introduction

According to the concept outlined in the Introduction, we accept the separation of variables of the electrons on inner shells in a divalent atom and associate a wave function $\psi_i(x)$ with each inner-shell electron. The index i specifies a one-electron state and, therefore, it denotes a set of four quantum numbers, the argument x denotes three components of the spatial coordinate and the spin coordinate of the electron. We suppose further that two valence electrons are described by the function $\varphi(x, x')$. Then a state of a divalent atom with $n+2$ electrons will be described by the function Ψ of $n+2$ variables $x_j\,(j = 1, 2, \ldots n+2)$ produced with the help of $n+1$ different functions. It is obvious that the Pauli principle will be fulfilled if we assume the function $\varphi(x, x')$ to be antisymmetric and take

$$\Psi(x_1, x_2, \ldots, x_{n+2}) = \sum_P (-1)^{[P]} \varphi(x_{\beta_1}, x_{\beta_2}) \cdot \\ \cdot \psi_1(x_{\beta_3}) \psi_2(x_{\beta_4}) \ldots \psi_n(x_{\beta_{n+2}}), \quad (1.1)$$

where the summation is done over all possible permutations P β_1, β_2, ..., β_{n+2} of numbers of the series $(1, 2, ...n+2)$, i.e., over all possible permutations of the electrons and the symbol $[P]$ is the parity of the permutation P.

Taking into account the antisymmetry of $\varphi(x, x')$ and dropping the factor 2, we can rewrite expression (1.1) as follows:

$$\Psi(x_1, x_2, \ldots, x_{n+2}) = \varphi(x_1, x_2) D^{12} + \sum_{k=3}^{n+2} (-1)^k \varphi(x_1, x_k) D^{1k} + \\ + \sum_{k=3}^{n+2} (-1)^{k+1} \varphi(x_2, x_k) D^{2k} + \sum_{k>i=3}^{n+2} (-1)^{k+i+1} \varphi(x_i, x_k) D^{ik}. \quad (1.2)$$

We denoted as

$$D^{rs} = D^{rs}(x_1, x_2, \ldots, x_{r-1}, x_{r+1}, \ldots, x_{s-1}, x_{s+1}, \ldots, x_{n+2})$$

the determinant of the wave functions of inner-shell electrons where the rows corresponding to the r-th and s-th electrons are dropped, i.e.,

$$D^{rs} = \begin{vmatrix} \psi_1(x_1), & \psi_2(x_1), & \ldots & \psi_n(x_1) \\ \psi_1(x_2), & \psi_2(x_2), & \ldots & \psi_n(x_2) \\ \ldots & \ldots & \ldots & \ldots \\ \psi_1(x_{r-1}), & \psi_2(x_{r-1}), & \ldots & \psi_n(x_{r-1}) \\ \psi_1(x_{r+1}), & \psi_2(x_{r+1}), & \ldots & \psi_n(x_{r+1}) \\ \ldots & \ldots & \ldots & \ldots \\ \psi_1(x_{s-1}), & \psi_2(x_{s-1}), & \ldots & \psi_n(x_{s-1}) \\ \psi_1(x_{s+1}), & \psi_2(x_{s+1}), & \ldots & \psi_n(x_{s+1}) \\ \ldots & \ldots & \ldots & \ldots \\ \psi_1(x_{n+2}), & \psi_2(x_{n+2}), & \ldots & \psi_n(x_{n+2}) \end{vmatrix}. \tag{1.3}$$

We suppose the one-electron functions $\psi_i(x)$ to be mutually orthogonal and normalized. Concerning the function $\varphi(x, x')$ we notice the following. The wave function Ψ of the whole atom is not changed if we add to $\varphi(x, x')$ an additional term that is linear with respect to the one-electron functions $\psi_i(x)$, since the extra terms appearing after this addition in expression (1.1) are the determinants with equal columns. Taking into account also the antisymmetry of the function Ψ, we can, therefore, replace the function $\varphi(x, x')$ in formulae (1.1) and (1.2) by the expression

$$\omega(x, x') = \varphi(x, x') + \sum_{i=1}^{n} [f_i(x')\psi_i(x) - f_i(x)\psi_i(x')] + \\ + \sum_{i,k=1}^{n} C_{ik}\psi_i(x)\psi_k(x'), \tag{1.4}$$

where $f_i(x)$ are arbitrary functions and C_{ik} comply with the condition $C_{ik} = -C_{ki}$.

We could omit the last summand in formula (1.4) and write down (1.4) as follows:

$$\omega(x, x') = \varphi(x, x') + \sum_{i=1}^{n} [f_i(x')\psi_i(x) - f_i(x)\psi_i(x')]. \tag{1.5}$$

In fact, equation (1.4) is obtained from (1.5) by the substitution

$$f_i(x) \to f_i(x) + \sum_{k=1}^{n} a_{ik}\psi_k(x), \tag{1.6}$$

with $C_{ik} = a_{ik} - a_{ki}$. We will not assume the coefficients C_{ik} to be equal to zero, having in mind to impose in the following some additional conditions on $f_i(x)$ (see equation (1.7) below).

For given $\varphi(x, x')$, we can choose the quantities C_{ik} and $f_i(x)$ to get the function $\omega(x, x')$ orthogonal to every function $\psi_i(x')$ identically for any x. For this, it is sufficient to take

$$f_i(x) = \int \varphi(x, x') \overline{\psi_i}(x') \, dx' \tag{1.7}$$

and

$$C_{ik} = \iint \varphi(x, x') \overline{\psi_i}(x) \psi_k(x') \, dx dx'. \tag{1.8}$$

Here and in the following the symbol of the integration over x denotes the integration over spatial coordinates and the summation over two components of the spin.

We denote as $\psi(x, x')$ the function $\omega(x, x')$ (1.4) when $f_i(x)$ and C_{ik} are subject to these conditions. Obviously, C_{ik} are the Fourier coefficients of the function $\varphi(x, x')$. If we complement the system of n one-electron functions $\psi_i(x)$ to a complete one and write the expansion of $\varphi(x, x')$ over the functions of this closed system

$$\varphi(x, x') = \sum_{i,k=1}^{\infty} C_{ik} \psi_i(x) \psi_k(x'), \tag{1.9}$$

then

$$\psi(x, x') = \sum_{i,k=n+1}^{\infty} C_{ik} \psi_i(x) \psi_k(x'). \tag{1.10}$$

Thus, the wave function Ψ of the whole atom will have the form

$$\Psi = \psi(x_1, x_2) + \sum_{k=3}^{n+2} (-1)^k \psi(x_1, x_k) D^{1k} + \sum_{k=3}^{n+2} (-1)^{k+1} \psi(x_2, x_k) D^{2k} + \\ + \sum_{k>i=3}^{n+2} (-1)^{k+i+1} \psi(x_i, x_k) D^{ik}, \tag{1.11}$$

where D^{rs} have the values (1.3) and functions $\psi_i(x)$ and $\psi(x, x')$ obey the conditions

$$\left.\begin{aligned} \int \overline{\psi_i}\,(x)\,\psi_k\,(x)\,dx &= \delta_{ik}, \\ \int \psi\,(x,x')\,\overline{\psi_i}\,(x')\,dx' &= 0, \\ \iint \overline{\psi}\,(x,x')\,\psi\,(x,x')\,dxdx' &= 2, \\ (i,k=1,2,\ldots n)\,. & \end{aligned}\right\} \tag{1.12}$$

Normalization of the function $\psi\,(x,x')$ is chosen so that

$$\int \overline{\Psi}\Psi\,(dx) = (n+2)!$$

$((dx) = dx_1 dx_2 \ldots dx_{n+2})$, as well as in case of the complete separation of the variables.

2 Calculation of Density Integral

Introduction

In order to get an expression for the total energy of the divalent atom, we have to calculate beforehand the integral

$$\int \ldots \int \overline{\Psi}\Psi'' dx_3 dx_4 \ldots dx_{n+2}, \tag{2.1}$$

where two primes over Ψ denote that two first arguments x_1 and x_2 of this function are replaced by x_1' and x_2'. When $x_1' = x_1$ and $x_2' = x_2$ this expression gives the density of the probability distribution for the coordinates of a pair of electrons of the atom. In order to calculate this integral, we write the determinants D^{ik} entering the wave function Ψ'' in the explicit form

$$\begin{aligned} D^{ik} = \sum_{\alpha} \varepsilon_\alpha \psi_{\alpha_1}\,(x_1')\,\psi_{\alpha_2}\,(x_2')\ldots\psi_{\alpha_{i-1}}\,(x_{i-1})\,\psi_{\alpha_i}\,(x_{i+1})\ldots \\ \ldots\psi_{\alpha_{k-2}}\,(x_{k-1})\,\psi_{\alpha_{k-1}}\,(x_{k+1})\ldots\psi_{\alpha_n}\,(x_{n+2})\,, \end{aligned} \tag{2.2}$$

where α is a set of indices $\alpha_1, \alpha_2, \ldots \alpha_n$ which is a permutation of the numbers $1, 2, \ldots n$. The factor ε_α is equal to plus or minus unity according to the parity of the permutation and the summation of all possible permutations is carried out. In a similar way, we write also the determinants entering the function Ψ. Then the integral (2.1) is reduced to three types of integrals, which are easily calculated because the majority of integrals are, in fact, equal to zero due to the orthogonality conditions (1.12). These integrals are written as follows:

$$\int\ldots\int \overline{\psi(x_1,x_2)\,D^{12}}\psi(x_1',x_2')\,D^{12}dx_3dx_4\ldots dx_{n+2} =$$

$$= n!\overline{\psi(x_1,x_2)}\psi(x_1',x_2')\,,$$

$$\int\ldots\int\sum_{k=3}^{n+2}(-1)^k\overline{\psi(x_1,x_k)\,D^{1k}}\sum_{m=3}^{n+2}(-1)^m\psi(x_1',x_m)D^{1m}dx_3dx_4\ldots dx_{n+2} =$$

$$= n!\varrho(x_2,x_2')\int\overline{\psi(x_1,x_k)}\psi(x_1',x_k)\,dx_k,$$

$$\int\ldots\int\sum_{k>i=3}^{n+2}(-1)^{i+k+1}\overline{\psi(x_i,x_k)D^{ik}}\sum_{r>s=3}^{n+2}\psi(x_r,x_s)D^{rs}dx_3dx_4\ldots dx_{n+2} =$$

$$= n!\left[\varrho(x_1,x_1')\,\varrho(x_2,x_2') - \varrho(x_1,x_2')\,\varrho(x_2,x_1')\right]. \tag{2.3}$$

In these formulae ϱ denotes the one-electron density matrix, i.e.,

$$\varrho(x,x') = \sum_{i=1}^{n}\overline{\psi_i(x)}\psi_i(x')\,. \tag{2.4}$$

Using these formulae, we get

$$\frac{1}{n!}\int\ldots\int\overline{\Psi}\Psi''dx_3dx_4\ldots dx_{n+2} = \overline{\psi(x_1,x_2)}\psi(x_1',x_2') +$$

$$+\varrho(x_1,x_1')\int\overline{\psi(x_2,x)}\psi(x_2',x)\,dx + \varrho(x_2,x_2')\int\overline{\psi(x_1,x)}\psi(x_1',x)\,dx -$$

$$-\varrho(x_1,x_2')\int\overline{\psi(x_2,x)}\psi(x_1',x)\,dx - \varrho(x_2,x_1')\int\overline{\psi(x_1,x)}\psi(x_2',x)\,dx +$$

$$+\varrho(x_1,x_1')\,\varrho(x_2,x_2') - \varrho(x_1,x_2')\,\varrho(x_2,x_1')\,. \tag{2.5}$$

We introduce the following notations:

$$\varrho_1(x_1, x_2; x_1', x_2') = \overline{\psi(x_1, x_2)}\psi(x_1', x_2'), \tag{2.6}$$

$$\varrho_1(x, x') = \int \overline{\psi(x, x'')}\psi(x', x'')\,dx''. \tag{2.7}$$

Then formula (2.5) is rewritten as

$$\frac{1}{n!}\int\ldots\int \overline{\Psi}\Psi'' dx_3 dx_4 \ldots dx_{n+2} = \varrho_1(x_1, x_2; x_1', x_2') +$$

$$+\begin{vmatrix} \varrho_1(x_1, x_1'), & \varrho_1(x_1, x_2') \\ \varrho(x_2, x_1'), & \varrho(x_2, x_2') \end{vmatrix} + \begin{vmatrix} \varrho(x_1, x_1'), & \varrho(x_1, x_2') \\ \varrho_1(x_2, x_1'), & \varrho_1(x_2, x_2') \end{vmatrix} +$$

$$+\begin{vmatrix} \varrho(x_1, x_1'), & \varrho(x_1, x_2') \\ \varrho(x_2, x_1'), & \varrho(x_2, x_2') \end{vmatrix}. \tag{2.8}$$

When we admit the complete separation of variables, the atomic wave function is constructed from $n+2$ different one-electron wave functions. In this case, the integral, which corresponds to the one of (2.1), is easily calculated:

$$\frac{1}{n!}\int\cdots\int \overline{\Psi}\Psi'' dx_3 dx_4 \ldots dx_{n+2} = \begin{vmatrix} R(x_1, x_1'), & R(x_1, x_2') \\ R(x_2, x_1'), & R(x_2, x_2') \end{vmatrix}, \tag{2.9}$$

where

$$R(x, x') = \sum_{i=1}^{n+2} \overline{\psi_i}(x)\,\psi_i(x'). \tag{2.10}$$

Each element of the density matrix can be obviously represented as a sum of two matrix elements

$$R(x, x') = \varrho(x, x') + \varrho_1^0(x, x'). \tag{2.11}$$

Here $\varrho(x, x')$ has the previous value (2.4) and

$$\varrho_1^0(x, x') = \sum_{i=n+1}^{n+2} \overline{\psi_i}(x)\,\psi_i(x'). \tag{2.12}$$

Then the determinant (2.9) is rewritten as a sum of four determinants

$$\begin{vmatrix} \varrho_1^0(x_1, x_1'), & \varrho_1^0(x_1, x_2') \\ \varrho_1^0(x_2, x_1'), & \varrho_1^0(x_2, x_2') \end{vmatrix} + \begin{vmatrix} \varrho_1^0(x_1, x_1'), & \varrho_1^0(x_1, x_2') \\ \varrho(x_2, x_1'), & \varrho(x_2, x_2') \end{vmatrix} +$$

$$+\begin{vmatrix} \varrho(x_1, x_1'), & \varrho(x_1, x_2') \\ \varrho_1^0(x_2, x_1'), & \varrho_1^0(x_2, x_2') \end{vmatrix} + \begin{vmatrix} \varrho(x_1, x_1'), & \varrho(x_1, x_2') \\ \varrho(x_2, x_1'), & \varrho(x_2, x_2') \end{vmatrix}. \tag{2.13}$$

Comparing this expression with the right-hand side of (2.8), we can formulate the following rule. In order to transform formula (2.13) into (2.8), which corresponds to incomplete separation of variables, we have to do the following replacements: the determinant that is composed only from the elements of the matrix ϱ_1^0 should be substituted for the matrix element ϱ_1 defined by equation(2.6); in the remaining determinants the matrix elements ϱ_1^0 should be replaced by the corresponding matrix elements of ϱ_1 in equation (2.7). Similar rules can be formulated also for the transition from the complete separation of variables to an incomplete separation for atoms with an arbitrary number of valence electrons.

If we set in equation (2.8) $x_1' = x_1, x_2' = x_2$ and then integrate it over x_1 and x_2, we see that the resulting integrals of separate terms in the right-hand side of (2.8) will be equal to $2 + 2n + 2n + n\,(n-1) = (n+1)\,(n+2)$. Separate terms in this formula have a simple interpretation. The first term up to a factor gives a probability that a pair of electrons is valent. The second and third terms together give a probability that one electron of the pair is valent and another is an inner-shell electron. The last term gives a probability that both electrons are from the inner-shell. A simple combinatorial calculation of the probabilities mentioned above gives the same result if we attribute to inner-shell states equal statistical weights and to the two-electron valence state the weight which is two times larger.

Strictly speaking, the probability interpretation of the integrals of separate terms in formula (2.8) still gives no right to consider the integrands themselves as the probability densities. Therefore, relation between formula (2.8) (at $x_1' = x_1$ and $x_2' = x_2$) and the theorem of the probability addition, natural at first sight, is hardly correct.

We can do the integration over x_2 and x_1 in two steps. We put first in (2.8) $x_2' = x_2$ keeping x_1' different from x_1 and make integration over x_2. Denoting as Ψ' the function Ψ where x_1 is replaced by x_1' we get the formula

$$\frac{1}{n+1!}\int\ldots\int \overline{\Psi}\Psi' dx_3 dx_4 \ldots dx_{n+2} = \varrho_1\left(x_1, x_1'\right) + \varrho\left(x_1, x_1'\right). \quad (2.14)$$

When $x_1' = x_1$ this expression gives a probability density for the coordinate distribution of an electron in the atom. Then, integrating over x_1, we get for the normalization integral the value

$$\int\ldots\int \overline{\Psi}\Psi dx_1 dx_2 \ldots dx_{n+2} = (n+2)! \quad (2.15)$$

which has been already given at the end of §1.

3 Dependence of Wave Functions on Spin[1]

Introduction

In order to single out the dependence of the wave functions $\psi(x, x')$ and $\psi_i(x)$ $(i = 1, 2 \dots n)$ on spin variables we use the fact that the energy operator does not depend on the spin explicitly. The wave function Ψ for a system of k electrons in a central field can be defined as an antisymmetric function, which is a common eigenfunction of the energy operator, the operators of the total and orbital angular momenta, the spin and z-th component of the total angular momentum, i.e.,

$$H\Psi = E\Psi, \tag{3.1}$$

$$\mathbf{M}^2\Psi = \hbar^2 j(j+1)\Psi, \tag{3.2}$$

$$\mathbf{m}^2\Psi = \hbar^2 l(l+1)\Psi, \tag{3.3}$$

$$\mathbf{s}^2\Psi = s(s+1)\Psi, \tag{3.4}$$

$$M_z\Psi = \hbar m\Psi. \tag{3.5}$$

The total angular momentum operator is $\mathbf{M} = \mathbf{m} + \hbar\mathbf{s}$, where $\mathbf{m}$ and $\hbar\mathbf{s}$ are the operators of the orbital angular momentum and the spin. The quantum numbers s, j and m for a given l and for k electrons can take the following values:

$$\left.\begin{aligned} s &= \tfrac{k}{2}, \tfrac{k}{2} - 1, \dots \tfrac{1}{2} \quad \text{or} \quad 0, \\ j &= l + s, l + s - 1, \dots |l - s|, \\ m &= j, j - 1, \dots - j. \end{aligned}\right\} \tag{3.6}$$

From the commutation relations for the operators of components of the angular momentum, we can write the following system of the first-order equations [3]:

$$(M_x + iM_y)\psi_{jm} = \hbar\alpha_{jm}\psi_{jm+1}, \tag{3.7}$$

$$(M_x - iM_y)\psi_{jm+1} = \hbar\overline{\alpha}_{jm}\psi_{jm}, \tag{3.8}$$

where

$$\overline{\alpha}_{jm}\alpha_{jm} = (j - m)(j + m + 1). \tag{3.9}$$

[1] Some of the results obtained in this section are well known. Nevertheless, we give this derivation not only for the completeness but also because they are usually derived either in a more complicated or in a less rigorous way. (*Authors*)

If the integral of $|\psi_{jm}|^2$ does not depend on m, then $\overline{\alpha}_{jm}$ is a complex conjugate to α_{jm}, which justifies the chosen notation. The system of equations (3.7)–(3.8) together with equation (3.5) leads to equation(3.2). Besides, it connects the eigenfunctions of the total angular momentum for a fixed value of j and different values of m.

The previous formulae are derived only from the commutator relations for M_x, M_y and M_z; therefore, they are also valid for the orbital angular momentum operators and the spin.

Consider now a function $\chi_{s\nu}^{(\lambda)}(\sigma_1, \sigma_2, \dots \sigma_k)$, which is a common eigenfunction of the operators $\mathbf{s}^2$ and s_z:

$$\mathbf{s}^2 \chi_{s\nu}^{(\lambda)} = s(s+1)\chi_{s\nu}^{\lambda}; \quad s = \frac{k}{2}, \frac{k}{2} - 1, \dots \frac{1}{2} \quad \text{or} \quad 0; \tag{3.10}$$

$$s_z \chi_{s\nu}^{(\lambda)} = \nu \chi_{s\nu}; \quad \nu = s, s-1, \dots - s. \tag{3.11}$$

Each variable σ_r of the function $\chi_{s\nu}^{(\lambda)}$ takes two values ± 1; then σ_r corresponds to σ_{rz}, i.e., to the doubled value of the spin component of the r-th particle calculated in the units of $\hbar$. An action of the spin operators on functions of these variables is represented as follows:

$$s_x \chi(\sigma_1, \sigma_2, \dots \sigma_k) = \frac{1}{2} \sum_{r=1}^{k} \chi(\sigma_1, \sigma_2, \dots \sigma_{r-1}, -\sigma_r, \sigma_{r+1}, \dots \sigma_k),$$

$$s_y \chi(\sigma_1, \sigma_2, \dots \sigma_k) = -\frac{i}{2} \sum_{r=1}^{k} \sigma_r \chi(\sigma_1, \sigma_2, \dots \sigma_{r-1}, -\sigma_r, \sigma_{r+1}, \dots \sigma_k),$$

$$s_z \chi(\sigma_1, \sigma_2, \dots \sigma_k) = \frac{1}{2} \left(\sum_{r=1}^{k} \sigma_r \right) \chi(\sigma_1, \sigma_2, \dots \sigma_k). \tag{3.12}$$

Using these formulae, it is easy to show that the operator $\mathbf{s}^2$ has the following expression:

$$\mathbf{s}^2 = s_x^2 + s_y^2 + s_z^2 = -\frac{k^2}{4} + k + \sum_{r<s}^{k} P_{rs}, \tag{3.13}$$

where P_{rs} is the permutation operator of σ_r and σ_s.

In the general case of k electrons, the common eigenvalues of the operators $\mathbf{s}^2$ and s_z are degenerate, so the multiplicity of the degeneracy is defined by the number λ_s which is equal to

$$\lambda_s = \binom{k}{\frac{k}{2} - s} - \binom{k}{\frac{k}{2} - s - 1}. \tag{3.14}$$

Therefore, the function $\chi_{s\nu}^{(\lambda)}$ has one more index λ besides the indices s, ν, which run the values $\lambda = 1, 2, \ldots \lambda_s$. But in the case of one- and two-electron problems, which we will be interested in, $\lambda_s = 1$ and the additional index is not necessary.

We construct the spin functions in this case.

According to equations (3.10) and (3.11) for a one-electron problem s and ν have the values $s = \frac{1}{2}$ and $\nu = \pm\frac{1}{2}$ and the corresponding wave functions will be

$$\left.\begin{aligned} \chi_{\frac{1}{2}\frac{1}{2}}(\sigma) &= \tfrac{1}{2}(1+\sigma), \\ \chi_{\frac{1}{2}-\frac{1}{2}}(\sigma) &= \tfrac{1}{2}(1-\sigma). \end{aligned}\right\} \tag{3.15}$$

It is easy to see that these functions are normalized and that they correspond to the value $\alpha_{\frac{1}{2}-\frac{1}{2}} = 1$. In the two-electron spin problem the $\mathbf{s}^2$ operator according to (3.12) is expressed as

$$\mathbf{s}^2 = 1 + P_{12}, \tag{3.16}$$

where P_{12} is the transposition operator of σ_1 and σ_2. In this case the quantum number s has the values 1 and 0 and, as a consequence, the number ν takes the values $-1, 0, 1$, and 0, correspondingly. A well-known result follows immediately from the expression for the operator $\mathbf{s}^2$, namely, the function $\chi_{00}(\sigma_1, \sigma_2)$ has to be antisymmetric and all three functions $\chi_{1\nu}(\sigma_1, \sigma_2), (\nu_2 = -1, 0, 1)$ have to be symmetric relative to a transposition of σ_1 and σ_2. As well as for the one-electron problem, using these symmetry properties one can easily construct the spin functions avoiding the tedious general prescriptions. Being obtained in this way, the analytic expressions for the normalized functions are written as follows:

$$\begin{aligned} \chi_{00}(\sigma_1, \sigma_2) &= \frac{1}{2\sqrt{2}}(\sigma_1 - \sigma_2), \\ \chi_{10}(\sigma_1, \sigma_2) &= \frac{1}{2\sqrt{2}}(1 - \sigma_1\sigma_2), \\ \chi_{1-1}(\sigma_1, \sigma_2) &= \frac{1}{4}(1-\sigma_1)(1-\sigma_2), \\ \chi_{11}(\sigma_1, \sigma_2) &= \frac{1}{4}(1+\sigma_1)(1+\sigma_2). \end{aligned} \tag{3.17}$$

We return now to the problem of the separation of spin variables in a general solution of equations (3.1)–(3.5) in the two-electron case.

Together with the function $\chi_{s\nu}(\sigma_1, \sigma_2)$, we also consider the function $\psi_{l\mu}(\mathbf{r}_1, \mathbf{r}_2)$, which satisfies the Schrödinger equation and the equations

$$\mathbf{m}^2 \psi_{\lambda\mu} = \hbar^2 l (l+1) \psi_{\lambda\mu}, \tag{3.18}$$

$$m_z \psi_{\lambda\mu} = \hbar\mu\psi_{\lambda\mu} \qquad (\mu = l, l-1, \ldots -l). \tag{3.19}$$

The wave function $\psi_{jm}^{(ls)}(x_1, x_2)$ of the two-electron system satisfying equations (3.1)–(3.5) will be expressed in terms of the functions $\chi_{s\nu}(\sigma_1, \sigma_2)$ and $\psi_{l\mu}(\mathbf{r}_1, \mathbf{r}_2)$ in the following way:

$$\psi_{jm}^{(ls)}(x_1, x_2) = \sum_{\mu+\nu=m} c_{jm}^{(ls)}(\mu, \nu) \psi_{l\mu}(\mathbf{r}_1, \mathbf{r}_2) \chi_{s\nu}(\sigma_1, \sigma_2). \tag{3.20}$$

The wave function (3.7) will be antisymmetric for a transposition of x_1 and x_2 if for $s = 0$ $\psi_{l\mu}(\mathbf{r}_1, \mathbf{r}_2)$ is symmetric and for $s = 1$ it is antisymmetric with respect to the transposition of the arguments $\mathbf{r}_1$ and $\mathbf{r}_2$.

The sum in the right-hand side of formula (3.20) satisfies equations (3.1), (3.3), (3.4) and (3.5) for any set of coefficients $c_{j\mu}^{(ls)}(\mu, \nu)$; these coefficients will be defined if one requires the function to satisfy also equation (3.2).

Using formulae (3.7) and (3.8), one can reduce this condition to the set of the following equations:

$$\alpha_{jm-1} c_{jm}^{(ls)}(\mu, \nu) = \alpha_{j\mu-1} c_{jm-1}^{(ls)}(\mu - 1, \nu) + \alpha_{s\nu-1} c_{jm-1}^{(ls)}(\mu, \nu - 1),$$

$$\overline{\alpha_{jm}} c_{jm}^{(ls)}(\mu, \nu) = \overline{\alpha_{j\mu}} c_{jm+1}^{(ls)}(\mu + 1, \nu) + \overline{\alpha_{s\nu}} c_{jm+1}^{(ls)}(\mu, \nu + 1), \tag{3.21}$$

where $\mu + \nu = m$.

From the orthogonality and normalization conditions of the function $\psi_{jm}^{(ls)}(x_1, x_2)$

$$\iint \overline{\psi_{jm}^{(ls)}}(x_1, x_2) \psi_{j'm}^{(ls)}(x_1, x_2) \, dx_1 dx_2 = \delta_{jj'}, \tag{3.22}$$

and from the orthogonality and normalization of the functions $\psi_{l\mu}(\mathbf{r}_1, \mathbf{r}_2)$ and $\chi_{s\nu}(\sigma_1, \sigma_2)$

$$\iint \overline{\psi_{l\mu}}(\mathbf{r}_1, \mathbf{r}_2) \psi_{l'\mu'}(\mathbf{r}_1, \mathbf{r}_2) \, dx_1 dx_2 = \delta_{ll'} \delta_{\mu\mu'}, \tag{3.23}$$

$$\sum_{\sigma_1, \sigma_2} \overline{\chi_{s\nu}}(\sigma_1, \sigma_2) \chi_{s'\nu'}(\sigma_1, \sigma_2) = \delta_{ss'} \delta_{\nu\nu'}, \tag{3.24}$$

we get the following equations for $c_{jm}^{(l,s)}(\mu,\nu)$:

$$\sum_{\mu}\overline{c_{jm}^{(ls)}}(\mu,m-\mu)\,c_{jm}^{(ls)}(\mu,m-\mu)=\delta_{jj'},\tag{3.25}$$

which show that the coefficients $c_{jm}^{ls}(\mu,m-\mu)$ constitute a unitary matrix with rows indicated by the numbers j and columns indicated by the numbers μ. As a consequence of these equations it follows

$$\sum_{j}\overline{c_{jm}^{(ls)}}(\mu,m-\mu)\,c_{jm}^{(ls)}(\mu',m-\mu')=\delta_{\mu\mu'}.\tag{3.26}$$

According to formula (3.20), the two-electron wave function of a singlet state ($s=0$) is represented by a single term

$$\psi_{jm}^{(l0)}(x_1,x_2)=\psi_{l\mu}(\mathbf{r}_1,\mathbf{r}_2)\,\chi_{00}(\sigma_1,\sigma_2)\,,\tag{3.27}$$

where $j=l,\mu=m$ and $\psi_{l\mu}(\mathbf{r}_1,\mathbf{r}_2)$ has to be symmetric for a transposition of $\mathbf{r}_1$ and $\mathbf{r}_2$. We drop the coefficient $c_{jm}^{(l0)}(\mu,0)$ at $\psi_{l\mu}\chi_{0,0}$ because it can be equated to unity due to equation (3.25). With the exception of some special cases, the functions of a triplet state ($s=1$) are represented, in general, as a sum of three terms, which correspond to the values $\nu=-1,0,1$ with antisymmetric $\psi_{l\mu}(\mathbf{r}_1,\mathbf{r}_2)$.

In particular cases this sum is reduced to one term (i.e., for $l=0$ or for $m=j$) or to two terms.

Similar to (3.20), the one-electron functions are represented by the sum of two terms

$$\psi_{jm}^{(l\frac{1}{2})}(x)=\sum_{\nu=\pm\frac{1}{2}}c_{jm}^{(l\frac{1}{2})}(m-\nu,\nu)\,\psi_{lm-\nu}(\mathbf{r})\,\chi_{1/2\nu}(\sigma)\,,\tag{3.28}$$

where $j=l\pm\frac{1}{2}$.

The same equations (3.21) and unitary conditions (3.25)-(3.26) are valid for coefficients $c_{jm}^{(ls)}(\mu,m-\mu)$ of the one-electron functions.

The formulae similar to (3.20) exhibit the so-called vector model and they are valid for the vector coupling of any two angular momentum operators. Formulae (3.7)–(3.9), (3.21) and (3.25)–(3.26) also have the same generality. Using these formulae one can define, for example, spin functions of two-electron states if the one-electron spin functions are known. Thus, in order to obtain $\chi_{00}(\sigma_1,\sigma_2)$, we calculate the corresponding coefficients $c_{00}^{(\frac{1}{2}\frac{1}{2})}(\frac{1}{2},\nu)$ and get for this function the following expression:

$$\chi_{00}(\sigma_1,\sigma_2)=\frac{1}{\sqrt{2}}\left\{\chi_{\frac{1}{2}\frac{1}{2}}(\sigma_1)\,\chi_{\frac{1}{2}-\frac{1}{2}}(\sigma_2)-\chi_{\frac{1}{2}-\frac{1}{2}}(\sigma_1)\,\chi_{\frac{1}{2}\frac{1}{2}}(\sigma_2)\right\}. \tag{3.17a}$$

It is easy to see that due to equation (3.15) this expression coincides with the formula (3.17) for $\chi_{00}(\sigma_1,\sigma_2)$ obtained earlier.

4 Determination of Dependence on Spin Variables in the Matrix Elements of the Electron Density

Introduction

In the following calculations, we restrict ourselves by the ground and, therefore, singlet state of a divalent atom. In order to determine explicitly the dependence on spin variables in formulae (2.5) and (2.14), we consider elements of the density matrices ϱ and ϱ_1, defined by formulae (2.4), (2.6) and (2.7). Substituting the function $\psi_{jm}^{(l0)}$, which is defined by formula (3.27) as $\psi(x,x')$ in formula (2.6), we get

$$\varrho_1(x_1,x_2;x_1',x_2')=\overline{\psi_{l\mu}}(\mathbf{r}_1,\mathbf{r}_2)\,\psi_{l\mu}(\mathbf{r}_1',\mathbf{r}_2')\,\overline{\chi_{0,0}}(\sigma_1,\sigma_2)\,\chi_{0,0}(\sigma_1',\sigma_2'). \tag{4.1}$$

Assuming then $\sigma_1'=\sigma_1,\sigma_2'=\sigma_2$ and making summation over σ_1 and σ_2, due to the normalization of function $\chi_{0,0}(\sigma_1,\sigma_2)$, we get for the quantity

$$\varrho_1(\mathbf{r}_1,\mathbf{r}_2;\mathbf{r}_1',\mathbf{r}_2')=\sum_{\sigma_1,\sigma_2}\varrho_1(\mathbf{r}_1\sigma_1,\mathbf{r}_2\sigma_2;\mathbf{r}_1'\sigma_1,\mathbf{r}_2'\sigma_2) \tag{4.2}$$

the following expression:

$$\varrho_1(\mathbf{r}_1,\mathbf{r}_2;\mathbf{r}_1',\mathbf{r}_2')=\overline{\psi_{l\mu}}(\mathbf{r}_1,\mathbf{r}_2)\,\psi_{l\mu}(\mathbf{r}_1',\mathbf{r}_2'). \tag{4.3}$$

In order to extract the dependence on spins of the matrix ϱ_1 defined by formula (2.7), we set $x_2'=x_2$ in expression (4.1) and integrate it over x_2 taking $\chi_{0,0}(\sigma_1',\sigma_2')$ from the corresponding formula (3.17a). Using relation (3.15) for one-electron spin functions, due to

$$\overline{\chi_{\frac{1}{2},\frac{1}{2}}}(\sigma)\,\chi_{\frac{1}{2},\frac{1}{2}}(\sigma')+\overline{\chi_{\frac{1}{2},-\frac{1}{2}}}(\sigma)\,\chi_{\frac{1}{2},-\frac{1}{2}}(\sigma')=\frac{1}{2}(1+\sigma\sigma')=\delta_{\sigma\sigma'}, \tag{4.4}$$

we get the following expression:

$$\varrho_1(x,x')=\frac{1}{2}\delta_{\sigma\sigma'}\int\overline{\psi_{l\mu}}(\mathbf{r},\mathbf{r}'')\,\psi_{l\mu}(\mathbf{r}',\mathbf{r}'')\,d\tau'', \tag{4.5}$$

where $d\tau$ is the differential of the ordinary volume.

Setting here $\sigma' = \sigma$ and making summation over σ, we get

$$\varrho_1(x, x') = \int \overline{\psi_{l\mu}}(\mathbf{r}, \mathbf{r}'')\, \psi_{l\mu}(\mathbf{r}', \mathbf{r}'')\, d\tau''. \tag{4.6}$$

Let us turn to a derivation of the known expression for the matrix element of the one-electron density ϱ of inner-shell electrons (formula (2.4)). The inner-shell electrons in our case form closed shells; hence, it is sufficient to calculate a matrix element for one shell, i.e., for equivalent electrons

$$\varrho^{(l)}(x, x') = \sum_{j,m} \overline{\psi_{jm}^{(ls)}}(x)\, \psi_{jm}^{(ls)}(x'), \tag{4.7}$$

and then to make summation of the obtained expression over all closed shells, in other words, to make summation over all values n and l of the principal and azimuth quantum numbers.

We substitute the expression (3.28) in (4.7) and make first summation over j and next we sum it over μ and ν also using the property (3.26) and formula (4.4). Then we get

$$\varrho^{(l)}(x, x') = \delta_{\sigma\sigma'} \sum_{\mu=-l}^{l} \overline{\psi_{l\mu}}(\mathbf{r})\, \psi_{l\mu}(\mathbf{r}'). \tag{4.8}$$

Finally, we obtain

$$\varrho(x, x') = \sum_{n,l} \sum_{\mu=-l}^{l} \overline{\psi_{l\mu}}(\mathbf{r})\, \psi_{l\mu}(\mathbf{r}')\, \delta_{\sigma\sigma'}. \tag{4.9}$$

Setting here $\sigma' = \sigma$ and making the summation over σ we receive

$$2\varrho(\mathbf{r}, \mathbf{r}') = 2\sum_{n,l} \sum_{\mu=-l}^{l} \overline{\psi_{l\mu}}(\mathbf{r})\psi_{l\mu}(\mathbf{r}'). \tag{4.10}$$

The summation over μ in formulae (4.8) and (4.9) is easily performed due to the well-known addition theorem for spherical harmonics.

Substituting the obtained expressions for matrix elements in formula (2.5) and (2.14), we separate, in this way, their dependence on spin variables. Setting further in new expression of the integral (2.5)

$\sigma_1' = \sigma_1,\ \sigma_2' = \sigma_2$ and summing up over σ_1 and σ_2 we get

$$\frac{1}{n!}\sum_{\sigma_1\sigma_2}\int\ldots\int \overline{\Psi}\left(\mathbf{r}_1\sigma_1,\mathbf{r}_2\sigma_2,\ldots \mathbf{r}_{n+2}\sigma_{n+2}\right)\times$$

$$\times\,\Psi\left(\mathbf{r}_1'\sigma_1,\mathbf{r}_2'\sigma_2,\ldots \mathbf{r}_{n+2}\sigma_{n+2}\right)dx_2\ldots dx_{n+2} =$$

$$= \varrho_1\left(\mathbf{r}_1,\mathbf{r}_2;\mathbf{r}_1',\mathbf{r}_2'\right) + 2\varrho_1\left(\mathbf{r}_1,\mathbf{r}_1'\right)\varrho\left(\mathbf{r}_2,\mathbf{r}_2'\right) + 2\varrho\left(\mathbf{r}_1,\mathbf{r}_1'\right)\varrho_1\left(\mathbf{r}_2,\mathbf{r}_2'\right) -$$

$$-\varrho_1\left(\mathbf{r}_1,\mathbf{r}_2'\right)\varrho\left(\mathbf{r}_2,\mathbf{r}_1'\right) - \varrho\left(\mathbf{r}_1,\mathbf{r}_2'\right)\varrho_1\left(\mathbf{r}_2,\mathbf{r}_1'\right) + 4\varrho\left(\mathbf{r}_1,\mathbf{r}_1'\right)\varrho\left(\mathbf{r}_2,\mathbf{r}_2'\right) -$$

$$-2\varrho\left(\mathbf{r}_1,\mathbf{r}_2'\right)\varrho\left(\mathbf{r}_2,\mathbf{r}_1'\right). \qquad (4.11)$$

Some matrix elements that form this formula are defined in equations (4.3), (4.6) and (4.10).

Setting further $\sigma_1' = \sigma_1$ in (2.14) and summing up over σ_1, we have

$$\frac{1}{(n+1)!}\sum_{\sigma_1}\int\ldots\int \overline{\Psi}\left(\mathbf{r}_1\sigma_1,\ldots \mathbf{r}_{n+2}\sigma_{n+2}\right)\times$$

$$\times\Psi\left(\mathbf{r}_1'\sigma_1,\mathbf{r}_2\sigma_2,\ldots \mathbf{r}_{n+2}\sigma_{n+2}\right)dx_2dx_3\ldots dx_{n+2} =$$

$$= \varrho_1\left(\mathbf{r}_1,\mathbf{r}_1'\right) + 2\varrho\left(\mathbf{r}_1,\mathbf{r}_1'\right). \qquad (4.12)$$

In the following, we use a notation for Schrödinger wave functions similar to that in Sections 1 and 2 for functions with the spin; namely, we will denote for brevity the set of quantum numbers defining a one-electron function of the spatial coordinates as i, where $i = 1, 2, \ldots \frac{1}{2}n$ (here n is the number of inner-shell electrons), i.e.,

$$\psi_{l\mu}\left(\mathbf{r}\right) \rightarrow \psi_i\left(\mathbf{r}\right), \qquad (4.13)$$

and we will drop indices of the wave functions of valent-shell electrons, i.e.,

$$\psi_{l\mu}\left(\mathbf{r}_1,\mathbf{r}_2\right) \rightarrow \psi\left(\mathbf{r}_1,\mathbf{r}_2\right). \qquad (4.14)$$

Let us separate the spin factors in the two-electron wave function (1.4) considered in Section 1. In the original function $\varphi\left(x,x'\right)$, describing the state of two valence electrons, the spin part, obviously, has to be the same as in the function $\psi\left(x,x'\right)$, which is orthogonal to one-electron functions of internal states, i.e.,

$$\varphi\left(x,x'\right) = \varphi\left(\mathbf{r},\mathbf{r}\prime\right)\chi_{0,0}\left(\sigma,\sigma'\right), \qquad (4.15)$$

where $\varphi(\mathbf{r}, \mathbf{r}')$ is symmetric.

Assuming conditions (1.7)–(1.8), we replace $\omega(x, x')$ by $\psi(x, x')$ in formula (1.4) according to its definition in Section 5. Inserting the one-electron functions (3.28) in (1.4), (1.7), (1.8), we make summation in (1.4) using again the unitarity conditions (3.26).

Finally, after cancelation of the spin function $\chi_{0,0}(\sigma, \sigma')$ and the passage to the simplified notation of functions (4.13) and (4.14) formula (1.4) will be written as

$$\psi(\mathbf{r}, \mathbf{r}') = \varphi(\mathbf{r}, \mathbf{r}') - \sum_{i=1}^{n/2} \left[f_i(\mathbf{r}) \psi_i(\mathbf{r}') + f_i(\mathbf{r}') \psi_i(\mathbf{r})\right] + \\ + \sum_{i,k=1}^{n/2} a_{ik} \psi_i(\mathbf{r}) \psi_k(\mathbf{r}'), \quad (4.16)$$

where

$$f_i(\mathbf{r}) = \int \overline{\psi_i}(\mathbf{r}') \varphi(\mathbf{r}, \mathbf{r}') \, d\tau' \quad (4.17)$$

and

$$a_{ik} = \iint \overline{\psi_i}(\mathbf{r}) \overline{\psi_k}(\mathbf{r}') \varphi(\mathbf{r}, \mathbf{r}') \, d\tau d\tau'. \quad (4.18)$$

An obvious equation

$$a_{ik} = a_{ki}$$

follows from the symmetry of $\varphi(\mathbf{r}, \mathbf{r}')$.

Concluding this section, we write down the normalization and orthogonality conditions for Schrödinger wave functions. As a result of spin separation in formulae (1.12), we obtain

$$\left.\begin{aligned} \iint \overline{\psi}(\mathbf{r}, \mathbf{r}') \psi(\mathbf{r}, \mathbf{r}') \, d\tau d\tau' &= 2, \\ \int \overline{\psi}(\mathbf{r}, \mathbf{r}') \psi_i(\mathbf{r}') \, d\tau' &= 0, \\ \int \overline{\psi_i}(\mathbf{r}) \psi_k(\mathbf{r}) \, d\tau &= \delta_{ik}, \\ \left(i, k = 1, 2, \ldots \frac{n}{2}\right). & \end{aligned}\right\} \quad (4.19)$$

5 Expression for the Energy of the Atom and Equations for Wave Functions

Introduction

We find now an expression for the total energy of the atom. We have

$$W = \frac{1}{(n+2)!}\int \overline{\Psi} H \Psi \,(dx)\,, \tag{5.1}$$

where H is the energy operator, Ψ is the wave function of the atom and

$$(dx) = dx_1 dx_2 \dots dx_{n+2}.$$

The energy operator in the atomic units has the form

$$H = \sum_{i=1}^{n+2} H_i + \sum_{i>k=1}^{n+2} \frac{1}{r_{ik}}, \tag{5.2}$$

where $H_i = -\frac{1}{2}\Delta_i - \frac{\zeta}{r_i}$ is the energy operator in the unscreened field of the nucleus.

Substituting this expression for H in formula (5.1), we get

$$W = \frac{1}{(n+2)!}\left\{\sum_{i=1}^{n+2}\int \overline{\Psi} H_i \Psi \,(dx) + \sum_{i>k=1}^{n+2}\int \overline{\Psi}\frac{1}{r_{ik}}\Psi\,(dx)\right\}. \tag{5.3}$$

By virtue of the symmetry of the product $\overline{\Psi}\Psi$ with respect to a permutation of the coordinates of all electrons, both all $n+2$ terms of the single sum and $\frac{1}{2}(n+2)(n+1)$ terms of the double sum are equal to each other, and consequently

$$W = \frac{1}{(n+1)!}\int \overline{\Psi} H_1 \Psi\,(dx) + \frac{1}{2n!}\int \overline{\Psi}\frac{1}{r_{12}}\Psi\,(dx)\,. \tag{5.4}$$

The integration over all variables, on which the operators in the integrands do not depend (i.e., over $x_2, x_3, \dots x_{n+2}$ in the first integral and over $x_3, x_4, \dots x_{n+2}$ in the second one), as well as the summation over σ_1 in the first and over σ_1 and σ_2 in the second integrals were carried out in the previous section.

Using now the derived formulae (4.11) and (4.12), we get the following expression for the total energy of a divalent atom:

$$W = \iint \overline{\psi}(\mathbf{r}_1, \mathbf{r}_2)\left[-\frac{1}{2}\Delta_1 - \frac{\zeta}{r_1}\right]\psi(\mathbf{r}_1, \mathbf{r}_2)\,d\tau_1 d\tau_2 +$$

$$+\frac{1}{2}\iint \overline{\psi}(\mathbf{r}_1,\mathbf{r}_2)\frac{1}{r_{12}}\psi(\mathbf{r}_1,\mathbf{r}_2)\,d\tau_1 d\tau_2 +$$
$$+2\sum_{i=1}^{n/2}\int \overline{\psi_i}(\mathbf{r}_1)\left[-\frac{1}{2}\Delta_1-\frac{\zeta}{r_1}\right]\psi_i(\mathbf{r}_1)\,d\tau_1 +$$
$$+2\iint\left[\varrho(\mathbf{r}_1,\mathbf{r}_1)\varrho(\mathbf{r}_2,\mathbf{r}_2)-\frac{1}{2}\left|\varrho(\mathbf{r}_1,\mathbf{r}_2)\right|^2\right]\frac{1}{r_{12}}d\tau_1 d\tau_2 +$$
$$+2\iiint \overline{\psi}(\mathbf{r}_2,\mathbf{r})\psi(\mathbf{r}_2,\mathbf{r})\frac{1}{r_{12}}\varrho(\mathbf{r}_1,\mathbf{r}_1)\,d\tau d\tau_1 d\tau_2 -$$
$$-\iiint \overline{\psi}(\mathbf{r}_2,\mathbf{r})\psi(\mathbf{r}_1,\mathbf{r})\frac{1}{r_{12}}\varrho(\mathbf{r}_1\mathbf{r}_2)\,d\tau d\tau_1 d\tau_2, \tag{5.5}$$

where, according to formula (4.10),

$$\varrho(\mathbf{r},\mathbf{r}')=\sum_{i=1}^{n/2}\overline{\psi_i}(\mathbf{r})\psi_i(\mathbf{r}')$$

and integration is done only over the spatial coordinates.

Taking into account the physical meaning of separate summands in (5.05), we can conveniently write it as follows:

$$W=W_1+W_2+W_{12}. \tag{5.6}$$

The expression

$$W_1=\iint \overline{\psi}(\mathbf{r}_1,\mathbf{r}_2)\left[-\frac{1}{2}\Delta_1-\frac{\zeta}{r_1}+\frac{1}{2}\frac{1}{r_{12}}\right]\psi(\mathbf{r}_1,\mathbf{r}_2)\,d\tau_1 d\tau_2 \tag{5.7}$$

gives the kinetic energy, the energy of the attraction to the nucleus and the interaction energy of valence electrons, whereas

$$W_2=2\sum_{i=1}^{n}\int \overline{\psi_i}(\mathbf{r}_1)\left[-\frac{1}{2}\Delta_1-\frac{\zeta}{r_1}\right]\psi_i(\mathbf{r}_1)\,d\tau_1 +$$
$$+2\iint\left[\varrho(\mathbf{r}_1,\mathbf{r}_1)\varrho(\mathbf{r}_2,\mathbf{r}_2)-\frac{1}{2}\left|\varrho(\mathbf{r}_1,\mathbf{r}_2)\right|^2\right]\frac{1}{r_{12}}d\tau_1 d\tau_2 \tag{5.8}$$

gives the same quantities for inner-shell electrons; finally, the remaining terms of formula (5.5), i.e., the expression

$$W_{12}=2\iiint \overline{\psi}(\mathbf{r}_2,\mathbf{r}_3)\psi(\mathbf{r}_2,\mathbf{r}_3)\frac{1}{r_{12}}\varrho(\mathbf{r}_1,\mathbf{r}_1)\,d\tau_1 d\tau_2 d\tau_3 -$$
$$-\iiint \overline{\psi}(\mathbf{r}_2,\mathbf{r}_3)\psi(\mathbf{r}_1,\mathbf{r}_3)\frac{1}{r_{12}}\varrho(\mathbf{r}_1\mathbf{r}_2)\,d\tau_1 d\tau_2 d\tau_3, \tag{5.9}$$

give the interaction energy of the inner-shell electrons and the valence ones.

Having now the complete expression for the energy, from the variation principle, we can derive the integro-differential equations for the wave functions. In this way, we should take into account the proper symmetry of the two-electron function $\psi(\mathbf{r}, \mathbf{r}')$. It is clear that if the integrand has the same symmetry as this function, the resulting equation will automatically preserve the symmetry. Therefore, we symmetrize beforehand W_1 and W_{12}, i.e., in the expression for the total energy, we set

$$W_1 = \frac{1}{2} \iint \overline{\psi}(\mathbf{r}_1, \mathbf{r}_2) \left[H_1 + H_2 + \frac{1}{r_{12}} \right] \psi(\mathbf{r}_1, \mathbf{r}_2) \, d\tau_1 d\tau_2 \quad (5.10)$$

and

$$\begin{aligned} W_{12} = & \iiint \overline{\psi}(\mathbf{r}_1, \mathbf{r}_2) \psi(\mathbf{r}_1, \mathbf{r}_2) \left[\frac{1}{r_{13}} + \frac{1}{r_{23}} \right] \varrho(\mathbf{r}_3, \mathbf{r}_3) \, d\tau_1 d\tau_2 d\tau_3 - \\ & - \frac{1}{2} \iiint \overline{\psi}(\mathbf{r}_1, \mathbf{r}_2) \left[\psi(\mathbf{r}_3, \mathbf{r}_1) \frac{1}{r_{23}} \varrho(\mathbf{r}_3, \mathbf{r}_2) + \right. \\ & \left. + \psi(\mathbf{r}_3, \mathbf{r}_2) \frac{1}{r_{13}} \varrho(\mathbf{r}_3 \mathbf{r}_1) \right] d\tau_1 d\tau_2 d\tau_3. \end{aligned} \quad (5.11)$$

Furthermore, the variation equation

$$\delta W = 0$$

has to be treated together with the additional conditions (4.19) of orthogonality and normalization, i.e., with the conditions

$$J_{ik} = \int \overline{\psi_i}(\mathbf{r}) \psi_k(\mathbf{r}) \, d\tau = \delta_{ik}, \quad \left(i, k = 1, 2 \ldots \frac{n}{2} \right) \quad (5.12)$$

$$J = \iint \overline{\psi}(\mathbf{r}, \mathbf{r}') \psi(\mathbf{r}, \mathbf{r}') \, d\tau d\tau' = 2. \quad (5.13)$$

The second condition in (4.19) will be written in the following way:

$$J_i = \iint \left[\lambda_i(\mathbf{r}) \psi_i(\mathbf{r}') + \lambda_i(\mathbf{r}') \psi_i(\mathbf{r}) \right] \overline{\psi}(\mathbf{r}, \mathbf{r}') \, d\tau d\tau' = 0, \quad (5.14)$$

where $\lambda_i(\mathbf{r})$ are arbitrary functions.

Let us write the expression

$$\delta\left[W+\lambda J+\sum_{i=1}^{n/2}\left(J_i+\overline{J}_i\right)+2\sum_{i,k=1}^{n/2}\lambda_{ik}J_{ik}\right]=0 \tag{5.15}$$

and vary the functions $\psi(\mathbf{r}_1,\mathbf{r}_2)$ and $\psi_i(\mathbf{r}_1)$, taking into account that the operators in the integrands are self-adjoint. Then, equating to zero the coefficients at the variations $\delta\overline{\psi}(\mathbf{r}_1,\mathbf{r}_2)$ and $\delta\overline{\psi_i}(\mathbf{r}_1)$, we get one equation for the two-electron function and $n/2$ equations for the wave functions of the inner-shell electrons.

The equation for two-electron function $\psi(\mathbf{r}_1,\mathbf{r}_2)$ is written as follows:

$$\begin{aligned}-&\frac{1}{2}\left(\Delta_1+\Delta_2\right)\psi(\mathbf{r}_1,\mathbf{r}_2)+\left[-\frac{\zeta}{r_1}-\frac{\zeta}{r_2}+\frac{1}{r_{12}}+2\int\varrho(\mathbf{r}_3,\mathbf{r}_3)\frac{1}{r_{23}}d\tau_3+\right.\\&\left.+2\int\varrho(\mathbf{r}_3,\mathbf{r}_3)\frac{1}{r_{13}}d\tau_3\right]\psi(\mathbf{r}_1,\mathbf{r}_2)-\int\psi(\mathbf{r}_3,\mathbf{r}_2)\frac{1}{r_{13}}\varrho(\mathbf{r}_3,\mathbf{r}_1)\,d\tau_3-\\&-\int\psi(\mathbf{r}_3,\mathbf{r}_1)\frac{1}{r_{23}}\varrho(\mathbf{r}_3,\mathbf{r}_2)\,d\tau_3+2\lambda\psi\psi(\mathbf{r}_1,\mathbf{r}_2)+\\&+2\sum_{i=1}^{n/2}\left[\lambda_i(\mathbf{r}_1)\psi_i(\mathbf{r}_2)+\lambda_i(\mathbf{r}_2)\psi_i(\mathbf{r}_1)\right]=0.\end{aligned} \tag{5.16}$$

This equation can be conveniently rewritten in the form

$$\begin{aligned}&L_{12}\psi(\mathbf{r}_1,\mathbf{r}_2)+2\lambda\psi(\mathbf{r}_1,\mathbf{r}_2)+\\&\qquad+2\sum_{i=1}^{n/2}\left[\lambda_i(\mathbf{r}_1)\psi_i(\mathbf{r}_2)+\lambda_i(\mathbf{r}_2)\psi_i(\mathbf{r}_1)\right]=0,\end{aligned} \tag{5.17}$$

where we denoted as $L_{12}\psi(\mathbf{r}_1,\mathbf{r}_2)$ the remaining terms in (5.16) so that L_{12} is, in fact, an integro-differential operator.

The parameter λ and the functions $\lambda_i(\mathbf{r})$ enter the equation as Lagrange multipliers.

We get the quantities 2λ calculating the integral

$$\frac{1}{2}\int\overline{\psi}(\mathbf{r}_1,\mathbf{r}_2)\,L_{12}\psi(\mathbf{r}_1,\mathbf{r}_2)\,d\tau_1d\tau_2.$$

It is easy to see that because of (5.10)–(5.11) this expression coincides with the sum W_1+W_{12} and, therefore, the value -2λ, which corresponds to the final solution of equation (5.16), gives the value of the total energy of the valence electrons in the field of the atomic core. The equation for the wave function that describes the state of the electron specified by the index i is written as:

$$-\frac{1}{2}\Delta_1\psi_i(\mathbf{r}_1)+\left[-\frac{\zeta}{r_1}+2\int\varrho(\mathbf{r}_2,\mathbf{r}_2)\frac{1}{r_{12}}d\tau_2+\right.$$

$$\left.+\iint\overline{\psi}(\mathbf{r}_2,\mathbf{r})\,\psi(\mathbf{r}_2,\mathbf{r})\frac{1}{r_{12}}d\tau d\tau_2\right]\psi_i(\mathbf{r}_1)-\int\psi_i(\mathbf{r}_2)\frac{1}{r_{12}}\varrho(\mathbf{r}_2,\mathbf{r}_1)\,d\tau_2-$$

$$-\frac{1}{2}\iint\overline{\psi}(\mathbf{r}_2,\mathbf{r})\,\psi(\mathbf{r}_1,\mathbf{r})\frac{1}{r_{12}}\psi_i(\mathbf{r}_2)\,d\tau d\tau_2+$$

$$+\int\overline{\lambda_i(\mathbf{r})}\psi(\mathbf{r}_1,\mathbf{r})\,d\tau+\sum_{i=1}^{n/2}\lambda_{ik}\psi_k(\mathbf{r}_1)=0. \qquad (5.18)$$

With the help of an orthogonal transformation of one-electron functions leaving invariant the one-electron density, one can bring the Lagrange multipliers λ_{ik} to the diagonal form $\lambda_{ik}=\lambda_i\delta_{ik}$.

Multiplying the equation on the left by $\overline{\psi_i}(\mathbf{r}_1)$ and integrating it over the coordinates, one can define an expression for the parameter λ_i, which corresponds to the solution of equation (5.18) finite everywhere. Defining in this way the corresponding parameters in the equations for other $\frac{n}{2}-1$ functions of the internal electrons, we get

$$-\sum_{i=1}^{n/2}\lambda_i=\frac{1}{2}(W_{12}+W_2')+W_2'',$$

where W_2' denotes the kinetic energy and the energy of the electron attraction to the nucleus and W_2'' denotes the energy of the electron interaction and, hence,

$$W_2'+W_2''=W_2.$$

The expression for the arbitrary Lagrange functions $\lambda_i(\mathbf{r})$ in the system of equations (5.16) and (5.18) follows from the comparison of equation (5.17) for the two-electron function $\psi(\mathbf{r}_1,\mathbf{r}_2)$ with the same equation but being derived in another way. In that way, we take into account by means of the Lagrange multipliers only the conditions (5.12) and (5.13), i.e., we consider the variation

$$\delta\left(W+\lambda J+2\sum_{i,k=1}^{n/2}\lambda_{ik}J_{ik}\right)=0. \qquad (5.19)$$

Then the variation $\delta\psi(\mathbf{r}_1,\mathbf{r}_2)$, which enters this equation, cannot be considered independent; it has to ensure the orthogonality conditions of the function $\psi(\mathbf{r},\mathbf{r}')$ to all functions $\psi(\mathbf{r})$. For that, we use formula (4.16), which gives an expression for $\psi(\mathbf{r}_1,\mathbf{r}_2)$ in terms of $\varphi(\mathbf{r}_1,\mathbf{r}_2)$, and take

$$\begin{aligned}\delta\overline{\psi}(\mathbf{r}_1,\mathbf{r}_2) &= \delta\overline{\varphi}(\mathbf{r}_1,\mathbf{r}_2) - \int \varrho(\mathbf{r}_1,\mathbf{r})\,\delta\overline{\varphi}(\mathbf{r}_2,\mathbf{r})\,d\tau - \\ &- \int \varrho(\mathbf{r}_2,\mathbf{r})\,\delta\overline{\varphi}(\mathbf{r}_1,\mathbf{r})\,d\tau + \iint \varrho(\mathbf{r}_1,\mathbf{r})\,\varrho(\mathbf{r}_2,\mathbf{r}')\,\delta\overline{\varphi}(\mathbf{r},\mathbf{r}')\,d\tau\,d\tau',\end{aligned} \tag{5.20}$$

where $\delta\overline{\varphi}(\mathbf{r}_1,\mathbf{r}_2)$ is restricted only by the symmetry condition but is arbitrary in all other respects. Inserting this expression for $\delta\overline{\psi}(\mathbf{r}_1,\mathbf{r}_2)$ in (5.19), we obtain an equation for the two-electron function $\psi(\mathbf{r}_1,\mathbf{r}_2)$ if we equate to zero the coefficient at an arbitrary variation $\delta\overline{\varphi}(\mathbf{r}_1,\mathbf{r}_2)$. We will get

$$\begin{aligned}&L_{12}\psi(\mathbf{r}_1,\mathbf{r}_2) + 2\lambda\psi(\mathbf{r}_1,\mathbf{r}_2) + \iint \psi(\mathbf{r}_3,\mathbf{r}_4)\,\frac{1}{r_{34}}\varrho(\mathbf{r}_3,\mathbf{r}_2)\,\varrho(\mathbf{r}_4,\mathbf{r}_1)\,d\tau_3 d\tau_4 \\ &- \int \varrho(\mathbf{r}_3,\mathbf{r}_1)\,\frac{1}{r_{23}}\psi(\mathbf{r}_3,\mathbf{r}_2)\,d\tau_3 - \int \varrho(\mathbf{r}_3,\mathbf{r}_2)\,\frac{1}{r_{13}}\psi(\mathbf{r}_3,\mathbf{r}_1)\,d\tau_3 - \\ &- \int \psi(\mathbf{r}_3,\mathbf{r}_2)\,H_3\varrho(\mathbf{r}_3,\mathbf{r}_1)\,d\tau_3 - \int \psi(\mathbf{r}_3,\mathbf{r}_1)\,H_3\varrho(\mathbf{r}_3,\mathbf{r}_2)\,d\tau_3 - \\ &- \iint \left[2\varrho(\mathbf{r}_3,\mathbf{r}_3)\,\varrho(\mathbf{r}_4,\mathbf{r}_1) - \varrho(\mathbf{r}_4,\mathbf{r}_3)\,\varrho(\mathbf{r}_3,\mathbf{r}_1)\right]\frac{1}{r_{34}}\psi(\mathbf{r}_2,\mathbf{r}_4)\,d\tau_3 d\tau_4 - \\ &- \iint \left[2\varrho(\mathbf{r}_3,\mathbf{r}_3)\,\varrho(\mathbf{r}_4,\mathbf{r}_2) - \varrho(\mathbf{r}_4,\mathbf{r}_3)\,\varrho(\mathbf{r}_3,\mathbf{r}_2)\right]\frac{1}{r_{34}} \times \\ &\qquad \times\psi(\mathbf{r}_1,\mathbf{r}_4)\,d\tau_3 d\tau_4 = 0,\end{aligned} \tag{5.21}$$

where L_{12} is the same operator as in equation (5.17). Comparing this equation with (5.21), we get the following expression for an arbitrary Lagrange function $\lambda_i(\mathbf{r})$:

$$\begin{aligned}2\lambda_i(\mathbf{r}_1) &= -\int \psi(\mathbf{r}_2,\mathbf{r}_1)\left[H_2 + \frac{1}{r_{12}}\right]\overline{\psi_i}(\mathbf{r}_2)\,d\tau_2 + \\ &+\frac{1}{2}\iint \psi(\mathbf{r}_2,\mathbf{r}_3)\,\psi_i(\mathbf{r}_2)\,\varrho(\mathbf{r}_3,\mathbf{r}_1)\,\frac{1}{r_{23}}d\tau_2 d\tau_3 - \\ &- \iint \left[2\varrho(\mathbf{r}_2,\mathbf{r}_2)\,\psi_i(\mathbf{r}_3) - \varrho(\mathbf{r}_3,\mathbf{r}_2)\,\psi_i(\mathbf{r}_2)\right]\frac{1}{r_{23}} \times \\ &\qquad \times\psi(\mathbf{r}_1,\mathbf{r}_2)\,d\tau_2 d\tau_3.\end{aligned} \tag{5.22}$$

Using this expression, we can determine the integral that contains the function $\lambda_i(\mathbf{r})$ in equation (5.18) for one-electron functions.

This integral is equal to

$$\int \overline{\lambda_i}(\mathbf{r}_1)\,\psi(\mathbf{r}_1,\mathbf{r}_2)\,d\tau_2 =$$

$$-\frac{1}{2}\iint \overline{\psi}(\mathbf{r}_2,\mathbf{r}_3)\,\psi(\mathbf{r}_1,\mathbf{r}_2)\left[H_3+\frac{1}{r_{23}}\right]\psi_i(\mathbf{r}_3)\,d\tau_2 d\tau_3 -$$

$$-\iiint\left[\varrho(\mathbf{r}_3,\mathbf{r}_4)\,\psi_i(\mathbf{r}_3)-\frac{1}{2}\varrho(\mathbf{r}_3,\mathbf{r}_4)\,\psi_i(\mathbf{r}_2)\right]\times$$

$$\times\frac{1}{r_{34}}\overline{\psi}(\mathbf{r}_2,\mathbf{r}_4)\,\psi(\mathbf{r}_1,\mathbf{r}_2)\,d\tau_4 d\tau_2 d\tau_3. \qquad (5.23)$$

After exclusion of the terms containing $\lambda_i(\mathbf{r})$ with the help of this formula, equations (5.18) represent a system of equations for functions $\psi(\mathbf{r}_1,\mathbf{r}_2)$ and $\psi_i(\mathbf{r})$. A solution of this system can, in principle, be obtained by the method of successive approximations. To this end, as a starting approximation, one can take, for instance, a solution of the reduced system in the approximation of the complete separation of variables for the function $\psi(\mathbf{r}_1,\mathbf{r}_2)$. It is easy to see that it will be an ordinary approximation of the self-consistent field with exchange.

However, practically a numerical integration of the system (5.18) is complicated by the necessity to tabulate the functions of several variables. Therefore, in order to determine the wave functions it is more practical to apply the direct variational methods looking for the minimum of the expression (5.4) for the complete energy of an atom. An example of the calculation of a divalent atom using the incomplete separation of variables will be presented in a separate paper.

References

1. D.R. Hartree and W. Hartree, Proc. Roy. Soc. **A 150**, 9, 1935;
   V. Fock and M. Petrashen, Sow. Phys. **8**, 359, 1935. (See also [34-1] in this book. (*Editors*))
2. E. Hylleraas, Zs. Phys. **71**, 739, 1931.
3. V.A. Fock, JETP **10**, 383, 1940.

Leningrad
Spectroscopy Laboratory
the USSR Academy of Sciences,
Leningrad University
Physical Institute

Translated by Yu. Yu. Dmitriev

40-2
On the Wave Functions of Many-Electron Systems[1]

V.A. Fock

Received 12 August 1940

JETP **10**, 961, 1940

Introduction

From the very first applications of quantum mechanics to many-electron problems it appeared necessary to state the symmetry of the Schrödinger wave function corresponding to a certain energy level of an atomic system.

The starting points here are, on the one hand, the Pauli principle (meant as a requirement of antisymmetry of the total wave function depending on coordinates and spins) and, on the other hand, the fact that neglecting relativistic corrections, one can consider the energy operator to be independent on spin. Starting from these basic assumptions and using the group theory, some authors, in particular Hund [1] and Wigner [2], deduced certain symmetry properties of the Schrödinger wave functions. Hund introduced the notions of "the symmetry and antisymmetry characters" of the wave functions and showed that in the case of n electrons it is possible, by means of wave functions belonging to the same energy level, to construct that one, which would be antisymmetric with respect to k coordinates and $n-k$ coordinates;[2] subdivision of coordinates on more than two antisymmetric groups is impossible. However, the question remained not settled: is the wave function antisymmetry with respect to two groups of coordinates the sufficient condition or only the necessary one and is any other condition needed to construct the total antisymmetric function with the help of that coordinate function? Therefore, the construction method of this latter function was not given.

---

[1]Reported on April 26, 1940 at the Session of the Phys.-Math. Section of the USSR Academy of Sciences.

[2]We mean an electron coordinate as a set of three spatial electron coordinates. (*V. Fock*)

The problem was considered more completely but also in a more abstract way by Wigner who investigated the irreducible representation of the permutation group for many-electron spin functions. Refusing to formulate the explicit symmetry conditions for the coordinate wave function, Wigner replaced them by the following requirement: the functions, belonging to the same energy level should be transformed in accordance with a certain irreducible representation of the permutation group. This representation is rather simply connected with the above-mentioned representation obtained by the spin functions. But an abstract character of this Wigner rule makes its application in practical calculations very complicated because it does not give any indications how to obtain the coordinate wave function with the required properties.

Since the works by Wigner and Hunds there is no essential progress in this field. Therefore, one has to conclude that the important question on the ways to construct a wave function with spin from wave functions without spin and the symmetry property of the latter is still unresolved.

The aim of this work is to fill this gap. Here we shall give the explicit expression of the total wave function of the n-electron system through a single Schrödinger wave function and point out, also in an explicit manner, the sufficient and necessary symmetry conditions for the latter. The solution of this problem (without the spatial symmetry) and our results obtained for the vector model [3] will allow us to solve the problem also for the case of spherical symmetry.

In our work we do not resort to the group theory at all and do not use its methods. However, after the results have been obtained by our method it is not difficult to state their connection with relations derived by the group theory.

1 Spin and General Expression of its Eigenfunctions

First let us consider the spin operator of the one-electron problem.

We shall prescribe for the spin variable two values $\sigma = +1$ and $\sigma = -1$ and treat the two-component wave function as a function of this variable. Omitting the dependence on the coordinates, we shall denote this function by $f(\sigma)$; the first component will be $f(+1)$, the second $f(-1)$. The action of Pauli's matrices

$$\sigma_x = \begin{pmatrix} 0 & 1 \\ 1 & 0 \end{pmatrix}; \quad \sigma_y = \begin{pmatrix} 0 & -i \\ i & 0 \end{pmatrix}; \quad \sigma_z = \begin{pmatrix} 1 & 0 \\ 0 & -1 \end{pmatrix} \tag{1.1}$$

on this function can be written in the following way:

$$\begin{aligned} \sigma_x f(\sigma) &= f(-\sigma), \\ \sigma_y f(\sigma) &= -i\sigma f(-\sigma), \\ \sigma_z f(\sigma) &= \sigma f(\sigma). \end{aligned} \tag{1.2}$$

From the third equation of (1.2) it is seen that the variable σ is nothing else but an eigenvalue of σ_z.

These formulae are easily generalized to the many-electron case. In this case, we have a function

$$f = f(\sigma_1, \sigma_2, \ldots \sigma_l, \ldots \sigma_n), \tag{1.3}$$

on which the operators $\sigma_{lx}, \sigma_{ly}, \sigma_{lz}$ corresponding to l-electron act as

$$\begin{aligned} \sigma_{lx} f &= f(\sigma_1, \ldots \sigma_{l-1}, -\sigma_l, \sigma_{l+1}, \ldots \sigma_n), \\ \sigma_{ly} f &= -i\sigma_l f(\sigma_1, \ldots \sigma_{l-1}, -\sigma_l, \sigma_{l+1}, \ldots \sigma_n), \\ \sigma_{lz} f &= \sigma_l f(\sigma_1, \ldots \sigma_{l-1}, -\sigma_l, \sigma_{l+1}, \ldots \sigma_n). \end{aligned} \tag{1.4}$$

From this determination of the individual electron operators it is easy to obtain the result of the action on the function f of the following operators:

$$\begin{aligned} s_x &= \frac{1}{2}(\sigma_{1x} + \sigma_{2x} + \ldots + \sigma_{nx}), \\ s_y &= \frac{1}{2}(\sigma_{1y} + \sigma_{2y} + \ldots + \sigma_{ny}), \\ s_z &= \frac{1}{2}(\sigma_{1z} + \sigma_{2z} + \ldots + \sigma_{nz}). \end{aligned} \tag{1.5}$$

Operators s_x, s_y, s_z correspond (in units of h) to the components of the total spin of n electrons. They obey the commutation relations

$$\begin{aligned} s_y s_z - s_z s_y &= i s_x, \\ s_z s_x - s_x s_z &= i s_y, \\ s_x s_y - s_y s_x &= i s_z. \end{aligned} \tag{1.6}$$

The operator of the square of the total spin is equal to

$$\mathbf{s}^2 = \frac{1}{4}\sum_l (\sigma_{lx}^2 + \sigma_{ly}^2 + \sigma_{lz}^2) + \frac{1}{2}\sum_{k,l} (\sigma_{kx}\sigma_{lx} + \sigma_{ky}\sigma_{ly} + \sigma_{kz}\sigma_{lz}) \tag{1.7}$$

or, since every term in the first sum is equal to unity,

$$\mathbf{s}^2 = \frac{3}{4}n + \frac{1}{2}\sum_{k,l}(\sigma_{kx}\sigma_{lx} + \sigma_{ky}\sigma_{ly} + \sigma_{kz}\sigma_{lz}). \tag{1.8}$$

To obtain the result of application of operator $\mathbf{s}^2$ to the function f let us first transform the relation

$$\frac{1}{2}(\sigma_{kx}\sigma_{lx} + \sigma_{ky}\sigma_{ly})f(\ldots\sigma_{k-1},\sigma_k,\sigma_{k+1},\ldots\sigma_{l-1},\sigma_l,\sigma_{l+1},\ldots) = \\ = \frac{1}{2}(1-\sigma_k\sigma_l)f(\ldots\sigma_{k-1},-\sigma_k,\sigma_{k+1},\ldots\sigma_{l-1},-\sigma_l,\sigma_{l+1},\ldots), \tag{1.9}$$

following from (1.4). This expression is equal to zero if $\sigma_l = \sigma_k$ and is not zero if $\sigma_l + \sigma_k = 0$. In these, the only possible, two cases it takes the same values as those of the expression

$$\frac{1}{2}(\sigma_{kx}\sigma_{lx} + \sigma_{ky}\sigma_{ly})f = \frac{1}{2}(1-\sigma_k\sigma_l)P_{kl}f, \tag{1.10}$$

where P_{kl} is the permutation operator of spin variables σ_k and σ_l so that

$$P_{kl}f(\ldots\sigma_{k-1},\sigma_k,\sigma_{k+1},\ldots\sigma_{l-1},\sigma_l,\sigma_{l+1},\ldots) = \\ = f(\ldots\sigma_{k-1},\sigma_k,\sigma_{k+1},\ldots\sigma_{l-1},\sigma_l,\sigma_{l+1},\ldots). \tag{1.11}$$

Further, we have an obvious relation

$$\frac{1}{2}(1+\sigma_{kz}\sigma_{lz})f = \frac{1}{2}(1+\sigma_k\sigma_l)P_{kl}f, \tag{1.12}$$

since it differs from zero only in the case of $\sigma_k = \sigma_l$, when permutation P_{kl} does not change anything.

Summing (1.10) and (1.12), one gets

$$\frac{1}{2}(1+\sigma_k\sigma_l)f = P_{kl}f. \tag{1.13}$$

Originally this equality was derived by Dirac using another less elementary method.

Substituting (1.13) into (1.8) and remembering that the number of terms in the double sum in (1.8) is equal to $\frac{1}{2}n(n-1)$, one gets the following final expression for $\mathbf{s}^2$:

$$\mathbf{s}^2 = n - \frac{n^2}{4} + \sum_{k<l} P_{kl}. \tag{1.14}$$

This expression, in fact, is identical to that obtained by Dirac in his book [4].

Let us find the general form of eigenfunctions of the operator $\mathbf{s}^2$. For that, let us consider the result of the application of this operator to a certain function of a partial form, namely, to the function

$$F_{\alpha_1\alpha_2\ldots\alpha_k} = F(\sigma_{\alpha 1}, \sigma_{\alpha_2}, \ldots \sigma_{\alpha_k} \mid \sigma_{\alpha k+1}, \ldots \sigma_{\alpha_n}). \tag{1.15}$$

We suppose this function to be symmetric with respect to both the permutations of the arguments $\sigma_{\alpha_1} \ldots \sigma_{\alpha_k}$ and the arguments $\sigma_{\alpha_{k+1}} \ldots \sigma_{\alpha_n}$. The permutation $P_{\alpha_i\alpha_j}$ of the arguments σ_{α_i} and σ_{α_j} in this function gives

$$P_{\alpha_i\alpha_j} F_{\alpha_1\ldots\alpha_k} = F_{\alpha_1\ldots\alpha_k} \qquad (i \le k, j \le k), \tag{1.16}$$

because it is evident that the function remains unchanged if both α_i and α_j are contained among its subscripts. Since the number of such subscripts is k, the number of corresponding terms in the operator $\mathbf{s}^2$ will be $\frac{1}{2}k(k-1)$.

Further, it will be

$$P_{\alpha_i\alpha_j} F_{\alpha_1\ldots\alpha_k} = F_{\alpha_1\ldots\alpha_k} \qquad (i \ge k+1, j \ge k+1), \tag{1.17}$$

because the function $F\alpha_1 \ldots \alpha_k$ remains unchanged also in the case when both the numbers α_i and α_j are not among its subscripts. The corresponding terms of the sum in $\mathbf{s}^2$ will be $\frac{1}{2}(n-k)(n-k-1)$.

Finally, if there is α_i among the subscripts, but not α_j, then

$$P_{\alpha_i\alpha_j} F_{\alpha_1\ldots\alpha_i\ldots\alpha_k} = F_{\alpha_1\ldots\alpha_{i-1}\alpha_j\alpha_{i+1}\ldots\alpha_k} \qquad (i \le k, j \ge k+1). \tag{1.18}$$

Thus, the result of the application of operator $\mathbf{s}^2$ to the function $F_{\alpha_1\ldots\alpha_k}$ will be

$$\begin{aligned} \mathbf{s}^2 F_{\alpha_1\ldots\alpha_k} &= [n - \frac{n^2}{4} + \frac{1}{2}k(k-1) + \frac{1}{2}(n-k)(n-k-1)] F_{\alpha_1\ldots\alpha_k} + \\ &+ \sum_{i=1}^{k} \sum_{j=k+1}^{n} F_{\alpha_1\ldots\alpha_{i-1}\alpha_j\alpha_{i+1}\ldots\alpha_k}. \end{aligned} \tag{1.19}$$

Let us consider the sum

$$\sum_{j=k+1}^{n} F_{\alpha_1\ldots\alpha_{i-1}\alpha_j\alpha_{i+1}\ldots\alpha_k}. \tag{1.20}$$

Here the subscript α_j runs the values

$$\alpha_j = \alpha_{k+1}, \alpha_{k+2}, \ldots \alpha_n, \tag{1.21}$$

i.e., all the values of the subscripts which do not appear among others (besides α_j) except α_i. If this exception is not made, the additional term equal to $F_{\alpha_1\ldots\alpha_k}$ is added to the sum (1.20) and it is transformed into

$$F^*_{\alpha_1\ldots\alpha_{i-1}\alpha_i+1\ldots\alpha_k} = \sum_{\alpha} F_{\alpha_1\ldots\alpha_{i-1}\alpha\alpha_{i+1}\ldots\alpha_k}, \tag{1.22}$$

where the summation subscript now runs all the values not equal to that of the other subscripts (i.e., all values in (1.21) besides the value of α_i). Therefore, the value of F^* will depend only on its subscripts and be a symmetric function of them (i.e., of corresponding spin variables). The original sum (1.20) will be, apparently,

$$\sum_{j=k+1}^{n} F_{\alpha_1\ldots\alpha_{i-1}\alpha_j\alpha_{i+1}\ldots\alpha_k} = F^*_{\alpha_1\ldots\alpha_{i-1}\alpha_{i+1}\ldots\alpha_k} - F_{\alpha_1\ldots\alpha_k}. \tag{1.23}$$

This expression should be substituted into (1.19) and the result be summed over i from 1 to k. Using the equality

$$\begin{aligned} n - \frac{n^2}{4} + \frac{1}{2}k(k-1) + \frac{1}{2}(n-k)(n-k-1) - k = \\ = (\frac{n}{2} - k)(\frac{n}{2} - k + 1), \end{aligned} \tag{1.24}$$

one gets

$$\mathbf{s}^2 F_{\alpha_1\ldots\alpha_k} = (\frac{n}{2} - k)(\frac{n}{2} - k + 1)F_{\alpha_1\ldots\alpha_k} + \sum_{i=1}^{k} F^*_{\alpha_1\ldots\alpha_{i-1}\alpha_{i+1}\ldots\alpha_k}. \tag{1.25}$$

Now it is easy to find the general form of the eigenfunctions of the operator $\mathbf{s}^2$. For that, one needs to construct a linear combination of equations (1.25) to eliminate the terms with F^*. We denote the coefficients of this linear combination by $a_{\alpha_1\ldots\alpha_k}$. These values are symmetric with respect to their subscripts; therefore, their number is the binomial coefficient $\binom{n}{k}$. To cancel the terms with F^*, we subject our coefficients to the conditions

$$\sum_{\alpha} a_{\alpha\alpha_2\alpha_3\ldots\alpha_k} = 0. \tag{1.26}$$

The number of such relations is obviously equal to $\binom{n}{k-1}$. This number must be less than the number of coefficients. Consequently, it should be

$$\binom{n}{k} > \binom{n}{k-1}, \tag{1.27}$$

whence it follows

$$k \leq \frac{n}{2} \tag{1.28}$$

(an equality can be only for the even n). Thus, the value of

$$\frac{n}{2} - k = s \tag{1.29}$$

has to be a non-negative number. This number will be integer for even n and half-integer for odd n. If the coefficients are known, we can construct the function

$$\chi(\sigma_1, \sigma_2, \ldots \sigma_n) = \sum_{\alpha_1 \ldots \alpha_k} a_{\alpha_1 \ldots \alpha_k} F_{\alpha_1 \ldots \alpha_k}(\sigma_1, \sigma_2, \ldots \sigma_n), \tag{1.30}$$

which, as we shall show now, will obey the equation

$$\mathbf{s}^2 \chi = s(s+1)\chi, \tag{1.31}$$

where s has a value of (1.29). Indeed, applying (1.25) for individual terms of the sum (1.30), we shall have

$$\begin{aligned} \mathbf{s}^2 \chi - (\frac{n}{2} - k)(\frac{n}{2} - k + 1)\chi = \sum_{\alpha_1 \ldots \alpha_k} a_{\alpha_1 \ldots \alpha_k} F^*_{\alpha_2 \alpha_3 \ldots \alpha_k} + \ldots \\ \ldots + \sum_{\alpha_1 \ldots \alpha_k} a_{\alpha_1 \ldots \alpha_k} F^*_{\alpha_1 \ldots \alpha_{i-1} \alpha_{i+1} \ldots \alpha_k} + \ldots \\ \ldots + \sum_{\alpha_1 \ldots \alpha_k} a_{\alpha_1 \ldots \alpha_k} F^*_{\alpha_1 \ldots \alpha_{k-1}}. \end{aligned} \tag{1.32}$$

In each sum in the right-hand side, we can do the first summation over the variable, which does not enter the corresponding F^*, namely: in the first sum, we do over α_1, in the second we do over α_2 and so on, and in the last sum we do over α_k. By the condition (1.26) of the coefficients $a_{\alpha_1, \ldots \alpha_k}$, the first summation gives zero each time and we obtain equation (1.31). Thus, we obtained the general expression of the eigenfunction of the operator $\mathbf{s}^2$ as a sum (1.30) with the coefficients obeying condition (1.26).

Starting from this general expression, one would first construct the common eigenfunctions of the operators $\mathbf{s}^2$ and s_z, which are independent of the coordinates, and then come to the total wave function depending both on the coordinates and spins. These total wave functions would be obtained as linear combinations of spin functions with coefficients depending on coordinates.

Denoting the total function by Ψ, we should have

$$\Psi(r_1\sigma_1, r_2\sigma_2, \ldots r_n\sigma_n) = \sum_\lambda \psi^{(\lambda)}(r_1 r_2 \ldots r_n)\chi^{(\lambda)}(\sigma_1\sigma_2\ldots\sigma_n), \quad (1.33)$$

where $\chi^{(\lambda)}$ are linearly independent common eigenfunctions of the operators $\mathbf{s}^2$ and s_z and give the irreducible representation of the permutation group, while the coefficients $\Psi^{(\lambda)}(r_1, r_2, \ldots r_n)$ are the Schrödinger wave functions.

However, this usual method to represent total wave function applied, in particular, by Wigner is in practice hardly suitable because of the difficulties connected with the construction and investigation of the Schrödinger wave functions.

But to obtain the total wave function, one needs not each term of the sum to obey such a heavy condition as the individual terms of the sum (1.33) do. It is sufficient for the total sum to obey the suitable conditions.

Based on this remark, in the next section we shall give a simpler and more explicit method to construct the Schrödinger wave function. Our treatment will be absolutely independent of the previous considerations. In particular, we shall give a new proof of the fact that our wave function is an eigenfunction of the operator $\mathbf{s}^2$. Thus, all the above will serve as an introduction that facilitates understanding of the further results and the methods by which they will be obtained.

2 The Symmetry Properties of the Schrödinger Wave Function

In this section, we formulate three basic symmetry properties of the Schrödinger wave function. A derivation and proof of these properties will be given in the next section. Consider the Schrödinger function

$$\psi = \psi(r_1 r_2 \ldots r_k \mid r_{k+1} r_{k+2} \ldots r_n), \quad (2.1)$$

depending only on coordinates $r_1, r_2, \ldots r_n$ of an n-electron system. The word "coordinate" is used here in the same sense as in Introduction, so each letter "r_i" denotes a set of three spatial coordinates x_i, y_i, z_i of an n-th electron. To simplify typesetting, we do not use bold script.

Let the function ψ obey the following symmetry conditions:

1) ψ is antisymmetric with respect to the first k arguments (which stand to the left of the line), for example,

$$\psi(r_2 r_1 r_3 \ldots r_k \mid r_{k+1} r_{k+2} \ldots r_n) = -\psi(r_1 r_2 r_3 \ldots r_k \mid r_{k+1} r_{k+2} \ldots r_n); \tag{2.2}$$

2) ψ is antisymmetric with respect to the last $n-k$ arguments (which stand to the right of the line), for example,

$$\psi(r_1 r_2 \ldots r_k \mid r_{k+2} r_{k+1} r_{k+3} \ldots r_n) = \\ = -\psi(r_1 r_2 \ldots r_k \mid r_{k+1} r_{k+2} r_{k+3} \ldots r_n); \tag{2.3}$$

3) ψ possesses the cyclic symmetry, which can be expressed in the form

$$\psi(r_1 \ldots r_{k-1} r_k \mid r_{k+1} r_{k+2} \ldots r_n) = \\ = \psi(r_1 \ldots r_{k-1} r_{k+1} \mid r_k r_{k+2} \ldots r_n) + \ldots \\ \ldots + \psi(r_1 \ldots r_{k-1} r_{k+l} \mid r_{k+1} \ldots r_{k+l-1} r_k r_{k+l+1} \ldots r_n) + \ldots \\ \ldots + \psi(r_1 \ldots r_{k-1} r_n \mid r_{k+1} \ldots r_{n-1} r_k). \tag{2.4}$$

Here we assume that the number of arguments to the right of the line is greater than that to the left of the line or equals it:

$$n - k \geq k, \tag{2.5}$$

since otherwise function ψ is identically equal to zero, as it follows from (2.2), (2.3), (2.4).

We called property 3 expressed by (2.4) the cyclic symmetry because permutations of arguments in (2.4) can be reduced (by using properties 1 and 2) to the multiple application of a single cyclic permutation. Indeed, let us consider a general term in the right-hand side of (2.4). The argument r_k stands there at the l-th position after the line. By using antisymmetry, one can shift this argument to the last position. For that one needs to do $n-k-l$ transpositions of adjacent arguments. Whence

$$\psi(r_1 \ldots r_{k-1} r_{k+l} \mid r_{k+1} .. r_{k+l-1} r_k r_{k+l+1} \ldots r_n) = \\ = (-1)^{n-k-l} \psi(r_1 \ldots r_{k-1} r_{k+l} \mid r_{k+1} \ldots r_{k+l-1} r_{k+l+1} \ldots r_n r_k). \tag{2.6}$$

Then the first argument after the line can be transposed to the last position. As a result, the function gets a factor $(-1)^{n-k-1}$ and the argument r_{k+2} takes the first position. In the same way one can (successively) transfer arguments $r_{k+2}, r_{k+3}, \ldots, r_{k+l-1}$ from the first position (after the line) to the last one. And every time the function gets the factor $(-1)^{n-k-1}$. As a result, a general term in (2.04) will be written in the form

$$\psi(r_1 \ldots r_{k-1} r_{k+l} \mid r_{k+1} \ldots r_{k+l-1} r_k r_{k+l+1} \ldots r_n) =$$
$$= -(-1)^{l(n-k)} \psi(r_1 \ldots r_{k-1} r_{k+l} \mid r_{k+l+1} \ldots r_n r_k \ldots r_{k+l-1}) -$$
$$-(-1)^{l(n-k)} P^l \psi, \tag{2.7}$$

where P^l is the cyclic permutation of arguments r_k, r_{k+1}, $\ldots$, r_n, repeated l times:

$$P = \begin{pmatrix} r_k & r_{k+1} & \ldots & r_{n-1} & r_n \\ r_{k+1} & r_{k+2} & \ldots & r_n & r_k \end{pmatrix}. \tag{2.8}$$

Thus, condition (2.4) can be written in two different forms depending on whether $n-k$ is even or odd, namely,

$$\{1 + P + P^2 + \ldots + P^{n-k}\}\psi = 0 \tag{2.9}$$

if $n-k$ is even, and

$$\{1 - P + P^2 - \ldots - P^{n-k}\}\psi = 0 \tag{2.10}$$

if $n-k$ is odd. Whence it is seen that the condition considered can be, with good reason, named the cyclic symmetry one.

For $n = 2, k = 1$ it is reduced to the demand that the wave function must be symmetric with respect to coordinates.

To understand the physical meaning of the cyclic symmetry condition, let us consider the form it takes when complete separation of variables becomes possible, i.e., when the wave function of the system ψ can be expressed by the wave functions of individual electrons.

To satisfy the first two conditions, we shall write ψ in the form of a product

$$\psi(r_1 \ldots r_{k-1} r_k \mid r_{k+1} r_{k+2} \ldots r_n) = \Psi^{(1)} \Psi^{(2)}, \tag{2.11}$$

where $\Psi^{(1)}$ and $\Psi^{(2)}$ are determinants

$$\Psi^{(1)} = \begin{vmatrix} \psi_1(r_1) & \ldots & \psi_1(r_k) \\ \ldots\ldots & \ldots & \ldots\ldots \\ \psi_k(r_1) & \ldots & \psi_k(r_k) \end{vmatrix}, \tag{2.12}$$

$$\Psi^{(2)} = \begin{vmatrix} \psi_{k+1}(r_{k+1}) & \ldots & \psi_{k+1}(r_n) \\ \ldots\ldots\ldots\ldots & \ldots & \ldots\ldots\ldots \\ \psi_n(r_{k+1}) & \ldots & \psi_n(r_n) \end{vmatrix}. \tag{2.13}$$

We denote the cofactors (minors with suitable signs) of these determinants by

$$\Psi_{ij}^{(1)} = \frac{\partial \Psi^{(1)}}{\partial \psi_i(r_i)}; \qquad \Psi_{ij}^{(2)} = \frac{\partial \Psi^{(2)}}{\partial \psi_i(r_i)}. \tag{2.14}$$

Let us substitute these expressions into the cyclic symmetry condition of the form (2.4). If in the left-hand side of (2.4) we expand the determinant (2.12) by elements of the last column, but leave the determinant (2.13) unchanged, then we shall get an equality

$$\sum_{i=1}^{k} \Psi_{ik}^{(1)} \psi_i(r_k) \Psi^{(2)} = \sum_{l=1}^{n-k} \sum_{i=1}^{k} \Psi_{ik}^{(1)} \psi_i(r_{k+l}) \sum_{j=k+1}^{n} \Psi_{jk+l}^{(2)} \psi_j(r_k). \tag{2.15}$$

Here summation over l corresponds to that in the right-hand side of (2.4). Making this summation and using a property of determinants, one gets

$$\sum_{i=1}^{k} \psi_i \psi_i(r_{k+l}) \Psi_{jk+l}^{(2)} = \{\Psi^{(2)}\}_{\psi_j \to \psi_i}, \tag{2.16}$$

where the right-hand part is the result of substitution of function ψ_j by function ψ_i in determinant $\Psi^{(2)}$.

Using this equality, we can write the previous equation in the form

$$\sum_{i=1}^{k} \Psi_{ik}^{(1)} \psi_i(r_k) \Psi^{(2)} = \sum_{i=1}^{k} \Psi_{ik}^{(1)} \sum_{j=k+1}^{n} \psi_j(r_k) \{\Psi^{(2)}\}_{\psi_j \to \psi_i} . \tag{2.17}$$

This equality has to be an identity with respect to arguments r_1, r_2, ..., r_k (which enter only $\Psi_{ik}^{(1)}$) and, consequently, with respect to $\Psi_{ik}^{(1)}$ because these values are linearly independent.[3] Consequently, the individual terms of the sum over i in both sides of (2.17) must be equal to each other.

[3]A determinant

$$D = \begin{vmatrix} \psi_1(r_1) & \ldots & \psi_1(r_{k-1}) & c_1 \\ \psi_1(r_2) & \ldots & \psi_2(r_{k-1}) & c_2 \\ \ldots\ldots & \ldots & \ldots\ldots & \ldots \\ \psi_k(r_1) & \ldots & \psi_k(r_{k-1}) & c_k \end{vmatrix}$$

with non-vanishing constants $c_1, \ldots, c_k$ cannot be identically equal to zero with respect to $r_1, \ldots, r_{k-1}$. (*V. Fock*)

We get the equality

$$\psi_i(r_k)\Psi^{(2)} = \sum_{j=k+1}^{n} \psi_j(r_k)\{\Psi^{(2)}\}_{\psi_j \to \psi_i} \qquad (i = 1, 2, 3, \ldots k). \tag{2.18}$$

Since argument r_k does not enter $\Psi^{(2)}$, the relations

$$\psi_i(r) = \sum_{j=k+1}^{n} a_{ij}\psi_j(r) \qquad (i = 1, 2, 3, \ldots k) \tag{2.19}$$

hold with the constant coefficients a_{ij}. These relations can be interpreted in the following sense: *each function entering the small determinant has to be linearly dependent on the functions of the big determinant.*

Inversely, relation (2.18) and consequently condition (2.4) follow from (2.19). In fact, from (2.13) and (2.19) and due to the determinant property, we get

$$\{\Psi^{(2)}\}_{\psi_j \to \psi_i} = a_{ij}\Psi^{(2)}. \tag{2.20}$$

This relation shows that (2.18) after dividing by $\Psi^{(2)}$ is reduced to (2.19) and hence is also valid.

Thus, in the case of the complete separation of variables relation (2.19) is a sufficient and necessary condition for the product of determinants (2.11) to possess the cyclic symmetry.

By using the determinant property, one can change the entering functions by their linear combinations. It allows one to choose them, e.g., in such a way so as to get

$$\left.\begin{aligned} \psi_{n-k+1} &= \psi_1, \\ \psi_{n-k+2} &= \psi_2, \\ \ldots&\ldots\ldots \\ \psi_n &= \psi_k. \end{aligned}\right\} \tag{2.21}$$

These equalities can substitute (2.19) without the loss of generality. Then the determinant $\Psi^{(1)}$ will be constructed, as before, from functions $\psi_1, \psi_2, \ldots, \psi_k$, and the determinant $\Psi^{(2)}$ from functions $\psi_1, \psi_2, \ldots, \psi_{n-k}$. This can be treated if you like, in the sense that there are k electrons at orbits $\psi_1, \psi_2, \ldots, \psi_k$ with one spin and $n-k$ electrons at orbits $\psi_1, \psi_2, \ldots \psi_{n-k}$ with another one.

A representation of the wave function in the form of the product of two determinants of the considered type was used more than ten years ago in a work by Waller and Hartree [5] and in our first work on the self-consistent field equations with quantum exchange [6], though at that time the cyclic symmetry property had not been known yet.

3 Expression for Functions with Spin by Means of the Schrödinger Function

In Section 1 it was shown that the sum of the kind (1.30), where coefficients are connected by relations (1.26), satisfies (1.31), i.e., it is an eigenfunction of the square of the total spin. Evidently nothing changes in this statement if one considers the constant coefficients $a_{\alpha_1 \dots \alpha_k}$ in sum (1.30) to be functions of coordinates. These functions can be chosen so that the whole sum satisfies the Pauli principle. This simple idea will be put into the basis of the present section. However, our consideration here will be independent of the results of Section 1. They are needed mainly to clarify the origin of our initial formulae.

Let the Schrödinger function

$$\psi = \psi(r_1 r_2 \dots r_k \mid r_{k+1} r_{k+2} \dots r_n) \tag{3.1}$$

satisfy three symmetry conditions, namely, two antisymmetry conditions and the condition of the cyclic symmetry.

Let P be an arbitrary permutation of numbers from 1 to n:

$$P = \begin{pmatrix} 1 & 2 & \dots & n \\ \alpha_1 & \alpha_2 & \dots & \alpha_n \end{pmatrix}, \tag{3.2}$$

where a number i is replaced by a number α_i and let $\varepsilon(P)$ be a number equal to $+1$ if permutation P is even, and to -1 if permutation P is odd, i.e.,

$$\left.\begin{aligned} \varepsilon(P) &= +1, \qquad \text{if } P \text{ is even,} \\ \varepsilon(P) &= -1, \qquad \text{if } P \text{ is odd.} \end{aligned}\right\} \tag{3.3}$$

Let us introduce a set of new functions by the equality

$$\psi_{\alpha_1 \alpha_2 \dots \alpha_k}(r_1 r_2 \dots r_n) = \varepsilon(P)\psi(r_{\alpha_1} r_{\alpha_2} \dots r_{\alpha_k} \mid r_{\alpha_{k+1}} \dots r_{\alpha_n}), \tag{3.4}$$

i.e., expressing them by a single Schrödinger function.

The right-hand side of (3.4) depends on subscripts $\alpha_1, \dots, \alpha_k$ through $\varepsilon(P)$ and through arguments $r_{\alpha_1}, r_{\alpha_2}, \dots, r_{\alpha_n}$. Due to the antisymmetry of function (3.1) with respect to its arguments and due to the antisymmetry of $\varepsilon(P)$ with respect to subscripts included into permutation P, expression (3.4) will be symmetric with respect to subscripts $\alpha_1, \dots, \alpha_k$ and with respect to the other subscripts $\alpha_{k+1}, \alpha_{k+2}, \dots, \alpha_n$. Consequently, the number of functions (3.4) will be $\binom{n}{k}$. But not all of these functions will be independent. Indeed, due to the cyclic symmetry of the Schrödinger function, $\binom{n}{k-1}$ relations will take place:

$$\sum_{\alpha} \psi_{\alpha\alpha_2\alpha_3\ldots\alpha_k} = 0, \tag{3.5}$$

which can be proved easily if one writes the cyclic symmetry condition in the form of (2.4).[4]

Thus, the number of linearly independent functions (3.4) will be

$$\binom{n}{k} - \binom{n}{k-1} = N_k. \tag{3.6}$$

If one transposes simultaneously arguments r_α and r_β of ψ and subscripts α and β in the right-hand side of (3.4), then obviously it does not change, while the factor $\varepsilon(P)$ changes its sign. Besides, if we keep in mind the symmetry of the left-hand side of (3.4) to the subscripts, whence we get the following property of the function $\psi_{\alpha_1\ldots\alpha_k}$ with respect to permutation of its arguments.

Under permutation of arguments r_α and r_β the function $P(r_\alpha, r_\beta)$ changes the sign

$$P(r_\alpha,, r_\beta)\psi_{\alpha_1\ldots\alpha_k} = -\psi_{\alpha_1\ldots\alpha_k}, \tag{3.7}$$

if both subscripts α and β occur among its subscripts

$$\alpha, \beta = \alpha_1, \alpha_2, \ldots, \alpha_k, \tag{3.8}$$

or if none of them occurs there

$$\alpha, \beta = \alpha_{k+1}, \alpha_{k+2}, \ldots, \alpha_n, \tag{3.9}$$

and the function $\psi_{\alpha_1\ldots\alpha_k}$ turns into another function taken with the opposite sign:

$$P(r_\alpha,, r_\beta)\psi_{\alpha_1\ldots\alpha_k} = -\psi_{\alpha_1\ldots\alpha_{i-1}\beta\alpha_i\ldots\alpha_k} \tag{3.10}$$

if only one of subscripts α or β occurs among its subscripts, for example,

$$\left.\begin{aligned} \alpha &= \alpha_i = \alpha_1, \alpha_2, \ldots, \alpha_k, \\ \beta &= \alpha_{k+1}, \alpha_{k+2}, \ldots, \alpha_n. \end{aligned}\right\} \tag{3.11}$$

[4]Recalling the remark at the beginning of this section, it can be easily seen that condition (3.05) is nothing else but condition (1.26) for coefficients $a_{\alpha_1,\ldots,\alpha_k}$. It is the property of cyclic symmetry of the Schrödinger function itself that follows herefrom. (*V. Fock*)

Now let us introduce two arbitrary functions of the spin variable σ, namely,

$$u = u(\sigma); \qquad v = v(\sigma); \tag{3.12}$$

for brevity, we shall write

$$u_i = u(\sigma_i); \qquad v_i = v(\sigma_i). \tag{3.13}$$

Consider the function[5]

$$\Phi(r_1 u_1 v_1, r_2 u_2 v_2, \ldots, r_n u_n v_n) = \\ = \sum_{\alpha_1 \ldots \alpha_k} \psi_{\alpha_1 \ldots \alpha_k}(r_1 \ldots r_n) u_{\alpha_1} \ldots u_{\alpha_k} v_{\alpha_{k+1}} \ldots v_{\alpha_n}. \tag{3.14}$$

First of all we shall show that this function satisfies the Pauli principle, i.e., is antisymmetric with respect to the simultaneous transposition of coordinates and spins of two electrons. In fact, if one performs a transposition of coordinates r_α and r_β and of spins σ_α and σ_β,[6] then the terms of sum (3.14) can be split into three categories corresponding to the cases (3.8), (3.9) and (3.11). Each term, which satisfies conditions (3.8) or (3.9), changes a sign due to the antisymmetry of the coordinate factor and the symmetry of the spin factor. The terms that satisfy condition (3.11) can be joined in pairs (e.g., terms with subscripts $(\alpha_1 \ldots \alpha_{i-1} \alpha \alpha_{i+1} \ldots \alpha_k)$ and $(\alpha_1 \ldots \alpha_{i-1} \beta \alpha_{i+1} \ldots \alpha_k)$). As a result of the transposition considered one of these terms transforms into another (in virtue of (3.10)) taken with the opposite sign. Thus, the antisymmetry of function (3.14) with respect to simultaneous transposition of coordinates and spins is proved.

Let us recapitulate the basic features of function Φ.

1. *Linearity.* The function φ is linear and homogeneous with respect to every pair of variables u_i and v_i. In other words, it can be presented in n different forms

$$\Phi = A_i u_i + B_i v_i. \tag{3.15}$$

2. *Homogeneity.* The function Φ is homogeneous of degree k with respect to variables $u_1, u_2, \ldots, u_n$ and of degree $n-k$ with respect to

[5] From the comparison of (3.14) with the sum (1.30) it is clear that Φ is an eigenfunction of the operator $\mathbf{s}^2$ corresponding to the quantum number $s = n/2 - k$. However, we shall not use this result and obtain it in some other way. (*V. Fock*)

[6] Transposition of spins σ_α and σ_β brings, in accordance with (3.13), both transposition of u_α with u_β and v_α with v_β. (*V. Fock*)

variables $v_1, v_2, \ldots, v_n$. So, if we will consider these variables to be continuous parameters, we can write

$$u_1 \frac{\partial \Phi}{\partial u_1} + u_2 \frac{\partial \Phi}{\partial u_2} + \ldots + u_n \frac{\partial \Phi}{\partial u_n} = k\Phi; \tag{3.16}$$

$$v_1 \frac{\partial \Phi}{\partial v_1} + v_2 \frac{\partial \Phi}{\partial v_2} + \ldots + v_n \frac{\partial \Phi}{\partial v_n} = (n-k)\Phi. \tag{3.17}$$

3. *Functional property.* Function Φ satisfies the functional equation

$$\Phi(r_1, u_1 + \lambda v_1, v_1, \ldots, r_n, u_n + \lambda v_n, v_n) = \Phi(r_1, u_1, v_1, \ldots, r_n, u_n, v_n), \tag{3.18}$$

where λ is an arbitrary parameter. This equation is a consequence of (3.5). To prove it, let us write the left-hand side of (3.18) as

$$\Phi(r_1, u_1 + \lambda v_1, v_1, \ldots, r_n, u_n + \lambda v_n, v_n) =$$

$$= v_1 v_2 \ldots v_n \sum_{\alpha_1 \ldots \alpha_k} \psi_{\alpha_1 \ldots \alpha_k} (\frac{u_{\alpha_1}}{v_{\alpha_1}} + \lambda) \ldots (\frac{u_{\alpha_k}}{v_{\alpha_k}} + \lambda). \tag{3.19}$$

Here the coefficient of zero degree of λ evidently gives the right-hand side of (3.18). The coefficient of every positive degree of λ is represented as a sum of terms, each no longer containing k factors of the kind $\frac{u_{\alpha_i}}{v_{\alpha_i}}$ but less number. One can sum over subscripts $\alpha_1, \alpha_2, \ldots, \alpha_k$, which do not enter these factors and as a result, in virtue of (3.5), one gets zero. Thus, formula (3.18) is proved.

Consider now an expression that is obtained from (3.14) if one substitutes there not u_i by $u_i + \lambda v_i$ as in the previous case of (3.19), but v_i by $v_i + \zeta u_i$ where ζ is another arbitrary parameter.

We shall denote function (3.14) briefly by

$$\Phi = \Phi(r, u, v), \tag{3.20}$$

and the result of our substitution will be denoted as

$$\Phi(r, u, v + \zeta u) == \Phi(r_1, u_1, v_1 + \zeta u_1, \ldots, r_n, u_n, v_n + \zeta u_n). \tag{3.21}$$

Evidently, this is a polynomial of ζ. But the degree of this polynomial is not $n-k$, as it may be expected due to the homogeneity with respect to the third arguments (we shall call r_i as the first, u_i as the second and v_i as the third arguments). Indeed, subtracting from the second arguments the third ones divided by ζ we get in virtue of the functional property of Φ and the homogeneity with respect to the second arguments,

$$\Phi(r, u, v + \zeta u) = \Phi(r, -v/\zeta, v + \zeta u) = (-1)^k \zeta^{-k} \Phi(r, v, v + \zeta u). \tag{3.22}$$

Since both functions $\Phi(r, u, v + \zeta u)$ and $\Phi(r, v, v + \zeta u)$ are polynomials of ζ of the degree less than $n - k$, then from (3.21) it follows that in the first of them the highest degree is $n - 2k$, and in the second one the lowest degree is k.

Putting

$$\frac{n}{2} - k = s, \tag{3.23}$$

so that s is a non-negative integer or half-integer, we can write the polynomial (3.21) as the sum

$$\Phi(r, u, v + \zeta u) = \sum_{\nu=-s}^{\nu=+s} \zeta^{\nu+s} \varphi_{s\nu}(r, u, v), \tag{3.24}$$

where the summation subscript ν runs the values

$$\nu = -s, -s+1, \dots .s-1, s, \tag{3.25}$$

so that $s + \nu$ runs all integer values from 0 to $n - 2k$.

Let us prove that if in (3.24) one puts

$$u(\sigma) = \frac{1+\sigma}{2}; \qquad v(\sigma) = \frac{1-\sigma}{2} \tag{3.26}$$

and, correspondingly,

$$u_i = \frac{1+\sigma_i}{2}; \qquad v_i = \frac{1-\sigma_i}{2}, \tag{3.27}$$

then functions $\varphi_{s\nu}$ determined by the generation function (3.24) satisfy the system of equations

$$(s_x + is_y)\varphi_{s\nu} = (s + \nu + 1)\varphi_{s\nu+1}, \tag{3.28}$$

$$(s_x - is_y)\varphi_{s\nu+1} = (s - \nu)\varphi_{s\nu}, \tag{3.29}$$

$$s_z \varphi_{s\nu} = \nu \varphi_{s\nu}, \tag{3.30}$$

where s_x, s_y, s_z are spin operators introduced in Section 1.

The action of these operators on the function of spin variables is determined by formulae (1.4) and (1.5). Taking into account that the change of the sign of variables σ_i corresponds to transposition u_i with v_i we can write the result of action of operators $s_x + is_y, s_x - is_y, s_z$ on a function

$$f = f(u_1, v_1, u_2, v_2, \dots u_n, v_n) \tag{3.31}$$

in the form of

$$(s_x + is_y)f = \sum_{i=1}^{n} u_i f(\ldots u_{i-1}v_{i-1}v_i u_i u_{i+1}v_{i+1}\ldots), \tag{3.32}$$

$$(s_x - is_y)f = \sum_{i=1}^{n} v_i f(\ldots u_{i-1}v_{i-1}v_i u_i u_{i+1}v_{i+1}\ldots), \tag{3.33}$$

$$s_z f = -\frac{n}{2}f + \sum_{i=1}^{n} u_i f, \tag{3.34}$$

or

$$s_z f = \frac{n}{2}f - \sum_{i=1}^{n} v_i f. \tag{3.35}$$

Put here

$$f = \Phi(r, u, v + \zeta u). \tag{3.36}$$

This function as well as $\Phi(r, u, v)$ can be represented in n ways of the form

$$\Phi(r, u, v + \zeta u) = A_i u_i + B_i v_i \tag{3.37}$$

with some values of coefficients, different from those in (3.15), namely,

$$A_i = \left(\frac{\partial \Phi}{\partial u_i} + \zeta \frac{\partial \Phi}{\partial v_i}\right)_{v+\zeta u}, \tag{3.38}$$

$$B_i = \left(\frac{\partial \Phi}{\partial v_i}\right)_{v+\zeta u}. \tag{3.39}$$

Here Φ means $\Phi(r, u, v)$ and the change of v_i by $v_i + \zeta u_i$ is made after differentiation. If for every term of sums (3.32) and (3.33) one uses the corresponding expression of the form (3.37) for f, then the result of transposition of u_i and v_i is easily written, and we get

$$(s_x + is_y)\Phi(r, u, v + \zeta u) = \sum_{i=1}^{n} u_i(A_i v_i + B_i u_i). \tag{3.40}$$

But, due to (3.27), we have

$$u_i^2 = u_i; \qquad v_i^2 = v_i; \qquad u_i v_i = 0 \tag{3.41}$$

and, therefore,

$$(s_x + is_y)\Phi(r, u, v + \zeta u) = \sum_{i} B_i u_i. \tag{3.42}$$

Similarly we get

$$(s_x - is_y)\Phi(r, u, v + \zeta u) = \sum_i A_i v_i, \tag{3.43}$$

$$s_z\Phi(r, u, v + \zeta u) = -\frac{n}{2}\Phi(r, u, v + \zeta u) + \sum_i A_i u_i =$$

$$= \frac{n}{2}\Phi(r, u, v + \zeta u) - \sum_i B_i v_i. \tag{3.44}$$

Using (3.38), (3.39) for A_i and B_i one can express the sums in the right-hand sides of these equations by the following four sums:

$$\left(\sum_i u_i \frac{\partial \Phi}{\partial u_i}\right)_{v+\zeta u} = k\Phi(r, u, v + \zeta u); \tag{3.45}$$

$$\left(\sum_i v_i \frac{\partial \Phi}{\partial v_i}\right)_{v+\zeta u} = (n - k)\Phi(r, u, v + \zeta u); \tag{3.46}$$

$$\left(\sum_i u_i \frac{\partial \Phi}{\partial v_i}\right)_{v+\zeta u} = \frac{\partial}{\partial \zeta}\Phi(r, u, v + \zeta u); \tag{3.47}$$

$$\left(\sum_i v_i \frac{\partial \Phi}{\partial u_i}\right)_{v+\zeta u} = 0. \tag{3.48}$$

The last equation can be obtained easier by differentiation of (3.18) over the parameter λ.

We shall have

$$\sum_i B_i u_i = \frac{\partial}{\partial \zeta}\Phi(r, u, v + \zeta u); \tag{3.49}$$

$$\sum_i A_i v_i = [(n - 2k)\zeta - \zeta^2 \frac{\partial}{\partial \zeta}]\Phi(r, u, v + \zeta u); \tag{3.50}$$

$$\sum_i A_i u_i = (k + \zeta \frac{\partial}{\partial \zeta})\Phi(r, u, v + \zeta u); \tag{3.51}$$

$$\sum_i B_i v_i = (n - k - \zeta \frac{\partial}{\partial \zeta})\Phi(r, u, v + \zeta u). \tag{3.52}$$

Thus, it is evident that

$$\sum_i (iA_i u_i + B_i v_i) = n\Phi(r, u, v + \zeta u). \tag{3.53}$$

Substituting the expressions obtained into the right-hand sides of (3.42)–(3.44), we get

$$(s_x + is_y)\Phi(r,u,v+\zeta u) = \frac{\partial}{\partial\zeta}\Phi(r,u,v+\zeta u); \quad (3.54)$$

$$(s_x - is_y)\Phi(r,u,v+\zeta u) = (2s\zeta - \zeta^2\frac{\partial}{\partial\zeta})\Phi(r,u,v+\zeta u); \quad (3.55)$$

$$s_z\Phi(r,u,v+\zeta u) = (-s + \zeta\frac{\partial}{\partial\zeta})\Phi(r,u,v+\zeta u). \quad (3.56)$$

But we had the equality (3.24)

$$\Phi(r,u,v+\zeta u) = \sum_{\nu=-s}^{+s} \zeta^{\nu+s}\varphi_{s\nu}(r,u,v), \quad (3.24)$$

which gives an expression for $\Phi(r,u,v+\zeta u)$ in the polynomial form. Substituting this equality into (3.54)–(3.56) and comparing the coefficients at equal degrees of ζ, we obtain a fundamental system of equations

$$\left.\begin{aligned} (s_x + is_y)\varphi_{s\nu} &= (s+\nu+1)\varphi_{s\nu+1}; \\ (s_x - is_y)\varphi_{s\nu} &= (s-\nu)\varphi_{s\nu}; \\ s_z\varphi_{s\nu} &= \nu\varphi_{s\nu+1} \end{aligned}\right\} \quad (3.57)$$

for the eigenfunctions of the angular-momentum operators.

From this system it follows that the function $\varphi_{s\nu}$ and, consequently, the whole sum (3.24) is an eigenfunction of the square of the total spin angular-momentum operator with a quantum number $s = n/2 - k$.

Note that the ordinary spherical functions can be obtained with the help of the generation function, which satisfies the system of equations that is analogous to (3.54)–(3.56). Indeed, if we put

$$w = \frac{x}{2}(1-\zeta^2) - \frac{iy}{2}(1+\zeta^2) + z\zeta, \quad (3.58)$$

then any function of w will satisfy the Laplace equation. But if we take

$$f(w) = w^l, \quad (3.59)$$

we obtain a homogeneous harmonic polynomial of degree l, for which it is easy to prove the following equalities:

$$(m_x + im_y)w^l = \frac{\partial}{\partial\zeta}w^l; \quad (3.60)$$

$$(m_x - im_y)w^l = (2l\zeta - \zeta^2 \frac{\partial}{\partial \zeta})w^l; \tag{3.61}$$

$$m_z w^l = (\zeta \frac{\partial}{\partial \zeta} - l)w^l, \tag{3.62}$$

where m_x, m_y, m_z are operators of the orbital angular momentum of a particle expressed in units of $\hbar$:

$$\left.\begin{aligned} m_x &= -i(y\frac{\partial}{\partial z} - z\frac{\partial}{\partial y}), \\ m_y &= -i(z\frac{\partial}{\partial x} - x\frac{\partial}{z\partial}), \\ m_z &= -i(x\frac{\partial}{\partial y} - y\frac{\partial}{\partial x}). \end{aligned}\right\} \tag{3.63}$$

These operators satisfy the same commutation relations as s_x, s_y, s_z, and (3.60)–(3.62) coincide with (3.54)–(3.56). Putting, similar to (3.24),

$$w^l = \sum_{\mu - l}^{+l} \zeta^{l+\mu} Q_{l\mu}(x, y, z), \tag{3.64}$$

we obtain a system of equations for the harmonic polynomials $Q_{l\mu}$

$$\left.\begin{aligned} (m_x + im_y)Q_{l\mu} &= (l + \mu + 1)Q_{l\mu+1}; \\ (m_x - im_y)Q_{l\mu+1} &= (l - \mu)Q_{l\mu}; \\ m_z Q_{l\mu} &= \mu Q_{l\mu}, \end{aligned}\right\} \tag{3.65}$$

which coincides in the form with the system (3.59).

4 The Case of Spherical Symmetry

The energy operator H of a many-electron system is symmetric with respect to all electrons, i.e., it commutes with all the permutation operators P of coordinates and spins of electrons:

$$HP - PH = 0. \tag{4.1}$$

Moreover, if we neglect the relativistic corrections, then the operator H does not depend on spin explicitly. Thus, it follows:

$$\left.\begin{aligned} Hs_x - s_x H &= 0, \\ Hs_y - s_y H &= 0, \\ Hs_z - s_z H &= 0 \end{aligned}\right\} \tag{4.2}$$

(note that the operators s_x, s_y, s_z themselves commute with all P). From (4.1) and (4.2) it follows that the eigenfunction of the energy operator can be imposed by the two following conditions: first, the antisymmetry condition, and, second, the demand that it should be a common eigenfunction of $\mathbf{s}^2$ and s_z. To fulfill all these conditions, it is sufficient that (2.2), (2.3), (2.4) are satisfied. The Schrödinger function ψ (it defines the total function $\psi_{s\nu}$ by means of (3.4), (3.14), and (3.24)) is an eigenfunction of the energy operator

$$H\psi = E\psi. \tag{4.3}$$

Let us assume now that the energy operator shows the spherical symmetry. The necessary and sufficient condition for that is to fulfill the commutation relations

$$\left.\begin{aligned} HM_x - M_xH &= 0; \\ HM_y - M_yH &= 0; \\ HM_z - M_zH &= 0, \end{aligned}\right\} \tag{4.4}$$

where $\mathbf{M} = \mathbf{m} + \mathbf{s}$ is the total angular-momentum operator (in units of $\hbar$).

If furthermore the conditions (4.2) are fulfilled, then the following commutation relations take place:

$$\left.\begin{aligned} Hm_x - m_xH &= 0; \\ Hm_y - m_yH &= 0; \\ Hm_z - m_zH &= 0, \end{aligned}\right\} \tag{4.5}$$

where m_x, m_y, m_z are the total orbital angular-momentum components (in the previous section we denoted the corresponding operators for a single electron by these symbols).

The eigenfunction of the energy operator can be subjected, besides the antisymmetry conditions, to the following equations:

$$H\Psi^{ls}_{jm} = E\Psi^{ls}_{jm}; \tag{4.6}$$

$$\mathbf{m}^2\Psi^{ls}_{jm} = l(l+1)\Psi^{ls}_{jm}; \tag{4.7}$$

$$\mathbf{s}^2\Psi^{ls}_{jm} = s(s+1)\Psi^{ls}_{jm}; \tag{4.8}$$

$$\mathbf{M}^2\Psi^{ls}_{jm} = j(j+1)\Psi^{ls}_{jm}; \tag{4.9}$$

$$M_z\Psi^{ls}_{jm} = m\Psi^{ls}_{jm}. \tag{4.10}$$

The last two equations can be replaced by the system

$$\left.\begin{aligned}(M_x + iM_y)\Psi^{ls}_{jm} &= \alpha_{jm}\Psi^{ls}_{jm+1};\\ (M_x - iM_y)\Psi^{ls}_{jm+1} &= \beta_{jm}\Psi^{ls}_{jm};\\ M_z\Psi^{ls}_{jm} &= m\Psi^{ls}_{jm},\end{aligned}\right\} \tag{4.11}$$

where

$$\alpha_{jm}\beta_{jm} = (j-m)(j+m+1). \tag{4.12}$$

To construct the wave function satisfying all these conditions one can use the vector model and find first the function $\psi_{l\mu s\nu}$ obeying the following equations:

$$H\varphi_{l\mu s\nu} = E\varphi_{l\mu s\nu}; \tag{4.13}$$

$$\left.\begin{aligned}(m_x + im_y)\varphi_{l\mu s\nu} &= \alpha_{l\mu}\varphi_{l\mu+1s\nu};\\ (m_x - im_y)\varphi_{l\mu+1s\nu} &= \beta_{l\mu}\varphi_{l\mu s\nu};\\ m_z\varphi_{l\mu s\nu} &= \mu\varphi_{l\mu s\nu};\end{aligned}\right\} \tag{4.14}$$

$$\left.\begin{aligned}(s_x + is_y)\varphi_{l\mu s\nu} &= \alpha_{s\nu}\varphi_{l\mu s\nu+1};\\ (s_x - is_y)\varphi_{l\mu s\nu+1} &= \beta_{s\nu}\varphi_{l\mu s\nu};\\ s_z\varphi_{l\mu s\nu} &= \nu\varphi_{l\mu s\nu}.\end{aligned}\right\} \tag{4.15}$$

But it is easy to see that the construction of such a function is reduced to obtaining the Schrödinger function $\psi_{l\mu}$, satisfying, besides the symmetry requirements (2.2), (2.3), (2.4), the following equations:

$$H\psi_{l\mu s\nu} = E\psi_{l\mu s\nu}; \tag{4.16}$$

$$\left.\begin{aligned}(m_x + im_y)\psi_{l\mu s\nu} &= \alpha_{l\mu}\psi_{l\mu+1s\nu};\\ (m_x - im_y)\psi_{l\mu+1s\nu} &= \beta_{l\mu}\psi_{l\mu s\nu};\\ m_z\psi_{l\mu s\nu} &= \mu\psi_{l\mu s\nu}.\end{aligned}\right\} \tag{4.17}$$

Indeed, let us assume that the function ψ entering particularly (3.4), (3.14), (3.24) of the previous section satisfies (4.16), (4.17). If everywhere in Section 3 function ψ is meant as the function $\psi_{l\mu}$, then the generated function Φ obtained leads to the functions $\varphi_{s\nu} = \varphi_{l\mu s\nu}$, satisfying eqs. (4.13), (4.14), (4.15). The required function Ψ^{ls}_{jm} is obtained by a linear combination

$$\Psi^{ls}_{jm} = \sum_{\mu\nu} c^{ls}_{jm}(\mu,\nu)\varphi_{l\mu s\nu} \tag{4.18}$$

with the coefficients defined by the vector model [3].

Thus, we reduced the solution of (4.6)–(4.11) for the functions Ψ^{ls}_{jm} to that of a simpler system of equations (4.16), (4.17) for the functions $\psi_{l\mu}$, which do not depend on the spin variables.

Supplement

The mixed density in terms of wave functions

Let

$$\psi = \psi(x_1, x_2, \dots x_n), \tag{5.1}$$

be a wave function of n electrons, where x_i is a set of variables r_i, σ_i.

We shall call the function of $2p$ arguments

$$\begin{aligned}&\langle x_1 x_2 \dots x_p \mid \varrho_p \mid x'_1 x'_2 \dots x'_p\rangle = \\ &= \tfrac{1}{(n-p)!}\int \psi(x_1, \dots x_p, x_{p+1} \dots x_n)\overline{\psi}(x'_1, \dots x'_p, x_{p+1} \dots x_n)dx_{p+1} \dots dx_n\end{aligned} \tag{5.2}$$

the mixed density or the density matrix of the order p.

In particular, for $p = n$ we shall have

$$\langle x_1 x_2 \dots x_p \mid \varrho_p \mid x'_1 x'_2 \dots x'_p\rangle = \psi(x_1, \dots x_n)\overline{\psi}(x'_1, \dots x'_n). \tag{5.3}$$

It is more suitable to normalize the many-electron function as

$$\int \mid \psi(x_1, \dots x_n) \mid^2 dx_1 \dots dx_n = n!. \tag{5.4}$$

Under this condition we shall have

$$\varrho_0 = 1; \tag{5.5}$$

$$\int \langle x_1 \dots x_p \mid \varrho_p \mid x_1, \dots x_p\rangle dx_1 \dots dx_p = \frac{n!}{(n-p)!}. \tag{5.6}$$

Putting $x'_p = x_p$ in (5.2) and integrating over x_p, we obtain

$$\begin{aligned}&\int \langle x_1 \dots x_{p-1} x_p \mid \varrho_p \mid x'_1, \dots x'_{p-1} x_p\rangle dx_p = \\ &= (n-p+1)\langle x_1 \dots x_{p-1} \mid \varrho_{p-1} \mid x'_1 \dots x'_{p-1}\rangle.\end{aligned} \tag{5.7}$$

This formula gives relations between the mixed densities of adjacent orders.

The interaction energy is given by the sum of the terms related to the electrons taken pairwise and for calculation it is sufficient to know the mixed density of the second order.

On the other hand, in many practically important cases the character of the energy dependence on orbital and spin variables allows one to express a general matrix element by means of the element corresponding to a definite quantum number ν, e.g., $\nu = s$ or $\nu = -s$.

Therefore, we can restrict our consideration by the mixed density of the second order for the state $\nu = -5$. We shall express it by the Schrödinger wave function.

In accordance with the general formulae (3.14) and (3.24), the expression of function $\varphi_{s\nu}$ is of the simplest form for $\nu = -s$. Denoting for brevity $\varphi_{s,-s}$ by φ, we shall have

$$\varphi(r_1\sigma_1, r_2\sigma_2, \dots r_n\sigma_n) = 0, \tag{5.8}$$

if the number of positives among $\sigma_1, \sigma_2, \dots \sigma_n$ is not equal to

$$k = \frac{n}{2} - s. \tag{5.9}$$

If this number is equal to k and

$$\left.\begin{array}{lllll} \sigma_{\beta_1} & = \sigma_{\beta_2} & = \dots = & \sigma_{\beta_k} & = +1, \\ \sigma_{\beta_{k+1}} & = \sigma_{\beta_{k+2}} & = \dots = & \sigma_{\beta_n} & = -1, \end{array}\right\} \tag{5.10}$$

then

$$\varphi(r_1\sigma_1, r_2\sigma_2, \dots r_n\sigma_n) = \varepsilon(P)\psi(r_{\beta_1}r_{\beta_2}\cdots r_{\beta_k} \mid r_{\beta_{k+1}}\cdots r_{\beta_n}), \tag{5.11}$$

where $\varepsilon(P)$ is determined by (3.3). Under conditions (5.10), there will be obviously

$$\nu = \frac{1}{2}\sum_{i=1}^{n}\sigma_i = \frac{1}{2}k - \frac{1}{2}(n-k) = -s. \tag{5.12}$$

According to the general formula (5.2), for the mixed density we shall have

$$\langle r_1\sigma_1, r_2\sigma_2 \mid \varrho_2 \mid r_1'\sigma_1', r_2'\sigma_2'\rangle = $$
$$= \frac{1}{(n-2)!}\sum_{\sigma_3,\dots\sigma_n}\int \varphi\ (r_1\sigma_1, r_2\sigma_2, r_3\sigma_3\dots r_n\sigma_n)\cdot$$
$$\cdot\ \overline{\varphi}\ (r_1'\sigma_1', r_2'\sigma_2', r_3\sigma_3, \dots r_n\sigma_n)d\tau_3\dots d\tau_n. \tag{5.13}$$

In accordance with (5.8), the terms of this sum differ from zero only if the spin arguments of φ and $\overline{\varphi}$ satisfy the conditions of the kind (5.12)

$$\frac{1}{2}(\sigma_1 + \sigma_2 + \sigma_3 + \dots + \sigma_n) = \frac{1}{2}(\sigma_1' + \sigma_2' + \sigma_3 + \dots + \sigma_n), \tag{5.14}$$

whence

$$\frac{1}{2}(\sigma_1 + \sigma_2) = \frac{1}{2}(\sigma_1' + \sigma_2') = s_z. \tag{5.15}$$

The physical meaning of the value s_z introduced here is evidently a z-component of the two-electron spin. Consistently with the possible values of s_z, we can distinguish three cases:

$$s_z = 1, \qquad s_z = 0, \qquad s_z = -1.$$

For the first case ($s_z = 1$), we must put in (5.13):

$$\sigma_1 = \sigma_2 = +1, \qquad \sigma_1' = \sigma_2' = +1.$$

As to the values $\sigma_3, \sigma_4, \dots \sigma_n$, among them there will be $k-2$ equal to $+1$. According to (5.10), we denote the subscripts at the positive σ by β_i ($i = 1, 2, \dots k$). We shall have $\beta_1 = 1$, $\beta_2 = 2$, while $(\beta_3, \beta_4, \dots \beta k)$ is a set of mutually nonequal numbers from 3 to n. A number of such sets will be, evidently, $\binom{n-2}{k-2}$.

Inserting (5.11) into (5.13) and keeping in mind that the factor $\varepsilon(P)$ in the functions φ and $\overline{\varphi}$ is the same, we shall have

$$\langle r_1, 1, r_2, 1 \mid \varrho_2 \mid r_1', 1, r_2', 1\rangle =$$
$$= \frac{1}{(n-2)!}\sum_{(\beta_3\dots\beta_k)}\int \psi(r_1r_2r_{\beta_3}\cdots r_{\beta_k} \mid r_{k+1}\cdots r_n)\cdot$$
$$\cdot\ \overline{\psi}(r_1'r_2'r_{\beta_3}\cdots r_{\beta_k} \mid r_{k+1}\cdots r_n)d\tau_3\dots d\tau_n. \tag{5.16}$$

But all these integrals are equal to each other and their number is $\binom{n-2}{k-2}$. Therefore,

$$\langle r_1, 1, r_2, 1 \mid \varrho_2 \mid r_1', 1, r_2', 1\rangle =$$
$$= \frac{1}{(k-2)!(n-k)!}\int \psi(r_1 r_2 r_3 \ldots r_k \mid r_{k+1} \ldots r_n)\cdot$$
$$\cdot\ \overline{\psi}(r_1' r_2' r_3 \ldots r_k \mid r_{k+1} \ldots r_n) d\tau_3 \ldots d\tau_n. \tag{5.17}$$

For the case of $s_z = 0$, we get in a similar way

$$\langle r_1, 1, r_2, -1 \mid \varrho_2 \mid r_1', 1, r_2', -1\rangle =$$
$$= \frac{1}{(k-1)!(n-k-1)1}\int d\tau_3 \ldots d\tau_n \quad \psi(r_1 r_3 \ldots r_{k+1} \mid r_{k+2} \ldots r_n r_2)\cdot$$
$$\cdot\overline{\psi}(r_1' r_3 \ldots r_{k+1} \mid r_{k+2} \ldots r_n r_2'). \tag{5.18}$$

This expression corresponds to the values $\sigma_1 = 1, \sigma_2 = -1; \sigma_1' = 1, \sigma_2' = -1$. For other values of the spin arguments the expression of density is obtained from (5.18) by using its symmetry properties, namely,

$$\langle r_1, 1, r_2, -1 \mid \varrho_2 \mid r_1', -1, r_2', 1\rangle = -\langle r_1, 1, r_2, -1 \mid \varrho_2 \mid r_2', 1, r_1', -1\rangle, \tag{5.19}$$
$$\langle r_1, -1, r_2, 1 \mid \varrho_2 \mid r_1', 1, r_2', -1\rangle = -\langle r_2, 1, r_1, -1 \mid \varrho_2 \mid r_1', 1, r_2', -1\rangle, \tag{5.20}$$
$$\langle r_1, -1, r_2, 1 \mid \varrho_2 \mid r_1', -1, r_2', 1\rangle = \langle r_2, 1, r_1, -1 \mid \varrho_2 \mid r_2', 1, r_1', -1\rangle. \tag{5.21}$$

Finally, for the case of $s_x = -1$, we have

$$\langle r_1, -1, r_2, -1 \mid \varrho_2 \mid r_1', -1, r_2', -1\rangle =$$
$$= \frac{1}{k!(n-k-2)!}\int d\tau_3 \ldots d\tau_n \quad \psi(r_3 \ldots r_{k+2} \mid r_{k+3} \ldots r_n r_1 r_2)\cdot$$
$$\cdot\overline{\psi}(r_3 \ldots r_{k+2} \mid r_{k+3} \ldots r_n r_1' r_2'). \tag{5.22}$$

Knowing ϱ_2, one can easily calculate the density of the first order ϱ_1. The nonvanishing matrix elements will be, evidently, diagonal (with respect to spins) elements only, namely,

$$\langle r_1, 1 \mid \varrho_1 \mid r_1', 1\rangle =$$
$$= \frac{1}{(k-1)!(n-k)!}\int d\tau_2 \ldots d\tau_n \quad \psi(r_1 r_2 \ldots r_k \mid r_{k+1} \ldots r_n)\cdot$$
$$\cdot\overline{\psi}(r_1' r_2 \ldots r_k \mid r_{k+1} \ldots r_n), \tag{5.23}$$

$$\langle r_1, -1 \mid \varrho_1 \mid r_1', -1\rangle =$$
$$= \frac{1}{k!(n-k-1)!}\int d\tau_2 \ldots d\tau_n \quad \psi(r_2 \ldots r_{k+1} \mid r_{k+2} \ldots r_n r_1)\cdot$$
$$\cdot\overline{\psi}(r_2 \ldots r_{k+1} \mid r_{k+2} \ldots r_n r_1'). \tag{5.24}$$

It is possible to deduce that if the total spin is equal to zero (n is even, $k = \frac{n}{2}$), then the values of (5.23) and (5.24) are equal to each other, so that the density of the first order is a diagonal matrix with respect to spin variables.

Finally, if we construct ϱ_0 by the general rule (5.02), then we have

$$\varrho_0 = \frac{1}{k!(n-k)!}\int d\tau_1 \ldots d\tau_n \mid \psi(r_1 \ldots r_k \mid r_{k+1} \ldots r_n) \mid^2 . \tag{5.25}$$

Since this value should be equal to 1, the normalization condition of the Schrödinger coordinate functions (which are connected with total functions by (5.11)) follows from the normalization (5.4) of spin functions:

$$\int d\tau_1 \ldots d\tau_n \mid \psi(r_1 \ldots r_k \mid r_{k+1} \ldots r_n) \mid^2 = k!(n-k)!. \tag{5.26}$$

Note that the product of the two determinants (2.11) satisfies this condition if the single-electron functions are orthogonal and normalized to unity.

References

1. F. Hund, *Allgemeine Quantenmechanik des Atom- und Molekulbaues*, Handb. Phys. **24**, Berlin, 1933, 561.
2. E. Wigner, Gruppentheorie und ihre Anwendung auf die Quantenmechanik der Atomspektren, Braunschweig, 1931.
3. V. Fock, *New deduction of the vector model*, JETP **10**, 383, 1940.
4. P.A.M. Dirac, Principles of Quantum Mechanics, Oxford, 1930.
5. J. Waller and D.R. Hartree, *On the Intensity of Total Scattering of X-Rays*, Proc. Roy. Soc. London **A 124**, 119, 1929.
6. V. Fock, *Näherungsmethode zur Lösung des quantenmechanischen Mehrkörperproblems*, Zs. Phys. **61**, 126, 1930. (See also [30-2] in this book. (*Editors*))

Leningrad
Spectroscopy Laboratory
the USSR Academy of Sciences,
Physical Institute
Leningrad State University

Translated by E.D. Trifonov

43-1
On the Representation of an Arbitrary Function by an Integral Involving Legendre's Function with a Complex Index[1]

V.A. Fock

Received 12 April 1943

DAN **39**, 253, 1943

A good method for solving certain problems of the potential theory is to use the toroidal coordinates ϑ, φ, which are connected with the usual cylindrical coordinates $r = \sqrt{x^2 + y^2}$ and z by means of the formulae

$$r = \frac{\sinh \vartheta}{\cosh \vartheta - \cos \varphi}; \quad z = \frac{\sin \varphi}{\cosh \vartheta - \cos \varphi}. \tag{1}$$

If the problem to be solved shows axial symmetry, the potential Φ satisfying the Laplace equation in the physical space may be expressed, according to the formula

$$\Phi = \sqrt{2(\cosh \vartheta - \cos \varphi)}\Psi, \tag{2}$$

by a function Ψ satisfying the equation

$$\frac{\partial^2 \Psi}{\partial \vartheta^2} + \coth \vartheta \frac{\partial \Psi}{\partial \vartheta} + \frac{\partial^2 \Psi}{\partial \varphi^2} = 0. \tag{3}$$

This equation admits separation of variables. By putting

$$\Psi = p(\vartheta) s(\varphi) \tag{4}$$

one comes to the equations

$$\frac{d^2 p}{d\vartheta^2} + \coth \vartheta \frac{dp}{d\vartheta} + \left(\mu^2 + \frac{1}{4}\right) p = 0, \tag{5}$$

[1]This paper was motivated by the diffraction theory. Later on its results entered the mathematical background of the well-known Regge method in quantum scattering. (*Editors*)

$$\frac{d^2s}{d\varphi^2} - \mu^2 s = 0, \tag{6}$$

where μ^2 is a parameter. If, by the nature of the problem, the range of variation of ϑ is $0 \le \vartheta < \infty$, then the requirement that the solution be finite leads to positive values of the parameter μ^2. In the case of eq. (5) the solution that meets this requirement is the Legendre function of the first kind with the complex index $i\mu - \frac{1}{2}$ and with the argument $\cosh\vartheta$

$$p(\vartheta) = P_{i\mu-\frac{1}{2}}(\cosh\vartheta), \tag{7}$$

while $s(\varphi)$ can be put equal to

$$s(\varphi) = a(\mu)\cosh\mu\varphi + b(\mu)\sinh\mu\varphi. \tag{8}$$

From the particular solution of shape (4) a more general solution may be derived:

$$\Psi = \int_0^\infty P_{i\mu-\frac{1}{2}}(\cosh\vartheta)\{a(\mu)\cosh\mu\varphi + b(\mu)\sinh\mu\varphi\}d\mu. \tag{9}$$

The problem arises to determine the functions $a(\mu)$ and $b(\mu)$ from some boundary conditions for Ψ. When these conditions include the function ψ or a linear combination of Ψ and $\frac{\partial\Psi}{\partial\varphi}$ is given for two values of the coordinate φ, the problem evidently reduces to the determination of the function $f(\mu)$ from a given function $\psi(\mu)$, so as to have

$$\psi(x) = \int_0^\infty P_{i\mu-\frac{1}{2}}(x)f(x)d\mu \quad (1 \le x < \infty). \tag{10}$$

In other words, the problem is reduced to the inversion of the integral (10) and to the expansion of an arbitrary function by the Legendre functions with a complex index. It is just this problem that we are to consider in the present article.

1. *Some properties of the Legendre functions.* Before we proceed to the solution of our problem, it will be reasonable to point out certain properties of the Legendre functions that will be used in the sequel. We define Legendre functions of the first kind by means of Mehler's integral

$$P_{i\mu-\frac{1}{2}}(\cosh\vartheta) = \frac{2}{\pi}\int_0^\vartheta \frac{\cos\mu t dt}{\sqrt{2(\cosh\vartheta - \cosh t)}}. \tag{11}$$

Those of the second kind are then defined by means of the integral

$$Q_{i\mu-\frac{1}{2}}(\cosh\vartheta) = \int_\vartheta^\infty \frac{e^{-i\mu t}dt}{\sqrt{2(\cosh t - \cosh\vartheta)}}. \tag{12}$$

The two kinds are connected by the relation

$$Q_{-i\mu-\frac{1}{2}}(\cosh\vartheta) - Q_{i\mu-\frac{1}{2}}(\cosh\vartheta) = i\pi\tan\mu\pi P_{i\mu-\frac{1}{2}}(\cosh\vartheta). \tag{13}$$

From the last two formulae, we have

$$P_{i\mu-\frac{1}{2}}(\cosh\vartheta) = \coth\mu\pi\frac{2}{\pi}\int_{\vartheta}^{\infty}\frac{\sin\mu t dt}{\sqrt{2(\cosh t - \cosh\vartheta)}}. \tag{14}$$

Substituting in the integral (11) the expansion

$$\sqrt{\frac{\sinh\vartheta}{2(\cosh t - \cosh\vartheta)}} = \\ = \sqrt{\frac{\vartheta}{\vartheta^2 - t^2}}\left\{1 + \frac{1}{8\vartheta}\left(\coth\vartheta - \frac{1}{\vartheta}\right)(\vartheta^2 - t^2) + \ldots\right\}, \tag{15}$$

which is a power series in $\vartheta^2 - t^2$, and integrating term by term, we obtain for $P_{i\mu-\frac{1}{2}}$ an expression of the form

$$P_{i\mu-\frac{1}{2}} = \sqrt{\frac{\vartheta}{\sinh\vartheta}}\left\{J_0(\mu\vartheta) + \frac{1}{8\mu}\left(\coth\vartheta - \frac{1}{\vartheta}\right)J_1(\mu\vartheta) + \ldots\right\}. \tag{16}$$

Here the general term in the brackets has the shape

$$A_n = a_n(\vartheta)\frac{J_n(\mu\vartheta)}{(\mu\vartheta)^n}, \tag{17}$$

where $a_n(\vartheta)$ is a polynomial in ϑ and $\cot\vartheta$, which with $\vartheta \to 0$ behaves like ϑ^{2n} and with $\vartheta \to \infty$ like ϑ^n. The series (16) converges when $\vartheta < \vartheta_0$ and diverges when $\vartheta > \vartheta_0$, where $\vartheta_0 = 2\pi\sqrt{2} + \sqrt{5}$. However, if μ is great, it may be regarded as an asymptotic series fit to be used for all values of ϑ. The corresponding series for the Legendre functions of the second kind have the form

$$Q_{i\mu-\frac{1}{2}} \cong -\frac{\pi i}{2}\sqrt{\frac{\vartheta}{\sinh\vartheta}}\left\{H_0^{(2)}(\mu\vartheta) + \frac{1}{8\mu}\left(\coth\vartheta - \frac{1}{\vartheta}\right)H_1^{(2)}(\mu\vartheta) + \ldots\right\}, \tag{18}$$

$$Q_{-i\mu-\frac{1}{2}} \cong \frac{\pi i}{2}\sqrt{\frac{\vartheta}{\sinh\vartheta}}\left\{H_0^{(1)}(\mu\vartheta) + \frac{1}{8\mu}\left(\coth\vartheta - \frac{1}{\vartheta}\right)H_1^{(1)}(\mu\vartheta) + \ldots\right\}, \tag{19}$$

where $H_s^{(1)}$ and $H_s^{(2)}$ are Hankel's functions. These are always divergent series, but their coefficients are the same as in (16). Our series appear to be a novelty and are also applicable for the complex value of μ. Their advantage over the ordinary asymptotic series is that when μ is large, they can also be used with small values of ϑ down to $\vartheta = 0$, while the ordinary series cannot.

The inequalities

$$\left|P_{i\mu-\frac{1}{2}}(\cosh\vartheta)\right| < P_{-\frac{1}{2}}(\cosh\vartheta) < \frac{\vartheta}{2\sinh\frac{\vartheta}{2}}, \tag{20}$$

$$\left|Q_{i\mu-\frac{1}{2}}(\cosh\vartheta)\right| < Q_{-\frac{1}{2}}(\cosh\vartheta) < \frac{\pi}{2}\coth\frac{\vartheta}{4} \tag{21}$$

hold for real values of μ and ϑ (in (21) it is supposed that $\vartheta > 0$).

2. *Inversion of the integral.* Equation (10) may be rewritten as

$$\psi(\cosh\vartheta) = \int_0^\infty P_{i\mu-\frac{1}{2}}(\cosh\vartheta)f(\mu)d\mu. \tag{22}$$

When $P_{i\mu-\frac{1}{2}}(\cosh\vartheta)$ is replaced by its integral representation (11), the problem of finding the function $f(\mu)$ is reduced to the solution of Abel's equation followed by the Fourier transformation. In a purely formal way, we obtain

$$f(\mu) = \mu\tanh\mu\pi\int_0^\infty P_{i\mu-\frac{1}{2}}(\cosh\vartheta)\psi(\cosh\vartheta)\sinh\vartheta d\vartheta. \tag{23}$$

The formula gives the inversion of the integral (22).

The conditions under which (22) and (23) are valid can be formulated in the form of the two following theorems:

Theorem I. *If a function $f(\mu)$ given in the interval $(1 \leq x < \infty)$ is such that $\varphi(t) = 2\sinh\frac{t}{2}\psi(\cosh t)$ has its first derivative absolutely integrable over an infinite interval $(0 \leq t < \infty)$, while its second derivative is absolutely integrable over any finite interval, and if $\varphi(0) = 0$; $\varphi(\infty) = 0$, then $\psi(x)$ is representable in the form of the integral (10), where $f(\mu)$ is defined by*

$$f(\mu) = \mu\tanh\mu\pi\int_0^\infty P_{i\mu-\frac{1}{2}}(x)\psi(x)dx. \tag{24}$$

Theorem II. *If a function $f(\mu)$ is absolutely integrable over an infinite interval $(0 \leq \mu < \infty)$ and has its derivative absolutely integrable over any finite interval, and if $f(0) = 0$, then $f(\mu)$ is representable in the form of the integral (24), where $\psi(x)$ is defined by (10).*

The conditions of the first theorem may be extended and made less rigorous. In particular, one can establish

Theorem III. *If a function $\psi(x)$, where $(z \leq x < \infty)$, is such that there exists an integral*

$$\int_1^\infty \frac{|\psi(x)|dx}{\sqrt{x+1}}, \tag{25}$$

then for every point x, in whose neighborhood $\psi(x)$ has a bounded variation, there holds

$$\frac{1}{2}[\psi(x+0)+\psi(x-0)] = \int_0^\infty P_{i\mu-\frac{1}{2}}(x)f(\mu)d\mu, \tag{26}$$

where $f(\mu)$ is given by (24).

The way to prove Theorems I and II is by examining the course of computations, which has led us to the inversion of the integral (10). Theorem III arises while the structure of formulae (10) and (24) is studied in detail, formulae (16), (18) and (19) being made use of in this study.

Let us take an example before we finish. Put

$$\psi(x) = \frac{\sin\alpha}{[2(x+\cos\alpha)]^{3/2}} \qquad (-\pi < \alpha < \pi). \tag{27}$$

Then

$$f(\mu) = \mu\frac{\sinh\mu\alpha}{\cosh\mu\alpha} \tag{28}$$

and we have

$$\frac{\sin\alpha}{[2(x+cos\alpha)]^{3/2}} = \int_0^\infty P_{i\mu-\frac{1}{2}}(x)\frac{\sinh\mu\alpha}{\cosh\mu\alpha}\mu d\mu, \tag{29}$$

$$\frac{\sinh\mu\alpha}{\cosh\mu\alpha} = \tanh\mu\pi\int_1^\infty P_{i\mu-\frac{1}{2}}(x)\frac{\sin\alpha}{[2(x+\cos\alpha)]^{3/2}}dx,$$

which can be easily verified by direct computation.

Translated by V.V. Fock

47-1
On the Uncertainty Relation between Time and Energy

V. Fock and N. Krylov

(Received 29 May 1946)

JETP **17**, N 2, 93, 1947
J. Phys. USSR **11**, N 2, 112, 1947 (English version)

Introduction

The physical meaning of the uncertainty relation between time and energy appears to be not completely clear until now. This fact is not surprising since the interpretation of the time–energy relation is much more difficult than the interpretation of the similar relation between the coordinate and momentum. Indeed, the latter relation is easily derived from the quantum-mechanical formalism and does not require the examination of the course of the physical process in time. The relation between time and energy, however, essentially requires such an examination, and there arises a question whether Schrödinger's equation may be used or not in deriving this relation, the latter possibility being not excluded since the relation may be interpreted to correspond to a measurement act, which does not obey Schrödinger's equation.

L. Mandelstam and Ig. Tamm in their recent paper (The Uncertainty Relation between Energy and Time in Non-Relativistic Quantum Mechanics) [1] made a very interesting attempt to derive the above-mentioned uncertainty relation from the Schrödinger equation.

The aim of the present note is to analyze the relation derived by these authors. At first we shall consider the uncertainty relation in its usual (Bohr's) interpretation and compare it with the relation derived by Mandelstam and Tamm. We shall find that the meaning and regions of application of these two relations are entirely different. In the last section we shall establish a connection between the law of decay of an almost stationary state and the energy distribution function in this state and consider the question on the practical applicability of the Mandelstam–Tamm relation to this problem.

1 Relation Referring to the Measurement Act

The uncertainty relation referring to the measurement act (for brevity we shall call it Bohr's relation) may be written in the form

$$\Delta(E - E')\Delta t > \hbar. \tag{1}$$

Here E and E' are the values of the energy of an object before and after the measurement (before and after the interaction of the object with the measuring apparatus); $\Delta(E-E')$ is the absolute value of the uncertainty in the increase of the energy during the process of interaction, Δt – the uncertainty in the time moment when this process has taken place. The letter $\hbar$ in the right-hand part of (1) stands for Planck's constant divided by 2π (the exact value of the right-hand part is not essential as we have to deal only with the order of magnitude).

As is mentioned above, Bohr's relation (1) can be directly applied to the measurement act. The quantity Δt may be interpreted as the minimum duration of the measurement and $\Delta(E - E')$ as the error of the measurement. The question as to the detailed course of the interaction process between the object and the apparatus in an elementary measurement does not arise at all. Indeed, in order to check any statement concerning the process of interaction, one should insert into a given elementary measurement a number of other measurements, which is obviously impossible. We can attach only a conditional meaning to the notion "the course of the interaction process during the measurement." Namely, it is possible to replace the given direct measurement by another indirect one, in which the primary apparatus is considered as part of the system under examination and the measurement itself is made by another apparatus. Then the interaction between the object and the primary apparatus may be described quantum-mechanically and examined in detail by means of the new apparatus. But in this case the interaction with the new apparatus will be uncontrollable.

Thus, the inclusion of the initial apparatus into the quantum-mechanical system only displaces the boundary behind which the quantum-mechanical description is to be replaced by a classical one, and the question on the uncontrollable interaction of the apparatus with the object (the measure of which is Bohr's relation) arises again.

It is self-evident that in spite of the fact that the apparatus is described in a classical way, the uncertainty relations remain valid for it, too. Indeed, the use of the apparatus as a means of measurement is based on the knowledge of *those* quantities (relative to the apparatus), and with that degree of accuracy only, that uncertainty relation cannot be violated.

At any rate, in examining the direct measurement one cannot use Schrödinger's equation to describe the behavior of the object during the time of measurement. This follows, among other reasons, from the fact that the object does not constitute a closed mechanical system during the measurement.

It is clear, however, that the role played by the Schrödinger equation in the theory of measurement is not invalidated by these considerations. This equation must give the theory of the indirect measurements,[1] which permits us to follow in detail the interaction of the object with the "intermediate" part of the system (i.e., with that part which at first was included in the apparatus and was described classically and then, after the replacement of the direct measurement by the indirect one, is considered as part of the object under examination and described by the quantum-mechanical method). It must also secure the consistency of the results obtained in the two ways (by the direct measurement and by the indirect one) and, in particular, the impossibility to break Bohr's relation by such a replacement.

We come to the conclusion that *since Bohr's relation governs the act of the measurement, it cannot be obtained directly from the Schrödinger equation.* It represents itself a certain basic principle. The role of the Schrödinger equation in its foundation has to be reduced to the proof of its consistency with the quantum-mechanical method. On the other hand, this principle may be established or at least illustrated by means of an analysis of various examples and mental experiments as it has been done by Bohr.

We wish to emphasize once more the fundamental character of the Bohr relation regarding the act of the measurement. The direct measurement forms the necessary intermediate link between the mathematical scheme of quantum mechanics and the experiment. The estimation of the dependence of the measurement accuracy upon its duration is just given by the Bohr relation. All the quantum mechanics would lose ground without this relation as there would not be any possibility to compare its predictions with the experiment.

Relation (1) was given for the first time in the well-known paper by Landau and Peierls (Erweiterung des Unbestimmtheitsprinzips für die relativistische Quantentheorie) [2] dated 1931; these authors pointed

[1]The development of the theory of indirect measurements brilliantly initiated by Prof. Mandelstam in his lectures is far from being completed up to the present. (*V. Fock*)

out that the relation belongs to Bohr. However, the derivation of Bohr's relation given in the quoted paper appears to us to be fallacious, being based on a wrong idea.

This derivation is based on the consideration of the interaction process between the object-particle and the apparatus-particle with the help of the perturbation theory. The interaction process taking place during the measurement is thus supposed to obey Schrödinger's equation. From what has been said above it is clear, however, that this equation cannot be applied to a direct measurement act (to which Bohr's relation applies), since the apparatus-particle must be described classically. The use of the perturbation theory and the consideration of the transition probabilities would mean that both particles are described quantum-mechanically and are observed by means of some other apparatus, not including one of them. This is by no means equivalent to a direct measurement act. As stated above, the attempt to consider transition probabilities in a direct measurement (or to detail it in some other way) leads to a contradiction since it supposes the possibility to intercalate in the given measurement a number of intermediate measurements in such a way that the result should not be affected, and that is evidently impossible.

In this connection we have to discuss two assertions made by Landau and Peierls regarding the measurement of momentum and energy. First, they assert that in this case the law of conservation of momentum is to be applied as a strict law, while the law of conservation energy is to be applied as an approximate law only, valid up to quantities of the order $\hbar/\Delta t$. Second, they assert that in the most favorable case the values of the momentum and energy of the apparatus-particle before and after the measurement may be considered as exactly known. Both assertions appear to be erroneous. There is no doubt that both conservation laws are to be applied as rigorous laws of classical mechanics, since they constitute the only means of defining and of measuring the momentum and energy of a particle (and of any other object). As to the second assertion, the supposition that the energies of the instrument-particle before and after the collision are exactly known, however small the duration of the collision may be, would signify a violation of the Bohr time–energy relation for the instrument.

In the case of a free particle, Bohr's relation (1) is a consequence of the relation

$$|v' - v|\Delta p\Delta t > \hbar, \tag{2}$$

where v and v' are the velocities of the particle before and after the collision. In deriving (2) one has to use the approximate equality of the uncertainties Δp and $\Delta p'$ in the momentum values of the particle before and after the collision. But this equality may be deduced from the supposition that v and v' are small as compared with the velocities of the apparatus-particle, the exact knowledge of the momentum of the latter being unnecessary.

2 Relation Referring to the Motion of a Wave Packet

Following Mandelatam and Tamm we consider a conservative system obeying Schrödinger's equation with the energy operator H. We suppose that in a given state the energy of the system has no definite value, but possesses a known distribution function. (The system being conservative, the energy distribution function will be independent of the time.) We denote the mean value (mathematical expectation) of the energy by $\overline{H}$ and the mean square value of the difference $H-\overline{H}$ by $(\Delta H)^2 = \overline{(H-\overline{H})^2}$. The quantity ΔH is the standard of the energy H of the system.

We also consider some quantity R, relating to the same system and having an operator which does not depend on time explicitly. Let $\overline{R}$ and ΔR be the mean values and the standard of R in a given state. The quantities $\overline{R}$ and ΔR will be the functions of time.

As shown by Mandelstam and Tamm, it follows from Schrödinger's equation and Schwarz' inequality that the quantities ΔH, ΔR and $\overline{R}$ are related by the inequality

$$\Delta H \Delta R \geq \frac{\hbar}{2}\left|\frac{\partial \overline{R}}{\partial t}\right|. \tag{3}$$

Putting then

$$\Delta T = \frac{\Delta R}{\left|\dfrac{\partial \overline{R}}{\partial t}\right|}, \tag{4}$$

the preceding relation may be written in the form

$$\Delta H \Delta T \geq \frac{\hbar}{2}. \tag{5}$$

Owing to the constancy of ΔH the quantities ΔR and $\partial\overline{R}/\partial t$ in formulae (3) and (4) may be replaced by their time averages over some interval that does not affect the meaning of relation (5), however.

The quantity ΔT is called by the authors "the standard time." It is the time during which the mean value $\overline{R}$ is changed by an amount of the order of its standard.

Relation (5) may be interpreted in another way. Let us determine the time moment when the quantity R passes through a given value R_0 and the energy of the system at that moment.

If $\partial\overline{R}/\partial t$ is the rate of variation of the quantity R (or of its mean value $\overline{R}$), then the uncertainty ΔT in the time moment defined above is obviously connected with the uncertainty ΔR in the quantity R, by relation (4) (Fig. 1). On the other hand, as the uncertainty ΔH in the energy does not change with time, it may be referred to the moment when the quantity R passes through a given value R_0.

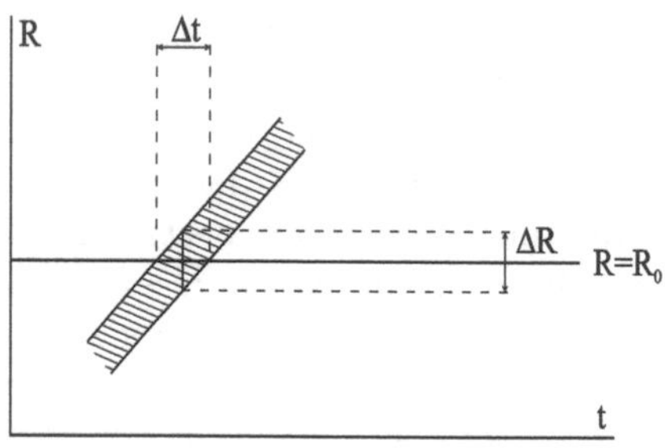

Fig. 1

In this interpretation, relation (5) connects the uncertainties in the energy and the time for a given value of R.

In this formulation, relation (5) was known earlier for the particular case when the quantity R is one of the rectangular coordinates x and the system under consideration is a free particle. This case corresponds to the passing of a wave packet through a given plane $x = x_0$ (see the first example by Mandelstam and Tamm). Relation (5) was usually derived (on assumption that the momentum components p_y and p_z parallel to the plane are known exactly) from the relations $\delta E = v_x \Delta p_x$ and $\Delta T = \Delta x v_x$ by multiplying them and making use of the Heisenberg relation

$$\Delta p_x \Delta x \geq \frac{\hbar}{2}. \tag{6}$$

In the case considered relation (5) is closely analogous to the spatial Heisenberg relations. Indeed, the Heisenberg relation connects the uncertainties in the momentum and the coordinate at a given time instant, while relation (5) connects the uncertainties in the energy and time for a given value of a coordinate.

The uncertainty relation discovered by Mandelstam and Tamm (for brevity we shall call it Mandelstam's relation) may be considered as a generalization of that previously known to an arbitrary quantity R and to an arbitrary conservative system.

The fact that it is possible to derive this generalized relation from the quantum–mechanical formalism is very interesting in itself. In addition, the formal derivation due to Mandelstam and Tamm has led to a clearer understanding of the physical meaning of the quantities involved.

3 Comparison of Bohr's Relation with That of Mandelstam

The physical meaning of the relation $\Delta H \Delta T \geq \frac{\hbar}{2}$ follows from the well-known statistical interpretation of the wave function, the wave packet and the Schrödinger equation.

The wave packet gives, as is well known, the time dependence of the distribution function for a given mechanical quantity and has no direct relation to the motion of an individual particle.

Since the wave packet model of a particle (as proposed initially by de Broglie and Schrödinger) turned out to be untenable and has been abandoned, the motion of a wave packet does not correspond to any physical process proceeding in time and cannot be observed in a single experiment made on an individual particle. Consequently, as Mandelstam's inequality refers to the motion of a wave packet, it cannot be applied to in individual experiment or to a measurement. On the contrary, Bohr's inequality refers just to such an experiment proceeding in time and performed on one individual object.

To plot a curve for any distribution function from observation data, we must have an estimate of the precision of the results of individual experiments. But this estimate has nothing to do with the form of the curve. The half-life period of an atomic nucleus may be as large as billions of years but the moment of its decay may be determined with an accuracy of a millionth of a second.

In the case of energy the estimate of the precision of an individual experiment is given by Bohr's inequality, while Mandelstam's relation refers to the properties of the distribution curves (and their dependence upon the time for the energy and for some other quantity R).

When speaking of individual particles or individual experiments, on the one hand, and of distribution curves, on the other hand, we do not oppose the notion of a single experiment to that of a series of experiments. Quite the contrary, we admit that the errors of individual experiments performed on individual particles have their own statistics, but that the states following from a given initial state by the Schrödinger equation have their own statistics. We only wish to emphasize that these two statistics have quite different objects and that their character depends on quite different circumstances.

If one considers a wave packet (to which Mandelstam's relation refers), then its statistics depends only on the initial state and on the form of the energy operator and is independent of any other circumstances. The statistics of the errors of individual experiments (to which Bohr's relation refers) does not depend upon the initial state, but depends upon the means used for the measurement, upon the duration of the measurement, etc.

Thus, the Bohr relation deals with the individual experiment (and its statistics) and the Mandelstam relation — with the initial state (and its statistics). This is their principal difference.

A more formal difference that makes evident that Mandelstam's relation has nothing to do with the real experiment manifests itself in the following. In the case of a wave packet the uncertainty ΔH in the energy of the system exists from the very beginning and does not change with time (neither does the energy change). In the case of a measurement the increase $E' - E$ in the energy takes place just during the time Δt, and $\Delta(E' - E)$ is the uncertainty in the value of this increase.

The preceding considerations may be applied to each of the Mandelstam and Tamm examples mentioned above.

Thus, the problem of the wave packet passing through a given plane is not equivalent to the problem of the measurement of the moment of time when the particle passes through the plane (the fixing of the passage being accompanied by an increase of its energy). The wave packet gives the statistics of the experiment which up to the moment of the experiment had moved in a constant (time-independent) external field, had preserved its energy and had not been subjected to any external influence. The Mandelstam relation refers to such experiments. The Bohr relation refers just to the real passage of a particle through the aperture in a diaphragm supplied with a shutter, the action of which is inevitably connected with an uncontrollable change $\Delta(E' - E)$ in the energy of the particle. The time ΔT in the Bohr relation may be connected in the given case with the duration of the action of the shutter. If we intended to describe the passage of the particle through the aperture more accurately we should account for the fact that the shutter with which the diaphragm is equipped is movable and that consequently the field in which the particle is moving depends on time and the system is not conservative. However, the quantum-mechanical description of the motion of a particle in a variable field leaves open the question as to the apparatus by means of which the measurements on such a particle are performed. At any rate we should stop considering the shutter as a measuring apparatus.

4 Connection between the Energy Distribution Function and the Law of Decay of a Given State

One of the applications of the Mandelstam relation refers to the problem of connection between the half-life period τ of a given state ψ_0 of the system and the uncertainty ΔH in the energy of this state.

More general results can be obtained by means of a theorem that establishes the connection between the law of decay of an almost-stationary state and the energy distribution function in this state.[2]

The derivation of this theorem is very simple.

Let Ψ_0 be the initial state of a system (for $t = 0$). What is the probability of finding the system in the initial state after time t had elapsed?

Denoting by x the coordinates (or those variables in terms of which the wave function is expressed) we write $\psi = \psi(x,t)$ and $\psi_0 = \psi(x,0)$. Let us expand ψ_0 in an integral involving proper functions $\psi_E(x)$ of the energy operator

$$\psi(x,0) = \int c(E)\psi_E(x)dE. \tag{7}$$

Then the state of the system at time t will be

$$\psi(x,t) = \int e^{-\frac{i}{\hbar}Et}c(E)\psi_E(x)dE. \tag{8}$$

The required probability $L(t)$ [3] will be equal to the squared modulus of the scalar product

$$p(t) = \int \overline{\psi}(x,0)\psi(x,t)dx, \tag{9}$$

which is equal to

$$p(t) = \int e^{-\frac{i}{\hbar}Et}\overline{c}(E)c(E)dE. \tag{10}$$

[2]The theorem is implicitly contained in some formulae derived by V. Fock in his course of lectures on quantum mechanics read in 1936/37 in the Leningrad University [3] (in a lecture on the passage of a particle through a potential barrier). (*Authors*)

[3]Here by $L(t)$ we denote the same quantity, which was denoted by $\overline{L}(t)$ in the paper by Mandelstam and Tamm. (*Authors*)

But the expression

$$dW(E) = w(E)dE = |c(E)|^2 dE \tag{11}$$

is the differential of the energy distribution function by the initial state (and therefore also for the state in any successive moment of time t). Therefore, the required probability is equal to

$$L(t) = |p(t)|^2 \tag{12}$$

with

$$p(t) = \int e^{-\frac{i}{\hbar}Et} w(E)dE = \int e^{-\frac{i}{\hbar}Et} dW(E). \tag{13}$$

Thus, we have obtained the following theorem.

The law of decay of the state ψ_0 depends only upon the energy distribution function

$$L(t) = \left| \int e^{-\frac{i}{\hbar}Et} dW(E) \right|^2 . \tag{14}$$

With a proper definition of the integral distribution function $W(E)$ formula (14) remains valid in the case of a discontinuous function $W(E)$ (point spectrum).

The following remarks concerning the theorem just proved should be added. First, the law of decay may be the same for two different states if the energy distribution functions of these states are the same. Second, in formula (14) for the decay probability the time t is reckoned from the moment when (for the very last time) it was stated that the atom (or the system) has not yet decayed; as to the state ψ_0 of the atom itself, it does not change. The situation may be expressed in such words: an atom does not grow old, but disintegrates suddenly. This conclusion holds for any law of decay, not necessarily an exponential one.

It is interesting to note that in formula (13) the Fourier transformation applies not to the probability amplitude (i.e., the wave function) as is usual in quantum mechanics, but to the probability itself. According to the terminology adopted in the probability theory, the function $p(t)$ is the *characteristic function* of the energy distribution.

From the properties of the Fourier transformation, a relation between the rate of decay and the smoothness of the distribution function follows.

We shall first examine the conditions that must be fulfilled in order that the decay takes place at all.

If the differential distribution function $w(E)$ exists, the integral distribution function is connected with $w(E)$ by the relation

$$W(E') - W(E) = \int_{E}^{E'} w(E)dE, \tag{15}$$

where E' and E are two arbitrary values of the energy. The above expression is the probability that the energy of the system has a value between E and E' (where $E' > E$). If there is no point spectrum, the function $W(E)$ will be continuous for any initial state. In the case when the point spectrum is also present the function $W(E)$ may be continuous only when all probabilities related to the point spectrum vanish in the initial state.

Suppose that the function $W(E)$ is continuous. It follows from relation (15) between $W(E)$ and $w(E)$ that the continuity of $W(E)$ is equivalent to the absolute integrability of $w(E)$ in its usual definition.[4] But if $w(E)$ is absolutely integrable then according to the Riemann–Lebesgue lemma the value of the integral (13) tends to zero when t increases to infinity. Thus, from the continuity of $W(E)$ it follows that

$$L(t) \to 0 \quad \text{for} \quad t \to 0. \tag{16}$$

On the other hand, according to a theorem on characteristic functions, the proof of which one can find in a book by S. Bernstein [4], condition (16) entails the continuity of $W(E)$.

We arrive at the conclusion that *the necessary and sufficient condition for the decay is the continuity of the integral energy distribution function.*

In many problems the energy distribution function satisfies much more stringent conditions than a simple continuity of $W(E)$. Thus in the problem of the escape of a particle from a potential well through a potential energy barrier, the probability density $w(E)$ will be a meromorphic function of the complex variable E (see Appendix). Since for real values of E the function $w(E)$ will be real, its poles will be situated symmetrically with respect to the real axis, the residues being complex conjugated quantities. Suppose that the pair of poles nearest to the real axis is

$$E = E_0 \pm i\Gamma \quad (\Gamma > 0). \tag{17}$$

[4]The usual definition of the absolute integrability includes the condition that the value of the integral should tend to zero with the region of the integration. According to this definition, the Dirac function $\delta(E - E_0)$ will not be absolutely integrable in the vicinity of $E = E_0$. (*V. Fock*)

Let the next pair of poles have an imaginary part Γ'. It is easy to see that if t is so large that

$$\left|\frac{\Gamma' - \Gamma}{\hbar} t\right| \gg 1, \tag{18}$$

then the value of the integral (13) will be determined by the residue relative to one of the two poles (17) while the other poles will be immaterial. Now if the function $w(E)$ had only one pair of poles, then we could write[5]

$$w(E) = \frac{1}{\pi} \frac{\Gamma}{(E - E_0)^2 + \Gamma^2}, \tag{19}$$

i.e., in this case we should come to the dispersion formula for the energy distribution.

Substituting (19) into (13), we obtain

$$p(t) = e^{-\frac{i}{\hbar}Et - \frac{\Gamma}{\hbar}|t|} \tag{20}$$

and, therefore,

$$L(t) = e^{-\frac{2\Gamma}{\hbar}|t|}. \tag{21}$$

Thus, *the usual exponential form of the law of decay follows from the general assumption that $w(E)$ is meromorphic in F*; this assumption may be substantiated by an analysis of the Schrödinger equation for the given problem.

In the problem of the escape of a particle from a potential well, the poles of $w(E)$ are the so-called "complex Gamow's eigenwerts." The order of magnitude of Γ in this problem is determined by the equation

$$\Gamma = \frac{2\hbar}{T} e^{-2S}, \tag{22}$$

where T is of the order of the period of oscillations of the particle in the potential well and the quantity S is the integral

$$S = \frac{1}{\hbar} \int \sqrt{2m(U - E)}dx \tag{23}$$

extended over the region where $U > E$ (the region of the well).

In this and in a more general problem on the decay of an excited, almost stationary state of an atomic system the order of magnitude of Γ may be determined from the perturbation theory, which gives

$$\gamma = \pi \left(|\langle E_0|H|k\rangle|^2 \frac{dk}{dE} \right)_{E=E_0}, \tag{24}$$

[5]We suppose here that $\Gamma \ll E_0 - E^*$ where E^* is the lower limit of integration in (13) (usually $E^* = 0$). (*V. Fock*)

where $\langle E_0|H|k\rangle$ is the matrix element of the perturbation energy, corresponding to the transition from the almost stationary state with energy E_0 into a state of the continuous spectrum specified by the parameter k and having the energy E.

The considerations which have led us to the dispersion formula (19) and to the exponential law of decay (21) permit us to make the following conclusion. The dispersion formula, which takes into account only one pair (or a finite number of pairs) of poles of the function $w(E)$, may be used only for the purpose of evaluation of integrals of the form (13) in the case of sufficiently large values of t. In order to calculate integrals of another type (for example, to calculate the quantum mechanical averages of powers of the energy, in particular of the standard ΔE), the dispersion formula cannot be used. Indeed, the value of the integral

$$\text{average } F(E) = \int F(E)w(E)dE \tag{25}$$

will be determined not only by that part of the range of integration which is close to $E = E_0$, but also by more distant parts or even by the behavior of $w(E)$ for very large values of E. In a most pronounced form this circumstance manifests itself in the fact that for some functions $F(E)$ the substitution in (25) of the dispersion formula (19) leads to a divergent integral.

Since the standard of the energy ΔE cannot be calculated by means of the dispersion formula (which, however, gives a correct law of the decay), one may conclude *that ΔE is not a characteristic quantity for the law of decay.*

On the other hand, for sufficiently small values of t a lower limit for $L(t)$ involving ΔE may be given. Putting

$$\int Ew(E)dE = E_0, \tag{26}$$

we introduce the function

$$p_1(t) = e^{\frac{i}{\hbar}E_0 t}P(t) = \int e^{\frac{i}{\hbar}(E_0 - E)t}w(E)dE. \tag{27}$$

We have then

$$p_1(0) = 1; \quad p_1'(0) = 0 \tag{28}$$

and the second derivative satisfies the inequality

$$|p_1''(t)| \le |p_1''(0)| = \frac{(\Delta E)^2}{\hbar^2} \tag{29}$$

for all values of t. Hence,

$$|p_1(t)| \geq 1 - \frac{(\Delta E)^2}{\hbar^2} \tag{30}$$

and, therefore,

$$L(t) \geq \left[1 - \frac{(\Delta E)^2}{\hbar^2} t^2\right]^2 \qquad \text{if} \quad t < \sqrt{2}\frac{\hbar}{\Delta E}. \tag{31}$$

A similar, but more accurate inequality

$$L(t) > \cos^2\left(\frac{\Delta E}{\hbar} t\right) \qquad \text{if} \quad t < \frac{\pi}{2}\frac{\hbar}{\Delta E} \tag{32}$$

is derived in the paper by Mandelstam and Tamm. The authors apply it to the estimation of the half-life period τ of a given state. Indeed, either the time τ is so large that inequalities (31), (32) cease to apply or they are still valid. In either case, we obtain

$$\tau \Delta E > \hbar\sqrt{2 - \sqrt{2}} == \hbar \cdot 2 \sin\frac{\pi}{8} = \hbar \cdot 0.7653 \tag{33}$$

from inequality (31) and

$$\tau \Delta E > \hbar \frac{\pi}{4} = \hbar \cdot 0.7854 \tag{34}$$

from inequality (32).

Inequality (34) [or (33)], although quite rigorous, is, however, of no practical use for the estimation of the half-life period. It might give a correct order of magnitude of τ only for those energy distribution functions for which ΔE is a characteristic parameter, for example, for the Gaussian distribution function

$$w(E) = \frac{1}{\sqrt{2\pi} \cdot \Delta E} \cdot \exp\left[-\frac{(E - E_0)^2}{2(\Delta E)^2}\right], \tag{35}$$

when the law of decay is of the form

$$L(t) = \exp\left[-\frac{(\Delta E)^2}{\hbar^2} t^2\right]. \tag{36}$$

In this case there is no half-life period in the strict sense; but if we define the half-life time by the condition $L(\tau) = 1/2p$ we get

$$\tau \Delta E = \hbar\sqrt{2} = \hbar \cdot 0.8326, \tag{37}$$

which is rather near to the lower limit given by inequality (34). However, this example is of a purely mathematical character. In the physical problems of the type considered (transitions from an almost stationary state into a state of continuous spectrum with the same energy) the function $w(E)$ is, as we have seen, of a quite different character and the energy standard if it exists at all, is not a specific quantity for the law of decay. In such problems the half-life period, if defined in terms of Γ by the relation

$$\tau\Gamma = \hbar \cdot 1/2 \log 2 = \hbar \cdot 0.3466, \tag{38}$$

turns out to be many times larger than its lower limit obtained from (34), since the quantity F is many times smaller than the energy standard Δ.

This situation may be examined more closely in the case of the problem of the potential well. At the attempt to calculate Δ for the almost-stationary state we meet with a characteristic difficulty which supports, however, our conclusion that ΔE is practically not connected with the law of decay. It is found that the value of ΔE sharply depends on the choice of the initial almost-stationary state (in our problem it depends on the values of the wave function outside the barrier), while the constant Γ is quite stationary for different admissible wave functions.

If we take a rectangular barrier and suppose that in the initial state the wave function outside the barrier is the analytical continuation of the function over the barrier, then for the quantity Γ a value is obtained of the order

$$\Gamma \sim (\Delta E)^2/E_0. \tag{39}$$

This value is many times smaller than ΔE. Indeed the quantity $(\Delta E)/E_0$ is in this case of the order of e^{-S}, while the quantity Γ/E_0 is of the order of e^{-2S} where e^{-S} is very small. Thus, the considered example supports our conclusion that inequalities of the form (33) or (34) cannot be applied to the estimation of the order of magnitude of the half-life period.

Appendix

The energy distribution function in the problem of the potential well

Let us denote by $f(r)$ the solution of the Schrödinger equation

$$-\frac{\hbar^2}{2m}\frac{d^2 f}{dr^2} + U(r)f = Ef \tag{A1}$$

satisfying the initial conditions

$$f(0) = 0; \quad f'(0) = 1. \tag{A2}$$

The function $f(r)$ belongs to the continuous spectrum, but is not normalized. The asymptotical expression for the function $f(r)$ at infinity is of the form

$$f(r) = B(E)e^{ikr} + \overline{B}(E)E^{-ikr}, \tag{A3}$$

where

$$k = (1/\hbar)\sqrt{2mE}. \tag{A4}$$

For the normalized proper function of the energy operator having a prescribed amplitude at infinity, we may take

$$f_E(r) = \frac{1}{\sqrt{2\pi}}\sqrt{\frac{dk}{dE}} \cdot \frac{1}{|B(E)|} \cdot f(r). \tag{A5}$$

To an almost stationary state such a value E_0 of the energy E corresponds, for which the quantity $B(E)$ becomes very small. Denote by a some point $r = a$ on the barrier, for example, the smaller root of the equation $U(r) = E_0$ and put

$$N = \int_0^a f^2(r)dr. \tag{A6}$$

The initial state may be described by the function

$$\begin{aligned} f_0(r) &= \frac{1}{\sqrt{N}} f(r) \qquad \text{for} \quad r < a, \\ f_0(r) &= \quad 0 \qquad\quad\ \ \text{for} \quad r > a, \end{aligned} \tag{A7}$$

or to be more exact by its value for $E = E_0$ (the function (A7) depends, but very little on E near $E = E_0$).

Using formula (11), we obtain for the energy distribution function in the almost stationary state considered the expression

$$w(E)dE = \frac{1}{2\pi} \cdot \frac{N}{|B(E)|^2} \cdot \frac{dk}{dE} dE. \tag{A8}$$

Here the quantity N depends but little upon E (or even is a constant if we put $F = E_0$ in $f(r)$ for $r < a$) and the main variation is due to the denominator $|B(E)|^2$.

Formula (A8) defines $w(E)$ for real values of E. For complex values of E this function is defined by analytical continuation.

As some arbitrariness is connected with the choice of the initial state, formula (A8) is not totally free from arbitrariness which may influence the value of the energy standard (but not that of the decay constant Γ). However, the general character of $w(E)$ as an analytical function of E is not essentially affected thereby and is correctly represented by the given expression.

Let us show that in the case when $U(r) = 0$ outside the barrier (for $r > b$, say) the quantity $kB(E)$ will be an integral transcendental function of k.

Putting $g = e^{-ikr}$, considering the expression $f''g - g''f$ and using the differential equation (Al) we shall get

$$2ikB(E) = f'(0) + \frac{2m}{\hbar^2}\int_0^b U(r)f(r)e^{-ikr}dr. \tag{A9}$$

Now it is known from the theory of the differential equations that for finite values of r the quantity f will be an integral transcendental function of E. As the integral in the right-hand side is taken in finite limits and as the factor e^{-ikr} is an integral function of k, the right-hand side of (A9) will also be an integral function of k, which was to be proved.

From the fact that $B(E)$ is an integral function it follows that the function $w(E)$ will be meromorphic — a conclusion that we have used in Section 4.

References

1. L. Mandelstam and Ig. Tamm, J. Phys. USSR **9**, 249, 1945.
2. L. Landau and R. Peierls, Zs. Phys. **69**, 56, 1931.
3. V.A. Fock, Lectures on Quantum Mechanics (read in 1936–37 in Leningrad University), Lithoprinted, Leningrad, 1937 (in Russian).
4. S.N. Bernstein, Theory of Probabilities, p. 346, Moscow–Leningrad, 1936 (in Russian).

Leningrad State University,
Physical Institute

Typeset by V.V. Fock

50-1
Application of Two-Electron Functions in the Theory of Chemical Bonds

V.A. Fock

Received 27 May 1950

DAN **73**, N 4, 735, 1950

The modern quantum theory of the valence [1] and the chemical bonds is entirely based on the application of one-electron functions. Meanwhile, by itself the idea of a bond which is built by a pair of electrons assumes such strong interaction between the electrons with opposite spins that any description of this interaction by means of two one-electron functions is somewhat artificial and cannot be exact. It is much more natural to associate with each saturated bond its own two-electron function and to preserve one-electron functions only for non-saturated bonds. (The electron functions are understood as the functions which depend only on spatial coordinates, but not on the spin.)

A possibility of such description will be proved if one finds a way to construct the wave function of all valence electrons from two-electron and one-electron functions which correspond to saturated and non-saturated bonds. The total wave function has to possess all the necessary symmetry properties and to represent a state of the molecule with a definite number of saturated and non-saturated bonds.

The symmetry properties of the coordinate wave function of a system of n electrons with a given spin was formulated for the first time in 1940 in our article about the wave functions of many-electron systems [2]. In this article a relation is demonstrated between the coordinate wave functions introduced there and the antisymmetric wave functions of the spatial and spin variables. In the same article, formulas are given for the mixed second-order density that is necessary for the calculation of the energy and other quantities.

The problem formulated here to construct the wave function, which corresponds to a given number of bonds, can be easily solved with the help of the formulas derived in our article.

Suppose here that a given system of n electrons has the resulting spin s. The number $k = \frac{n}{2} - s$ will be an integer and it is equal to the number of electron pairs with the compensated spin. If one considers the valence electrons that build the bonds, then k is the number of saturated bonds. The coordinate wave function of such a system will be written as follows:

$$\psi = \psi\left(r_1, r_2, \ldots, r_k \mid r_{k+1}, r_{k+2}, \ldots, r_n\right), \tag{1}$$

where $r_1, \ldots, r_n$ denotes the spatial coordinates (the radii-vectors) of n electrons where the first k arguments are separated from the other coordinates by the vertical line. We will also use the abridged notation, namely,

$$\psi = \psi\left(1, 2, \ldots, k \mid k+1, k+2, \ldots, n\right). \tag{2}$$

The function ψ has to
a) be antisymmetric relative to the first k arguments (to the left of the vertical line),
b) be antisymmetric relative to the last $n-k$ arguments (to the right of the vertical line) and
c) have the cycle symmetry, which is expressed by the equation

$$\psi\left(1, 2, \ldots, k \mid k+1, k+2, \ldots, n\right) = \\ = \sum_{l=1}^{n-k} \psi\left(k+l, 2, \ldots, k \mid k+1, \ldots, k+l-1, 1, k+i+1, \ldots, n\right). \tag{3}$$

The right-hand side of this equation consists of $n-k$ terms where the argument 1 (i.e., r_1) is set consecutively on the place of each of $n-k$ arguments to the right of the vertical line.

We have to express the wave function with this symmetry through k two-electron functions

$$\psi_1\left(r, r'\right), \psi_2\left(r, r'\right), \ldots, \psi_k\left(r, r'\right), \tag{4}$$

which correspond to the saturated bonds and through $n - 2k = 2s$ one-electron functions

$$\varphi_1\left(r\right), \varphi_2\left(r\right), \ldots, \varphi_{n-2k}\left(r\right), \tag{5}$$

which correspond to the non-saturated bonds. We denote through $\alpha_1, \alpha_2, \ldots, \alpha_k$ the numbers $1, 2, \ldots, k$, which are taken in an arbitrary order, and through P the corresponding transposition. Analogously, we denote $\beta_1, \beta_2, \ldots, \beta_k$ the numbers $k+1, k+2, \ldots, n$ in an arbitrary order and through Q we denote the corresponding transposition. Further, we set $\varepsilon(P)$ to $+1$ if the transposition P is even, and to -1 if the transposition P is odd; analogously we set $\varepsilon(Q)$ to the same value for the transposition Q. Using this notation, we obtain

$$\psi(1,2,\ldots,k \mid k+1,k+2,\ldots,n) = \\ = \sum_P \sum_Q \varepsilon(P)\,\varepsilon(Q)\,\psi_1(\alpha_1,\beta_1)\,\psi_2(\alpha_2,\beta_2)\ldots\psi_k(\alpha_k,\beta_k)\times \\ \times\varphi_1(\beta_{k+1})\,\varphi_2(\beta_{k+2})\ldots\varphi_{n-2k}(\beta_{n-k}). \quad (6)$$

It is directly seen from formula (6) that the function ψ satisfies the antisymmetry conditions a) and b). But, besides, it is possible to demonstrate that if the two-electron functions are symmetric relative to their arguments, then the condition of the cycle symmetry c) is also fulfilled. In order to prove this, it is more convenient to represent ψ as the sum of $k!$ determinants of the order $n-k$. We get

$$\psi(1,2,\ldots,k \mid k+1,k+2,\ldots,n) = \\ = \sum_P \begin{vmatrix} \psi_{\alpha_1}(1,k+1) & \ldots & \psi_{\alpha_1}(1,n) \\ \ldots & \ldots & \ldots \\ \psi_{\alpha_k}(k,k+1) & \ldots & \psi_{\alpha_k}(k,n) \\ \varphi_1(k+1) & \ldots & \varphi_1(n) \\ \ldots & \ldots & \ldots \\ \varphi_{n-2k}(k+1) & \ldots & \varphi_{n-2k}(n) \end{vmatrix}. \quad (7)$$

In this formula each determinant separately satisfies the cycle symmetry condition (in the case of the symmetric functions (4)). In order to prove this property, we consider one of the determinants from the sum (7), for example, the determinant with $\alpha_1 = 1, \alpha_2 = 2 \ldots, \alpha_k = k$ and denote it as Δ. We have then

$$\Delta = \begin{vmatrix} \psi_1(1,k+1) & \ldots & \psi_1(1,n) \\ \ldots & \ldots & \ldots \\ \psi_k(k,k+1) & \ldots & \psi_k(k,n) \\ \varphi_1(k+1) & \ldots & \varphi_1(n) \\ \ldots & \ldots & \ldots \\ \varphi_{n-2k}(k+1) & \ldots & \varphi_{n-2k}(n) \end{vmatrix}. \quad (8)$$

We further denote as Δ_{k+1} the determinant, which is obtained from Δ after the transposition of the argument 1 with the one of $k+l$. We now have to prove the equation

$$\Delta = \Delta_{k+1} + \Delta_{k+2} + \ldots + \Delta_n. \tag{9}$$

We note that in the determinant Δ_{k+l} the minor (i.e., the cofactor) of the element $\psi_1(k+l, k+m)$ is equal with the opposite sign to the minor of the element $\psi_1(k+m, k+l)$ in the determinant Δ_{k+m}. Therefore, if the function ψ_1 is symmetric, then in the sum $\Delta_{k+l}+\Delta_{k+m}$ the quantity $\psi_1(k+l, k+m) = \psi_1(k+m, k+l)$ is multiplied by zero and, hence, it can be replaced by zero.

One can verify that in the first rows of the determinants in the right-hand side of (9) all functions with the argument that differs from unity can be replaced by zeros. We denote as $\Delta'_{k+1}, \Delta'_{k+2}, \ldots, \Delta'_n$ the determinants obtained from $\Delta_{k+1}, \Delta_{k+2}, \ldots, \Delta_n$ after this replacement. In Δ'_{k+l} the only nonzero element is the entry of the first row $\psi_1(1, k+l)$. The values of other elements that are placed under it are nonessential and they can be chosen as in the determinant Δ. But after this replacement the equation

$$\Delta = \Delta'_{k+1} + \Delta'_{k+2} + \ldots + \Delta'_n \tag{10}$$

is, obviously, correct because it represents the expansion of the determinant Δ over the elements of the first row. From here, the validity of equation (9) follows. In this way, we proved the cycle symmetry property of the function ψ, which can be considered as the determinant Δ symmetrized relative to the arguments $1, 2, \ldots, k$. From (6) and (7), it is easy to see that if the function $\psi_m(r, r')$ is factorized and has the form

$$\psi_m(r, r') = \psi_m(r)\,\psi_m(r'), \tag{11}$$

i.e., is split in two equal functions, then the function ψ is reduced to the product of two determinants of the orders k and $n-k$. This is the case for an atom; then the function $\psi_m(r, r')$ represents an "orbit," which is occupied by two electrons. An expression for ψ as a product of two determinants was the initial representation for the wave function in our first article [3] on a self-consistent field method with the quantum exchange.

In quantum chemistry, one usually considers only the wave functions corresponding to the valence electrons. However, in order to achieve the accuracy of atomic calculations one should take into account not only valence, but all the electrons of a molecule. To a pair of inner electrons that occupies the same orbit, one could associate the wave function of the form (11) where in the first approximation one could consider $\psi_m(r)$ as to be known from the atomic calculations. In the complete function ψ only the functions of the valence electrons remain to be unknown. As they correspond to definite bonds, one can judge their qualitative character by the structural formula of the molecule. Theoretically speaking, for more exact determination of these functions one could use the complete function to write the energy of the system and then to apply the variation principle. It is obvious that this program is feasible only for very simple molecules. In more complicated cases one has to apply rather approximate, semi-empirical methods that are customary to the quantum chemistry, but in any case it is necessary in one way or another to take into account the impossibility for the valence electron to get to already occupied orbits of inner electrons.

Regardless of the possibility to do the practical calculations with the two-electron wave functions, an introduction of them gives the advantage of the visualization since each of these functions corresponds to a definite bond. In addition, double and multiple bonds can be taken into consideration; for this it is reasonable to suggest that one or several two-electron wave functions correspond to electrons belonging to the same pair of atoms.

References

1. H. Eyring, J. Walter and G.E. Kimball, Quantum Chemistry, New York, 1944.
2. V. Fock, JETP **10**, 961, 1940 (in Russian); V. Fock, Yubileinyi sbornik 30-let Oktjabr'skoi Revoljutsii; Izv. AN **1**, 255, 1957 (in Russian).
3. V. Fock, TOI **5**, 51, 1931; V. Fock, Zs. Phys. **61**, 126, 1930. (See [30-2] in this book. (*Editors*))

Translated by Yu. Yu. Dmitriev

54-1
On the Schrödinger Equation of the Helium Atom

V. Fock

Izv. AN **18**, 161, 1954
Det Kongelige Norske Videnskabers Selskabs
Forhandlinger **B 31**, 138, 1958
(Author's English version)

The wave equation for the helium atom in a state with vanishing angular momentum was first deduced by E. Hylleraas in his classical papers on the ground state of helium [1]. This equation shall be referred to, for the sake of brevity, as the Hylleraas equation. The Hylleraas equation corresponds to a mechanical system with three degrees of freedom; it is natural to take as independent variables the two distances of the electrons from the nucleus and the angle between their radii-vectors. To simplify calculations connected with the Ritz method, Hylleraas took as independent variables the three distances; the wave function to be varied was expanded in a power series of the three distances. Careful calculations made by Hylleraas showed a good agreement with the experimental values of the energy levels. It has been pointed out, however [2], that the exact solution of the Hylleraas equation does not possess an expansion of the supposed type, and the true form of the expansion was still to be found.

The energy levels obtained from the Hylleraas equation are to be corrected for the finite value of the nuclear mass, and also for the relativistic effects; but the more subtle the corrections introduced, the more accurate the calculated values of the "uncorrected" energy levels must be.

In connection with the increased subtlety of the corrections the need for a more accurate solution of the Hylleraas equation arose. It is obvious, however, that an increase in accuracy can only be obtained if expansions are used which actually satisfy the differential equation. The aim of the present paper is to exhibit the form of such an expansion and to give a method for calculating its successive terms.

1 The Hylleraas Equation and its Transformation

Taking as independent variables the two distances r_1 and r_2 of the electrons from the nucleus and the angle ϑ between the directions pointing from the nucleus to the two electrons, we write the Hylleraas equation in the form

$$\frac{\partial^2\psi}{\partial r_1^2}+\frac{2}{r_1}\frac{\partial\psi}{\partial r_1}+\frac{\partial^2\psi}{\partial r_2^2}+\frac{2}{r_2}\frac{\partial\psi}{\partial r_2}+\left(\frac{1}{r_1^2}+\frac{1}{r_2^2}\right)\Delta^*\psi+2(E-V)\psi=0. \quad (1.1)$$

Here the symbol $\Delta^*\psi$ denotes the operator

$$\Delta^*\psi=\frac{1}{\sin\vartheta}\frac{\partial}{\partial\vartheta}\left(\sin\vartheta\frac{\partial\psi}{\partial\vartheta}\right). \quad (1.2)$$

In the following we shall consider a more general equation that is of the same form as (1.1), but with $\Delta^*\psi$ denoting the Laplacian on the unit sphere:

$$\Delta^*\psi=\frac{1}{\sin\vartheta}\frac{\partial}{\partial\vartheta}\left(\sin\vartheta\frac{\partial\psi}{\partial\vartheta}\right)+\frac{1}{\sin^2\vartheta}\frac{\partial^2\psi}{\partial\vartheta^2}. \quad (1.3)$$

A physical meaning is attached only to solutions independent of the angle φ. The quantity V is the potential energy which is equal to

$$V=-\frac{Z}{r_1}-\frac{Z}{r_2}+\frac{1}{\sqrt{r_1^2+r_2^2-2r_1r_2\cos\vartheta}}. \quad (1.4)$$

All quantities are expressed here in atomic units. In order to include the case of helium-like atoms we put the nuclear charge equal to Z.

We now proceed to a change of independent variables, introducing in place of r_1, r_2, φ, ψ the quantities x, y, z, u defined by

$$\begin{aligned} x &= 2r_1r_2\sin\vartheta\cos\varphi,\\ y &= 2r_1r_2\sin\vartheta\sin\varphi,\\ z &= 2r_1r_2\cos\vartheta,\\ u &= r_1^2-r_2^2. \end{aligned} \quad (1.5)$$

Putting

$$R=\sqrt{x^2+y^2+z^2+u^2} \quad (1.6)$$

we have

$$R=r_1^2+r_2^2. \quad (1.7)$$

Let us build the four-dimensional Laplacian in Euclidean space with rectangular coordinates x, y, z, u:

$$\Box\psi = \frac{\partial^2\psi}{\partial\, x^2} + \frac{\partial^2\psi}{\partial\, y^2} + \frac{\partial^2\psi}{\partial\, z^2} + \frac{\partial^2\psi}{\partial\, u^2}. \tag{1.8}$$

It is easily verified that

$$4R\Box\psi = \frac{\partial^2\psi}{\partial r_1^2} + \frac{2}{r_1}\frac{\partial\psi}{\partial r_1} + \frac{\partial^2\psi}{\partial r_2^2} + \frac{2}{r_2}\frac{\partial\psi}{\partial r_2} + \left(\frac{1}{r_1^2} + \frac{1}{r_2^2}\right) + \Delta^*\psi. \tag{1.9}$$

Thus, the Hylleraas equation takes the form

$$2R\Box\psi + (E - V)\psi = 0. \tag{1.10}$$

In the new variables x, y, z, u the potential energy V is homogeneous of degree minus one half, namely,

$$V = -\frac{Z\sqrt{2}}{\sqrt{R+u}} - \frac{Z\sqrt{2}}{\sqrt{R-u}} + \frac{1}{\sqrt{R-z}}. \tag{1.11}$$

If we define the norm N of a function ψ according to

$$\Box N = 16\int |\psi|^2 r_1^2 r_2^2 dr_1 dr_2 \sin\vartheta d\vartheta d\varphi, \tag{1.12}$$

we shall have in the new variables

$$N = \int |\psi|^2 \frac{1}{R} dxdydzdu. \tag{1.13}$$

The energy level is the extremal value of the functional

$$W = \frac{2}{N}\int \left\{(\mathrm{grad}\psi)^2 + \frac{V}{2R}\psi^2\right\} dxdydzdu, \tag{1.14}$$

where we have put for brevity

$$(\mathrm{grad}\psi)^2 = \left(\frac{\partial\psi}{\partial x}\right)^2 + \left(\frac{\partial\psi}{\partial y}\right)^2 + \left(\frac{\partial\psi}{\partial z}\right)^2 + \left(\frac{\partial\psi}{\partial u}\right)^2. \tag{1.15}$$

In the last two equations the wave function ψ is to be real [3].

2 The Equations of Laplace and Poisson on a Four-dimensional Sphere

In studying the Hylleraas equation we shall use four-dimensional spherical coordinates $R, \alpha, \vartheta, \varphi$ defined by the equations

$$\begin{aligned} x &= R\sin\alpha\sin\vartheta\cos\varphi, \\ y &= R\sin\alpha\sin\vartheta\sin\varphi, \\ z &= R\sin\alpha\cos\vartheta, \\ u &= R\cos\alpha. \end{aligned} \tag{2.1}$$

Comparison with (1.5) gives

$$r_1 = \sqrt{R}\cos\frac{\alpha}{2}, \qquad r_2 = \sqrt{R}\sin\frac{\alpha}{2}. \tag{2.2}$$

Since $0 < \alpha < \pi$, both quantities r_1 and r_2 are positive as they ought to be.

In hyperspherical coordinates thus introduced, the four-dimensional Laplacian takes the form

$$\Box\psi = \frac{\partial^2\psi}{\partial R^2} + \frac{3}{R}\frac{\partial\psi}{\partial R} + \Box^*\psi, \tag{2.3}$$

where $\Box^*\varphi$ is the Laplacian on the four-dimensional sphere

$$\Box^*\psi = \frac{1}{\sin^2\alpha}\left\{\frac{\partial}{\partial\alpha}\left(\sin^2\alpha\frac{\partial\psi}{\partial\alpha}\right) + \Delta^*\psi\right\}, \tag{2.4}$$

$\Delta^*\psi$ denoting the quantity (1.3).

The eigenfunctions of the operator $\Box^*\psi$ may be called hyperspherical harmonics. These are well known [3]. We shall state here some of their fundamental properties. The equation

$$\Box^*\psi + (n^2 - 1)\psi = 0 \tag{2.5}$$

possesses solutions that are finite, one-valued and continuous everywhere on the four-dimensional unit sphere (i.e., for $0 \le \alpha \le \pi, 0 \le \vartheta \le \pi, 0 \le \varphi \le 2\pi$) only if n is an integer ($n = 1, 2, 3, \ldots$). To every integral value of n there correspond n^2 eigenfunctions, but only n of them are independent of φ. The latter functions may be written as

$$\psi = \Phi_{nl}(\alpha, \vartheta) = \Pi_l(n, \alpha)P_l(\cos\vartheta), \tag{2.6}$$

where P_l is Legendre's polynomial and $\Pi_l(n,\alpha)$ is a function that admits a representation either in the form of an integral

$$\Pi_l(n,\alpha) = \frac{M_l}{(\sin\alpha)^{l+1}} \int_0^\alpha \cos n\beta \frac{(\cos\beta - \cos\alpha)^l}{l!} d\beta, \tag{2.7}$$

or else in the form of a derivative

$$\Pi_l(n,\alpha) = \frac{(\sin\alpha)^l}{M_l} \frac{d^{l+1}(\cos n\alpha)}{d(\cos\alpha)^{l+1}}, \tag{2.8}$$

where

$$M_l^2 = n^2(n^2-1)\dots(n^2-l^2). \tag{2.9}$$

For a given n the number l takes the values $l = 0,1,2,\dots n-1$. The functions $\Pi_l(n,\alpha)$ are normalized according to the formula

$$\frac{2}{\pi}\int_0^\pi \Pi_l^2(n,\alpha)\sin^2\alpha d\alpha = 1. \tag{2.10}$$

Equation (2.5) admits solutions of the form

$$\psi = F(\alpha)P_l(\cos\vartheta) \tag{2.11}$$

also in the case $l \geq n$, as well as for non-integral values of n, but in these cases the solutions do not remain finite on the hypersphere and are thus not eigenfunctions.

The multiplier $F(\alpha)$ in (2.11) is a linear combination of $\Pi_l^*(n,\alpha)$ and $\Lambda_l^*(n,\alpha)$ where

$$\Pi_l^*(n,\alpha) = \frac{1}{(sin\alpha)^{l+1}} \int_0^\alpha \cos n\beta \frac{(\cos\beta - \cos\alpha)^l}{l!} d\beta \tag{2.12}$$

and

$$\Lambda_l^*(n,\alpha) = (\sin\alpha)^l \frac{d^{l+1}(\sin n\alpha)}{d(\cos\alpha)^{l+1}}. \tag{2.13}$$

We also have

$$\sin^2\alpha\left(\Pi_l^*\frac{d\Lambda_l^*}{d\alpha} - \Lambda_l^*\frac{d\Pi_l^*}{d\alpha}\right) = n. \tag{2.14}$$

The functions thus defined appear in the solution of the non-homogeneous equation

$$\Box^*\psi + (n^2-1)\psi = -f \tag{2.15}$$

if the right-hand side is expanded in a series of Legendre's polynomials.

In the following we have to solve non-homogeneous equations of the form (2.15), with integral and half-integral values of n.

If n has a half-integral value, the solution of (2.15) is uniquely determined for an arbitrary f-function (no additional conditions of the kind of orthogonality relations are imposed upon f). In this case a Green's function exists, using which we may write the solution in the form of a definite integral, namely

$$\psi(\alpha, \vartheta, \varphi) = \frac{1}{4\pi} \int f(\alpha', \vartheta', \varphi') \frac{\cos n\omega}{\sin \omega} d\Omega'. \tag{2.16}$$

Here ω denotes an angle $(0 < \omega < \pi)$ defined by

$$\cos \omega = \cos \alpha \cos \alpha' + \sin \alpha \sin \alpha' \cos \gamma, \tag{2.17}$$

where

$$\cos \gamma = \cos \vartheta \cos \vartheta' + \sin \vartheta \sin \vartheta' \cos (\varphi - \varphi'). \tag{2.18}$$

The symbol $d\Omega'$ in (2.16) denotes the expression

$$d\Omega' = \sin^2 \alpha' d\alpha' \sin \vartheta' d\vartheta' d\varphi'. \tag{2.19}$$

Formula (2.16) is written for the general case when the function f in the differential equation depends on the angle φ as well.

For integral values of n, in order that the non-homogeneous equations possess a finite and continuous solution, orthogonality is to be satisfied: the function f must be orthogonal to all solutions of the corresponding homogeneous equation. This gives n^2 conditions that may be written in the form

$$\int f(\alpha', \vartheta', \varphi') \frac{\cos n\omega}{\sin \omega} d\Omega' = 0; \tag{2.20}$$

this equation is to be understood as an identity in the quantities $\alpha, \vartheta, \varphi$ involved in ω. If f is independent of φ, only n conditions

$$\int f(\alpha, \vartheta) \Phi_{nl}(\alpha, \vartheta) \sin^2 \alpha \sin \vartheta d\vartheta = 0, \tag{2.21}$$

with Φ_{nl} defined by (2.6), are relevant, the other conditions being satisfied automatically.

If the above conditions are satisfied, a finite solution of equation (2.15) exists; this solution is not unique, however, since an additive hyperspherical harmonic of degree n remains arbitrary.

For integral values of n, there exists a generalized Green's function, of which the solution of equation (2.15) may be written in the form

$$\psi(\alpha, \vartheta, \varphi) = \frac{1}{4\pi^2} \int f(\alpha', \vartheta', \varphi')(\pi - \omega) \frac{\cos n\omega}{\sin \omega} d\Omega'. \tag{2.22}$$

The orthogonality relations (2.20) are supposed to be satisfied.

If the function f is expanded in a series of hyperspherical harmonics, the solution of equation (2.15) may be written at once. Supposing f to be independent of φ, we have

$$f = \sum_{p=1}^{\infty} \sum_{l=0}^{p-1} c_{pl} \Phi_{pl}(\alpha, \vartheta) \tag{2.23}$$

and

$$\psi = \sum_{p=1}^{\infty} \frac{1}{p^2 - n^2} \sum_{l=0}^{p-1} c_{pl} \Phi_{pl}(\alpha, \vartheta). \tag{2.24}$$

Owing to the orthogonality relations all the coefficients c_{pl} ($l = 0, 1, \ldots, p-1$) vanish if $p = n$; thus the expansion (2.24) for ψ does not contain terms with a vanishing denominator. An arbitrary hyperspherical harmonic of degree n may, of course, be added to the solutions (2.22) and (2.24).

3 Solution of the Hylleraas Equation for Finite Values of R

Let us transform the Hylleraas equation (1.10) to hyperspherical coordinates as independent variables (these coordinates were discussed in the preceding section). The expression $\square\psi$ is transformed by applying formula (2.3). The potential energy V being homogeneous of degree minus one half in the variables (2.1), it may be written in the form

$$V = \frac{U}{\sqrt{R}}, \tag{3.1}$$

where U is independent of R and depends only on α and ϑ. We have

$$U = -\frac{Z}{\cos\frac{\alpha}{2}} - \frac{Z}{\sin\frac{\alpha}{2}} + \frac{1}{\sqrt{1 - \sin\alpha \cos\vartheta}}. \tag{3.2}$$

Using expressions (2.3) and (3.1) for $\Box\psi$ and for V, the Hylleraas equation takes the form

$$R^2\frac{\partial^2\psi}{\partial R^2}+3R\frac{\partial\psi}{\partial R}+\Box^*\psi+(\frac{1}{2}ER-\frac{1}{2}U\sqrt{R})\psi=0. \tag{3.3}$$

In attempting to solve equation (3.3), it would be natural to try an expansion of the form

$$\psi=\psi_1+R^{\frac{1}{2}}\psi_{\frac{3}{2}}+R\psi_2+R^{\frac{3}{2}}\psi_{\frac{5}{2}}+\ldots+R^{n-1}\psi_n+\ldots, \tag{3.4}$$

i.e., an expansion arranged in integral and half-integral powers of R, the coefficients ψ_n being functions of α and ϑ. As a matter of fact, however, no such expansion exists if the ψ_n are to be functions finite and continuous on the hypersphere.

We consider, therefore, a more general expansion

$$\psi=\sum_{n=1,\frac{3}{2},\ldots}R^{n-1}\sum_{k=0}^{[n-1]}(\log R)^k\psi_{nk}, \tag{3.5}$$

which is arranged in powers of both R and $\log R$, the powers of R being integral and half-integral and that of $\log R$ only integral. The symbol $[n-1]$ in the upper limit of the k-summation denotes as usual the integral part of the number $n-1$. The number n takes all integral and half-integral values from 1 upward:

$$n=1,\ \frac{3}{2},\ 2,\ \frac{5}{2},\ldots \tag{3.6}$$

and, for a given n, the number k takes integral values from 0 to the integral part of $n-1$:

$$k=0,1,\ldots,\ [n-1]. \tag{3.7}$$

Let us write down some of the first members of the series (3.5) for ψ:

$$\begin{aligned}\psi &= \psi_{10}+R^{\frac{1}{2}}\psi_{\frac{3}{2}0}+R(\log R\cdot\psi_{21}+\psi_{20})+\\ &+ R^{\frac{3}{2}}(\log R\cdot\psi_{\frac{5}{2}1}+\psi_{\frac{5}{2}0})+\\ &+ R^2((\log R)^2\psi_{32}+\log R\cdot\psi_{31}+\psi_{30})+\ldots\ .\end{aligned} \tag{3.8}$$

With the help of the relation

$$\begin{aligned}&(R^2\frac{\partial^2}{\partial R^2}+3R\frac{\partial}{\partial R})R^{n-1}(\log R)^k=\\ &=R^{n-1}\{(n^2-1)(\log R)^k+2nk(\log R)^{k-1}+k(k-1)(\log R)^{k-2}\}\end{aligned} \tag{3.9}$$

we can easily calculate the result of inserting the expansion (3.5) into (3.3). Equating to zero the coefficient of $R^{n-1}(\log R)^k$ in the resulting expression, we obtain

$$\begin{aligned}\Box^*\psi_{n,k} + (n^2-1)\psi_{n,k} &= -2n(k+1)\psi_{n,k+1} - \\ -(k+1)(k+2)\psi_{n,k+2} &+ \frac{1}{2}U\psi_{n-\frac{1}{2}} - \frac{1}{2}E \quad \psi_{n-1,k}. \end{aligned} \tag{3.10}$$

In dealing with equations (3.10), it is to be remembered that the index k in $\psi_{n,k}$ is at most equal to $[n-1]$; otherwise the function k in $\psi_{n,k}$ is zero. We order equations (3.10) according to increasing values of n and, for a given n, according to decreasing values of k.

We now proceed to show that equations (3.10) may be solved consecutively, one by one, in the above-mentioned order, the choice of the solutions being essentially limited by the finiteness and uniformity condition.

Among equations (3.10) the first corresponds to the values $n = 1, k = 0$ and is of the form

$$\Box^*\psi_{10} = 0. \tag{3.11}$$

This is an equation of the type (2.05) with $n = 1$, i.e., an equation for hyperspherical harmonics. But for $n = I$ the only hyperspherical harmonic is a constant. We may thus put

$$\psi_{10} = 1. \tag{3.12}$$

To the value $n = 3/2$ there corresponds only one value of k, namely, $k = 0$. Thus, the set of equations with $n = 3/2$ reduces to a single equation

$$\Box^*\psi_{\frac{3}{2}0} + \frac{5}{4}\psi_{\frac{3}{2}0} = \frac{1}{2}U\psi_{10} \tag{3.13}$$

with U taken from (3.2). Since n is half-integral in this case, the solution of equation (3.13) is unique and does not involve any arbitrary constants. It is even possible to write this solution in an explicit form, namely,

$$\psi_{\frac{3}{2}0} = -Z(\cos\frac{\alpha}{2} + \sin\frac{\alpha}{2}) + \frac{1}{2}\sqrt{1-\sin\alpha\cos\alpha}. \tag{3.14}$$

For $n = 2$ we get two equations (for $k = 1$ and $k = 0$). They are of the form

$$\Box^*\psi_{20} + 3\psi_{20} = 0 \tag{3.15}$$

and

$$\Box^*\psi_{20} + 3\psi_{20} = -4\psi_{21} + \frac{1}{2}U\psi_{\frac{3}{2}0} - \frac{1}{2}E\psi_{10}. \tag{3.16}$$

The first of these is an equation for hyperspherical harmonics. There are two harmonics independent of φ:

$$\Phi_{20} = \cos\alpha \qquad \text{and} \qquad \Phi_{21} = \sin\alpha\cos\vartheta \tag{3.17}$$

(we take non-normalized functions). The function ψ_{21} will be a linear combination of the two functions (3.17), but the coefficients in this combination are not arbitrary. In fact, they are determined from the orthogonality conditions satisfied by the right-hand side of (3.16). Since these conditions involve just the same functions (3.17) that form the linear combinations, it is clear that there are as many conditions as there are coefficients to be found. The function ψ_{21} is thus uniquely determined.

The function ψ_{21} being known, the equation for ψ_{20} will be of the type (2.15), with the right-hand side f satisfying all orthogonality conditions required. The solution of this inhomogeneous equation may be found by one of the procedures indicated in Section 2. In this solution an additive linear combination of the hyperspherical harmonics (3.17) remains arbitrary; the function ψ_{20} thus involves two arbitrary constants.

Proceeding now to the value $n = 5/2$, we shall have (for $k = 1$ and $k = 0$) two equations

$$\Box^*\psi_{\frac{5}{2}1} + \frac{21}{4}\psi_{\frac{5}{2}1} = \frac{1}{2}U\psi_{21} \tag{3.18}$$

and

$$\Box^*\psi_{\frac{5}{2}0} + \frac{21}{4}\psi_{\frac{5}{2}0} = -5\psi_{\frac{5}{2}1} + \frac{1}{2}U\psi_{20} - \frac{1}{2}E\psi_{\frac{5}{2}1}. \tag{3.19}$$

These are inhomogeneous equations of the type (2.15), the value $n^2 - 1 = 21/4$ of the parameter not being an eigenvalue. We first obtain from (3.18) the function $\psi_{\frac{5}{2}1}$ and then insert it in (3.19), whence the other function $\psi_{\frac{5}{2}0}$ is obtained. The determination of both $\psi_{\frac{5}{2}1}$ and $\psi_{\frac{5}{2}0}$ is quite unique; no new constants are hereby introduced.

Proceeding further, we have for $n = 3$ three equations: for ψ_{32}, ψ_{31} and for ψ_{30}. The solution ψ_{32} of the first of these is a hyperspherical harmonic; the coefficients therein are to be determined subsequently from the orthogonality conditions for the right-hand side of the ψ_{31} equation. The function ψ_{31} is at first determined except for an additive hyperspherical harmonic, which is then to be found from the orthogonality conditions belonging to the ψ_{30} equation. Solving this, we obtain ψ_{30} except for a harmonic with three constants, and these are the only new constants introduced in ψ_{30}.

This process of solution may clearly be continued indefinitely, the only difficulties being of computational nature. For a half-integral n, all functions $\psi_{n,k}$ are found consecutively, beginning with that which has a maximum k, down to the function $\psi_{n,0}$ with $k = 0$, and no orthogonality conditions are required. For an integral n each function $\psi_{n,k}$ is at first determined only except for an additive hyperspherical harmonic, but the latter is then uniquely obtained from the orthogonality conditions satisfied by the right-hand side of the equation for the next function $\psi_{n,k-1}$. An additive hyperspherical harmonic remains arbitrary only in $\psi_{n,0}$.

It is to be expected that the series (3.5) obtained as a result of this process will be convergent for all finite values of R.

4 The Behavior of the Solution for Unlimited Values of R

Let us assume that for very large values of R the function ψ satisfying equation (3.3) is asymptotically equal to

$$\psi = Ae^{-\mu\sqrt{R}}, \tag{4.1}$$

where the quantity A increases not faster than some finite power of R and the coefficient μ in the exponent is independent of R. In the physical problem considered the function ψ, and consequently μ, is also independent of φ. Thus we have $\mu = \mu(\alpha, \vartheta)$. Inserting in (3.3) the expression (4.1) for μ, dividing by ψ and equating to zero the terms that are independent of R (and do not vanish at infinity), we obtain the following partial differential equation of the first order which must be satisfied by μ:

$$\left(\frac{\partial\mu}{\partial\alpha}\right)^2 + \frac{1}{\sin^2\alpha}\left(\frac{\partial\mu}{\partial\vartheta}\right)^2 + \frac{\mu^2}{4} + \frac{E}{2} = 0. \tag{4.2}$$

A particular solution of this equation is obviously the constant

$$\mu = \mu_0 \equiv \sqrt{-2E} \tag{4.3}$$

(E is negative in our case, and thus μ_0 is real). Putting

$$\mu = \mu_0 \cos\frac{\omega}{2} \tag{4.4}$$

we introduce in place of μ a new function $\omega(\alpha, \vartheta)$. Excluding the $\omega = 0$ that corresponds to the solution (4.03), we obtain the following equation for the function ω:

$$\left(\frac{\partial\omega}{\partial\alpha}\right)^2 + \frac{1}{\sin^2\alpha}\left(\frac{\partial\omega}{\partial\vartheta}\right)^2 = 1. \tag{4.5}$$

To proceed further with the determination of μ, we take recourse to physical considerations. Our system consists of two electrons and the nucleus. Consider the case when one of the distances of the electrons from the nucleus, r_1 or r_2, is very large and the other is finite. In this case we may speak of an outer and an inner electron. The outer electron is under the influence of the screened nuclear field, and the inner one under the influence of the Coulomb field of the nucleus. The states of the two electrons will be nearly independent; for the inner electron the wave function will be of a Coulomb type, while the outer electron will be described by a wave function that corresponds to a spherically symmetrical (though not Coulombian) field.

The fact that the states of the two electrons are independent means, in particular, that the quantity μ (and thus ω) will be independent of the angle ϑ. But if ω depends only on α, then, according to (4.5), it must be of the form

$$\omega = \alpha - \alpha_0, \tag{4.6}$$

where α_0 is a constant (there is no need to consider the other solution $\omega = \alpha_0 - \alpha$ since an inversion of sign in ω does not alter the value of μ). Putting

$$\mu_0 \cos\frac{\alpha_0}{2} = a, \quad \mu_0 \sin\frac{\alpha_0}{2} = b \tag{4.7}$$

we get

$$\mu = a\cos\frac{\alpha}{2} + b\sin\frac{\alpha}{2} \tag{4.8}$$

and thus the exponent in the expression (4.1) for the wave function is equal to

$$\mu\sqrt{R} = ar_1 + br_2. \tag{4.9}$$

The constants a and b are connected by the relation

$$a^2 + b^2 + 2E = 0 \tag{4.10}$$

that follows from (4.3) and (4.7).

Assuming that, for large values of r_2, the electron with the radius-vector r_1 is in the state of lowest energy, its wave function must be proportional to $\exp(-Zr_1)$. Consequently, the coefficient of r_1 in the exponent (4.9) must be equal to Z, and we have

$$a = Z. \tag{4.11}$$

Equation (4.10) then gives

$$b = \sqrt{2V_i}, \tag{4.12}$$

where

$$V_i = -\frac{1}{2}Z^2 - E > 0 \tag{4.13}$$

is the ionization potential of the outer electron.

Equating the two expressions (4.1) and (3.8) for ψ, we obtain for the quantity A a series of the form

$$\begin{aligned} A = {} & \psi_{10} + R^{\frac{1}{2}}(\psi_{\frac{3}{2}0} + \mu\psi_{10}) + R(\psi_{21}\log R + \psi_{20} + \mu\psi_{\frac{3}{2}0} + \frac{1}{2}\mu^2\psi_{10}) + \\ & + R^{\frac{3}{2}}\Big[(\psi_{\frac{5}{2}1} + \mu\psi_{21})\log R + \\ & + \psi_{\frac{5}{2}0} + \mu\psi_{20} + \frac{1}{2}\mu^2\psi_{\frac{3}{2}0} + \frac{1}{6}\mu^3\psi_{10}\Big] + \dots \,. \end{aligned} \tag{4.14}$$

An approximate value of A is obtained if one keeps in this series only the first few terms and neglects the rest. Inserting this approximate value in (4.1), we obtain for the wave function an expression that, on the one hand, satisfies in the vicinity of $R = 0$ the differential equation and, on the other hand, has a correct behavior at infinity.

In order to construct the solution that is physically relevant, we must take a linear combination of the functions (4.1) that is either symmetric or antisymmetric with respect to r_1 and r_2. Permutation of r_1 and r_2 corresponds to a change of α into $\pi - \alpha$; in particular, if

$$\mu(\alpha) = ar_1 + br_2 \tag{4.15}$$

then we have

$$\mu(\pi - \alpha) = ar_2 + br_1. \tag{4.16}$$

Consequently, the relevant combination of the functions (4.1) may be written as

$$\psi = A(R, \alpha, \vartheta)e^{-\sqrt{R}\mu(\alpha)} \pm A(R, \pi - \alpha, \vartheta)e^{-\sqrt{R}\mu(\pi-\alpha)}. \tag{4.17}$$

This expression contains the constants involved in the functions $\psi_{n,0}$ $(n = 1, 2, 3, \dots)$ and, besides them, the constants a and b in the exponent (except if they are fixed in advance according to (4.11) and (4.12)). The constants in the wave function (4.17) are to be determined from the variational principle.

In the case that a and b are fixed in advance, their values are to some extent arbitrary. But this arbitrariness is compensated for by the freedom of choice of the other constants, and the use of inaccurate values of a and b only makes the convergence slower, but does not alter the final result. It is to be borne in mind that in practical applications of the variational method the calculations are much simpler if the constants varied enter the wave function linearly and not exponentially.

Expression (4.17) for the wave function has the advantage of taking into account the singularity of the differential equation that corresponds to the triple collision ($R = 0$). It is, therefore, to be expected that the use of this expression should permit one to attain a relatively high accuracy with a small number of constants to be varied; in fact, the success of the Ritz method fully depends on the appropriate choice of the family of functions that are varied. The practical application of the process described in this paper may involve laborious calculations; nevertheless, the process turns out to be advantageous, since the use of functions that fit well the nature of the problem is a condition of the primary importance if high accuracy is desired.

References

1. F.A. Hylleraas, Zs. Phys. **48**, 469, 1928; Zs. Phys. **54**, 347, 1929.
2. J.H. Bakilett et al., Phys. Rev. **47**, 679, 1935.
3. V. Fock, Izv. AN **2**, 169, 1935; Zs. Phys. **98**, 145, 1935. (See [35-1] in this book. (*Editors*))

Leningrad State University
Physical Institute

Typeset by I.V. Komarov

57-1
On the Interpretation of Quantum Mechanics

V.A. Fock

Leningrad, 1957

UFN **52**, N 4, 461, 1957

The modern physics is characterized by deeper and deeper penetration into the laws of the world of atoms and other minute particles of matter. These laws require not only new experimental techniques to be developed, but also new notions to formulate them. To describe atomic phenomena new methods that are completely different from those used to investigate larger objects of our world are required. The new problem of description of atomic objects is of particular practical importance since the arising notions could be applied to other fields of knowledge. Therefore on the ground of the modern physics some questions of the philosophical character arise. We are going to consider them here.

As has happened several times in the history of physics, the mathematical part of the theory together with some formal prescriptions relating theory to experiment was developed much earlier than the corresponding physical notions. The technique of nonrelativistic mechanics free of internal contradictions was successfully applied to several particular problems of atomic physics; however, its physical interpretation remained unclear for some time. The necessity of a suitable interpretation of the mathematical technique of quantum mechanics arises more and more sharply.

1 Attempts of Classical Interpretation of the Wave Function and the Causes of their Inconsistency

The original point of view by de Broglie and Schrödinger tells us that the quantum mechanical wave function is a kind of field distributed in space that is similar to the electromagnetic field or other known kinds of fields. The stationary states of atoms correspond, according to Schrödinger, to eigenstates of this field. A little later de Broglie put forward a slightly different point of view according to which the space distributed field is a carrier of particles and defines their motion in the classical sense (a wave-pilot or more exactly a wave-guide). This point of view was discarded by de Broglie after a while, but he has returned to it 25 years later. Some works of Bohm are close to this point of view. He tried there to preserve the notion of a trajectory and to make it compatible with the formulae of the usual quantum mechanics by means of a "quantum potential" specially constructed in each particular situation. Some modifications of the "wave-pilot" point of view are the attempts by Vigier to apply in this case Einstein's idea of a particle as a singularity of a field; however, such attempts are not supported by any convincing mathematical arguments.

Although during the first years of the development of wave mechanics it was natural to try to interpret it in a classical spirit, it is not longer possible to say the same about the attempts of de Broglie and his followers made in recent years. The common feature of all these attempts is that they are extremely artificial and of no heuristic value: the authors did not try to solve new problem. On the contrary, the speculations were adjusted (in an inconvincing way) to the results known from quantum mechanics. Therefore the criterion of practice is entirely against such a scientific direction.

What are the features of quantum mechanics that do not allow us to interpret them in a classical spirit and consider the wave function as a distributed field similar to the classical one? Discarding for a while some more deep epistemological arguments one can indicate some formal reasons contradicting this interpretation. First, in the case of a complex system consisting of several particles the wave function depends not only on three coordinates, but on all degrees of freedom of the system. It is a function of a point of a multidimensional configuration space and not of a real physical space. Second, in quantum mechanics the canonical transformations analogous to the Fourier transformation are allowed and all transformed functions obtained in this way describe the same state and are equivalent to the original wave function expressed in terms of the coordinates. And it is not only the absolute value squared of the original function that has a physical meaning, but the squared absolute values of the transformed functions as well. Third, the many body problem (in particular the problem of several identical particles) has in quantum mechanics some features that do not allow us to reduce this problem either to the problem of several disjoint particles or to formulate it as a field problem in the ordinary three-dimensional space. Hence if a complex system possesses a wave function then it is impossible to assign wave functions to single particles. Moreover, in the case of identical particles satisfying the Pauli principle there exists a quantum interaction of a special kind irreducible to a force interaction in the ordinary three-dimensional space. Another kind of interaction that is also irreducible to the classical one exists between particles described by symmetric wave functions. Finally, not only for the case of identical particles but for a single particle as well the wave function does not always exist and does not always change according to the Schrödinger equation; under certain conditions it simply disappears or gets replaced by another one (the so-called reduction of a wavelet, see §11). It is obvious that such "momentary change" does not agree with the notion of a field.

The described features of quantum mechanics make in advance inconsistent all attempts to interpret the wave function in the classical spirit.

The relation between quantum mechanics and classical mechanics comes from another direction. It amounts to the correspondence principle according to which there exists a limiting case when quantum mechanical formulae, which can be compared to the experiment, tend to the classical ones. In this limiting case the quantities characteristic for a given mechanical system and having the dimension of action can be considered too much greater than the "action quantum" $\hbar$.[1] The correspondence principle was established by Bohr at the very start of the development of quantum mechanics and played a significant role in it.

2 Niels Bohr's Ideas and His Terminology

The true sense of the wave function and other notions of quantum mechanics started to be revealed a little in the works by Max Born about the statistical interpretation of quantum mechanics. The fundamental role of the notion of probability was clarified, although at the beginning it was not clear at all the probability of what we were talking about. The essential role in understanding this question as well as in the whole interpretation of quantum mechanics was played by the ideas of Niels Bohr that the quantum mechanical description of an object should be

[1] For a material particle in a time-independent field one can take $\frac{mv^2}{w}$ for this characteristic quantity. Here m is the mass of the particle, v its velocity and w acceleration. Since we are estimating only the order of magnitude one can replace v and w by their mean values (in any sense). If we take $\hbar$ for the Planck constant divided by 2π then the criterion of the applicability of classical mechanics to the motion of a material point writes down as $mv^3 \gg \hbar w$ (see V. Fock, UZ LGU **3**, 5–9, 1937 (in Russian)). (*V. Fock*)

compatible with the description of a classical observation (experimental device).

In his works devoted to the principal questions of quantum mechanics N. Bohr especially emphasizes the necessity to consider an experiment entirely and follow the experiment up to the indications of measuring devices. This idea is correct in the sense that in principle there should exist a possibility to follow the experiment up to the indications of measuring devices. However, the overestimation of the role of devices gives a ground to accuse Bohr in underestimation of the necessity of abstraction and forgetting that it is the properties of the micro object that are under study rather than indications of the devices. The properties of atomic objects, such as charge, mass, spin, form of the energy operator and of the particle interaction law, are, however, absolutely objective and can be considered abstractly from observation tools. On the other hand, the properties to be formulated require new quantum mechanical notions. In particular it concerns the many body problem.

The origin of confusion is also in the terminology used by Bohr. For example, he speaks about "uncontrollable interaction" although an interaction considered as a physical process is always controllable. Bohr needs to speak about "uncontrollability" only to hide the discrepancy arising at using classical notions out of the range of their applicability. Further, one can point out the interpretation by Bohr of the principles of "complementarity" and "causality" as controversial notions. If one understands these principles literally such contraposition is obviously incorrect. However, under the "complementarity principle" Bohr understands not only the Heisenberg relations but also all the characteristic differences between classical and quantum mechanics. Under the "causality principle" Bohr understands causality in the narrow mechanical sense — in the sense of the Laplace-type determinism. Therefore, in fact Bohr means the incompatibility of quantum mechanics with the Laplace-type determinism rather than with the causality principle in a more general sense. And then one can agree with him.

The causality principle in the general sense should be understood as the statement of the existence of the laws of nature and in particular those related to the general properties of space and time (finite speed of action propagation, impossibility to influence the past). Being understood in this way, quantum mechanics not only agrees with the causality principle but also gives some new content to it and enlarges it to probabilistic laws.

As I have an opportunity to make certain from personal discussions with Niels Bohr, his position is much closer to the materialistic one than it could seem from his papers on the principal question of quantum mechanics. First of all Bohr assumes that the nature should be taken as it is. He definitely expresses his disagreement with the positivistic point of view and completely agrees with the objective character of atomic objects. As for the terminology, Bohr is ready to get rid of the term "uncontrollable interaction" which he considers to be inadequate. Bohr agrees also that the general causality principle should be distinguished from the Laplace-type determinism and that only such determinism contradicts the laws of atomic physics.

3 Rejection of New Ideas as a Reaction to Their Positivistic Interpretation

The novelty of Bohr's ideas and their hardly understandable description using the terminology that is not always appropriate led to several misunderstandings and wrong interpretations of these ideas in the spirit of positivism. (It is to be mentioned that it is just the positivistic interpretation of these ideas that is usually understood under the notion of the "Copenhagen school.") The most extreme positivistic position is taken by P. Jordan; other more serious physicists such as M. Born, W. Heisenberg and others who shared such a point of view for a while now gradually get rid of it. For example, in one of the latest papers published in a volume devoted to the 70th birthday of Niels Bohr, W. Heisenberg accepts objectivity of the notion of the wave function.

The interpretation of Bohr's ideas in the spirit of positivism carried out by some of his followers gave rise to the reaction rejecting the new ideas on behalf of materialism (de Broglie, Bohm, Vigier and others). The main stimulus of the above-mentioned scientists that forced them to take a position of rejecting the usual probabilistic interpretation of quantum mechanics is the false belief that the probabilistic interpretation means the rejections of the objectivity of the micro world and of its laws, i.e., the rejection of the main assertion of materialism. According to the followers of the de Broglie school it is only the classical-type determinism that is compatible with materialism. For this reason they call their point of view "materialistic."

The fact that such understanding of materialism is narrow and therefore wrong seems doubtless to us. To force nature to obey just the deterministic form of laws, rejecting to assume their more general probabilistic form, means to be based on dogmas rather than on properties of nature itself. Such point of view is philosophically wrong. Therefore one should not be surprised by the failures of attempts to interpret quantum mechanics deterministically — the failures discussed in §1. On the other hand, repetition of such attempts and interest of nonspecialists in them because they are undertaken on behalf of materialism make the detailed analysis of new ideas from the point of view of materialistic philosophy absolutely necessary. Such analysis should doubtlessly lead to the conclusion that new ideas significantly enlarge the class of notions that materialistic philosophy deals with and that do not contradict its spirit.

4 Relativity to the Observation Tools

Let us try to demonstrate the main features of quantum mechanics distinguishing it from the classical one.

Properties of objects manifest themselves only in interaction with other objects, in particular with observation tools (devices). This is true in classical as well as in quantum physics. However, in classical physics it is allowed to neglect the influence of observation tools, or at least pay much less attention to it than it is possible in quantum physics. It becomes clear if we mention that the observation tools are always of the "human" scale, although the scale of objects that the classical physics, on the one hand, and the quantum physics, on the other hand, deal with are completely different: classical objects are in general of the same size as the observation tools though the quantum objects are much smaller.

In classical physics it is assumed that if one uses observation tools sufficiently carefully they cannot essentially influence the behavior of the object under study. Even if they do, such an influence can be taken into account by making some corrections. Therefore one can consider a classical situation as if the observation tools play no role and speak, for example, about the state of motion of an object without making reference to the observation tools and therefore giving an absolute character to the notion of "the state of motion." However, an element of relativity remains in this situation since we need to make a reference to a certain coordinate frame; from the point of view that we develop here it can be interpreted as accounting for the motion of observation tools. In quantum mechanics one needs to take into account not only the motion of observation tools, but their internal structure in some schematic way as well.

Therefore, classical description should be understood to be the one that is not relative to observation tools (if we neglect their motion). The accuracy of such a description is limited by the Heisenberg uncertainty relations. This accuracy is sufficient to describe the mechanical properties of relatively large objects, but it becomes insufficient for the description of micro world objects. Not only accuracy in the quantitative sense but rather formulation of qualitatively new properties of micro objects requires new methods of description, and first of all it is necessary to introduce a new element of relativity into their description — the relativity to the observation tools.

It is absolutely clear that relativity is not in contradiction with objectivity. Already in classical theory such a simple notion as a material point trajectory being completely objective is relative at the same time since it is well defined only if a coordinate frame is fixed. Similarly, in quantum physics the relativity to observation tools makes physical notions more precise and allows one to introduce new notions without depriving them of objectivity. Micro world objects are as real and their properties are as objective as in classical physics.

5 The Notion of a Device

In the previous section we have shown some general consequences of the simple and fundamental fact that the study of the atomic world is possible only using some larger objects that serve as observation tools (devices). Since the notion of a device plays an important role in our speculations we need to make it more precise. We can call "a device" such a technical construction that, on the one hand, is able to interact with the micro object and to react to it and, on the other hand, admits a classical description with accuracy sufficient for a given purpose (and therefore does not require further "observation tools"). It is to be mentioned here that it is absolutely inessential whether the device is constructed by a human being or it is just a collection of natural conditions manageable for observation of the micro object. The only thing that is important is that all these conditions as well as the observation tools should be describable classically themselves.

Having understood the term "device" in this way we are able to formulate the problem of the quantum-mechanical description as follows.

All properties of a micro object including just the quantum ones, i.e., such that the classical mechanics is insufficient to describe, should be characterized by the ability to influence a device admitting a classical description.

6 The Essence of the Wave–Particle Dualism

To manifest different properties of a micro object different external conditions are required. It may turn out that different kinds of external conditions are incompatible with each other. Consider, for example, the diffraction of electrons on a crystal. The regular pattern of the diffusing centers is necessary to obtain a clear diffraction pattern and therefore to manifest the wave properties of the electron. But just the same regularity is an obstacle to the exact space localization of the electron under diffraction: without disturbing the regularity of the centers it is impossible to find which particular one has scattered the electron. In the literature and in particular in the papers by N. Bohr and W. Heisenberg many other examples are discussed that can be considered as examples of incompatibility of external conditions necessary to manifest the corpuscular and the wave properties of the electron, respectively.

There exist conditions under which both corpuscular and wave properties of the electron manifest themselves, but in this case these properties are not very definite and sharp. For example, an electron bound in an atom has a wave function that is of the type of a stationary wave with the amplitude rapidly decreasing with the distance from the center of the atom. It just means that the electron is almost localized (the corpuscular property), but on the other hand it manifests the wave properties as well.

Therefore atomic objects under certain conditions manifest their wave properties and under other conditions the corpuscular ones; the conditions when both kinds of properties are manifest, however not sharply, are also possible. One can say that for an atomic object there exists a *potential possibility* to manifest itself either as a wave or as a particle or in an intermediate way depending on external conditions. It is just this potential possibility of manifesting different properties that is the content of the wave–particle dualism. Any other more literal understanding of this dualism as any model is wrong. In particular the model of a particle carried by a wave or a particle as a singularity of a field proposed by de Broglie and his followers is absolutely inadequate as was already shown in §1.

Moreover it is necessary to remember that the features of quantum mechanics (even of the nonrelativistic one) are not exhausted by the wave–particle dualism. Such electron properties as the spin or the quantum statistics (the Pauli principle) are not reducible to such a dualism; nevertheless they can be formulated in terms of quantum mechanics. The fundamental character of these properties follows even from the fact that they define the structure of electronic shells of atoms and therefore their optical and chemical properties.

The technique of quantum mechanics describing series of fundamental properties of atomic objects correctly finds a consistent interpretation only on the basis of the extended problem of the description of micro objects — just the one where their behavior is considered together with their interaction with observation tools.

7 Probabilistic Description of the Interaction between an Object and a Device

Speaking about the interaction between a micro object and a device we should distinguish between two sides of the story: on the one hand the interaction considered as a physical process and on the other hand the interaction as a connection between the quantum-mechanically describable part of the system (the micro object) and the part that can be described classically. In the former case the term "interaction" is used in a more direct way though in the latter case in a more conditional one. We are interested here mainly in the second aspect of the story since the very notion of a quantum-mechanical description should be based on the analysis of the interaction in the second sense.

What are the features of the interaction between a quantum object and a classical device?

To answer this question one should recall that both the external conditions of the object and the result of its interaction with the device should be described in the language of classical physics. From these classical data one should make a conclusion about the quantum characteristics of an atomic object.

Even if an atomic object is under given external conditions the result of its interaction with a device is not unambiguous. This result can not be certainly predicted from earlier observations independently of their precision level. It is only the probability of a given result that is well defined. The maximally complete result of a measurement is a probability distribution of the measured quantity rather than an exact value of it.

Of course it may happen that the probability of one particular value of the measured quantity (or of a narrow interval of values) is so much exceeding the probability of the other values that practically it is possible to assign a definite value to this quantity. In this case a precise or an almost precise prediction of the experiment results is possible. However, this situation is an exception or rather a particular case. The typical case taking place in quantum mechanics is the generic case giving just a probability distribution as a result of a measurement.

The fact that a higher precision of earlier measurements does not lead to an unambiguous prediction of the measurement result is of principal importance. This fact should be considered as an expression of the law of nature related to the properties of atomic objects and in particular to the wave-corpuscular dualism. Admitting this fact means the rejection of classical determinism and requires a new form of the causality principle.

Expression of the results of a series of measurements as a probability distribution is known for classical physics as well. But there the probabilities were considered as a kind of a "strange element," as a result of ignoring some unknown factors and averaging over unknown data. In classical physics a principal possibility to sort the objects under measurement beforehand in order to obtain a single value instead of the probability distribution was always assumed. Conversely, in quantum physics such sorting of atomic objects is not possible since these properties are such that measured quantities may have no definite values under certain conditions. In quantum physics the notion of the probability is the primary notion. It plays the fundamental role there and is closely related to the quantum mechanical notion of the state of an object.

8 Probabilistic Characteristic of the State of an Object

To study the properties of an atomic object the most important is to consider experiments allowing one to distinguish between three stages: preparing of an object, object's behavior under certain external conditions and just the measurement. According to that, one should emphasize three parts of the measuring device: the preparing part, the working part and the registering part. For example, for the electron diffraction on a crystal the preparing part is the source of a monochromatic electronic beam as well as diaphragms and other tools located before the crystal. The working part is just the crystal and the registering part is the photographic plate or an electronic counter.

Once this distinction has been made one can vary the last stage (the measurement) keeping the first two unchanged. This example is the most convenient to follow the physical interpretation of the quantum mechanical technique.

Varying the last stage of the experiment one can measure different quantities (e.g., energy of the particles, their velocities, their positions in space etc.) starting from a given initial state of the object. To each quantity there corresponds a series of measurements having a probability measure as a result.

All these measurement results can be described parametrically via a single wave function that does not depend on the final stage of the experiment and therefore is an objective characteristic of the object's state just before the final stage.

The state of an object described by the wave function is objective in the sense that it represents an objective (independent the observer) characteristic of the possibilities of one or another result of interaction between the object and the device. In the same sense it is related to a single object. But this objective state is not yet real in the sense that these possibilities for a given object have not been realized yet. The transition from something potentially possible to something real and actual happens at the last stage of the experiment. For the statistical characteristic of this transition, i.e., to obtain the probability distribution experimentally, a series of measurements is required and the probability distribution is a result of statistics applied to this series. This experimental probability distribution may be then compared with the theoretical one obtained from the wave function.

It is to be mentioned that although the result of the final stage of an experiment can be formulated classically, one can derive values of specifically quantum quantities out of it, such as the spin of a particle, the energy level of an atomic system etc. Therefore by statistically processing the experiment results one can obtain probability distributions for both quantities with classical analogues and specifically quantum quantities.

9 The Notions of Potentially Possible and Actual in Classical Physics

In classical deterministic physics the question of the transition between something potentially possible to something actual does not arise at all since the unambiguous predetermination of the sequence of events is postulated there. According to it anything possible becomes actual and it is not necessary to distinguish between these notions. The practical impossibility of predicting events is related only to the incompleteness of the initial data.

Such a deterministic point of view is not unavoidable logically but is rather a consequence of historical conditions and mainly of successes of celestial mechanics in the 18th and 19th centuries. The high accuracy of predictions of the motion of celestial objects generated a belief in mechanistic determinism (the Laplace determinism). As a result of this belief the deterministic point of view had spread over the whole physics (with thermodynamics perhaps being the only exception) and started to pretend to be the only scientific one. The success of the electromagnetic theory of light that put forward the notion of a field as a new physical reality has shown narrow sides of the mechanistic point of view, but, still, did not destroy the belief in determinism.

However everyday experience when one needs to distinguish between a possibility and its realization was against such determinism; but this experience was being rejected as a "nonscientific" one. Thermodynamics remained in the domain of "unsafe" physics from the point of view of determinism since one did not manage to make it agree with the latter. But the real crush of determinism took place with the development of quantum mechanics starting from the paper by A. Einstein on radiation theory (1916), where he first introduced *a priori* probabilities into physics.[2] The correct interpretation of the quantum-mechanical properties of atomic objects absolutely excludes the deterministic point of view. Quantum mechanics restores the rights of the difference between the potential possibility and the realization dictated by everyday life.

10 Probability and Statistics in Quantum Mechanics

The probabilistic character of quantum mechanics is doubtless, although almost nobody tries to object. However, the questions of what the probabilities correspond to, what statistical ensemble are they related to and whether quantum mechanics is a theory of single atomic objects or a theory of collections of such objects continue to be discussed, although at present they can be answered unambiguously.

During the first years of the development of quantum mechanics and the first attempts of its statistical interpretation physicists had not yet gotten rid of the interpretation of the electron as a classical material point. One would speak about an electron as if it were a particle with

[2]It is amazing that Einstein, who has done so much for the theory of quanta at the first stage of its development and introduced into physics *a priori* probabilities for the first time, later became an adversary of quantum mechanics and a supporter of determinism; he said several times half seriously that he could not believe that God plays dices (daß der liebe Gott würfelt). (*V. Fock*)

definite but unknown values of the coordinates and velocity. The Heisenberg relations were interpreted as inaccuracy relations rather than uncertainty relations. The absolute value squared of the wave function was interpreted as a probability density of a particle to have given coordinates (as if the coordinates were always defined). The absolute value squared of the wave function in the momentum space was interpreted analogously. Both probabilities (in the coordinate space and in the momentum space) were considered simultaneously as if the values of coordinates and momenta were compatible. The actual impossibility to measure them together expressed by the Heisenberg relations was interpreted under this consideration as a kind of paradox or a caprice of nature as if not everything existing is understandable.

All these difficulties drop out if we adopt completely the wave-corpuscular nature of the electron, clarify the essence of this duality and understand what the probabilities considered in quantum mechanics correspond to. In order not to repeat what has been clarified above, recall that the probabilities for different quantities obtained from the wave function correspond to different experiments and that they characterize not the behavior of a particle "itself," but rather its influence on a device of a given kind.

The question of what statistical ensemble corresponds to the probabilities was also a subject of discussion. It was L.I. Mandel'shtam (Collected Papers, vol. 5, p. 356) who was the first to pose this question. However he gave a wrong answer. Mandel'shtam is speaking about "micromechanical ensemble which the wave function belongs to" and calls it also "an electronic ensemble," underlying in this way that he considers a collection of suitably prepared micro objects. These starting points of Mandel'shtam contain some mistakes that are related to an insufficiently exact definition of the statistical ensemble. Let us correct them and give a more exact definition of the ensemble.

Imagine an infinite series of elements possessing the properties that can be used to sort these elements and observe a frequency of an element with a given property. If there exists a probability for an element to have a given property,[3] then the considered series of elements is a statistical ensemble.

What statistical ensemble can be considered in quantum mechanics?

[3]Existence of a definite probability is a hypothesis that can be introduced either *a priori* (for instance, from symmetry considerations) or starting from the permanence of the external conditions under which the physical realization of a given series of elements takes place. The hypothesis of the existence of probability is equivalent to the hypothesis that a given series of elements is a statistical ensemble. (*V. Fock*)

It is obvious that it should be an ensemble of elements that can be described classically since only to such elements one can assign definite values of sorting parameters. For this reason, a quantum object cannot be an element of an ensemble even if it is under such conditions that one can assign a wave function to it. Therefore it is impossible to speak about the "micromechanical" and "electronic" ensemble in the sense of Mandel'shtam.

The elements of statistical ensembles considered in quantum mechanics are not micro objects, but rather the results of experiments on them. A given experiment corresponds to a definite ensemble. Since the probability distributions for different quantities obtained from the wave function correspond to different experiments, they correspond to different ensembles as well. Therefore *the wave function does not correspond to any statistical ensemble.*

All the above can be illustrated by the following diagram:

	E	p	x	$\dots$
ψ_1				
ψ_2				
ψ_3				
$\vdots$				

Each cell of this diagram corresponds to a definite statistical ensemble with its own probability distribution. Each line contains ensembles obtained by measuring different quantities E, p, x of a given initial state. Each column contains ensembles obtained by measuring a given quantity for different states $\psi_1, \psi_2, \psi_3 \dots$.

A deeper reason for the impossibility to put a statistical ensemble into correspondence to a wave function is that the notion of a wave function corresponds to a potential possibility (to the experiments that have not yet been performed), although the notion of a statistical ensemble corresponds to the already performed experiments.

Attempts to reduce a wave function to a collection of micro objects were made by several authors. There was an opinion that the whole quantum mechanics is a theory of such collections of micro objects (ensembles), and the theory of single, individual objects is ostensibly nonexistent. This opinion was based on a misunderstanding of what the probability is. The probability of one or another behavior of an object under given external conditions is determined by the internal properties of a given individual object and given conditions; it is a numerical estimation of the potential possibility of one or another behavior of an object. This probability expresses itself in the relative number of realized cases of the given behavior of an object; this number is its measure. Therefore the probability corresponds to a single object and characterizes its potential possibilities; on the other hand, to determine its numerical value experimentally, one needs to have statistics of realizations of such possibilities, i.e., several repetitions of the experiment. It is obvious from the above that the probabilistic character of the theory does not exclude its correspondence to a single object. It is also true for quantum mechanics.

11 Forms of Expression of the Causality Principle in Quantum Mechanics

The quantum-mechanical notion of a state allows us to formulate the causality principle in the form applicable to atomic phenomena. According to quantum mechanics the wave function of an atomic system satisfies the wave equation that determines it unambiguously from its initial value (the Schrödinger equation). Therefore the law of probability change expressible via the wave function is determined. The wave equation allows us to solve nonstationary problems of quantum mechanics corresponding to experiments with stages separated in time. A typical example of such a problem is given by the problem of decay of an almost stationary state of an atomic system, in particular by the problem of ionization of an atom by an electric field. In principle, the problem of radioactive decay of an atom belongs to this class of problems as well.

In modern physics the causality principle is related not only to the impossibility of influencing the past, but also to the existence of the limiting speed of action propagation, which is equal to the speed of light in free space. Both these requirements are satisfied in quantum mechanics, although the nonrelativistic form of it (the Schrödinger theory) accounts for the limiting speed only in an indirect way — in a form of an additional requirement that all the considered velocities were much less than the limiting one. However, in all the relativistic generalizations of quantum mechanics the existence of the limiting speed is accounted for automatically. The relations implied by the causality principle and, in particular, the relations for the scattering amplitudes play an important role in quantum field theory.

In connection with the limiting speed of action propagation one should consider the question about the so-called "wavelet reduction." Under this, one understands the following. If one assumes that a final stage of one experiment is at the same time the initial stage of another one, then the wave function that gave the probability distribution of the results of the first experiment should be replaced by another one corresponding to the actual obtained result. Such a replacement happens at once; the change of the wave function does not satisfy the Schrödinger equation. It might look like (and this question has been in fact debated) that the sudden change of the wave function contradicts the finiteness of the speed of action propagation. However, it is easy to see that in this situation we are dealing not with the propagation of an action, but rather with a change of the question of probabilities. In the experiment, only one of the possible results prescribed by the wave function was realized. The change of the question of probabilities consists of accounting for the realized result, i.e., accounting for the new data. But to the new data a new wave function corresponds.

These speculations show how important it is to interpret the quantum mechanics to distinguish between something potentially possible and something actually realized. They also show in a completely transparent way that the wave function is not a real field and that its sudden change is not a physical process like a change of a field. A physical process is in fact related to an experiment but it influences the wave function indirectly by means of the requirement to reformulate the problem of probabilities.

Quantum mechanical understanding of causality drastically differs from the classical one, although the former is a natural generalization of the latter. The classical (Laplacian) determinism that we were speaking about in §9 can be defined as a point of view according to which the elaboration of observation methods, together with making the formulation of the laws of nature and their mathematical treatment more precisely, can in principle allow us to predict unambiguously the whole sequence of events. The study of the atomic world shows that the classical determinism not only disagrees with the laws of nature, but even does not allow us to formulate them with sufficient accuracy. This agreement takes place even in the case of the simplest elementary processes (quantum transitions). It means that the problem lies not in the complexity of an event, but rather in the unsuitability of the old ways of description. As we have already seen, the essential features of the new methods consist of the probabilistic character of the description according to which one should distinguish between something potentially possible and something realized, accounting for the relativity with respect to observation tools and, finally, in the new understanding of the causality principle according to which this principle corresponds directly only to probabilities, i.e., to something potentially possible rather than to actually realized events.

12 Philosophical Questions Put Forward by Quantum Mechanics

The development of new ideas put forward by the development of quantum physics requires the consideration of some philosophical questions and in particular the questions related to the analysis of the process of study. Such questions arise in connection with the above-mentioned impossibility to abstract oneself from observation tools when studying atomic objects. They are related also to the necessity to consider the probability as a fundamental notion and distinguish between something potentially possible and actually happening, which in turn is related to the formulation of the causality principle. Here one cannot restrict oneself to studies of classical heritage and collections of classic quotations, but rather one needs to approach these questions of human knowledge creatively. One should creatively develop dialectic materialism. On the other hand, one should remember that the ideas of atomic physics are radically new and it is impossible to get rid of them trying to reduce the story to the ideas that the quotations of the classics are ready for.

It is also incorrect to refer to the fact that the ideas of quantum mechanics are not the last words of science and that a satisfactory quantum field theory has not yet been constructed. Each theory, and quantum mechanics in particular, is only a relative truth; however, this is not the reason to reject its ideas and notions.

The physical notions will doubtlessly develop; however, it is clear even now that this development will proceed farther from the classical patterns rather than toward them. In particular the hopes for a return to a certain kind of classical determinism expressed by some followers of the de Broglie school have no grounds. He who tries on behalf of materialism to reject new ideas and to restore old ones gives a bad service to materialism.

A philosophical generalization of new ideas originally coming from atomic physics could be helpful for the development of other branches of science where the questions analogous to the ones already solved in quantum mechanics may arise.

The resolution of contradictions achieved in quantum mechanics between the wave and corpuscular nature of the electron, between the probability and causality, between the quantum description of an atomic object and the classical description of a device and finally between the properties of an individual object and the statistical behavior gives a series of brilliant examples of the application of dialectics to questions of natural science. This remains true independent of whether the dialectic method has been applied consciously or not. The achievement of quantum mechanics should become a strong stimulus to the development of dialectic materialism. Inclusion of new ideas to its treasury is the primary task of materialistic philosophy.

Translated by V. V. Fock

59-1
On the Canonical Transformation in Classical and Quantum Mechanics[1]

V. Fock

(Received 3. II. 1969)

UZ LGU **16**, 67, 1959,
Acta Phys. Acad. Sci. Hungaricae, **27**, 219, 1969
(Author's English version)

In his famous book on the principles of quantum mechanics Dirac[2] formulates (in §26) the following proposition: "... for a quantum dynamical system that has a classical analogue, unitary transformations in the quantum theory are the analogue of contact transformations in the classical theory." No detailed description of this analogy is, however, given in Dirac's book. In the following we intend to investigate this analogy in more detail.

Let $q_1, q_2, \dots q_n$; $p_1, p_2, \dots p_n$ be the original coordinates and momenta and $Q_1, Q_2, \dots Q_n$; P_1, $P_2, \dots P_n$ the transformed coordinates and momenta of a dynamical system with n degrees of freedom. We consider the case when the transformation function S depends on the old and new coordinates:

$$S = S(q_1, q_2, \dots q_n; Q_1, Q_2, \dots Q_n). \tag{1}$$

The contact transformation is defined by the relation between the differentials

$$\sum_{r=1}^{n} p_r dq_r - \sum_{r=1}^{n} P_r dQ_r \tag{2}$$

from which it follows

$$p_r = \frac{\partial S}{\partial q_r}; \qquad P_r = -\frac{\partial S}{\partial Q_r}. \tag{3}$$

The expressions for the quantities q, p in terms of Q, P and the inverse expressions are obtained by solving eqs. (3). The solution is always

[1]Dedicated to Prof. P. Gombás on his 60th birthday.

[2]P.A.M. Dirac. Quantum Mechanics, 4th ed., London and New York, 1958. This paper was included in subsequent Russian editions of the Dirac book. (*Editors*)

possible, since the determinant

$$D = \mathrm{Det}\frac{\partial^2 S}{\partial q_r \partial Q_s} \neq 0 \tag{4}$$

is assumed to be different from zero.

The contact transformation thus defined corresponds in quantum mechanics to a unitary transformation from the representation in which the quantities q are "diagonal" to the representation in which the quantities Q are "diagonal."[3]

This unitary transformation has the following form. Let $\psi_Q(q)$ be a complete set of simultaneous eigenfunctions of the operators $Q_1, \ldots Q_n$ expressed in the variables $q_1, \ldots q_n$. Let F be the operator undergoing the transformation. Then the kernel (or the matrix) of the transformed operator will have the form

$$\langle Q'|F^*|Q\rangle = \int \overline{\psi}_{Q'}(q) F \psi_Q(q) dq, \tag{5}$$

where dq denotes the product of the differentials

$$dq = dq_1 \ldots dq_n. \tag{6}$$

In Dirac's notation

$$\psi_Q(q) = \langle q|Q\rangle \tag{7}$$

and formula (5) takes the form

$$\langle Q'|F^*|Q\rangle = \int \langle Q'|q\rangle F \langle q|Q\rangle dq. \tag{8}$$

The eigenfunction $\psi_Q(q)$ can be considered as the kernel $\langle q|\overline{U}|Q\rangle$ of a unitary operator $\overline{U} = U^{-1}$ and formula (8) can be written as

$$F^* = UFU^{-1}. \tag{8*}$$

In the case $F = 1$, formula (5) reduces to the orthogonality condition, and the kernel of the unit operator expressed in variables Q must appear on its left-hand side, i.e., the expression

$$\langle Q'|1|Q\rangle = \delta_0(Q - Q') \equiv \delta(Q_1 - Q_1') \ldots \delta(Q_n - Q_n'), \tag{9}$$

where δ is the Dirac delta function.

[3]The set of variables $q_1, \ldots\ q_n$ will often be denoted by a single letter q; a similar meaning will be assigned to the symbol Q and also to p and P. (*V. Fock*)

In semi-classical approximation we may take as $\psi_Q(q)$ the quantity

$$\psi_Q(q) = c\sqrt{\left|\frac{\partial^2 S}{\partial q \partial Q}\right|} e^{(i/\hbar)S}. \tag{10}$$

We have put for brevity

$$\frac{\partial^2 S}{\partial q \partial Q} = Det \frac{\partial^2 S}{\partial q_r \partial Q_s} = D \tag{11}$$

and the absolute value of the determinant stands under the square-root in (10). The constant c equals

$$c = (2\pi\hbar)^{-n/2}. \tag{12}$$

Let us verify that the semi-classical functions (10) approximately satisfy the orthogonality relations. Inserting expressions (10) in the integral and taking $F = 1$, we obtain under the integral sign a rapidly varying exponential factor $\exp\left[\left(\frac{i}{\hbar}\right)(S - S')\right]$, where S' is the value of S with Q replaced by Q'. This factor ceases to be rapidly varying only if Q' is near Q; this is the condition that the integral should noticeably differ from zero. Consequently, the difference $S - S'$ in the exponent may be replaced by the expression

$$S - S' = -\sum_{r=1}^{n} (Q'_r - Q_r)\frac{\partial S}{Q_r} \tag{13}$$

or

$$S - S' = -\sum_{r=1}^{n} (Q'_r - Q_r)P_r, \tag{14}$$

where P_r has the value (3). Formula (14) may be briefly written as

$$S - S' = (Q' - Q)P. \tag{15}$$

In all factors of the exponential we may put $Q' = Q$. Then we have

$$\int \overline{\psi}_{Q'}(q)\psi_Q(q)dq = c^2 \int e^{\frac{i}{\hbar}(Q'-Q)P} \left|\frac{\partial^2 S}{\partial Q \partial q}\right| dq. \tag{16}$$

Now, if P_r has the value (3), the determinant under the integral sign in (16) is the Jacobian for the transformation from P to q, so that

$$\left|\frac{\partial^2 S}{\partial Q \partial q}\right| dq = dP_1 \ldots dP_n = dP. \tag{17}$$

Thus, formula (16) may be written as

$$\int \overline{\psi}_{Q'}(q)\psi_Q(q)dq = c^2 \int e^{\frac{i}{\hbar}(Q'-Q)P} dP. \tag{18}$$

But the remaining integral (multiplied by c^2) is simply the product (9) of the delta functions. We finally obtain

$$\int \overline{\psi}_{Q'}(q)\psi_Q(q)dq = \delta_0(Q - Q') \tag{19}$$

and the orthogonality condition (as well as the normalization condition) is thus satisfied.

We now consider the matrix for an arbitrary operator F expressed in terms of q_r and of $p_r = -i\hbar\partial/\partial q_r$.

Let

$$F = F(q,p) = F\left(q, -i\hbar\frac{\partial}{\partial q}\right). \tag{20}$$

When applied to the exponential function $\exp\left(\frac{i}{\hbar}S\right)$, the operator F yields approximately this function multiplied by $F(q, \partial S/\partial q)$. Thus

$$F\left(q, -i\hbar\frac{\partial}{\partial q}\right) e^{\left(\frac{i}{\hbar}S\right)} \cong e^{\left(\frac{i}{\hbar}S\right)} F\left(q, \frac{\partial S}{\partial q}\right). \tag{21}$$

A similar relation holds if we replace the exponential in (21) by the function $\psi_Q(q)$ which is approximately equal to (10). Accordingly, we may mean by F in formula (5) not the differential operator on the left-hand side of (21), but the function on the right-hand side of this equation. Putting, as before, $Q' = Q$ in all factors of the exponential, we obtain

$$\langle Q'|F^*|Q\rangle = c^2 \int F\left(q, \frac{\partial S}{\partial q}\right) e^{\frac{i}{\hbar}(S-S')} \left|\frac{\partial^2 S}{\partial Q \partial q}\right| dq. \tag{22}$$

As in formula (18), we take in (22) the quantities P as integration variables. Transforming to the said variables the function F, we shall have

$$F(q,p) = F(q(Q,P), \qquad p(Q,P)) = F^*(Q,P), \tag{23}$$

where p and P are the classical expressions (3). Using the approximate expression (15) for the quantity in the exponent, we may write

$$\langle Q'|F^*|Q\rangle = c^2 \int F^*(Q,P) e^{\frac{i}{\hbar}(Q'-Q)P} dP. \tag{24}$$

To evaluate this integral we note that the multiplication of the exponential in the integrand by P is equivalent to the application of the operator $-i\hbar\partial/\partial Q'$. We also use our previous statement (which led us to (16)) that in the factors of the exponential we may put $Q' = Q$. We may then write

$$\int F^*(Q,P)e^{\frac{i}{\hbar}(Q'-Q)P}dP = \int F^*\left(Q', -i\hbar\frac{\partial}{\partial Q'}\right)e^{\frac{i}{\hbar}(Q'-Q)P}dP. \quad (25)$$

Taking the operator F^* out of the integral sign and using (18) and (19), we obtain

$$\langle Q'|F|Q\rangle = F^*\left(Q', -i\hbar\frac{\partial}{\partial Q'}\right)\delta_0(Q-Q'), \quad (26)$$

and interchanging Q' and Q,

$$\langle Q|F|Q'\rangle = F^*\left(Q, -i\hbar\frac{\partial}{\partial Q}\right)\delta_0(Q-Q'). \quad (26^*)$$

Now, the result of applying the operator F^* to a given function $\Psi(Q)$ is, by definition,

$$F^*\Psi(Q) = \int (Q|F^*|Q')\Psi(Q')dQ'. \quad (27)$$

Using (26*) we obtain

$$F^*\Psi(Q) = F^*\left(Q', -i\hbar\frac{\partial}{\partial Q'}\right)\Psi(Q). \quad (28)$$

This equation gives (apart from terms depending on the order of factors in $F^*(Q,P)$) the form of the transformed operator as applied to the function $\Psi(Q)$.

Our calculations can be summarized as follows. The approximate equation (21) has been used twice. First, we made a transition from the operator $F(q, -i\hbar\partial/\partial q)$ to the function $F(q,p)$; this function was expressed by means of classical formulae in terms of new canonical variables Q, P. Second, we made the inverse transition from the resulting function $F = F^*(Q,P)$ to the operator $F^*(Q, -i\hbar\partial/\partial Q)$. This transition presented itself naturally when we applied the method of differentiation with respect to a parameter to the calculation of the integral for the matrix element of the operator considered. The results of our calculations can be stated as follows. Let the operator

$$F = F(q,p); \qquad p = -i\hbar\frac{\partial}{\partial q}, \quad (29)$$

first expressed in variables q, then be transformed, by means of a unitary transformation, to new variables Q. When expressed like (29), the resulting operator F^* will have the form

$$F^* = F^*(Q, P); \qquad P = -i\hbar\frac{\partial}{\partial Q}. \tag{30}$$

Suppose that the unitary transformation is performed by means of eigenfunctions having in classical approximation the form (10) (so that their phase is $\frac{1}{\hbar}S(q, Q)$). Then the form of F^* can be obtained from that of F (if one disregards the order of the factors) by means of a purely algebraic transformation expressed by the formulae

$$F(q, p) = F^*(Q, P), \tag{31}$$

$$p = \frac{\partial S}{\partial q}; \qquad P = -\frac{\partial S}{\partial Q}, \tag{32}$$

where S is the function involved in the phase of the unitary transformation. These formulae represent a contact transformation of classical mechanics.

We see that one can speak not only of an analogy between unitary and contact transformations, but also of an approximate equality between the corresponding quantum mechanical and classical expressions.

Typeset by I. V. Komarov

Index

Milton Keynes UK
Ingram Content Group UK Ltd.
UKHW021929071024
449327UK00022B/1742